Springer Series in Materials Science

Volume 266

Series editors

Robert Hull, Troy, USA
Chennupati Jagadish, Canberra, Australia
Yoshiyuki Kawazoe, Sendai, Japan
Richard M. Osgood, New York, USA
Jürgen Parisi, Oldenburg, Germany
Udo W. Pohl, Berlin, Germany
Tae-Yeon Seong, Seoul, Republic of Korea (South Korea)
Shin-ichi Uchida, Tokyo, Japan
Zhiming M. Wang, Chengdu, China

The Springer Series in Materials Science covers the complete spectrum of materials physics, including fundamental principles, physical properties, materials theory and design. Recognizing the increasing importance of materials science in future device technologies, the book titles in this series reflect the state-of-the-art in understanding and controlling the structure and properties of all important classes of materials.

More information about this series at http://www.springer.com/series/856

Claudia Cancellieri · Vladimir N. Strocov
Editors

Spectroscopy of Complex Oxide Interfaces

Photoemission and Related Spectroscopies

 Springer

Editors
Claudia Cancellieri
Empa—Swiss Federal Laboratories
 for Materials Science and Technology
Dübendorf
Switzerland

Vladimir N. Strocov
Spectroscopy of Novel Materials Group
Paul Scherrer Institute
Villigen
Switzerland

ISSN 0933-033X   ISSN 2196-2812 (electronic)
Springer Series in Materials Science
ISBN 978-3-319-74988-4   ISBN 978-3-319-74989-1 (eBook)
https://doi.org/10.1007/978-3-319-74989-1

Library of Congress Control Number: 2018931429

© Springer International Publishing AG, part of Springer Nature 2018
This work is subject to copyright. All rights are reserved by the Publisher, whether the whole or part of the material is concerned, specifically the rights of translation, reprinting, reuse of illustrations, recitation, broadcasting, reproduction on microfilms or in any other physical way, and transmission or information storage and retrieval, electronic adaptation, computer software, or by similar or dissimilar methodology now known or hereafter developed.
The use of general descriptive names, registered names, trademarks, service marks, etc. in this publication does not imply, even in the absence of a specific statement, that such names are exempt from the relevant protective laws and regulations and therefore free for general use.
The publisher, the authors and the editors are safe to assume that the advice and information in this book are believed to be true and accurate at the date of publication. Neither the publisher nor the authors or the editors give a warranty, express or implied, with respect to the material contained herein or for any errors or omissions that may have been made. The publisher remains neutral with regard to jurisdictional claims in published maps and institutional affiliations.

Printed on acid-free paper

This Springer imprint is published by the registered company Springer International Publishing AG part of Springer Nature
The registered company address is: Gewerbestrasse 11, 6330 Cham, Switzerland

Preface

This volume deals with recent important achievements of photoemission spectroscopy on buried Transition Metal Oxide (TMO) interfaces. The idea of this monograph began at the APS conference in Denver, Colorado in 2014, when presented there for the first time was the case that electronic structure of buried interface was disclosed with a high-resolution angle-resolved soft X-ray photoemission experiment at the ADRESS beamline of the Swiss Light Source. Performing an ARPES experiment not in-situ, without surface preparation, and on an electron system buried behind a few-nm-thick overlayer was a very challenging and ambitious project. The combination of advanced instrumentation with high energy resolving power, the excellent quality of samples produced after many years of growth parameters optimization (already initiated in 2007 by the Triscones research group in Geneva), and the team of motivated people made possible to achieve the results presented in this book.

In this monograph, we want to report on the latest progresses in photoemission spectroscopy techniques which allowed the investigation of buried complex oxide interfaces. While the conventional ARPES experiment focuses on the surface properties, surfaces are hardly construction elements of solid-state electronics, staying rather a playground for fundamental physics. The interfaces, on the contrary, represent the core of nowadays electronics and spintronics, and can therefore be characterized by the famous phrase "the interface is the device" by the Nobel laureate Herbert Kroemer. Selected examples of complex material systems, with particular attention to the paradigm system of oxide electronic, $LaAlO_3/SrTiO_3$ two-dimensional electron system (2DES), will be given together with quantitative theoretical description which often demands the advanced methods to be validated. Although photoemission spectroscopy is the main technique considered in this book, a special consideration is given also to the description of the growth of the oxides and the influence of the growth parameters on the transport properties and electronic structure.

In Chap. 1, an introduction to ARPES will be given stressing specifically the employ of high energy photoemission to probe interfaces. In Chaps. 2 and 3, a detailed description of the LAO/STO interface will be provided, including the effect of the growth parameters on the electronic and transport properties. Chapters 4–7 are devoted to the photoemission spectroscopy techniques (XPS, ARPES, and standing wave photoemission) to investigate complex oxide interfaces and bare TMO surfaces to extract the valence state, the band structure, and the Fermi surface. Chapters 8 and 9 present the theoretical work done on TMO interfaces within the DFT models further extended to include dynamical effects. Chapters 10 and 11 deal with two other spectroscopy techniques, XAS and RIXS, applied to buried interface and to magnetic materials.

We are very grateful to all the authors who participated in the realization of this book with their valuable chapter contributions. We enjoyed to compile and put together the excellent chapters prepared with care and attention by different authors in different laboratories all around the world.

Zürich, Switzerland
Claudia Cancellieri
Vladimir N. Strocov

Contents

1 Introduction: Interfaces as an Object of Photoemission Spectroscopy 1
C. Cancellieri and Vladimir N. Strocov
1.1   Introduction 1
1.2   Basics of Photoemission Spectroscopy: Energetics and Momentum Resolution 4
1.3   Spectral Function and Many-Body Interactions in Photoemission 5
    1.3.1   Matrix Element Effects 7
1.4   High-Energy Photoemission as a Probe for Buried Systems 8
    1.4.1   Probing Depth of High Energy Photoemission 10
    1.4.2   ARPES Instrumentation 11
References 14

2 The LaAlO$_3$/SrTiO$_3$ Interface: The Origin of the 2D Electron Liquid and the Fabrication 17
S. Gariglio and C. Cancellieri
2.1   Origin of the 2D Electron Liquid 17
    2.1.1   Introduction 17
    2.1.2   The Interface 18
    2.1.3   The Polar Discontinuity 20
    2.1.4   The Polar Catastrophe 21
    2.1.5   Beyond a Perfect Interface 22
2.2   The Growth of Crystalline LaAlO$_3$/SrTiO$_3$ Heterostructures 24
    2.2.1   PLD Growth 24
    2.2.2   The Role of LAO Deposition Conditions 26
    2.2.3   The Role of LAO Stoichiometry 29
2.3   Conclusions 31
References 31

3 Transport Properties of TMO Interfaces 37
A. M. R. V. L. Monteiro, A. D. Caviglia and N. Reyren
- 3.1 Introduction 37
- 3.2 Evidence for Multi-band Conduction from Magnetotransport ... 40
  - 3.2.1 Anisotropic Magnetotransport 40
  - 3.2.2 Universal Lifshitz Transition 40
- 3.3 Ground State of the LaAlO$_3$/SrTiO$_3$: Superconductivity and Magnetism 41
- 3.4 Nanopatterning 45
- 3.5 Other Paths of Exploration 46
  - 3.5.1 Spintronics 46
  - 3.5.2 Diode Effects, Circuits and Sensors 47
- References 48

4 ARPES Studies of Two-Dimensional Electron Gases at Transition Metal Oxide Surfaces 55
Siobhan McKeown Walker, Flavio Y. Bruno and Felix Baumberger
- 4.1 Introduction 55
  - 4.1.1 Metallic Subbands at the Surface of an Insulator 57
- 4.2 Origin of Surface 2DELs in TMOs 58
  - 4.2.1 UV Induced Oxygen Vacancies 58
- 4.3 2DEL Subband Structure at the (001), (110) and (111) Surfaces of SrTiO$_3$ 61
  - 4.3.1 SrTiO$_3$ (001) 61
  - 4.3.2 SrTiO$_3$ (111) and (110) surface 2DELs 69
- 4.4 Surface 2DELs in Other Transition Metal Oxides 71
  - 4.4.1 KTaO$_3$ 72
  - 4.4.2 Anatase TiO$_2$ 73
- 4.5 Many-Body Interactions in TMO 2DELs 74
- 4.6 Discussion 77
- References 79

5 Photoelectron Spectroscopy of Transition-Metal Oxide Interfaces 87
M. Sing and R. Claessen
- 5.1 Introduction 87
- 5.2 Depth Profiling 88
- 5.3 Band Bending and Offset 91
- 5.4 Role of Oxygen Vacancies 95
- 5.5 Conclusions and Outlook 103
- References 103

6 Electrons and Polarons at Oxide Interfaces Explored by Soft-X-Ray ARPES ... 107
Vladimir N. Strocov, Claudia Cancellieri and Andrey S. Mishchenko
- 6.1 Soft-X-Ray ARPES: From Bulk Materials to Interfaces and Impurities ... 108
  - 6.1.1 Virtues and Challenges of Soft-X-Ray ARPES ... 108
  - 6.1.2 Experimental Technique ... 110
  - 6.1.3 Application Examples ... 111
- 6.2 k-Resolved Electronic Structure of LAO/STO ... 117
  - 6.2.1 Resonant Photoemission ... 118
  - 6.2.2 Electronic Structure Fundamentals: Fermi Surface, Band Structure, Orbital Character ... 120
  - 6.2.3 Doping Effect on the Band Structure ... 121
- 6.3 Electron-Phonon Interaction and Polarons at LAO/STO ... 123
  - 6.3.1 Basic Concepts of Polaron Physics ... 123
  - 6.3.2 Polaronic Nature of the LAO/STO Charge Carriers ... 130
- 6.4 Oxygen Vacancies at LAO/STO ... 134
  - 6.4.1 Signatures of Oxygen Vacancies in Photoemission ... 134
  - 6.4.2 Tuning the Polaronic Effects ... 136
  - 6.4.3 Interfacial Ferromagnetism ... 137
  - 6.4.4 Phase Separation ... 138
- 6.5 Prospects ... 140
- 6.6 Conclusions ... 142
- References ... 143

7 Standing-Wave and Resonant Soft- and Hard-X-ray Photoelectron Spectroscopy of Oxide Interfaces ... 153
Slavomír Nemšák, Alexander X. Gray and Charles S. Fadley
- 7.1 Basic Principles of Resonant Standing-Wave Soft- and Hard-X-ray Photoemission (SXPS, HXPS) and Angle-Resolved Photoemission (ARPES) ... 153
  - 7.1.1 Standing-Wave Photoemission and Resonant Effects ... 153
  - 7.1.2 Near-Total Reflection Measurements ... 156
  - 7.1.3 ARPES in the Soft- and Hard-X-ray Regimes ... 157
- 7.2 Applications to Various Oxide Systems and Spintronics Materials ... 159
  - 7.2.1 Overview of Past Studies—Standing Waves from Multilayer Reflection ... 159
  - 7.2.2 Overview of Past Studies—Standing Waves from Atomic-Plane Reflection ... 160
  - 7.2.3 Multilayer Standing-Wave Soft- and Hard-X-ray Photoemission and ARPES from the Interface Between a Half-Metallic Ferromagnet and a Band Insulator: $La_{0.67}Sr_{0.33}MnO_3/SrTiO_3$ ... 160

	7.2.4	Multilayer Standing-Wave Soft X-ray Photoemission and ARPES Study of the Two-Dimensional Electron Gas at the Interface Between a Mott Insulator and a Band Insulator: GdTiO$_3$/SrTiO$_3$	163
	7.2.5	Multilayer Standing-Wave Photoemission Determination of the Depth Distributions at a Liquid/Solid Interface: Aqueous NaOH and CsOH on Fe$_2$O$_3$	167
	7.2.6	Near-Total Reflection Measurement of the Charge Accumulation at the Interface Between a Ferroelectric and a Doped Mott Insulator: BiFeO$_3$/(Ca$_{1-x}$Ce$_x$)MnO$_3$	169
	7.2.7	Atomic-Plane Bragg Reflection Standing-Wave Hard X-ray ARPES: Element- and Momentum-Resolved Electronic Structure of GaAs and the Dilute Magnetic Semiconductor Ga$_{1-x}$Mn$_x$As	170
7.3	Conclusions and Future Outlook		176
References			177

8 Ab-Initio Calculations of TMO Band Structure 181
A. Filippetti

- 8.1 Fundamentals of 2DEG Formation in SrTiO$_3$/LaAlO$_3$ According to Ab-Initio Calculations 181
  - 8.1.1 Ab-Initio Description of Polar Catastrophe: Role of Polar Discontinuity and Band Alignment 182
  - 8.1.2 Ab-Initio Description of Non-stoichiometric Mechanisms: Oxygen Doping and Cation Mixing 186
- 8.2 Band Structure and Related Properties: Spectroscopy, Thermal, and Transport Properties 190
  - 8.2.1 Basic Aspects of Band-Structure Calculation at the STO/LAO Interface 190
  - 8.2.2 Calculated Band Structures and Comparison with Spectroscopy 192
  - 8.2.3 Thermoelectric and Transport Properties: The Bloch-Boltzmann Approach 196
  - 8.2.4 Transport and Thermoelectric Properties of STO/LAO 200
  - 8.2.5 Analysis of Phonon-Drag in 2D Systems 203
- 8.3 Conclusions 207

References 208

9 Dynamical Mean Field Theory for Oxide Heterostructures 215
O. Janson, Z. Zhong, G. Sangiovanni and K. Held
- 9.1 Introduction 215
- 9.2 Steps of a DFT + DMFT Calculation Illustrated
  by SVO/STO Heterostructures 218
  - 9.2.1 Motivation for DFT + DMFT: The Electronic
    Structure of Bulk $SrVO_3$ 218
  - 9.2.2 Workflow of a DFT + DMFT Calculation 220
  - 9.2.3 Comparison with the Experiment 224
- 9.3 Applications 226
  - 9.3.1 Titanates 226
  - 9.3.2 Nickelates 227
  - 9.3.3 Vanadates 230
  - 9.3.4 Ruthenates 232
- 9.4 Conclusion and Outlook 234
- References 236

10 Spectroscopic Characterisation of Multiferroic Interfaces 245
M.-A. Husanu and C. A. F. Vaz
- 10.1 Introduction 245
  - 10.1.1 Intrinsic Multiferroics and Heterostructures 247
  - 10.1.2 Types of Interfacial Coupling 248
- 10.2 Introduction to PES and XAS: Differences
  and Complementarities 251
  - 10.2.1 X-ray Photoelectron Spectroscopy (XPS) 251
  - 10.2.2 X-ray Absorption Spectroscopy (XAS) 253
- 10.3 Photoemission Studies of Multiferroic Interfaces 254
  - 10.3.1 Probing Interface Region with Core-Level PES 254
  - 10.3.2 Schottky Barrier Height and Band Alignment
    at the Interface 257
  - 10.3.3 Angle-Resolved Photoelectron Spectroscopy
    of Multiferroic Interfaces 261
- 10.4 X-ray Absorption Spectroscopy Studies of Multiferroic
  Systems 264
  - 10.4.1 L-edge XAS Studies of Multiferroics
    and Heterostructures 265
  - 10.4.2 K-edge X-ray Absorption Spectroscopy
    of Multiferroics 268
  - 10.4.3 X-ray Photoemission Electron Microscopy
    (XPEEM) 270
- 10.5 Conclusions and Outlook 274
- References 275

11 Oxides and Their Heterostructures Studied with X-Ray Absorption Spectroscopy and Resonant Inelastic X-Ray Scattering in the "Soft" Energy Range 283
M. Salluzzo and G. Ghiringhelli
- 11.1 Introduction 283
- 11.2 Introduction to X-Ray Absorption Spectroscopy and Resonant Inelastic X-Ray Scattering 285
  - 11.2.1 X-Ray Absorption Spectroscopy 286
  - 11.2.2 X-Ray Linear Dichroism 288
  - 11.2.3 X-Ray Magnetic Circular Dichroism 291
- 11.3 Introduction to Resonant Inelastic X-Ray Scattering 293
- 11.4 XAS and RIXS of Oxide Heterostructures 296
  - 11.4.1 XAS on Cuprate Heterostructures 297
  - 11.4.2 RIXS of Cuprate Heterostructures 303
  - 11.4.3 XAS and RIXS of $LaAlO_3/SrTiO_3$ Heterostructures ... 305
- 11.5 Perspectives and Conclusions 311
- References 312

Index 315

Contributors

Felix Baumberger University of Geneva, Geneva, Switzerland; Swiss Light Source, Paul Scherrer Institut, Villigen, Switzerland

Flavio Y. Bruno University of Geneva, Geneva, Switzerland

Claudia Cancellieri Empa, Swiss Federal Laboratories for Materials Science and Technology, Laboratory for Joining Technologies and Corrosion, Dübendorf, Switzerland

A. D. Caviglia Kavli Institute of Nanoscience Delft University of Technology, Delft, The Netherlands

R. Claessen Physikalisches Institut and Röntgen Center for Complex Material Systems (RCCM), Universität Würzburg, Würzburg, Germany

Charles S. Fadley Department of Physics, University of California Davis, Davis, CA, USA; Lawrence Berkeley National Laboratory, Materials Sciences Division, Berkeley, CA, USA

A. Filippetti CNR-IOM Cagliari and Physics Department, University of Cagliari, Monserrato, Italy

S. Gariglio Department of Quantum Matter Physics, University of Geneva, Genève 4, Switzerland

G. Ghiringhelli CNR-SPIN and Dipartimento di Fisica, Politecnico di Milano, Milan, Italy

Alexander X. Gray Department of Physics, Temple University, Philadelphia, USA

K. Held Institute for Solid State Physics, TU Wien, Vienna, Austria

M.-A. Husanu Paul Scherrer Institut, Villigen PSI, Villigen, Switzerland; National Institute of Materials Physics, Magurele, Romania

O. Janson Institute for Solid State Physics, TU Wien, Vienna, Austria

Siobhan McKeown Walker University of Geneva, Geneva, Switzerland

Andrey S. Mishchenko RIKEN Center for Emergent Matter Science (CEMS), Saitama, Japan

A. M. R. V. L. Monteiro Kavli Institute of Nanoscience Delft University of Technology, Delft, The Netherlands

Slavomír Nemšák Peter-Grünberg-Institut PGI-6, Forschungszentrum Jülich GmbH, Jülich, Germany

N. Reyren Unité Mixte de Physique, CNRS, Thales, Univ. Paris-Sud, Université Paris-Saclay, Palaiseau, France

M. Salluzzo CNR-SPIN, Complesso MonteSantangelo via Cinthia, Naples, Italy

G. Sangiovanni Institut Für Theoretische Physik Und Astrophysik, Universität Würzburg, Würzburg, Germany

M. Sing Physikalisches Institut and Röntgen Center for Complex Material Systems (RCCM), Universität Würzburg, Würzburg, Germany

Vladimir N. Strocov Swiss Light Source, Paul Scherrer Institute, Villigen-PSI, Switzerland

C. A. F. Vaz Paul Scherrer Institut, Villigen PSI, Villigen, Switzerland

Z. Zhong Max Planck Institute for Solid State Physics, Stuttgart, Germany

Chapter 1
Introduction: Interfaces as an Object of Photoemission Spectroscopy

C. Cancellieri and Vladimir N. Strocov

Abstract In this short introductory chapter, basic concepts of photoemission techniques will be given. In particular, the importance of some parameters like probing depth, energy and momentum resolution will be tackled by comparing photoemission experiments in different photon energy ranges from ultraviolet to soft and hard X-rays. Buried system i.e. interfaces could be probed only by using high energy photoemission. Apart from the band structure resolved in electron momentum **k**, the photoemission technique directly probes the electron spectral function encoding information about how particles are dressed by their interactions with the remainder of the system. Many body effects and electron correlation can in this way be accessed, in particular, the electron-phonon interaction affecting electron mobility. Finally, the instrumental development of photoemission is described in connection with its scientific perspective.

1.1 Introduction

Angle-resolved photoelectron spectroscopy (ARPES) is a worldwide spread experimental technique to explore a variety of solid-state systems. Used in many laboratory and synchrotron facilities, this technique allows access to electronic band structure and many-body interactions in condensed matter physics. Historically the first photoelectric effect experiments that revealed the interaction of light with solids, were performed by Heinrich Hertz (Karlsruhe) and Wilhelm Hallwachs (Dresden) in 1887. At that time, the maximum kinetic energy of the photoelectrons could be determined under vacuum conditions by the retarding-field technique [1]. However, only several

---

C. Cancellieri (✉)
Empa, Swiss Federal Laboratories for Materials Science and Technology,
Laboratory for Joining Technologies and Corrosion,
Dübendorf, Switzerland
e-mail: claudia.cancellieri@empa.ch

V. N. Strocov
Paul Scherrer Institut, SLS, Villigen, Switzerland
e-mail: vladimir.strocov@psi.ch

© Springer International Publishing AG, part of Springer Nature 2018
C. Cancellieri and V. Strocov (eds.), *Spectroscopy of Complex Oxide Interfaces*, Springer Series in Materials Science 266,
https://doi.org/10.1007/978-3-319-74989-1_1

years after, the Einsteins formula was experimentally confirmed by observing a linear dependence between kinetic energy and photon energy hν. Progress in theory, and by rather diverse, but important instrumental improvements have occurred and still develops today. On the way of historical development of photoemission as the spectroscopic technique [2], various kinds of magnetic and electrostatic analyzers were developed in order to increase resolution of the photoelectron energy analysis. A large variety of photon sources became available, and at the same time physical and chemical surface condition become better controlled such that surface and bulk effects could be disentangled. Despite the huge instrumentation and technology improvement of nowadays setups, a photoemission experiment for spectroscopic purposes is basically performed in the same way as more than 100 years ago. The ARPES experiment is sketched in Fig. 1.1. Photons from a monochromatic light source—which can be produced either by a laboratory source like gas discharge lamp or X-ray tube, or by a synchrotron radiation facility—are shined on a sample. The photoelectrons,

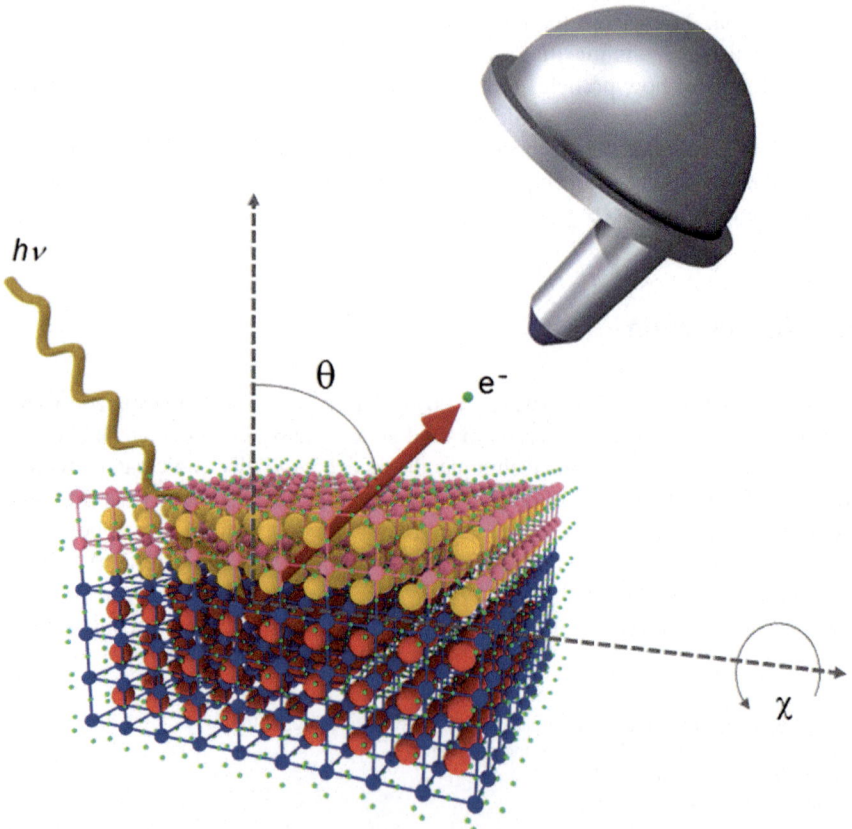

Fig. 1.1 The schematic representation of high energy ARPES. The detected electron travel through an overlayer material. Courtesy of M. Lopes, University of Geneva

liberated by the photoelectric effect, are then analyzed with respect to their emission angle and kinetic energy by an electrostatic analyzer. Photoemission as the spectroscopic technique was pioneered in the early 1960s by several groups, in particular by the group of Spicer (Stanford) [3], who measured the first ultraviolet (UV) photoelectron valence band spectrum on copper and conceived the three step models. Electronic surface states were first reported in 1962 for Si(111) [4].

The use of synchrotron radiation for a photoemission experiment has become very popular not only for a bigger data throughput in comparison with the laboratory UV or X-ray sources but also because many dedicated synchrotron facilities have become available worldwide. The main advantage of these radiation sources is the possibility to properly and accurately select the photon energy by a monochromator from a continuous spectrum over a wide range of energies. Other important advantages of the synchrotron light include variable polarization, high intensity and brightness, small photon spot on the sample, and the possibility of time-resolved experiments on a very short timescale, below the nanosecond range. These advantages made possible many experiments and pioneering findings otherwise unattainable using laboratory-sources. However, the comparatively simple and cheap laboratory sources are still widely used for many applications.

Photoemission spectroscopy can be performed on different classes of materials with a variety of physical properties like magnetism, superconductivity, 2D and 3D conductivity which can result in complicated band structures and correlation effects. The possibility to access to electron correlation effects should lead to a detailed understanding of many intriguing aspects of many-body physics that is fundamental for the development and application of complex functional materials for future generations of electronic and spintronic devices.

Complex oxides comprises mostly transition metal oxides (TMO) qualify nowadays as unique and diverse class of solids and a hot topic of interest for applications. The TMO materials exhibit a broad range of significant electronic properties ranging from ferroelectricity to metal-insulator transitions as well as from magnetism to superconductivity. Many of these compounds display structural instabilities, strong electronic correlations, and complex phase diagrams with competing ground states [5]. These features are quite desirable from an engineering perspective since they could prove instrumental in the design of novel ultrasensitive sensors with a strong response to small stimuli. Engineered structures of transition metal oxides therefore seem the ideal platform to explore interfacial effects that could possibly lead to new physics and material phases. In this respect, the conducting layer at the interface between $LaAlO_3$ (LAO) and $SrTiO_3$ (STO) that was observed by Ohtomo and Hwang a decade ago [6] has attracted a considerable amount of attention. Many efforts have been dedicated to the detailed investigation of this system, using a variety of techniques worldwide resulting in excellent works and interesting findings which make the LAO/STO the representative system for complex oxides interfaces. LAO/STO it is also one of the "protagonist" of this monograph devoted to various spectroscopic techniques applied to TMO interfaces.

1.2 Basics of Photoemission Spectroscopy: Energetics and Momentum Resolution

Photoemission spectroscopy (PES) is a well-known technique which measures electronic structure of a solid through the external photoelectric effect. If incident photon energy is higher than work function of the material, electrons from the top few or a few tens of atomic layers near the surface will be stimulated into vacuum with kinetic energy E_{kin}. Binding energy E_B of electrons back in the solid relative to the Fermi level E_F can be determined from the following equation:

$$E_B = E_{kin} - h\nu - \Phi \qquad (1.1)$$

where Φ is work function of the material and $h\nu$ photon energy. The simplest photoemission process consists of photoexcitation of an electron from the K-shell, i.e. the 1s core-level, of an atom. The corresponding photoemission spectrum consist of a single line with a binding energy that increases with the atomic number [7].

Angle-Resolved Photoemission Spectroscopy (ARPES) makes one step further and measures, in addition to energy of photoelectrons, their momenta rendering into band dispersions $E(\mathbf{k})$. The momentum selectivity of ARPES is based on the fact that photoexcitation in the solid conserves electron momentum $\mathbf{k}$. Furthermore, owing to the translation symmetry of the crystal surface, the surface-parallel momentum component $k_\parallel$ of the outgoing photoelectron is equal to its $k_\parallel$ back in the valence band. It can therefore be determined from the photoelectron polar emission angle θ (measured from the surface normal, see Fig. 1.1):

$$k_\parallel = \frac{\sqrt{2m(h\nu - \phi - E_B)}}{\hbar} \sin\theta \qquad (1.2)$$

The surface-perpendicular momentum is distorted by the crystal potential alternation at the surface, but in principle can be recovered assuming free-electron dispersion of photoexcited electron states. In the case of 2D states like the surface and interface states or electron states in layered materials such as graphene, it is possible to ignore $k_\perp$. Then the band dispersion $E(k_\parallel)$ is simply determined as E_B depending on $k_\parallel$ defined by the two above equations. The ARPES spectral intensity is related to so-called spectral function $A(\mathbf{k}, \omega)$, which contains fingerprints of all many-body interactions in the solid including electron-phonon coupling and electronic correlations discussed in Sect. 1.3, modulated by the photoexcitation matrix element.

A typical image of ARPES intensity rendered from the raw experimental (E_{kin}, θ) coordinates to $(E_B, \mathbf{k})$ is shown in Fig. 1.2a. Analysis of the ARPES data often employs so-called energy distribution curves (EDC) representing energy dependence of ARPES intensity for fixed $\mathbf{k}$, (b). The EDC peaks identify the band dispersions, (c). For metals, one can trace points in $\mathbf{k}$-space where the bands cross E_F. The manifold

1 Introduction: Interfaces as an Object of Photoemission Spectroscopy

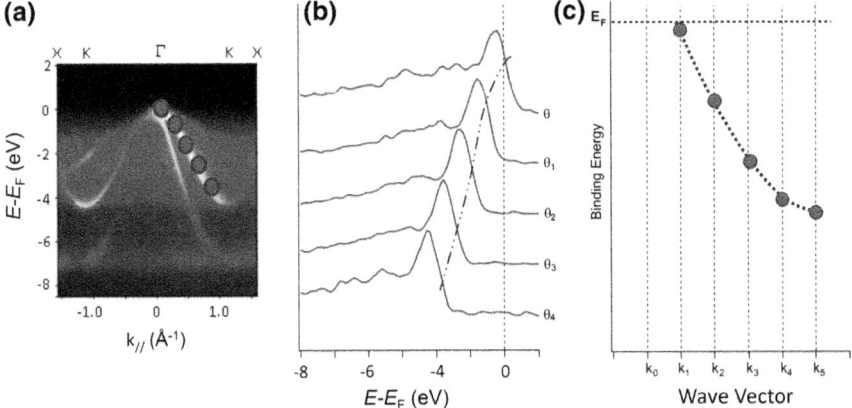

Fig. 1.2 An example of ARPES data measured on annealed Ag (100) cleaned surface (unpublished): **a** Photoemission intensity rendered from the measured (E_{kin}, θ) to $(E_B, \mathbf{k})$ coordinates using the 1.1 and 1.2; **b** Corresponding series of energy distribution curves (EDCs) crossing the Fermi level, and **c** band dispersions and **d** Fermi surface map

of these **k**-points defines the Fermi surface. Mapping of photoemission intensity at E_F as a function of two emission angles θ and χ (Fig. 1.2), rendering into two $k_\parallel$, produces the experimental Fermi surface contours.

1.3 Spectral Function and Many-Body Interactions in Photoemission

The power of ARPES as a spectroscopic technique is however not exhausted by informing only the electron energy levels and their distribution in **k**-space. If the photoelectron excitation and removal from the system are fast enough compared to the electron relaxation time (so-called sudden approximation which can be questioned only at very low excitation energies) the energy- and **k**-resolved photoemission intensity can be expressed as the following:

$$I(\omega, \mathbf{k}) = \left|M_{fi}\right|^2 F(\omega) A(\omega, \mathbf{k}) \tag{1.3}$$

where M_{fi} is the photoemission matrix element coupling the final state $|f>$ with the initial sate $<i|$, the Fermi-Dirac function $F(\omega) = \frac{1}{1+e^{\frac{\omega}{kT}}}$ restricts photoemission to the occupied initial states, and $A(\omega, \mathbf{k})$ is the one-particle spectral function. The latter, being essentially a probability to find an electron with energy ω and momentum k, informs the whole spectrum of electron interactions with other electrons, plasmons,

phonons, spin excitations, etc. in the system. While the reader can find exhaustive reviews of the many-body aspects of ARPES elsewhere—see, for example, [8, 9] and the references therein—we will here recap the most basic concepts. A typical shape of $A(\omega, \mathbf{k})$ is sketched in Fig. 1.3. A non-interacting electron system (where the electron interaction is not actually neglected but approximated by interaction with an average effective potential) is characterized by $A(\omega, \mathbf{k})$ shown in Fig. 1.3a where each k-state is a δ-peak corresponding to a non-interacting particle having infinite lifetime. Each state has the normalized weight $Z = 1$, and the occupation number is either 1 below the Fermi vector k_F or 0 above k_F. In the interacting electron case, however, the excitations in the system such as electrons or holes get dynamically screened by other electrons and finally dissipate, which results in shifting of their energy compared to the non-interacting case and in their finite lifetime. Mathematically, this characteristic property of the interacting system is expressed with a concept of a (non-local operator of) complex self-energy

$$\Sigma(\omega, \mathbf{k}) = \Sigma'(\omega, \mathbf{k}) + i\Sigma''(\omega, \mathbf{k}) \tag{1.4}$$

whose real part $\Sigma'(\omega, \mathbf{k})$ describes the energy shift relative to the non-interacting case and imaginary part $\Sigma''(\omega, \mathbf{k})$ the finite lifetime. The corresponding $A(\omega, \mathbf{k})$ takes the form

$$\mathbf{A}(\omega, \mathbf{k}) = \frac{-1}{\pi} \frac{\Sigma''(\omega, \mathbf{k})}{\left[\omega - \varepsilon(\mathbf{k}) - \Sigma'(\omega, \mathbf{k})\right]^2 + \left[\Sigma''(\omega, \mathbf{k})\right]^2} \tag{1.5}$$

where $\varepsilon(\mathbf{k})$ is the non-interacting band dispersion. Such screened excitations in the interacting electron system are called quasiparticles (QPs). The sketch of $A(\omega, \mathbf{k})$ in this case is shown in Fig. 1.3b. The electron interactions shift the QP peaks in energy by $\Sigma'(\omega, \mathbf{k})$ compared to the non-interacting case and broaden them to an energy width equal to $\Sigma''(\omega, \mathbf{k})$. The broadening results in fractional occupation number of states near k_F. ARPES studies of the QP peak lineshape focus on mechanisms of electron-electron correlation which can follow distinctly different models of the electron system ranging from the canonical Fermi liquid to so-called marginal Fermi liquid in strongly correlated systems such as high-temperature superconductors and exotic Tomonaga–Luttinger liquid in one-dimensional systems [8, 10]. Figure 1.3b illustrates that $A(\omega, \mathbf{k})$ can embed, apart from the QP peak, additional satellites at higher binding energy. These satellites arise from entanglement of electrons with charge-neutral bosonic excitations such as plasmons, spinons, excitons, phonons, etc. [11]. As in any case $A(\omega, \mathbf{k})$ obeys the sum rule $\int_{-\infty}^{\infty} A(\omega, \mathbf{k})d\omega = 1$, the satellites steal spectral weight from the QP peak which acquires then a fractional normalized weight $Z_k < 1$. We find examples of electron entanglement with bosons in the conventional superconductivity, where electron-phonon coupling mediates formation of the Cooper pairs, or in possible scenarios of high-temperature superconductivity mechanisms involving electron coupling to spin excitations. For oxide systems with their ionic character and easy structural transformations, particularly important is

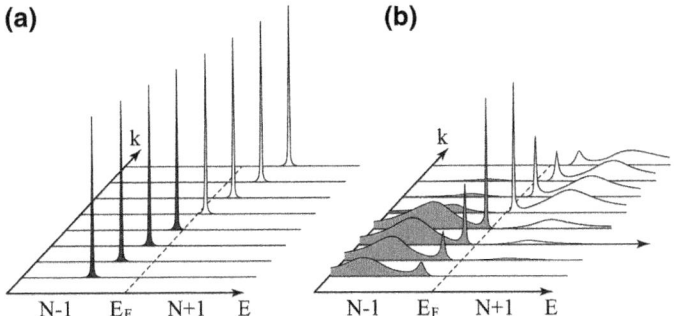

Fig. 1.3 One-electron spectral function $A(\omega, \mathbf{k})$ for a non-interacting (**a**) and interacting (**b**) electron systems. Electron-electron interaction shifts and broadens the $A(\omega, \mathbf{k})$ peaks, and electron-boson interaction forms a satellite structure. Figure adapted from [8]

formation of so-called polarons [12] when, in response to strong electron-phonon coupling, electrons drag behind them a local lattice distortion significantly reducing their mobility (see the Chap. 6). Therefore, ARPES studies of the satellite structures deliver information about the electron entanglement with bosonic modes crucial for a wealth of physical phenomena.

1.3.1 Matrix Element Effects

As we have seen above, the photoemission response of $A(\omega, \mathbf{k})$ is modulated by the matrix element which, in the framework of one-step theory of photoemission [13], is found as $M_{fi} = \langle f | \mathbf{A} \cdot \mathbf{p} | i \rangle$ where the final and initial states are coupled by a product of the electromagnetic field vector potential $\mathbf{A}$ and momentum operator $\mathbf{p}$. The experimental geometry and photon energy dependence of M_{fi} can be used constructively in the ARPES experiment. For example, one can perform symmetry analysis of the valence states by choosing the experimental geometry in such a way that the photoelectrons are detected in a mirror plane of the sample. In this case p-polarization of the incident photons (the electric field vector E parallel to the mirror plane) excites only the valence states having even symmetry, and s-polarization (E perpendicular to the mirror plane) only having odd symmetry relative to the mirror plane [8, 14, 15]. Furthermore, if we neglect the polarization effects expressed by $\mathbf{A} \cdot \mathbf{p}$, then M_{fi} appears simply as a scalar product $\langle f | i \rangle$ of the final and initial sate wave functions. Tuning photon energy varies $\langle f |$ and therefore its overlap with $| i \rangle$, which allows us to either silence or boost photoemission response of certain valence states to discriminate them from the rest of the valence band. Examples of such constructive use of the matrix elements can be found in the Chaps. 5 and 6.

1.4 High-Energy Photoemission as a Probe for Buried Systems

An important parameter to describe surface sensitivity of the photoemission experiment is its probing depth defined by mean free path (MFP) of photoelectrons escaping from the solid. In Fig. 1.4, the so-called "universal curve" of the electron MFP is shown as a function of kinetic energy [16]. The best method to increase the probing depth and simultaneously increase electron momentum resolution seems to reduce $h\nu$ It seems then natural to consider a UV laser as a low-energy light source for the ARPES experiments. However, the pulse length of a regular laser is very short and the number of photons in each pulse is very high, that brings in space charge effects deteriorating the energy resolution. Apart from instrumental limitations, a fundamental drawback of ARPES experiments with UV laser sources is their limited intercept in **k**-space coming from small $h\nu$, see 1.2. Moreover, the "universal curve" compiled mostly for metals does not take into account elastic scattering of photoelectrons in the low energy final states [17] and thus largely overestimates MFP for oxide and insulating materials. In the higher energy UV photoelectron spectroscopy (UPS) energy range the MFP dramatically decreases. This translates into the fact that only a thin layer of the sample is probed in UPS, indicated in Fig. 1.4, and the surface preparation is fundamental. Clearly the UPS cannot be applied to investigate buried interfaces but only to bare surfaces which are properly prepared prior to the experiment. Chapter 4 is dedicated to the photoemission of bare TMO surfaces like STO. In order to overcome the surface sensitivity of the UPS, $h\nu$ has to be pushed to the opposite side of the "universal curve" at higher energy values. The soft and

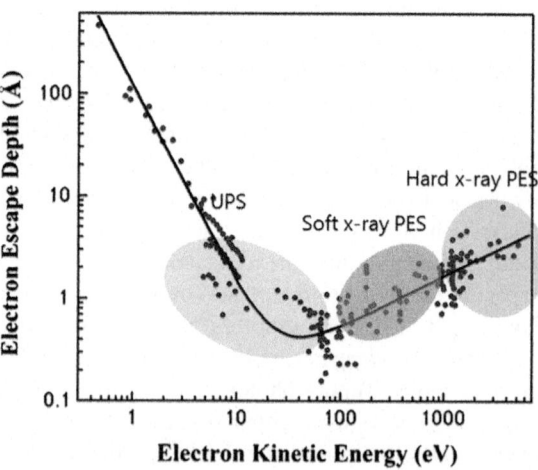

Fig. 1.4 Universal curve of photoemission with different PES energy ranges indicated. Adapted from [16]

hard X-ray regions indicated in Fig. 1.4, represents the spectroscopic ranges where the enhanced bulk sensitivity allows access to buried systems such as interfaces, heterostructures, and impurities (see Fig. 1.1 as a representative scheme of photoemission on buried systems). Moreover, surface preparation and control is less critical in this range of photon energies, making the ARPES experiment more practical.

The fundamental advantage of soft-X-ray ARPES (SX-ARPES) with $h\nu$ around 1 keV compared to the conventional UV-ARPES, as we have seen above, is an increase of the photoelectron MFP and thus probing depth [16]. SX-ARPES virtues and challenges will be discussed in detail in Chap. 6, and here we briefly summarize them. For the merits, we can mention:

- Probing depth: the use of soft X-rays makes this technique less surface sensitive than UV-photoemission
- 3D momentum resolution: the increase of MFP results in improvement of the intrinsic $k_\perp$ resolution [18].
- Chemical specificity achieved using resonant photoemission: the SX-ARPES energy range goes through the L-absorption edges of transition metals and M-edges of the rare earths.

Challenges are however present:

- Reduction of the valence band (VB) photoexcitation cross-section by a few orders of magnitude compared to the UV-ARPES energy range. This can only be overpowered with advanced synchrotron radiation instrumentation delivering high photon flux [19].
- Relaxation of **k**-selectivity because of the atomic thermal motion.
- Energy resolution roughly proportional to $h\nu$.

Hard X-ray photoelectron spectroscopy (HAXPES) in the multi-keV range can be beneficial for further increase of the MFP to 5–10 nm, delivering true bulk sensitivity. This virtue goes along however with dramatic reduction of the VB photoexcitation cross section and relaxation of **k** resolution, resulting in loss of coherent spectral and, on the high-energy end of HAXPES, the recoil effects that smear not only the **k** but also energy definition [20]. Although **k** resolved studies have been demonstrated for some materials, the main focus of HAXPES is **k**-integrated core level studies. HAXPES studies on buried interfaces and heterostructures, including depth-resolved experiments with standing X-ray waves, are discussed in Chaps. 5 and 7.

This volume focuses mainly on SX-ARPES combining significant increase of the probing depth compared on the UV with good **k** resolution and still acceptable VB cross-section. The use of cutting edge instrumentation, combining high energy and **k** resolution with high photon flux and detection efficiency, is essential for studying the electronic structures of buried interfaces and strongly correlated electron systems with different surface and bulk electronic structures.

1.4.1 Probing Depth of High Energy Photoemission

1.4.1.1 GaAs behind As protective layer

A representative example of the penetrating ability of SX-ARPES is the work on GaAs buried behind a capping layer of amorphous As layer with a thickness of 1 nm [21, 22] which protected the GaAs underlayer against oxidization in atmosphere. Although such capping layer can be removed by annealing up to 180 °C this complicates the photoemission experiment. The problem is overcome by the use of SX-ARPES with photon energies up to 1 KeV which allows the observation of bulk dispersions in GaAs underlayer through the capping layer, an obviously impossible case for the conventional UV-ARPES.

In Fig. 1.5 three experimental ARPES images acquired with increasing photon energy are shown. The $h\nu$ values were selected to stay at the Γ-X line of the Brillouin zone (BZ). The low-energy image in panel (b) evidences no band dispersion and it is dominated by the structureless signal from the amorphous As overlayer. Increasing the energy of the incident photons allows to access the GaAs underlayer as the electron MFP is also increased. In panels (c) and (d) the ARPES images gradually develop astonishingly clear dispersive structures formed by the coherent photoelectrons crossing the amorphous layer. These band structures can be distinguished in the canonical manifold of the light-hole (LH), heavy-hole (HH), and spin-orbit split (SO) bands of GaAs seen through the amorphous capping layer.

1.4.1.2 LaAlO$_3$/SrTiO$_3$ Interface

New physical phenomena, not anticipated from the properties of individual bulk constituents, can arise at interfaces and heterostructures of strongly correlated TMOs. The new functionalities associated can have large impact and relevance for future device applications. One such system is the LaAlO$_3$/SrTiO$_3$ (LAO/STO) interface, one of the main subjects of this book. The use of ARPES technique to investigate this kind of buried systems, achievable nowadays by virtue of the cutting edge instrumentation and last generation synchrotrons, allows direct access to their **k**-resolved electronic band structure. In this case SX-ARPES allows penetration through the LAO layer which is at least 4 u.c. (1.5 nm) thick in order to achieve the two-dimensional electron system (2DES) at the LAO/STO interface (Fig. 1.6). However, the 2DES signal is extremely small. The use of resonant photoemission with its elemental and chemical state specificity [24], can overcome the low intensity signal, increasing specifically the signal of the interface Ti^{3+} ions whose valency contrasts them with the Ti^{4+} ones from the STO bulk. The long sought for **k**-resolved information about the interfacial Fermi surface, band structure and orbital composition is obtained in this SX-ARPES experiment. The detailed account of SX-ARPES experiments, including identification of polaronic nature of the interface charge carriers, will be given in Chaps. 5 and 6. In Chaps. 2 and 3 the growth parameters and the transport

1 Introduction: Interfaces as an Object of Photoemission Spectroscopy

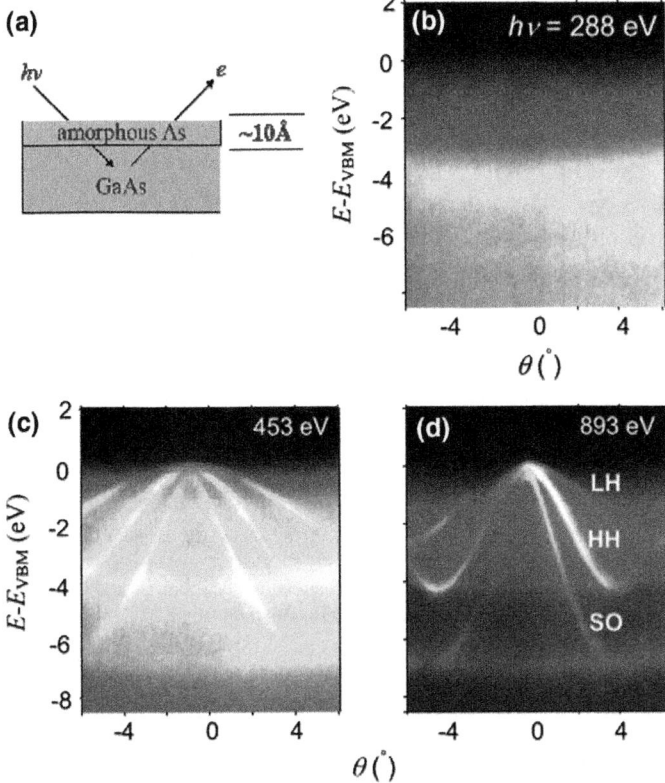

Fig. 1.5 Soft X-ray ARPES of As-protected GaAs: **a** GaAs(100) samples protected by amorphous As capping layer; (**b–d**) ARPES images along the Γ-K-X line of the BZ measured with the indicated hν. Development of the GaAs band structure signal with hν evidences the increase of the electronic MFP. Adapted from [21]

properties of the LAO/STO interface will be presented. For a comprehensive photoemission investigation of the bare STO surface the reader is referred to Chap. 4. Density-functional and many-body theoretical picture of the LAO/STO electronic band structure will be described in Chaps. 8 and 9, respectively.

1.4.2 ARPES Instrumentation

The enormous progress in ARPES instrumentation seen over the last decades is coming from dramatic improvement of the light sources, on one side, and of the ARPES analyser on the other side. Although routine ARPES experiments can be performed with laboratory sources such as X-ray tubes, UPS gas-discharge lamps and lasers, the most advanced experiments require synchrotron radiation sources (see [25] for

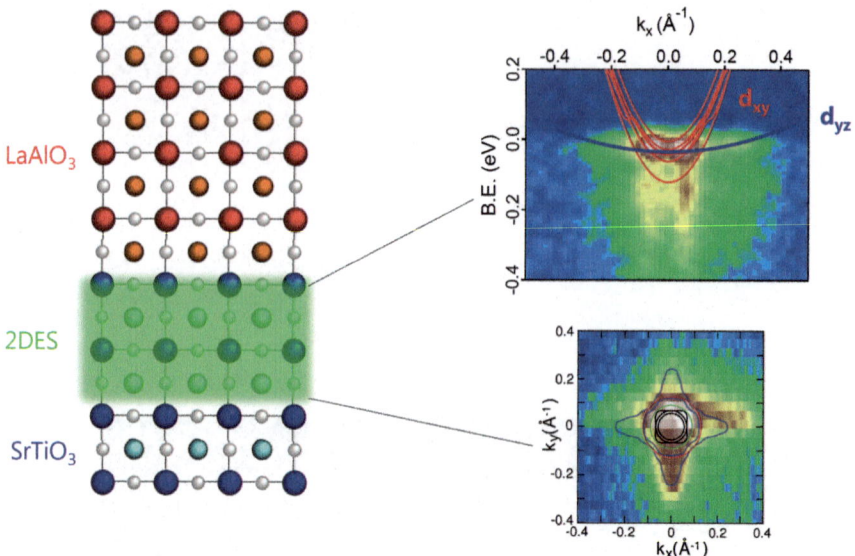

Fig. 1.6 Schematic representation of the ARPES signal collected on buried LAO/STO interface. Fermi surfaces and band dispersion can be accessed. Adapted from [23]

entries). Starting from the 3rd generation machines, the synchrotrons produce light in straight sections of the ring using wigglers or undulators where movement of relativistic electrons through a periodic array of magnets produces intense coherent radiation. These sources bring advantages of high photon flux, tunable photon energy and polarization, and small focused spot size on the sample. The photon flux is particularly important for ARPES of buried interfaces with soft and hard X-rays because of dramatic reduction of the valence band photoexcitation cross-section at high photon energies [26] and, naturally, attenuation of photoelectrons in the overlayer. On the photoelectron analyser side, the general evolution trend of the ARPES instrumentation has been development of dispersive and imaging electron optics to achieve multichannel detection in an extended region of photoelectron energies E and emission angles θ [27–29]. A modern high-resolution ARPES analyzer—the hemispherical analyser (HSA)—is sketched in Fig. 1.7. Energy dispersion and focusing of electrons, performed by the hemisphere, combine in this analyzer with imaging of emission angles onto the entrance slit of the hemisphere, performed by the electrostatic lens. This combination creates in the HSA focal plane a 2D image of photoelectron intensity as a function of E and θ. The dramatic efficiency gain achieved with this multichannel concept has been instrumental for the recent progress of ARPES. The most recent development has been the angle-resolved time-of-flight (ARTOF) analyzer [30] which allows multichannel acquisition of already 3D photoelectron intensity data as a function of E and two orthogonal emission angles θ_x and θ_y.

1 Introduction: Interfaces as an Object of Photoemission Spectroscopy

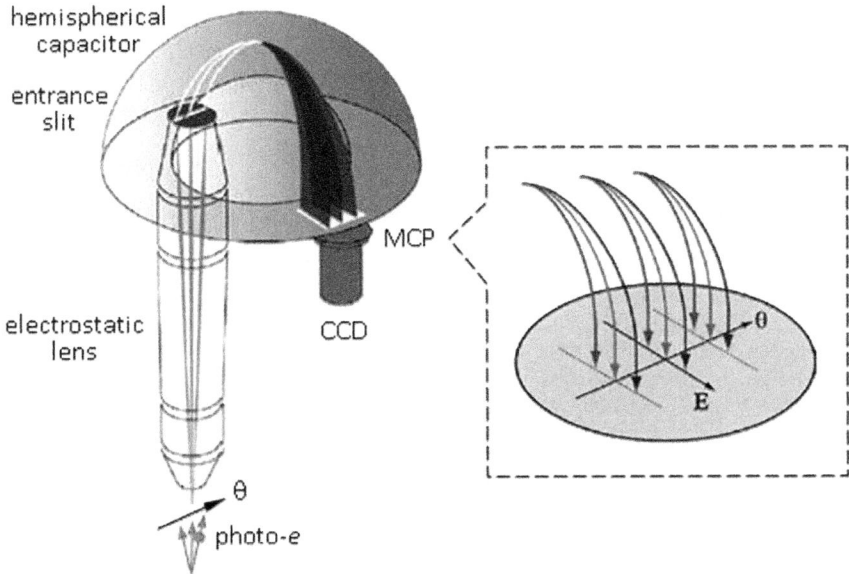

Fig. 1.7 Scheme of a hemispherical electron analyzer. Electron energy dispersion in the hemispherical capacitor combines with angular imaging in the electrostatic lens, which creates a 2D image of photoemission intensity as a function of energy E and emission angle θ (adopted from [28])

Detection of the photoelectron spin in the ARPES experiment, in addition to energy and momentum, plays a crucial role in investigation of many physical phenomena, ranging from the obvious example of magnetism, via novel materials for spintronics applications such multiferroic Rashba semiconductors [31] to high temperature superconductivity (see a recent review [32] and the references therein). On the technical side, however, spin-resolved ARPES (SARPES) suffers from an immense intensity loss of at least 2 orders of magnitude associated with spin resolution and, until very recently, detecting photoelectrons only in one energy and angular channel. Most recently, however, a few concepts of multichannel SARPES detectors have been demonstrated. They use either spin-polarized reflectivity of low-energy electrons [28, 33] or Mott scattering of high-energy electrons [15] as spin-selective processes, which are combined with imaging-type electron optics to deliver the full 2D image of spin asymmetry as a function of photoelectron E and θ. A dramatic efficiency gain achieved with these multichannel instruments promises to push SARPES from essentially surface science applications in the UV photon energy range to buried interfaces and heterostructures in the soft- and hard-X-ray ranges.

Concluding remarks We have therefore seen that high-energy photoemission has just broken out as the experimental technique delivering previously unthinkable k-resolved information about the electronic structure of buried interfaces and heterostructures such as their Fermi surface, band structure and spectral function. The following chapters develop in detail all the subjects outlined in this introductory chapter. The reader is referred to them for each particular topic of interest.

References

1. H. Bonzel, C. Kleint, On the history of photoemission. Prog. Surf. Sci. **48**(1), 179 (1995), https://doi.org/10.1016/0079-6816(95)93425-7, http://www.sciencedirect.com/science/article/pii/0079681695934257
2. F. Reinert, S. Hfner, Photoemission spectroscopy-from early days to recent applications. New J. Phys. **7**(1), 97 (2005), http://stacks.iop.org/1367-2630/7/i=1/a=097
3. W.E. Spicer, C.N. Berglund, d band of copper. Phys. Rev. Lett. **12**, 9–11 (1964), https://doi.org/10.1103/PhysRevLett.12.9, http://link.aps.org/doi/10.1103/PhysRevLett.12.9
4. G.W. Gobeli, F.G. Allen, Direct and indirect excitation processes in photoelectric emission from silicon. Phys. Rev. **127**, 141–149 (1962), https://doi.org/10.1103/PhysRev.127.141, https://link.aps.org/doi/10.1103/PhysRev.127.141
5. D.I. Khomskii, *Transition Metal Compounds*, (Cambridge University Press, Cambridge 2014), https://doi.org/10.1017/CBO9781139096782, https://www.cambridge.org/core/books/transition-metal-compounds/037907D3274F602D84CFECA02A493395
6. A. Ohtomo, H. Hwang, A high-mobility electron gas at the LaAlO3/SrTiO3 heterointerface (vol 427, pg 423, 2004). Nature **441**(7089), 120 (2006), https://doi.org/10.1038/nature04773
7. S. Hüfner, Photoelectron spectroscopy: principles and applications, in *Advanced Texts in Physics*. (Springer, 2003), https://books.google.ch/books?id=WfOw6jP9-oIC
8. A. Damascelli, Z. Hussain, Z.X. Shen, Angle-resolved photoemission studies of the cuprate superconductors. Rev. Mod. Phys. **75**, 473–541 (2003), https://doi.org/10.1103/RevModPhys.75.473, http://link.aps.org/doi/10.1103/RevModPhys.75.473
9. S. Kevan (ed.), *Studies in Surface Science and Catalysis*, vol. 74 (Elsevier 1992), https://doi.org/10.1016/S0167-2991(08)61767-X, http://www.sciencedirect.com/science/article/pii/S016729910861767X
10. L. Perfetti, S. Mitrovic, M. Grioni, Fermi liquid and non-fermi liquid spectral lineshapes in low-dimensional solids. J. Electron Spectrosc. Relat. Phenom. **127**(12), 77–84 (2002), https://doi.org/10.1016/S0368-2048(02)00175-5, http://www.sciencedirect.com/science/article/pii/S0368204802001755. IWASES 5 Special Issue
11. G.D. Mahan, Many-particle physics [Elektronische Ressource], in *Physics of Solids and Liquids* (Springer, Boston, MA 2000), http://search.ebscohost.com/login.aspx?direct=true&AuthType=ip&db=cat04420a&AN=LIB.gbv744982855&site=eds-live
12. A.S. Alexandrov, J.T. Devreese, Advances in polaron physics, in *Springer series in solid-state sciences*, vol. 159 (Springer, Berlin 2010), http://search.ebscohost.com/login.aspx?direct=true&AuthType=ip&db=cat04420a&AN=LIB.swissbib305572687&site=eds-live
13. P.J. Feibelman, D.E. Eastman, Photoemission spectroscopy—correspondence between quantum theory and experimental phenomenology. Phys. Rev. B **10**, 4932–4947 (1974), https://doi.org/10.1103/PhysRevB.10.4932, http://link.aps.org/doi/10.1103/PhysRevB.10.4932
14. J. Hermanson, Final-state symmetry and polarization effects in angle-resolved photoemission spectroscopy. Solid State Communic. **22**(1), 9–11 (1977), https://doi.org/10.1016/0038-1098(77)90931-0, http://www.sciencedirect.com/science/article/pii/0038109877909310
15. V.N. Strocov, V.N. Petrov, J.H. Dil, Concept of a multichannel spin-resolving electron analyzer based on Mott scattering. J. Synchrotron Radiat. **22**(3), 708–716 (2015), https://doi.org/10.1107/S160057751500363X, https://dx.doi.org/10.1107/S160057751500363X

16. M.P. Seah, W.A. Dench, Quantitative electron spectroscopy of surfaces: a standard data base for electron inelastic mean free paths in solids. Surf. Interface Anal. **1**(1), 2–11 (1979), https://doi.org/10.1002/sia.740010103, https://doi.org/10.1002/sia.740010103
17. V.N. Strocov, P. Blaha, H.I. Starnberg, M. Rohlfing, R. Claessen, J.M. Debever, J.M. Themlin, Three-dimensional unoccupied band structure of graphite: very-low-energy electron diffraction and band calculations. Phys. Rev. B **61**, 4994–5001 (2000), https://doi.org/10.1103/PhysRevB.61.4994, http://link.aps.org/doi/10.1103/PhysRevB.61.4994
18. V. Strocov, Intrinsic accuracy in 3-dimensional photoemission band mapping. J. Electron Spectrosc. Relat. Phenom. **130**(1–3), 65–78 (2003). https://doi.org/10.1016/S0368-2048(03)00054-9. http://www.sciencedirect.com/science/article/pii/S0368204803000549
19. V.N. Strocov, T. Schmitt, U. Flechsig, T. Schmidt, A. Imhof, Q. Chen, J. Raabe, R. Betemps, D. Zimoch, J. Krempasky, X. Wang, M. Grioni, A. Piazzalunga, L. Patthey, High-resolution soft X-ray beamline adress at the swiss light source for resonant inelastic X-ray scattering and angle-resolved photoelectron spectroscopies. J. Synch. Rad. **17**(5), 631–643 (2010). https://doi.org/10.1107/S0909049510019862
20. S. Suga, A. Sekiyama, High energy photoelectron spectroscopy of correlated electron systems and recoil effects in photoelectron emission. European Phys. J. Spec. Top. **169**(1), 227–235 (2009), https://doi.org/10.1140/epjst/e2009-00997-4, https://dx.doi.org/10.1140/epjst/e2009-00997-4
21. M. Kobayashi, I. Muneta, T. Schmitt, L. Patthey, S. Ohya, M. Tanaka, M. Oshima, V.N. Strocov, Digging up bulk band dispersion buried under a passivation layer. Appl. Phys. Lett. **101**(24), 242103 (2012), https://doi.org/10.1063/1.4770289, http://aip.scitation.org/doi/abs/10.1063/1.4770289
22. M. Kobayashi, I. Muneta, Y. Takeda, Y. Harada, A. Fujimori, J. Krempaský, T. Schmitt, S. Ohya, M. Tanaka, M. Oshima, V.N. Strocov, Unveiling the impurity band induced ferromagnetism in the magnetic semiconductor (Ga, Mn)As. Phys. Rev. B **89**, 205204 (2014), https://doi.org/10.1103/PhysRevB.89.205204, http://link.aps.org/doi/10.1103/PhysRevB.89.205204
23. C. Cancellieri, M.L. Reinle-Schmitt, M. Kobayashi, V.N. Strocov, P.R. Willmott, D. Fontaine, P. Ghosez, A. Filippetti, P. Delugas, V. Fiorentini, Doping-dependent band structure of $LaAlO_3/SrTiO_3$ interfaces by soft X-ray polarization-controlled resonant angle-resolved photoemission. Phys. Rev. B **89**, 121412 (2014). https://doi.org/10.1103/PhysRevB.89.121412, http://link.aps.org/doi/10.1103/PhysRevB.89.121412
24. S.L. Molodtsov, M. Richter, S. Danzenbächer, S. Wieling, L. Steinbeck, C. Laubschat, Angle-resolved resonant photoemission as a probe of spatial localization and character of electron states. Phys. Rev. Lett. **78**, 142–145 (1997), https://doi.org/10.1103/PhysRevLett.78.142, http://link.aps.org/doi/10.1103/PhysRevLett.78.142
25. P. Willmott, An introduction to synchrotron radiation [Elektronische Daten]: techniques and applications. (Wiley, Chichester, 2011), http://search.ebscohost.com/login.aspx?direct=true&AuthType=ip&db=cat04420a&AN=LIB.swissbib12226536X&site=eds-live
26. J. Yeh, I. Lindau, Atomic subshell photoionization cross sections and asymmetry parameters: $1 \leq z \leq 103$. At. Data Nucl. Data Tables **32**(1), 1–155 (1985), https://doi.org/10.1016/0092-640X(85)90016-6, http://www.sciencedirect.com/science/article/pii/0092640X85900166
27. N. Mårtensson, P. Baltzer, P. Brühwiler, J.O. Forsell, A. Nilsson, A. Stenborg, B. Wannberg, A very high resolution electron spectrometer. J. Electron Spectrosc. Relat. Phenom. **70**(2), 117–128 (1994), https://doi.org/10.1016/0368-2048(94)02224-N, http://www.sciencedirect.com/science/article/pii/036820489402224N
28. S. Suga, C. Tusche, Photoelectron spectroscopy in a wide h region from 6 ev to 8 kev with full momentum and spin resolution. J. Electron Spectrosc. Relat. Phenom. **200**, 119–142 (2015), https://doi.org/10.1016/j.elspec.2015.04.019, http://www.sciencedirect.com/science/article/pii/S0368204815000912. Special Anniversary Issue: Volume 200

29. B. Wannberg, Electron optics development for photo-electron spectrometers. Nuclear Instruments and Methods in Physics Research Section A: Accelerators, Spectrometers, Detectors and Associated Equipment **601**(1–2), 182–194 (2009). https://doi.org/10.1016/j.nima.2008.12.156. http://www.sciencedirect.com/science/article/pii/S0168900208020238. Special issue in honour of Prof. Kai Siegbahn
30. G. Öhrwall, P. Karlsson, M. Wirde, M. Lundqvist, P. Andersson, D. Ceolin, B. Wannberg, T. Kachel, H. Dürr, W. Eberhardt, S. Svensson, A new energy and angle resolving electron spectrometer—first results. J. Electron Spectrosc. Relat. Phenom. **183**(1–3), 125–131 (2011), https://doi.org/10.1016/j.elspec.2010.09.009, http://www.sciencedirect.com/science/article/pii/S0368204810002045. Electron Spectroscopy Kai Siegbahn Memorial Volume
31. J. Krempasky, S. Muff, F. Bisti, M. Fanciulli, H. Volfová, A.P. Weber, N. Pilet, P. Warnicke, H. Ebert, J. Braun, F. Bertran, V. Volobuev, J. Minár, G. Springholz, J.H. Dil, V. Strocov, Entanglement and manipulation of the magnetic and spin-orbit order in multiferroic Rashba semiconductors (2016), arXiv:1606.00241
32. N. Mannella, Measuring spins in photoemission experiments: old challenges and new opportunities. Synchrotron Radiat. News **27**(2), 4–13 (2014), https://doi.org/10.1080/08940886.2014.889548, https://dx.doi.org/10.1080/08940886.2014.889548
33. M. Kolbe, P. Lushchyk, B. Petereit, H.J. Elmers, G. Schönhense, A. Oelsner, C. Tusche, J. Kirschner, Highly efficient multichannel spin-polarization detection. Phys. Rev. Lett. **107**, 207601 (2011), https://doi.org/10.1103/PhysRevLett.107.207601, http://link.aps.org/doi/10.1103/PhysRevLett.107.207601

Chapter 2
The LaAlO$_3$/SrTiO$_3$ Interface: The Origin of the 2D Electron Liquid and the Fabrication

S. Gariglio and C. Cancellieri

Abstract This chapter discusses the formation of the 2D electron liquid at the LAO/STO interface. The first part presents the theoretical proposals aimed at explaining the origin of the charge at the interface. The second part focuses on the importance of the growth techniques and parameters like temperature, oxygen pressure, post-deposition annealing and their influence on the electronic transport properties.

2.1 Origin of the 2D Electron Liquid

2.1.1 Introduction

The progress in the deposition of epitaxial oxides thin films achieved in the last 20 years allows currently the fabrication of oxide heterostructures with a quality equivalent to the one reached in III–V semiconductors. This control at the atomic level has enabled the discovery of a set of fascinating properties and phenomena in complex oxides such that today novel materials are first designed by first-principles calculations before their fabrication. The discovery of a conducting interface between two insulating perovskites, LaAlO$_3$ and SrTiO$_3$, instead, happened as a surprise: stacking together two insulating materials with large energy gaps (>3 eV) is not expected to produce conductivity at their interface. This finding stems from the ability to prepare perovskite single crystals with atomically smooth surfaces, chemically and thermally treated to offer only one chemical termination for the deposition of

---

S. Gariglio (✉)
Department of Quantum Matter Physics, University of Geneva,
24 Quai Ernest-Ansermet, 1211 Genève 4, Switzerland
e-mail: stefano.gariglio@unige.ch

C. Cancellieri
Empa, Laboratory for Joining technologies and corrosion,
Dübendorf, Switzerland
e-mail: claudia.cancellieri@empa.ch

© Springer International Publishing AG, part of Springer Nature 2018
C. Cancellieri and V. Strocov (eds.), *Spectroscopy of Complex Oxide Interfaces*, Springer Series in Materials Science 266,
https://doi.org/10.1007/978-3-319-74989-1_2

an oxide layer which is grown unit-cell by unit-cell. In this chapter we present the LaAlO$_3$/SrTiO$_3$ system, the models proposed to explain the presence of free carriers at its interface, before reviewing in detail the growth of the LaAlO$_3$ layer and the influence of the film deposition parameters on the physical properties of the interface.

2.1.2 The Interface

SrTiO$_3$ stands as one of the most remarkable examples of complex oxides. Having a perovskite structure (see Fig. 2.1), it displays a variety of physical properties upon small changes in its stoichiometry that put this system at the center of an intense research, both fundamental and application-oriented. In its stoichiometric form, it is an insulating compound with a large dielectric constant ($\varepsilon_r \sim 300$ at room temperature) which tends to diverge lowering the temperature [1]. Quantum fluctuations are thought to suppress a ferroelectric order [2], which can establish if one replaces O^{16} with O^{18} [3, 4]. The compound has a 3.2 eV gap between an O 2p valence band and a Ti 3d conduction band. Electron doping by substitution of La^{3+} for Sr^{2+}, Nb^{5+} for Ti^{4+} or by creating oxygen vacancies (SrTiO$_{3-\delta}$) leads to a metallic state for one of the lowest known doping level (10^{16} cm^{-3}) [5]. This can be explained, in a model of shallow donors, as a consequence of the large dielectric constant ε_r which lowers the donor binding energy below the μeV at 4.2 K. Once the doping level reaches 10^{18} cm^{-3}, a superconducting state appears [6]: the critical temperature, T_c, depends on doping and has a dome-shaped behaviour, reaching a maximum value of ~400 mK [7]. For an electron doping of 10^{21} cm^{-3}, superconductivity disappears. Along this continuous increase of carriers, the low temperature charge mobility is reduced from ~20000 cm^2 V^{-1} s^{-1} at the insulator to metal transition to few hundreds cm^2 V^{-1} s^{-1}

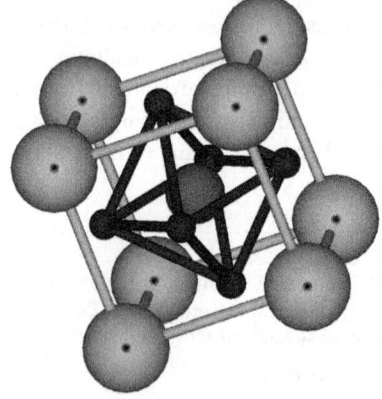

Fig. 2.1 Representation of the SrTiO$_3$ perovskite unit cell in its cubic phase. Sr atoms are green, O in red while Ti in gray (color figure online)

for carrier densities of 10^{20} cm^{-3}. This decrease is due to the ionized impurity scattering which scales with the dopants introduced into the material [8]. At higher temperatures, the mobility is limited by scattering with several phonon modes, both optical and acoustic, a phenomenon that is currently well described.

If one looks along the crystallographic direction [0 0 1], the crystal appears as a sequence of Sr^{2+}O^{2-} and Ti^{4+}(O^{2-})$_2$ atomic planes, each being charge neutral according to the cation nominal valence state. The cutting and polishing procedure to obtain (0 0 1) surfaces results in a mixed termination, both SrO and TiO$_2$ cation planes exposed to air. This is due to the miscut angle (from 0.1° up to 1°) between the polishing (optical) plane and the crystallographic plane. In 1994, Kawasaki et al. proposed to use a wet solution of HF-buffered NH$_4$F with a pH = 4.5 [9] to dissolve the SrO and hence obtain a single TiO$_2$ termination. This approach, perfected during the years using different chemical solutions and thermal treatments [10], produces surfaces with atomically smooth TiO$_2$ terraces, usually between 200 nm and 1 μm wide, separated by steps one-unit-cell high (i.e. 3.905 Å at room temperature). This chemical termination is a requisite for the formation of the 2DEL: when a LaAlO$_3$ layer thicker than 3 unit cells (u.c.) is epitaxially deposited on top, interface conductivity can be observed [11]; changing the termination to SrO results in an insulating system [12]. The crystal structure of the heterostructure LaAlO$_3$/SrTiO$_3$ is shown in a schematic view in Fig. 2.2.

LaAlO$_3$ is also a perovskite insulator, with a gap of 5.6 eV between the O 2p valence band and the La 4f conduction band. At room temperature it has a rhombohedral symmetry with an equivalent pseudo-cubic lattice parameter of 3.7911 Å. Looking along the [0 0 1] pseudocubic direction, the crystal is a stack of La^{3+}O^{2-} and Al^{3+}O$_2^{2-}$ atomic planes, each charged, positively and negatively respectively, of 1 electron charge per u.c. As a consequence, the (0 0 1) LaAlO$_3$/SrTiO$_3$ heterostructures presents a polar discontinuity.

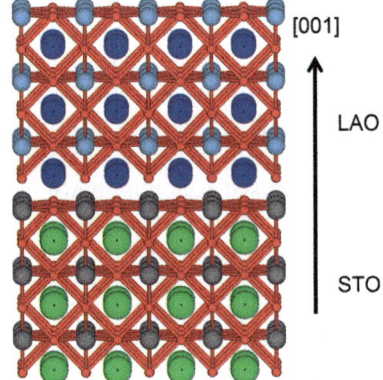

Fig. 2.2 Schematic view of the LaAlO$_3$/SrTiO$_3$ interface. The [0 0 1] direction is oriented along the growth direction, the [0 1 0] axis points towards the reader, while the [1 0 0] points to the left of the page. *Source* Drawing produced by VESTA [61]

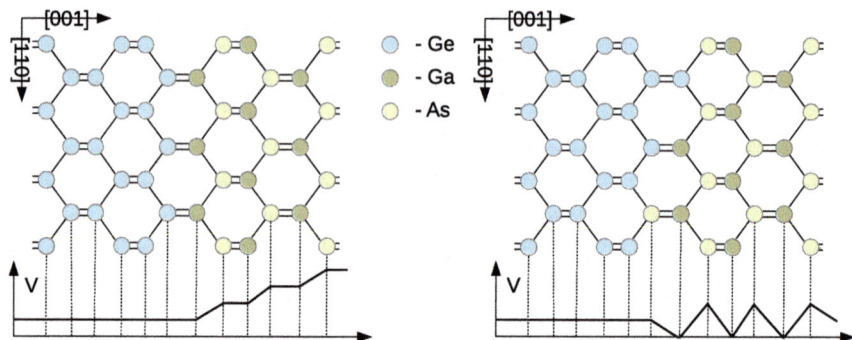

Fig. 2.3 A (0 0 1) heterojunction between Ge and GaAs. Left: a sharp interface produces a surface polarization and consequently the rise of an electric potential in the right side of the junction. Right: the same heterojunction with two transitions planes, the first is 1/4 As, the second is 3/4 Ga while the remaining atoms are Ge. This charge configuration eliminates the build up of the electrical potential. Image adapted from [13]

2.1.3 The Polar Discontinuity

The growth of heterojunctions between III–V semiconductors like GaAs and Si/Ge (IV) already illustrated the issues of polar discontinuities between materials possessing the same crystal structure and nearly an exact lattice match but different surface polarities. Along the [0 0 1] direction, Ge atomic planes are neutral while each Ga or As plane, parallel to the interface, is charged. The consequence of this charge configuration is an electric potential in GaAs which develops, with the layer thickness, an average gradient and a fluctuating component (see Fig. 2.3). The GaAs layer can be viewed as a parallel plate capacitor, where the charge accumulation at the interface and surface generates the average gradient. This potential cannot be sustained by the material since it leads to a gradient over few atoms that is greater than the band gap, leading to a Zener break-down and a charge transfer to the interface. There is currently no experimental evidence of such free carrier density at the interface neither of a significant difference between polar (prepared on (0 0 1) surfaces) and non-polar (prepared on (1 1 0) surfaces) heterojunctions. The system may prefer to roughen the interface, intermixing Ge, As and Ga atoms in some transition layers (the intermixed region is shown in the right panel of Fig. 2.3) before stacking the polar GaAs layers. This configuration leads to an oscillating potential in the GaAs, but avoids the diverging potential [13].

The $LaAlO_3/SrTiO_3$ interface bears many similarities to the GaAs-Ge system. At the oxide interface, however, an electron gas is observed and electrical conductivity is measured, despite both systems being insulators with a large band gap. Conduction could come from the transition layers between $LaAlO_3$ and $SrTiO_3$, where the stoichiometry is altered. Transmission electron microscopy [14, 15] and surface X-ray diffraction [16, 17] indeed show an interface region of two unit cells where the composition is mixed. This alloying could lead to a cation doping of $SrTiO_3$

(La^{3+} on Sr^{2+} or Al^{3+} on Ti^{4+}) and hence be the origin of the metallic state. Similar intermixing is however also reported for interfaces prepared on SrO-terminated substrates, where the conductivity is absent.

2.1.4 The Polar Catastrophe

To quantify the effect of the polar discontinuity, we can estimate the voltage gradient that develops across a layer of $LaAlO_3$. In the framework of the modern theory of polarization [18], $LaAlO_3$ has a formal polarization $P^0_{LAO} = e/2S = 0.529$ cm^{-2}, where S is the unit-cell area in the plane of the interface. Assuming a $LaAlO_3$ dielectric constant $\varepsilon_{LAO} \sim 24$, the potential drop across one unit cell (lattice parameter $a = 3.7911$ Å) is $\delta V_{LAO} = P^0_{LAO} a / \varepsilon_0 \varepsilon_{LAO} \sim 0.9$ V. If not screened by surface/interface charges, this potential leads to a substantial bending of the $LaAlO_3$ bands. This is shown in Fig. 2.4[1]: we note that after few unit cells of $LaAlO_3$, its valence band reaches the energy level of the bottom of the $SrTiO_3$ conduction band. This is the condition for a Zener breakdown: electrons tunnel from the $LaAlO_3$ valence band to the $SrTiO_3$ conduction band. This phenomenon occurs in a Zener diode under reverse bias: the applied external field tilts the electronic bands until a direct tunneling transfers electrons from occupied states into empty states sitting at the same energy level but on the other side of the barrier.

In a polar oxide interface, this is the intrinsic charge transfer mechanism that has been predicted by first-principles calculations [19–21]. The critical thickness t_c can be handy calculated, considering the band alignment shown in Fig. 2.4: $t_c = E_g(STO)/\delta V_{LAO} = 3.2$ eV/0.9 eV/u.c.=3.5 u.c.. The experimental observation of such critical thickness, firstly reported by Thiel et al. [22] and successively confirmed by many groups, has for long been a strong fact in favor of the Zener breakdown model. Other predictions of the Zener breakdown scenario deriving from first-principles calculations have, however, not been verified: the amount of screening charge, for instance, should increase progressively with the $LaAlO_3$ thickness, reaching a density of 3×10^{14} cm^{-2} for $LaAlO_3$ layers thicker than 10 u.c. These differences have motivated further theoretical work to include surface defects and adsorbates as origins of charges. A set of experiments [23, 24] has revealed the sensitivity of the interface conduction to the surface state: applying a voltage to the surface with an atomic force microscope tip allows the conductance to be switch on locally in an insulating sample; this metallic regime remains stable for some time, without any applied voltage.

---

[1]Given the similar O $2p$ nature of the valence bands of $SrTiO_3$ and $LaAlO_3$, the tops of these bands are aligned in energy at the interface.

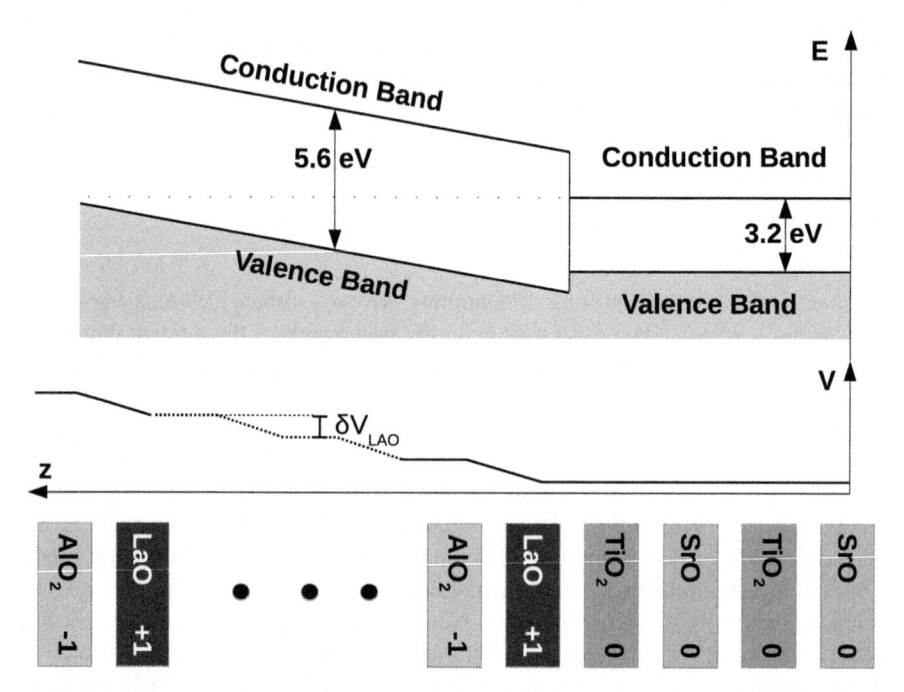

Fig. 2.4 Representation of the band alignment configuration at the LaAlO$_3$/SrTiO$_3$ interface. The electric field inside LaAlO$_3$ bends the layer electronic bands by $\sim$1 eV/u.c.: at a critical thickness electrons tunnel from the LaAlO$_3$ surface into SrTiO$_3$, where they form the electron liquid

2.1.5 Beyond a Perfect Interface

Real materials are rarely as perfect as the ones simulated by ab-initio theory: different kinds of defects enter into a thin film during its fabrication, such as cation or oxygen vacancies, dislocations, impurities, ... These are often overlooked in the description of physical phenomena, as their contribution seldom determines the properties of a compound. In the present case, the presence of an electric field may nevertheless change these considerations. Since samples are exposed to air, the surface can be covered by adsorbates like water: chemical processes, confined to the surface, can transform bound charges into free-carrier charges which can then move to the buried interface upon the effect of the electric field. Bristowe et al. proposed, as a relevant example, the formation of oxygen vacancy at the surface of LaAlO$_3$ whereby O^{2-} anions transform into O$_2$ molecules, releasing two electrons that tunnel to the interface [25] (Fig. 2.5).

This is not the only type of defects that could be promoted by the electric field. Ab-initio calculations of the interface system for different LaAlO$_3$ layer thicknesses reveal that cation intermixing lowers the polar discontinuity, similarly to the case of semiconductors previously discussed. In the form of antisite defects, Ti^{4+} on Al^{3+} sites in LaAlO$_3$ can provide an extra electron to Al on Ti sites in SrTiO$_3$,

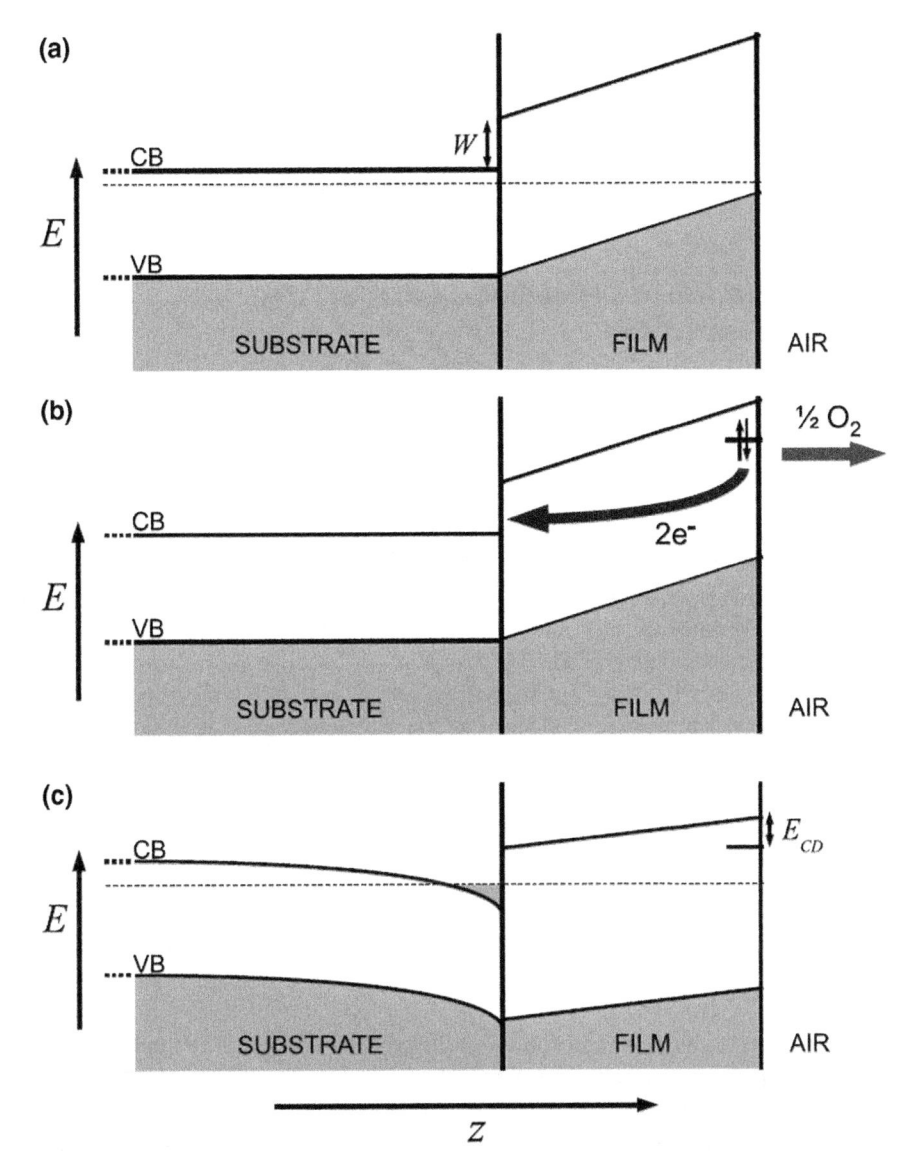

Fig. 2.5 Schematic band diagram of an interface between a polar film and a nonpolar substrate along the normal direction z. **a** The pristine system under the critical film thickness. **b** The creation of a donor state at the surface via a redox reaction and subsequent electron transfer. **c** The reconstruction reduces the film's electric field. Reprint from [25]

creating a dipole field that counteracts the polar field [26]. At a critical thickness, the formation of surface oxygen vacancies becomes the leading doping mechanism: first-principles theory shows that at 4 u.c. of LaAlO$_3$, the formation energy of oxygen vacancies becomes negative and these surface defects form spontaneously. This scenario predicts that, at such thickness, the charge transfer to the interface is complete.

2.2 The Growth of Crystalline LaAlO$_3$/SrTiO$_3$ Heterostructures

The previous discussion on the origin of the electron liquid has revealed the importance of the control of the materials for the realization of a conducting LaAlO$_3$/SrTiO$_3$ interface.

Different deposition techniques achieve high quality LaAlO$_3$ epitaxial thin films on SrTiO$_3$ substrates. Most of the studies reported in literature were carried out on samples fabricated by Pulsed Laser Deposition (PLD). Samples grown by on-axis RF sputtering were reported to be non conductive [27], while the use of off-axis geometry granted the fabrication of conducting interfaces [28]. The main issue encountered when growing LAO/STO interfaces with sputtering techniques is the La/Al stoichiometry [27, 29, 30]. We will present in Sect. 2.2.3, the importance of the La/Al ratio in determining conducting interfaces. Molecular beam epitaxy (MBE) has also been recently used to grow this system on Si wafers [31] and to study the effect of the layer stoichiometry on the interface properties. In the following sections, we review the growth conditions for the different techniques and the effect of these parameters on the physics of the interface.

2.2.1 PLD Growth

PLD represents the most used technique for the growth of crystalline complex oxides since it combines a large flexibility in the growth parameters (substrate temperature, growth atmosphere) with in-situ diagnostic tools [33]. One major limitation, in particular for a widespread use in industrial production, is the size of the material flux, around few mm^2. Efforts to bridge this technological gap and render this technique applicable to large wafers are underway [34]. A common issue to physical deposition techniques is the achievement of the correct oxidation state for the compound. In particular, dealing with transition metals which have several valence states, different oxide phases can indeed coexist and compete: considering, for instance, vanadium oxide, we can have V$_2$O$_3$ (V^{+3}), VO$_2$ (V^{4+}) and V$_2$O$_5$ (V^{5+}). Moreover, to achieve high quality crystalline growth, high temperatures are often required. For these reasons, a layer can be deposited in low oxygen pressure and annealed in oxygen in a post-deposition treatment. This procedure is for instance common in high temperature cuprate superconductors to obtain metallic and superconducting films. Most of the studies reported on LaAlO$_3$/SrTiO$_3$ interfaces consider samples grown by PLD. With this technique were indeed produced the first conducting interfaces described in the pioneering work of Ohtomo and Hwang [11]. From this report onwards, the role of the growth conditions has been investigated in many laboratories, achieving nowadays a large knowledge on the growth processes. At the University of Geneva, we use a PLD system, designed by Pascal Co., Ltd., equipped with Reflection High Energy Electron Diffraction (RHEED), a KrF excimer laser and a Nd-YAG heating source. The target-substrate distance is set at 95 mm and the LaAlO$_3$ target is a single crystal as well as a ceramic pellet. The LaAlO$_3$ layers were grown onto TiO$_2$-terminated [0 0 1] oriented SrTiO$_3$ substrates in oxygen pressure of 10^{-4} mbar at a substrate temperature T = 800 °C using a laser fluence 0.6 J/cm^2 with a repetition rate of 1 s. After growth, the samples are in-situ annealed in an oxygen pressure of 200 mbar for 1 h keeping the temperature at 500 °C, before being cooled down to room temperature in the same atmosphere. Typical RHEED oscillations of the specular intensity and a diffraction pattern of the LaAlO$_3$ grown layer are shown in Fig. 2.6, panel a and b. The oscillations indicate a layer-by-layer growth, allowing us to control very precisely the thickness of the LaAlO$_3$ layer. Films grown in these conditions show atomically flat surfaces (for thicknesses lower than 20 u.c.). A topography image of a layer surface is shown in Fig. 2.6c) where we recognize the characteristic step-and-terrace morphology of the underneath vicinal STO substrate.

In order to obtain the 2DEL at the interface, the LaAlO$_3$ layer must have a thickness strictly higher than 3 u.c. (∼1.1 nm): a sharp change of interface conductance

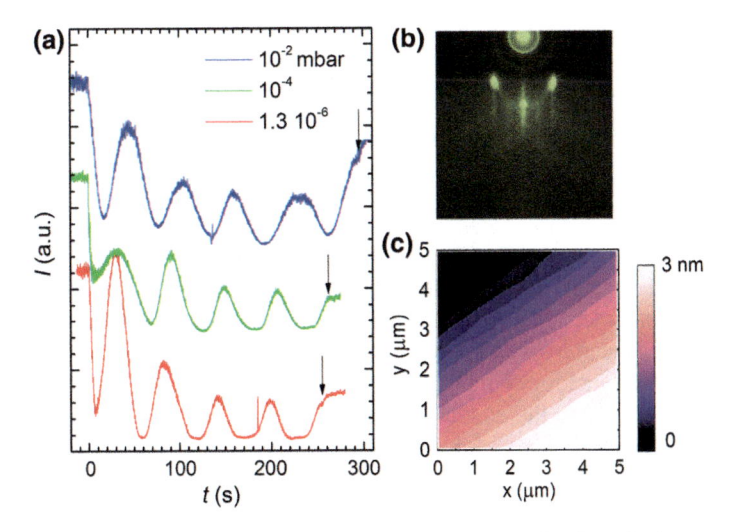

Fig. 2.6 a RHEED intensity oscillations as a function of time revealing layer-by-layer growth for three different oxygen deposition pressures. (Figure adapted from [32]) b RHEED pattern after LAO deposition. c Atomic Force Microscope topographic image of the surface of an 11 u.c. LaAlO$_3$ film

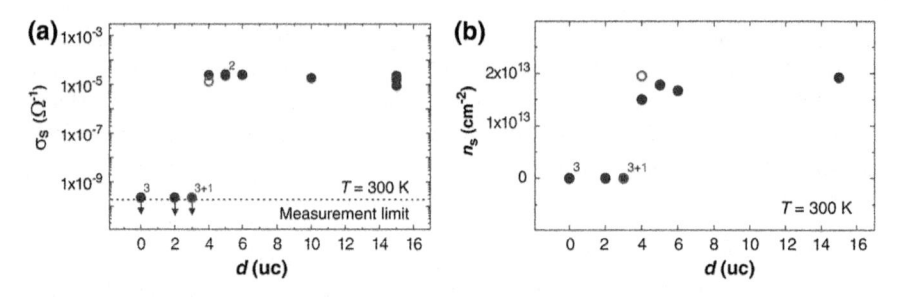

Fig. 2.7 Influence of LaAlO$_3$ thickness on the electronic properties of the LAO/STO interfaces. **a** Sheet conductance and **b** carrier density of the heterostructures plotted as a function of the number of their LAO unit cells from [22]. These results have been reproduced in many laboratories around the world

occurs at 4 u.c., as observed firstly by Thiel et al. [22] and then confirmed in several laboratories around the world.

Figure 2.7 displays the conductance σ_S (panel a) and the sheet carrier density n_S (panel b) as a function of the LaAlO$_3$ thickness: at 4 u.c. we note the sharp insulator to metal transition and the corresponding jump in carrier density. As discussed in Sect. 2.1.4, these experimental findings point towards an intrinsic doping mechanism.

The electronic properties of this system have been found to be highly sensitive to the preparation conditions. These include the oxygen partial pressure during and after the deposition, the growth temperature, the layer thickness and the laser fluence. In the next section, we address the role of oxygen pressure and substrate temperature during deposition.

2.2.2 The Role of LAO Deposition Conditions

Oxygen Pressure

The role of oxygen pressure during the growth of LaAlO$_3$ layers has being explored by many groups [35–39], due to the particular sensitivity of the properties of the system to this parameter. The STO is indeed known to become conducting if oxygen vacancies are induced by a high temperature anneal in high vacuum [40, 41] or by illumination with synchrotron light [42]; each oxygen vacancy introduces 2 electrons which can contribute to the conduction. So far, structural characterizations do not reveal a noticeable influence of the oxygen pressure used during the deposition or the annealing process on the crystalline quality of the structures [32]; the transport data, on the contrary, strongly depend on the deposition atmosphere.

Figure 2.8 shows the dramatic effects of the oxygen pressure on the sheet conductance of samples with 2/3 u.c. thick LaAlO$_3$ layers. Samples grown at different oxygen pressures and not annealed (green diamond) show a sheet conductance that scales inversely with the growth pressure. Samples grown at 10^{-6} mbar and not post-

Fig. 2.8 Room temperature sheet conductance as a function of the oxygen growth pressure for samples with a 2–3 u.c. LAO thickness; the gray region illustrates the scattering in the room temperature sheet conductance observed for LAO films 5–10 u.c. thick. (Figure adapted from [32]) (color figure online)

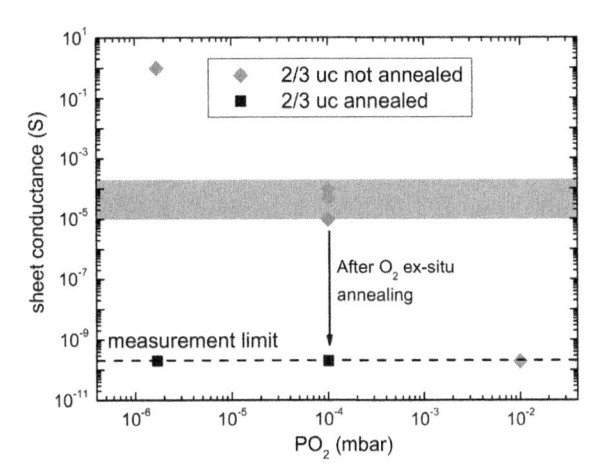

annealed display, at room temperature, a sheet resistance of $1\,\Omega/\square$ and a carrier density of $\approx 10^{17}$ e/cm^2 [32]. Considering a maximum doping of 1 electron per Ti, i.e. a volume doping of 1.7×10^{22} cm^{-3}, the charge would occupy a 100 nm-thick layer. These values suggest that this is a bulk conduction due to oxygen vacancies created inside the SrTiO$_3$ substrate. A visual confirmation of such bulk doping has been provided by measurements of the vertical extension of the conducting region performed with a conducting atomic force microscope (C-AFM) [38]: interfaces prepared in 10^{-5} mbar of oxygen reveal an extension of hundreds of micrometers. The effect of an in-situ anneal performed in an oxygen pressure of 200 mbar after the deposition is evident looking at Fig. 2.8: we note that all samples (red squares) become insulating, independently on the growth pressure! We remind that the thickness of the LaAlO$_3$ layer of these samples is below the critical thickness to observe surface conductivity. For interfaces grown in a 10^{-4} mbar oxygen atmosphere and in-situ annealed at 0.2 bar, we indeed observe the occurrence of a metallic state at four and higher LaAlO$_3$ unit cells. Such samples, post-annealed in 300 mbar of oxygen, have a charge confinement less than 7 nm at room temperature, as observed by C-AFM. These results point to the importance of the post-deposition annealing oxygen treatment to suppress oxygen vacancies created during the deposition. Oxygen deficient samples, i.e. interfaces not annealed, show electronic band structure different from the one of annealed samples: in-gap defect states are observed for non-annealed LAO/STO interfaces as well as a different polaronic effect is detected in oxygen deficient samples compared to fully oxygenated interfaces; these differences will be discussed in Chap. 6 on photoemission experiments.

The oxygen pressure used during the deposition has also an effect on the growth mode: at low oxygen pressure (below 10^{-3} mbar), the LaAlO$_3$ film grows layer-by-layer while at high oxygen pressure (above 10^{-3} mbar), the plasma plume is thermalized before impinging on the substrate and the deposition occurs following an island growth mode [43].

Growth Temperature

Charge mobility is a key ingredient for electronic applications as well as for the observation of quantum effects like Shubnikov de Haas oscillations. LAO/STO interfaces typically display mobilities of 500–1000 cm^2/Vs at low temperature [44, 45]. Different studies have attempted to pinpoint the origin of the scattering mechanisms that limit the electron mobility at the interface. In a charge transfer scenario, free carriers at the interface are not generated by ionized dopants; since these limit the charge mobility μ in bulk SrTiO$_3$ [8], one could expect that μ and n_s are not linked for this quantum well. Experimentally, the LaAlO$_3$/SrTiO$_3$ interface mobility equals the one of SrTiO$_3$ for the same doping level. Thiel et al. investigated the role of microstructural defects on the conductivity of the 2DEL by introducing edge dislocations at the junction of a bi-crystal. By changing the angle of the crystallographic planes that meet at the junction, they could vary the density of these dislocations and observe a dramatic reduction of the conductance. The analysis of the data indicates that the line defects affect the transport over a length larger than their structural size [46]. Since the conductivity occurs at the surface of SrTiO$_3$, one could expect a relevant role of the step edges on the scattering rate. The proximity of the charge distribution to the interface could indeed be an important factor in defining the mobility (this point will be discussed further in the next chapter). To study the effect of surface scattering induced by step edges, Fix and coworkers measured the transport properties of a series of interfaces prepared on SrTiO$_3$ substrates whose miscut angle (the angle between the optical and crystallographic planes) was varied. In this way, the density of step edges could be modified and its variation linked to the observed change in the electron mobility [47]. Another scattering mechanism could come from the charges sitting at the LaAlO$_3$ surface that could be only partially screened by the layer. In an experiment to probe the contribution of this scattering to the 2DEL mobility, Bell et al. [48] observed instead a dramatic decrease of the electron mobility of nearly two orders of magnitude increasing the LaAlO$_3$ thickness from 5 to 25 unit cells.

In Geneva, Fête et al. discovered that the growth temperature plays a crucial role in determining the carrier density and the mobility of the 2DEL.

The transport properties of 5 u.c. thick LaAlO$_3$/SrTiO$_3$ samples grown at different temperatures are shown in Fig. 2.9. One notes that the samples show very different numbers of carriers (n_{2D}) and (Hall) mobilities (μ_H) for the three growth temperatures. The samples grown at 650 °C exhibit a linear Hall effect which is nearly temperature independent. Considering a single channel conduction, the carrier density can be estimated as 2×10^{12} cm^{-2}, one of the lowest reported for a conducting interface. The low carrier density value and the low sheet resistance measured at low temperature yield an estimate of the mobility of few thousand cm^2/(Vs) (Fig. 2.9b) for these samples, one of the highest values observed. For a growth temperature of 800 and 900 °C, the Hall resistance R_{xy} displays a more complex behavior, changing from a linear dependence in magnetic field above 100 K to a non-linear function of the magnetic field below 100 K. This behavior suggests the contribution of two channels to the conduction: for samples grown at 900 °C transport data can be described by a band with low carrier density ($\approx 10^{12}$ e/cm^2, see Fig. 2.9a) and high mobility

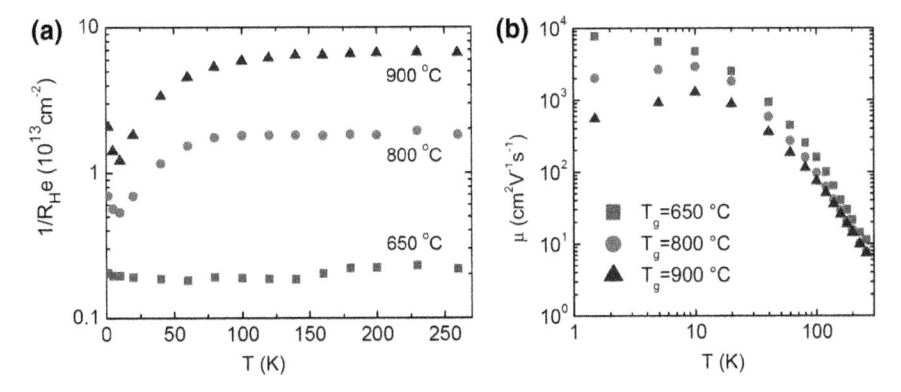

Fig. 2.9 a Temperature dependence of the inverse Hall constant measured at $B \rightarrow 0$ (low carrier density channel) for interfaces grown at different temperature (symbols) and according to the model described in the text (dashed line); **b** μ_H as a function of temperature for samples grown at 900 °C (wine triangles), 800 °C (gold dots) and 650 °C (blue squares). Figure adapted from [49] (color figure online)

(in the range of 1000–2000 cm^2/Vs) and a band with higher carrier density ($\approx 6 \times 10^{13}$ e/cm^2) and low mobility (in the range of 100 cm^2/Vs) [50, 51]. The presence of carriers from several bands, i.e. carriers with different mobilities (i.e. masses) and densities, had been predicted by ab-initio calculation [19, 21, 52, 53] (please refer to Chaps. 8, 9 for a discussion of the theoretical and experimental band structure for the 2DEL).

2.2.3 The Role of LAO Stoichiometry

PLD is often described as a deposition technique allowing a stoichiometric transfer between the target and the substrate. Recent work on the plume composition [55] and on the film off-stoichiometry [56, 57] has revealed that the gas pressure in the chamber and the pulsed laser fluence are crucial parameters to achieve the correct film stoichiometry. Concerning the LaAlO$_3$ layer, the ratio between La and Al cations was shown to depend on the laser fluence used during the growth: it varies from 0.88 to 1.15 when the laser fluence changes from 0.7 to 1.9 J/cm^2. As a consequence, the LaAlO$_3$ layer structure and the interface conductivity are modified [54, 58]. In Fig. 2.10 we see the effect of the La/Al off-stoichiometry on the number of carriers and sheet resistance: a 2 orders of magnitude change in the carrier density and a metal to insulator transition when the La/Al ratio is different than 1 are shown. Conducting interfaces were found only in slightly Al-rich (or La-deficient) samples. Sato et al. [54] proposed that the film off-stoichiometry changes the balance between atomic and electronic reconstructions, the former resolving the polar discontinuity without doping electrons. The effect of laser energy density combined with different

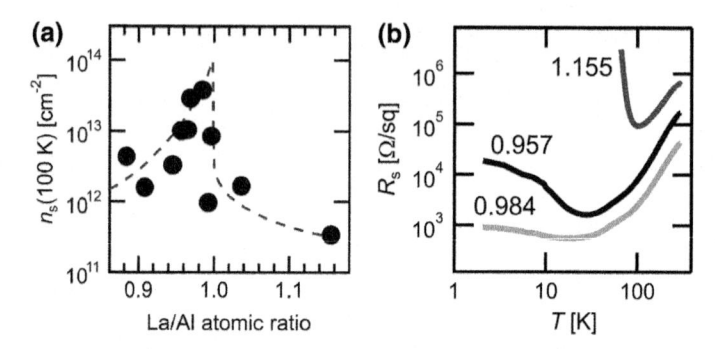

Fig. 2.10 a Sheet carrier density of LAO/STO samples at 100 K as a function of cation ratio. **b** Temperature dependence of the sheet resistance of the LaAlO$_3$/SrTiO$_3$ samples with different film cation ratios. Figure adapted from [54]

oxygen pressures was studied by Golalikhani et al. [59] on 100 nm thick LaAlO$_3$. The depositions were carried out at laser energy densities ranging from 0.7 to 2 J/cm^2, and oxygen pressures between 10^{-4} and 0.3 Torr. For oxygen pressures lower than 10^{-2} Torr, the films are La-rich; to achieve the correct film stoichiometry, the oxygen pressure was found to be about 10^{-1} Torr.

A careful study of the role of the LaAlO$_3$ stoichiometry on the interface properties has been performed by Warusawithana et al. comparing two deposition techniques, oxide molecular beam epitaxy (o-MBE) and PLD [60]. Taking advantage of the control of the separated evaporation of La and Al in a o-MBE approach, they could determine that interface conductivity occurs only in a well defined La/Al ratio. Figure 2.11

Fig. 2.11 Room temperature sheet resistance measured in a four-point geometry of La$_{(1-\delta)}$Al$_{(1+\delta)}$O$_3$/(001) SrTiO$_3$ interfaces is plotted as a function of the La/Al ratio determined by Rutherford backscattering spectrometry. A sharp jump in sheet resistance is observed at La/Al = 0.97 ± 0.03, consistently in all three samples. (Figure adapted from [60])

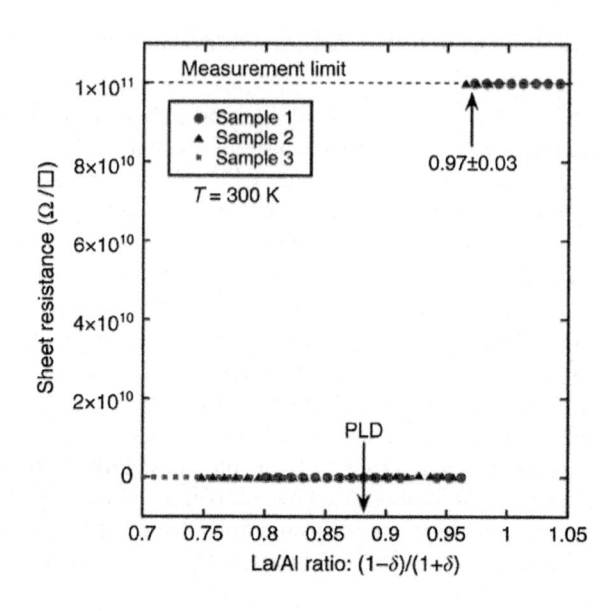

illustrates this point: only samples that are Al-rich conduct while stoichiometric and La-rich samples are insulating. Measuring the composition of conducting samples prepared by PLD in standard conditions (films were deposited from stoichiometric, single crystalline targets in an oxygen atmosphere of 2×10^{-5} mbar at $\sim 800\,°C$ and cooled to room temperature in 400 mbar of O_2 [22]), the authors observed an off-stoichiometry with a La/Al ratio of 0.97. These results are not in contradiction with the polar catastrophe model: in insulating (La-rich) samples, the predicted and observed local accumulation of cation vacancies on the B-site in the vicinity of the interface acts to remove the diverging potential, whereas this is accomplished in conducting (Al-rich) samples by an electronic reconstruction that forms the 2DEL. These results demonstrate that one needs to take into account the LAO film stoichiometry, in addition to the oxygen vacancy effects, when studying the 2DEL properties at the $LaAlO_3/SrTiO_3$ interfaces.

2.3 Conclusions

The discussion on the origin of the 2DEL at the interface between two insulating oxides is yet not settled. Despite a large amount of experimental evidence has cumulated pointing to the polar discontinuity as a driving mechanism for the presence of the free charges at the interface, the difficulty to achieve and observe p doping in the $LaAlO_3$ layer, as expected in a Zener breakdown scenario, suggests that surface defects are the source of the charges. This scenario poses a challenge to be detected by experimentalists.

References

1. K. Müller, H. Burkard, $SrTiO_3$: An intrinsic quantum paraelectric below 4 K. Phys. Rev. B **19**, 3593 (1979). https://doi.org/10.1103/PhysRevB.19.3593
2. D. Bäuerle, D. Wagner, M. Wöhlecke, B. Dorner, H. Kraxenberger, Soft modes in semiconducting $SrTiO_3$: II. The ferroelectric mode. Zeitschrift für Physik B Conden. Mat. **38**, 335 (1980). https://doi.org/10.1007/BF01315325
3. M. Itoh, R. Wang, Y. Inaguma, T. Yamaguchi, Y.J. Shan, T. Nakamura, Ferroelectricity Induced by Oxygen Isotope Exchange in Strontium Titanate Perovskite. Phys. Rev. Lett. **82**, 3540–3543 (1999). https://doi.org/10.1103/PhysRevLett.82.3540
4. S.E. Rowley, L.J. Spalek, R.P. Smith, M.P.M. Dean, M. Itoh, J.F. Scott, G.G. Lonzarich, S.S. Saxena, Ferroelectric quantum criticality. Nat. Phys. **10**, 367 (2014). https://doi.org/10.1038/nphys2924
5. A. Spinelli, M.A. Torija, C. Liu, C. Jan, C. Leighton, Electronic transport in doped $SrTiO_3$: Conduction mechanisms and potential applications. Phys. Rev. B **81**, 155110 (2010). https://doi.org/10.1103/PhysRevB.81.155110
6. J.F. Schooley, W.R. Hosler, M.L. Cohen, Superconductivity in semiconducting $SrTiO_3$. Phys. Rev. Lett. **12**, 474 (1964). https://doi.org/10.1103/PhysRevLett.12.474

7. C.S. Koonce, M.L. Cohen, J.F. Schooley, W.R. Hosler, E.R. Pfeiffer, Superconducting transition temperatures of semiconducting SrTiO$_3$. Phys. Rev. **163**, 380 (1967). https://doi.org/10.1103/PhysRev.163.380
8. A. Verma, A.P. Kajdos, T.A. Cain, S. Stemmer, D. Jena, Intrinsic mobility limiting mechanisms in lanthanum-doped Strontium Titanate. Phys. Rev. Lett. **112**, 216601 (2014). https://doi.org/10.1103/PhysRevLett.112.216601
9. M. Kawasaki, K. Takahashi, T. Maeda, R. Tsuchiya, M. Shinohara, O. Ishiyama, T. Yonezawa, M. Yoshimoto, H. Koinuma, Atomic control of the SrTiO$_3$ crystal surface. Science **266**, 1540 (1994). https://doi.org/10.1126/science.266.5190.1540
10. G. Koster, B.L. Kropman, G.J.H.M. Rijnders, D.H.A. Blank, H. Rogalla, Quasi-ideal strontium titanate crystal surfaces through formation of strontium hydroxide. Appl. Phys. Lett. **73**, 2920 (1998). https://doi.org/10.1063/1.122630
11. A. Ohtomo and H. Y. Hwang. A high-mobility electron gas at the LaAlO3/SrTiO3 heterointerface. Nature 427, 423 (2004). http://doi.org/10.1038/nature02308
12. J. Nishimura, A. Ohtomo, A. Ohkubo, Y. Murakami, and M. Kawasaki. Controlled Carrier Generation at a Polarity-Discontinued Perovskite Heterointerface. Jpn. J. Appl. Phys. 43, L1032 (2004). https://doi.org/10.1143/JJAP.43.L1032
13. W. Harrison, E. Kraut, J. Waldrop, and R. Grant. Polar heterojunction interfaces. Phys. Rev. B 18, 4402 (1978).https://doi.org/10.1103/PhysRevB.18.4402
14. C. Cantoni, J. Gazquez, F. Miletto Granozio, M. P. Oxley, M. Varela, A. R. Lupini, S. J. Pennycook, C. Aruta, U. Scotti di Uccio, P. Perna, and D. Maccariello. Electron Transfer and Ionic Displacements at the Origin of the 2D Electron Gas at the LAO/STO Interface: Direct Measurements with Atomic-Column Spatial Resolution. Adv. Mater. 24, 3952 (2012). https://doi.org/10.1002/adma.201200667
15. N. Nakagawa, H. Y. Hwang, and D. A. Muller. Why some interfaces cannot be sharp. Nat. Mater. 5, 204 (2006). https://doi.org/10.1038/nmat1569
16. P. Willmott, S. Pauli, R. Herger, C. Schlepütz, D. Martoccia, B. Patterson, B. Delley, R. Clarke, D. Kumah, C. Cionca, Y. Yacoby, Structural basis for the conducting interface between LaAlO$_3$ and SrTio$_3$. Phys. Rev. Lett. **99**, 155502 (2007). https://doi.org/10.1103/PhysRevLett.99.155502
17. M. Salluzzo, S. Gariglio, D. Stornaiuolo, V. Sessi, S. Rusponi, C. Piamonteze, G.M. De Luca, M. Minola, D. Marré, A. Gadaleta, H. Brune, F. Nolting, N.B. Brookes, G. Ghiringhelli, Origin of interface magnetism in BiMnO$_3$/SrTiO$_3$ and LaAlO$_3$/SrTiO$_3$ heterostructures. Phys. Rev. Lett. **111**, 087204 (2013). https://doi.org/10.1103/PhysRevLett.111.087204
18. M. Stengel, D. Vanderbilt, Berry-phase theory of polar discontinuities at oxide-oxide interfaces. Phys. Rev. B **80**, 241103 (2009). https://doi.org/10.1103/PhysRevB.80.241103
19. Z. Popović, S. Satpathy, R. Martin, Origin of the two-dimensional electron gas carrier density at the LaAlO$_3$ on SrTiO$_3$ Interface. Phys. Rev. Lett. **101**, 256801 (2008). https://doi.org/10.1103/PhysRevLett.101.256801
20. R. Pentcheva, W.E. Pickett, Electronic phenomena at complex oxide interfaces: insights from first principles. J. Phys. Condens. Mater. **22**, 043001 (2010). https://doi.org/10.1088/0953-8984/22/4/043001
21. P. Delugas, A. Filippetti, V. Fiorentini, D.I. Bilc, D. Fontaine, Ph. Ghosez, Spontaneous 2-Dimensional carrier confinement at the n-Type SrTiO$_3$/LaAlO$_3$ interface. Phys. Rev. Lett. **106**, 166807 (2011). https://doi.org/10.1103/PhysRevLett.106.166807
22. S. Thiel, G. Hammerl, A. Schmehl, C.W. Schneider, J. Mannhart, Tunable quasi-two-dimensional electron gases in oxide heterostructures. Science **313**, 1942 (2006). https://doi.org/10.1126/science.1131091
23. C. Cen, S. Thiel, G. Hammerl, C.W. Schneider, K.E. Andersen, C.S. Hellberg, J. Mannhart, J. Levy, Nanoscale control of an interfacial metal-insulator transition at room temperature. Nat. Mater. **7**, 298 (2008). https://doi.org/10.1038/nmat2136
24. Y. Xie, Y. Hikita, Ch. Bell, H.Y. Hwang, Control of electronic conduction at an oxide heterointerface using surface polar adsorbates. Nat. Commun. **2**, 494 (2011). https://doi.org/10.1038/ncomms1501

25. N.C. Bristowe, P.B. Littlewood, E. Artacho, Surface defects and conduction in polar oxide heterostructures. Phys. Rev. B **83**, 205405 (2011). https://doi.org/10.1103/PhysRevB.83.205405
26. L. Yu, A. Zunger, A polarity-induced defect mechanism for conductivity and magnetism at polarnonpolar oxide interfaces. Nat. Commun. **5**, 5118 (2014). https://doi.org/10.1038/ncomms6118
27. I.M. Dildar, M. Neklyudova, Q. Xu, H.W. Zandbergen, S. Harkema, D. Boltje, J. Aarts, Growing LaAlO$_3$/SrTio$_3$ interfaces by sputter deposition. AIP Advances **5**, 067156 (2015). https://doi.org/10.1063/1.4923285
28. J.P. Podkaminer, T. Hernandez, M. Huang, S. Ryu, C.W. Bark, S.H. Baek, J.C. Frederick, T.H. Kim, K.H. Cho, J. Levy, M.S. Rzchowski, C.B. Eom, Creation of a two-dimensional electron gas and conductivity switching of nanowires at the LaAlO$_3$/SrTiO$_3$ interface grown by 90° off-axis sputtering. Appl. Phys. Lett. **103**, 071604 (2013). https://doi.org/10.1063/1.4817921
29. I.M. Dildar, D.B. Boltje, M.H.S. Hesselberth, J. Aarts, Q. Xu, H.W. Zandbergen, S. Harkema, Non-conducting interfaces of LaAlO$_3$/SrTiO$_3$ produced in sputter deposition: The role of stoichiometry. Appl. Phys. Lett. **102**, 121601 (2013). https://doi.org/10.1063/1.4798828
30. L. Qiao, T. Droubay, T. Varga, M. Bowden, V. Shutthanandan, Z. Zhu, T.C. Kaspar, S. Chambers, Epitaxial growth, structure, and intermixing at the LaAlO$_3$/SrTiO$_3$ interface as the film stoichiometry is varied. Phys. Rev. B **83**, 085408 (2011). https://doi.org/10.1103/PhysRevB.83.085408
31. J.W. Park, D.F. Bogorin, C. Cen, D.A. Felker, Y. Zhang, C.T. Nelson, C.W. Bark, C.M. Folkman, X.Q. Pan, M.S. Rzchowski, J. Levy, C.B. Eom, Creation of a two-dimensional electron gas at an oxide interface on silicon. Nat. Commun. **1**, 94 (2010). https://doi.org/10.1038/ncomms1096
32. C. Cancellieri, N. Reyren, S. Gariglio, A.D. Caviglia, A. Fête, J.-M. Triscone, Influence of the growth conditions on the LaAlO$_3$/SrTiO$_3$ interface electronic properties. EPL **91**, 17004 (2010). https://doi.org/10.1209/0295-5075/91/17004
33. R. Eason (ed), Pulsed Laser Deposition of Thin Films: Applications-Led Growth of Functional Materials. Wiley (2007). https://doi.org/10.1002/0470052120
34. D.H.A. Blank, M. Dekkers, G. Rijnders, Pulsed laser deposition in Twente: from research tool towards industrial deposition. J. Phys. D. Appl. Phys. **47**, 034006 (2014). https://doi.org/10.1088/0022-3727/47/3/034006
35. G. Herranz, M. Basletic, M. Bibes, C. Carrétéro, E. Tafra, E. Jacquet, K. Bouzehouane, C. Deranlot, A. Hamzic, J.-M. Broto, A. Barthélémy, A. Fert, High Mobility in LaAlO$_3$/SrTiO$_3$ Heterostructures: Origin, Dimensionality, and Perspectives. Phys. Rev. Lett. **98**, 216803 (2007). https://doi.org/10.1103/PhysRevLett.98.216803
36. A. Brinkman, M. Huijben, M. van Zalk, J. Huijben, U. Zeitler, J. C. Maan, W. G. van derWiel, G. Rijnders, D. H. A. Blank, H. Hilgenkamp, Magnetic effects at the interface between nonmagnetic oxides. Nat. Mater. **6**, 493 (2007). https://doi.org/10.1038/nmat1931
37. W. Siemons, G. Koster, H. Yamamoto, W. A. Harrison, G. Lucovsky, T. H. Geballe, D. H. A. Blank, M. R. Beasley, Origin of Charge Density at LaAlO$_3$ on SrTiO$_3$ Heterointerfaces: Possibility of Intrinsic Doping. Phys. Rev. Lett. **98**, 196802 (2007). https://doi.org/10.1103/PhysRevLett.98.196802
38. M. Basletic, J.-L. Maurice, C. Carrétéro, G. Herranz, O. Copie, M. Bibes, E. Jacquet, K. Bouzehouane, S. Fusil, A. Barthélémy, Mapping the spatial distribution of charge carriers in LaAlO$_3$/SrTiO$_3$ heterostructures. Nat. Mater. **7**, 621 (2008). https://doi.org/10.1038/nmat2223
39. A. Kalabukhov, R. Gunnarsson, J. Borjesson, E. Olsson, T. Claeson, D. Winkle, Effect of oxygen vacancies in the SrTiO$_3$ substrate on the electrical properties of the LaAlO$_3$/SrTiO$_3$ interface. Phys. Rev. B **75**, 121404 (2007). https://doi.org/10.1103/PhysRevB.75.121404
40. N. C. Plumb, M. Salluzzo, E. Razzoli, M. Månsson, M. Falub, J. Krempasky, C. E. Matt, J. Chang, M. Schulte, J. Braun, H. Ebert, J. Minár, B. Delley, K.-J. Zhou, T. Schmitt, M. Shi, J. Mesot, L. Patthey M. Radović, Mixed Dimensionality of Confined Conducting Electrons in the Surface Region of SrTiO$_3$. Phys. Rev. Lett. **113**, 086801 (2014). https://doi.org/10.1103/PhysRevLett.113.086801

41. A. F. Santander-Syro, O. Copie, T. Kondo, F. Fortuna, S. Pailhès, R. Weht, X. G. Qiu, F. Bertran, A. Nicolaou, A. Taleb-Ibrahimi, P. Le Fèvre, G. Herranz, M. Bibes, N. Reyren, Y. Apertet, P. Lecoeur, A. Barthélémy, M. J. Rozenberg, Two-dimensional electron gas with universal subbands at the surface of SrTiO$_3$. Nature **469**, 189 (2011). https://doi.org/10.1038/nature09720

42. W. Meevasana, P. D. C. King, R. H. He, S.-K. Mo, M. Hashimoto, A. Tamai, P. Songsiriritthigul, F. Baumberger Z.-X. Shen, Creation and control of a two-dimensional electron liquid at the bare SrTiO$_3$ surface. Nat. Mater. **10**, 114 (2011). https://doi.org/10.1038/nmat2943

43. M. Huijben, A. Brinkman, G. Koster, G. Rijnders, H. Hilgenkamp, D. H. A. Blank, Structure-Property Relation of SrTiO$_3$/LaAlO$_3$ Interfaces. Adv. Mater. **21**, 1665 (2009). https://doi.org/10.1002/adma.200801448

44. N. Reyren, S. Thiel, A. D. Caviglia, L. Fitting Kourkoutis, G. Hammerl, C. Richter, C. W. Schneider, T. Kopp, A.-S. Rüetschi, D. Jaccard, M. Gabay, D. A. Muller, J.-M. Triscone, J. Mannhart, Superconducting interfaces between insulating oxides. Science **317**, 1196 (2007). https://doi.org/10.1126/science.1146006

45. S. Gariglio, N. Reyren, A. D. Caviglia, J.-M. Triscone. Superconductivity at the LaAlO$_3$/SrTiO$_3$ interface. J. Phys.: Condens. Matter **21**, 164213 (2009). https://doi.org/10.1088/0953-8984/21/16/164213

46. S. Thiel, C. Schneider, L. Kourkoutis, D. Muller, N. Reyren, A. Caviglia, S. Gariglio, J.-M. Triscone, J. Mannhart, Electron Scattering at Dislocations in LaAlO$_3$/SrTiO$_3$ Interfaces. Phys. Rev. Lett. **102**, 046809 (2009). https://doi.org/10.1103/PhysRevLett.102.046809

47. T. Fix, F. Schoofs, Z. Bi, A. Chen, H. Wang, J. L. MacManus-Driscoll, M. G. Blamire, Influence of SrTiO$_3$ substrate miscut angle on the transport properties of LaAlO$_3$/SrTiO$_3$ interfaces. Appl. Phys. Lett. **99**, 022103 (2011). https://doi.org/10.1063/1.3609785

48. C. Bell, S. Harashima, Y. Hikita, H. Y. Hwang, Thickness dependence of the mobility at the LaAlO$_3$/SrTiO$_3$ interface. Appl. Phys. Lett. **94**, 222111 2009. https://doi.org/10.1063/1.3149695

49. A. Fête, C. Cancellieri, D. Li, D. Stornaiuolo, A. D. Caviglia, S. Gariglio, J.-M. Triscone, Growth-induced electron mobility enhancement at the LaAlO$_3$/SrTiO$_3$ interface. Appl. Phys. Lett. **106**, 051604 (2015). https://doi.org/10.1063/1.4907676

50. M. Ben Shalom, A. Ron, A. Palevski, Y. Dagan, Shubnikov-De Haas Oscillations in SrTiO$_3$/LaAlO$_3$ Interface. Phys. Rev. Lett. **105**, 206401 (2010). https://doi.org/10.1103/PhysRevLett.105.206401

51. S. Lerer, M. Ben Shalom, G. Deutscher, and Y. Dagan. Low-temperature dependence of the thermomagnetic transport properties of the SrTiO/LaAlO3 interface. Phys. Rev. B 84, 075423 (2011). https://doi.org/10.1103/PhysRevB.84.075423

52. W.-J. Son, E. Cho, B. Lee, J. Lee, S. Han, Density and spatial distribution of charge carriers in the intrinsic n-type LaAlO$_3$-SrTiO$_3$ interface. Phys. Rev. B **79**, 245411 (2009). https://doi.org/10.1103/PhysRevB.79.245411

53. R. Pentcheva, W. E. Pickett. Charge localization or itineracy at LaAlO$_3$/SrTiO$_3$ interface: hole polarons, oxygen vacancies, and mobile electrons. Phys. Rev. B **74**, 035112 (2006). https://doi.org/10.1103/PhysRevB.74.035112

54. H. K. Sato, C. Bell, Y. Hikita, H. Y. Hwang. Stoichiometry control of the electronic properties of the LaAlO$_3$/SrTiO$_3$ heterointerface. Appl. Phys. Lett. **102**, 251602 (2013). https://doi.org/10.1063/1.4812353

55. C. Xu, S. Wicklein, A. Sambri, S. Amoruso, M. Moors, R. Dittmann. Impact of the interplay between nonstoichiometry and kinetic energy of the plume species on the growth mode of SrTiO$_3$ thin films. J. Phys. D. Appl. Phys. **47**, 034009 (2014). https://doi.org/10.1088/0022-3727/47/3/034009

56. T. Ohnishi, M. Lippmaa, T. Yamamoto, S. Meguro, H. Koinuma. Improved stoichiometry and misfit control in perovskite thin film formation at a critical fluence by pulsed laser deposition. Appl. Phys. Lett. **87**, 241919 (2005). https://doi.org/10.1063/1.2146069

57. Y. Tokuda, S. Kobayashi, T. Ohnishi, T. Mizoguchi, N. Shibata, Y. Ikuhara, T. Yamamoto, Growth of Ruddlesden-Popper type faults in Sr-excess SrTiO$_3$ homoepitaxial thin films

by pulsed laser deposition. Appl. Phys. Lett. **99**, 173109 (2011). https://doi.org/10.1063/1.3656340
58. E. Breckenfeld, N. Bronn, J. Karthik, A. R. Damodaran, S. Lee, N. Mason, L. W. Martin. Effect of growth induced (non)stoichiometry on interfacial conductance in LaAlO$_3$/SrTiO$_3$. Phys. Rev. Lett. **110**, 196804 (2013). https://doi.org/10.1103/PhysRevLett.110.196804
59. M. Golalikhani, Q. Y. Lei, G. Chen, J. E. Spanier, H. Ghassemi, C. L. Johnson, M. L. Taheri, X. X. Xi, Stoichiometry of LaAlO$_3$ films grown on SrTiO$_3$ by pulsed laser deposition. J. Appl. Phys. **114**, 027008 (2013). https://doi.org/10.1063/1.4811821
60. M. P. Warusawithana, C. Richter, J. A. Mundy, P. Roy, J. Ludwig, S. Paetel, T. Heeg, A. A. Pawlicki, L. F. Kourkoutis, M. Zheng, M. Lee, B. Mulcahy, W. Zander, Y. Zhu, J. Schubert, J. N. Eckstein, D. A. Muller, C. Stephen Hellberg, J. Mannhart, D. G. Schlom. LaAlO$_3$ stoichiometry is key to electron liquid formation at LaAlO$_3$/SrTiO$_3$ interfaces. Nat. Commun. **4**, 2351 (2013). https://doi.org/10.1038/ncomms3351
61. K. Momma, F. Izumi, VESTA 3 for three-dimensional visualization of crystal, volumetric and morphology data. J. Appl. Crystallogr. **47**, 1272–1276 (2011). https://doi.org/10.1107/S0021889811038970

Chapter 3
Transport Properties of TMO Interfaces

A. M. R. V. L. Monteiro, A. D. Caviglia and N. Reyren

Abstract Phenomena that are absent of bulk TMO compounds can emerge at their interfaces when they are grown on top of each-other. A prototypical example of such emerging states is found at the LaAlO$_3$/SrTiO$_3$ interface, which also attracted most of the initial interest for this new field of research (in the TMO context). Here we review some properties of this peculiar interface as investigated by transport measurements allowing the studies of different effects such as magnetism, superconductivity or Rashba effect; hence indirectly accessing the band structures studied by the methods presented in the rest of the book.

3.1 Introduction

Electronic dc transport is a fundamental tool for the study of TMO interfaces, providing complementary information to spectroscopic techniques. We will illustrate this fact in the case of the LaAlO$_3$ films grown on top of (0 0 1)-oriented TiO$_2$-terminated SrTiO$_3$ substrate. A two-dimensional system (2DES) is found at this particular interface. Electrostatic field effect experiments have proven to be particularly valuable as they allowed tuning of the carrier density in a very sensitive region of the phase diagram in which TMO interfaces undergo quantum phase transitions (insulator to metal/superconductor) accompanied by various changes in electronic properties. Here we will discuss changes in (1) mobility and carrier localization, (2) spin-orbit

---

A. M. R. V. L. Monteiro (✉) · A. D. Caviglia
Kavli Institute of Nanoscience Delft University of Technology,
Lorentzweg 1, 2628 CJ Delft, The Netherlands
e-mail: A.M.Monteiro@tudelft.nl

A. D. Caviglia
e-mail: A.Caviglia@tudelft.nl

N. Reyren
Unité Mixte de Physique, CNRS, Thales, Univ. Paris-Sud,
Université Paris-Saclay,
91767 Palaiseau, France
e-mail: nicolas.reyren@cnrs-thales.fr

© Springer International Publishing AG, part of Springer Nature 2018
C. Cancellieri and V. Strocov (eds.), *Spectroscopy of Complex Oxide Interfaces*, Springer Series in Materials Science 266,
https://doi.org/10.1007/978-3-319-74989-1_3

coupling, (3) capacitance, (4) polar order and domain wall conductivity and (5) thermopower. Field effect on superconductivity is discussed in a dedicated section. The purpose of this discussion is to highlight some of the insight on oxide interfaces acquired during the past 10 years through transport and field effect experiments.

In 2006, 2 years after the discovery of the conducting $LaAlO_3/SrTiO_3$ (LAO/STO) interfaces, Thiel et al. reported the first results on the effects of electrostatic gating in this system [1]. It was shown that (a) a critical thickness of 4 u.c. of LAO exists for conductivity and (b) a 3 u.c. sample can be made conducting (and reversibly turned insulating) at room temperature by means of the electrostatic field effect, using a back gate geometry (the STO substrate acts as a gate dielectric). The same approach was used to determine the influence of gating on the superconducting properties. In 2008, it was demonstrated that the electrostatic field effect can lead to an on/off switching of superconductivity, uncovering a complex phase diagram with a non-superconducting phase, a quantum critical point, underdoped and overdoped 2D superconducting regions [2].

Mobility and carrier localization. C. Bell et al. showed that gating in LAO/STO leads to a large change in carrier mobility [3], measured through Hall effect experiments. In the underdoped region of the phase diagram a mobility of the order 10^2 cm^2/Vs is observed. Its magnitude continuously increases up to several 10^3 cm^2/Vs as the system is brought into the overdoped region by means of electrostatic gating. It was argued that this effect is related to a variation in the spatial confinement of the electronic wave functions in the out-of-plane direction. Around the same time, carrier localization mechanisms were considered. In the non-superconducting state it was shown that carriers undergo weak localization, evidenced by a negative quantum correction to the conductivity, correction which is suppressed by magnetic fields, meaning that a negative magnetoresistance is observed.

Spin-orbit coupling. In 2010 it was shown by means of magnetotransport experiments, that spin-orbit coupling undergoes large changes throughout the phase diagram of the system [4]. As the system enters a gate voltage range corresponding to the underdoped superconducting regime, a steep rise in spin-orbit coupling is observed, leading to a spin-splitting of the Fermi surface up to $\sim$10 meV. A similar correlation between spin-orbit coupling and superconductivity is observed also in the overdoped regime [5]. The magnitude of the spin splitting is comparable to the Fermi energy, indicating that spin-orbit coupling is a dominant energy scale of the system. This leads to various interesting magnetotransport effect, including conductance oscillations with respect to the angle between the magnetic field and the current vector [6] and a complex evolution of the Shubnikov-de Haas oscillations with gating [7, 8]. More recently, Boltzmann transport calculations have shown that spin-orbit-induced modifications of the Fermi surface can also account for the large in-plane magnetoresistance observed in LAO/STO [9].

Capacitance enhancement. The electrostatic field-effect can be used in order to bring the electron system on the verge of strong carrier localization. In this regime, it was shown that top-gated LAO/STO exhibits a very large enhancement of capacitance [10], attributed to a negative electron compressibility, arising from correlation or disorder effects. It was argued that these effects offer a route for reducing operating voltages in field effect transistors.

Lattice effects: polar order and domain wall conductivity. Further insight into gating and electron localization in LAO/STO was acquired in 2013 by Rössle et al. [11]. Using ellipsometry and x-ray diffraction experiments, they showed that a 1 μm thick region of the STO substrate, undergoes a polar structural phase transition at temperatures below 50 K under the application of a negative gate voltage. This was evidenced by the electric-field induced splitting of an infrared active phonon mode observed only in the tetragonal phase of STO, representing antiphase rotations of oxygen octahedra. A second evidence presented was the electric-field induced satellite peaks observed around a specific Bragg x-ray reflection, representing spatial modulations of the polar order. It was argued that the electron localization observed at LAO/STO in field effect experiments is either influenced or even induced by the polar order: this phase transition strongly reduces the lattice polarizability of STO at the interfaces. This in turns reduces the dielectric screening and enhances the effect of disorder leading to a tighter confinement and a decrease in mobility. Additional indications for polar order in LAO/STO, in the absence of gating, are provided by electron microscopy experiments [12, 13]. The data points at polar order developing even at room temperature at much sharper (atomic) length scales, involving a combination of octahedral rotation and polar displacements in both LAO and STO. The effect of electrostatic gating on these short-scale atomic displacements remains to be investigated.

A second class of lattice effects relevant to transport and gating experiments in LAO/STO pertains to the tetragonal ferroelastic domains formed below 105 K. By means of a scanning single electron transistor technique, Honig et al. [14] demonstrated that the electrostatic landscape of LAO/STO is a direct map of the tetragonal domains of STO, with the local potential exhibiting ∼1 mV steps at the domain boundaries between in-plane and out-of-plane oriented domains. As the LAO/STO interface is gated, these domains move by ∼1 μm/V driven by either anisotropic electrostriction or direct coupling to polar walls. Importantly for transport experiments, Kalisky et al. [15] have shown, using a scanning magnetometry technique, that these domain boundaries constitute enhanced conduction paths. This can be understood as a combined effect of enhanced carrier density and mobility at the domain walls. Motivated by these findings, the investigation of nanoscale properties of LAO/STO remains to this date a frontier area of research that is discussed below in a separate section.

Thermopower. In 2010 Pallecchi et al. [16] considered the Seebeck effect of LAO/STO under the application of gating down to 77 K. The data is consistent with a tightly confined layer with a 2D density of states. Electrostatic gating was found to change the carrier density as well as the width of the confinement. More recently, the same authors have considered thermopower at low temperature and at gating fields on the verge of carrier depletion [17]. They observe a remarkably high thermopower (10^5 μV/K) oscillating as a function of the gate voltage, attributed to a periodic density of states arising from localized states.

3.2 Evidence for Multi-band Conduction from Magnetotransport

At high carrier densities (several 10^{13} cm^{-2}), magnetotransport at the LAO/STO interface displays a complex evolution in magnetic field, which has been attributed to the presence and occupation of several electronic bands [3, 7, 18–20]. Evidence of the existence of several bands has also been provided by Nernst effect measurements [21]. A growing body of theoretical models have been proposed to explain the multiband transport, making room to accommodate for its peculiar gate voltage dependence [3, 21–25]. The electron spatial distribution in the confinement potential has been a key ingredient, providing an explanation for the existence of bands with different mobilities, accessible at different electrostatic doping levels set by the application of a gate-voltage.

3.2.1 Anisotropic Magnetotransport

From an extensive list of exotic properties, one of the most surprising experimental observations is the peculiar anisotropy of magnetotransport under externally applied magnetic fields of large magnitude. When the field is applied in the plane of the 2DES, a large negative magnetoresistance is observed, showing a dramatic bell-shaped drop in resistance with respect to its zero-field value [9, 25, 26]. This negative magnetoresistance is extremely sensitive to the angle of the applied magnetic field, vanishing when the field is slightly tilted out of the plane. Furthermore, a strong, approximately six-fold, anisotropy in transport is observed with respect to the angle of the applied field within the plane of the 2DES [25].

At low temperatures, Shubnikov-de Haas oscillations in the longitudinal resistivity have been observed [7, 20], from which the extracted carrier density is one order of magnitude smaller than that extracted from the Hall effect. To date, the origin of this discrepancy is missing a clear explanation. At small magnetic fields applied perpendicularly to the plane of the 2DES, the magnetoresistance gradually changes sign as a function of gate-voltage from negative (WL) to positive (WAL) [4, 6], originating from spin-orbit coupling with a rather large energy scale. Interestingly enough, superconductivity was shown to emerge at the same gate-voltage that strong spin-orbit coupling sets in [4, 5].

3.2.2 Universal Lifshitz Transition

For samples displaying high-mobility, a simple yet effective model was put forward to explain the observed transport. Experiments have shown [24] that, for this type of samples, there is a critical carrier density below which the Hall voltages are linear in magnetic field, indicating that transport follows a single-band behavior. Above this critical density, Hall curves become non-linear, which is consistent with a two-band transport. This transition is observed to occur at a critical carrier density of $\sim 1.6 \times 10^{13}$ cm^{-2} for several samples with different thicknesses and mobilities. This apparent universality hints that the observed transition is not disorder-driven [27], but instead has its origin in intrinsic properties of the 2DES.

This scenario predicts that the critical density corresponds to a Lifshitz transition between the population of a single, light along the conduction plane, d_{XY} band and the additional population of two heavy bands: the d_{XZ} and the d_{YZ}. The reason for the difference in their mobilities can be understood from a simple geometrical argument: both the d_{XZ} and d_{YZ} bands have one pair of lobes pointing out-of-plane, while both lobes of the d_{XY} are in plane, shifting this band to a lower energy. Gate-dependent angle-resolved photoemission spectroscopy measurements recently corroborated this scenario [28]. This band picture provides an elegant explanation for the sudden appearance of spin-orbit interactions as a function of gate-voltage. Spin-orbit interactions should be most prominent where the bands are degenerate, which corresponds exactly to the energy where the heavy bands start being populated, i.e., the Lifshitz transition. In turn, the influence of Rashba spin-orbit coupling is also peaked at the Lifshitz point, because it is directly proportional to the atomic spin-orbit coupling. The resulting band structure, introduced by Ruhman et al. [29], has been pivotal to explain various magnetotransport phenomena [6, 9, 30, 31].

3.3 Ground State of the LaAlO$_3$/SrTiO$_3$: Superconductivity and Magnetism

In 2007, shortly after the discovery of the 2DES at the LAO/STO interface, magnetic effects were reported in this system, based on the observation of hysteretic magneto-resistances at 0.3 K [32]. This was especially exciting, as magnetism is not present in any of the bulk components, revealing new states emerging from "boring band insulators" by their combination. At the same time, it was discovered that the 2DES is also superconducting below about 0.2 K [33]. Moreover, these measurements were compatible with a two-dimensional (2D) superconducting system, with a peculiar type of transition, as it was later confirmed by other experiments [2, 5, 34]. Both phenomena being thought to be antagonist, and indeed not being observed in the same samples in these first years, it triggered a debate about the "true nature" of the ground state. Possible explanations rely on the presence of oxygen vacancies: On one hand, it was suggested that the superconductivity state is obtained when the carrier concentration is increased due to their presences; on the other hand it was also suggested that the same vacancies were responsible for titanium polarization. Several years later, coexistence of superconductivity and magnetism was reported [35, 36], but some techniques indicate that superconductivity and magnetism might be spatially separated [37], or occurring in different electronic bands [38]. Due to the difficulties inherent to the pulsed laser deposition (note however that superconducting samples have also been fabricated using growth by molecular beam epitaxy [39]), the extreme sensitivity of the $SrTiO_3$ to oxygen defects, and the variation of $SrTiO_3$ substrate qualities (number of defects, chemical content, etc.), different groups might conclude differently simply because they have different samples.

This section will focus on three topics, first on magnetism, then on superconductivity and its two-dimensional nature, and finally on the modulation of the LAO/STO properties by electrostatic field effect.

The first indications of "magnetic effects" [32] were relying on hysteresis in the magnetoresistance curves (see Fig. 3.1a). The interpretation of the curves (which were also depending on the sweep rate of the magnetization) could not give a clear picture of the mechanisms at play. Magnetoresistance and Hall effect remained a technique of investigation of the magnetic effect, and qualitative behavior could be reproduced considering two conducting bands in parallel, one of them containing magnetic impurities [38]. Other effects related to the Rashba effect at the interface might be responsible for magnetic-like effects [40]. Macroscopic magnetic measurements were also performed, but the magnetic volume and the associated moments being so weak or diluted, totalizing a few $nA\,m^2$ at most, their interpretation must be extremely cautious [35, 41]. Alternatively, x-ray magnetic circular dichroism at the Ti $L_{2,3}$-edge (the magnetism being potentially found in the $3d$ band of Ti [42], even though it was also predicted to occur at the $LaAlO_3$ surface [43]) gives more direct evidence of "intrinsic" (not related to impurities) magnetism situated at the interface, or at least close to it. Some groups reported [44] the observation of such dichroism, other did not find any (see e.g. [45] or supplementary of [46]). It was also reported that oxygen vacancies seem to play a major role in the observation of the Ti dichroic signal in LAO/STO [47]. Local probe, precisely a micro-SQUID, allowed to observe localized and disconnected dipole patches ($<3\,\mu m$ and $\sim 10^7 \mu_B$) [37] (Fig. 3.1b) situated probably near the interface and sensitive to the tip pressure [48]. This last observation suggested a role of strain related to defects and to step edges. Magnetic force microscopy may have also revealed some magnetic patches [49]. Finally, LAO/STO has been used as an electrode in a magnetic tunnel junction with Co as second electrode [50]. The tunnel magnetoresistance signal changes with gate voltage, suggesting a connection with the 2DES, but results are varying with thermal cycling, pointing again to a phenomenon related to defects or domain walls appearing at the $SrTiO_3$ cubic-tetragonal structural phase transition at about 105 K [51]. It has also been suggested that the ground state could be a long-wavelength spiral [52]. The magnetic effects at the LAO/STO interface are hence still matter of research to understand the phenomenon in more details.

The $SrTiO_3$ is known to be superconducting in bulk systems when it is oxygen deficient [53, 54] or if it is doped by substitution of Ti by Nb or Sr by La [55]. It was hence natural to think that chemical doping could explain this observation. Interestingly, the growth of superconducting very thin films ($<10\,nm$) of cation-doped $SrTiO_3$, either by La or Nb, was failing until a new strategy was adopted: Growing a "delta-doped" $SrTiO_3$ avoids band-bending effects at the surface and hence allowed the fabrication of extremely thin (5.5 nm) layers of doped $SrTiO_3$ which exhibit very similar

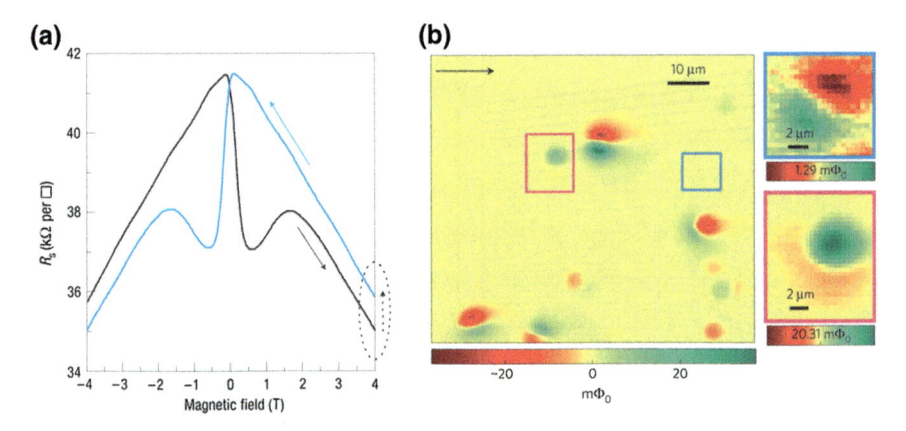

Fig. 3.1 Magnetic effects at the LAO/STO interface. **a** First hints of magnetic properties were revealed by hysteretic magneto-resistance curves at low temperature [32]. **b** Later, local SQUID measurements exposed the presence of local dipoles totalizing typically 10^7 Bohr magnetons [37]

superconducting properties than the LAO/STO system [56]. When LaAlO$_3$ is grown on top of SrTiO$_3$, similar band-bending might occur, but probably not strong enough to insure conductivity due to the differences in workfunctions only [57]. For bulk Nb-doped SrTiO$_3$, the superconductivity shows two gaps [58], but this was not observed in the LAO/STO system [59]. The coupling mechanism could be a rather conventional BCS electron-phonon, despite the low carrier density. Interestingly, on the other hand, a pseudogap temperature has also been found, indicating that the 2D superconductivity at the LAO/STO interface might share some ingredients with high-temperature superconductors [59, 60].

The two-dimensional character of the superconductivity in LAO/STO manifests itself in several aspects. We detail two of them: the resistive transition in temperature and the anisotropies of the critical magnetic fields. For 2D materials where a Berezinskii-Kosterlitz-Thouless (BKT) transition is expected, a finite resistance appears at temperature at which the thermally activated vortex-antivortex pairs unbind. In the LAO/STO case, a more complex case of the melting of a vortex-antivortex lattice could replace the conventional BKT mechanism, and is indeed quantitatively agreeing with observations [33]. The resistance as a function of temperature follows a characteristic law close to the BKT transition temperature T_{BKT} and the current-voltage curves exhibit a power-law $V = I^a$, the a coefficient taking the value 3 at T_{BKT}, as it was observed now is several groups [2, 33, 61]. Deviations from these laws are associated with finite size effects [62]. A somewhat more direct evidence for the 2D nature of the superconducting state is the anisotropy of the critical magnetic fields [34]: Applying the field in-plane or out-of-plane leads to markedly different effect. Considering a superconductor with a magnetic field applied in-plane, if its thickness is lower than its coherence length, the wave function amplitude cannot vary over it, and hence the superconductivity is not destroyed before the field-associated energy goes beyond the pairing energy. This happens in

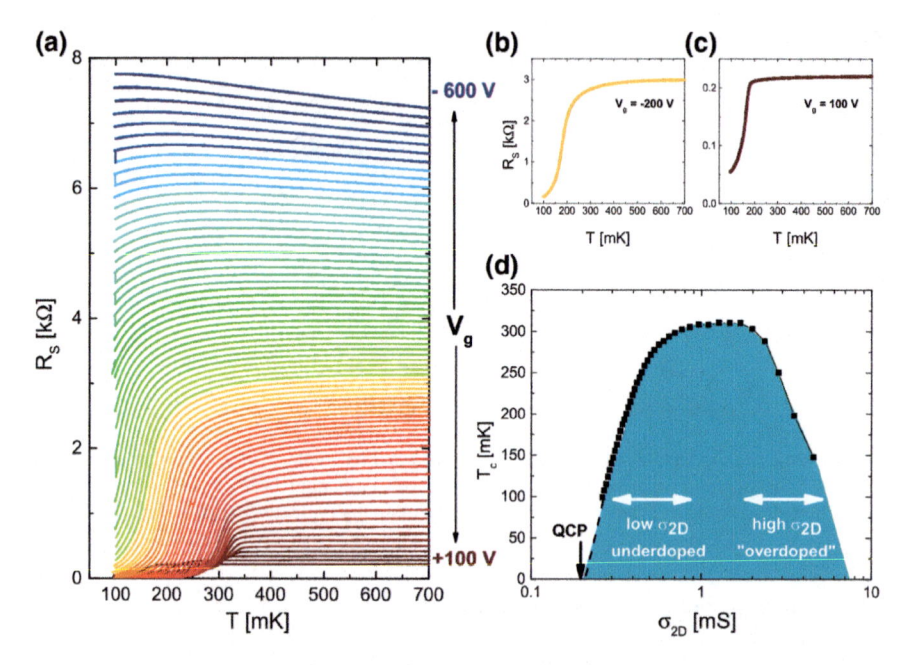

Fig. 3.2 Superconductivity and its modulation in the LAO/STO system: **a** Resistive transition as a function of the gate voltage, displaying a remarkable tunability and a superconductivity-insulator transition near the lowest doping; **b**, **c** zoom on the high concentration curves; **d** resulting phase diagram with the superconducting T_c dome as a function of the normal state sheet conductance, which reflects the carrier concentration. Figure from [66]

BCS systems at the Clogston-Chandrasekhar paramagnetic limit [63, 64] given by $\mu_0 H_{CC} = \Delta(0)/(\sqrt{2}\mu_B)$, where $\mu_0 H_{CC}$ is the applied field, $\Delta(0)$ is the gap energy at $T = 0$, μ_B is the Bohr magnetron and a gyomagnetic ratio of 2 is assumed. It has been observed that this limit is overcome by a factor 5 [5, 34]. This is an extremely strong indication of the 2D nature of the superconductivity in LAO/STO and it also reveals that spin-orbit effects or other corrections such Fulde-Ferrell-Larkin-Ovchinikov type of condensate [65] must be taken into account for a precise quantitative description of the observed critical fields.

Owing to the very low carrier concentration (of the order of 10^{13} cm^{-2}) of the 2DES, the extremely large dielectric constant ($\sim 10^4 \, \varepsilon_0$) of the SrTiO$_3$ substrate at low temperature, and its two dimensional nature, electrostatic field effect using the SrTiO$_3$ substrate as a gate dielectric is very efficient to modulate the properties of LAO/STO. In particular, it has been possible to tune the superconducting state and reveal a dome-like shape of the critical temperature as a function of the gate voltage, or the carrier concentration [2, 60, 67]. "Top gates" (without SrTiO$_3$) lead to similar properties, even though the mechanism is cleaner, it is harder to realize [67–69]. Side gates taking again advantage of the huge dielectric constant of SrTiO$_3$ have also been used [70, 71]. Finally, it is noteworthy that the adsorbates at the surface of LaAlO$_3$ influence the properties of the 2DES [72]. Whatever the doping technique, a superconducting dome is found: the critical temperature first increases with the carrier number, it culminates at about 0.3 K and then reduces as the carrier number is further increased, as visible in Fig. 3.2. At the low concentration edge of the dome, a quantum critical point (QCP) is found, corresponding to a superconductor to insulator transition [2, 66]. Explaining this dome shape is rather complicated as many parameters are at play. First the volume carrier density does not scale linearly with the areal carrier density, which is the quantity that is modulated with gate voltages (without mentioning extra difficulties stemming from trapped states in $SrTiO_3$ [73]): Indeed due to the strong field-dependence of the $SrTiO_3$ dielectric constant, and the electrical potential well shape, the 2DES can actually expand as carrier concentration is reduced! Second, as the areal carrier concentration is increased, different conduction bands get populated, and associated to the different bands is a more or less strong Rashba spin-orbit coupling. As mentioned above, as the Rashba coefficient strongly increases (several folds), the superconducting transition appears and culminates [4]. The correlation between Rashba and superconductivity might be more than coincidental...

The ground state properties of LAO/STO are hence still a matter of active research! But whatever the nature of the "true" ground state, the superconductivity that can be modulated by gate voltage or by geometrical constriction establishes a very unique test system to understand two dimensional, or even one-dimensional superconductivity, and, who knows, even help to uncover the mysteries of high temperature superconductivity.

3.4 Nanopatterning

Patterning of the interfacial 2DES is crucial to the realization of functional electrical devices. When compared to their semiconductor counterparts, where the 2DES is typically buried hundreds of nanometers below the surface, the 2DES at the LAO/STO interface offers the exciting possibility of extremely reduced dimensions, since it lives only a few nanometers below the surface. However, producing high quality nanoscale structures at the LAO/STO interface has proven challenging due to inherent stoichiometric and structural intricacies associated with complex oxides [74]. Here, we make a brief overview of the main approaches to patterning LAO/STO and the progress in creating functional devices in this system.

Conventional photo- and e-beam lithographic techniques have been extensively used to laterally define structures by locally controlling the thickness of the crystalline LAO layer [75]. The STO substrate is patterned prior to the LAO thin film deposition and, after development, an amorphous LAO layer is deposited. After lift-off, the STO substrate is cleared in the areas protected by the resist, thus yielding conducting regions upon epitaxial LAO growth. The areas covered by the amorphous layer remain insulating. Figure 3.3b shows a microbridge realized by means of this technique. In certain cases, to ensure that resist residue does not disrupt the

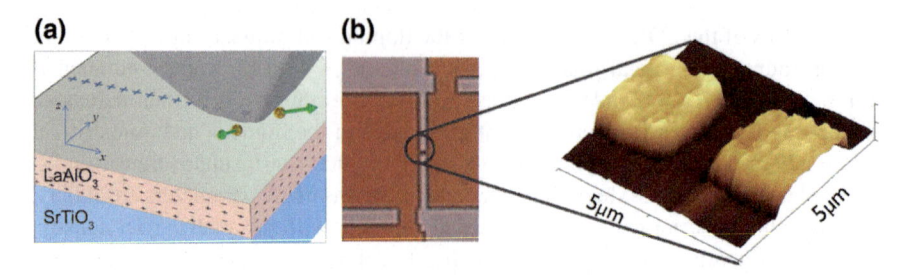

Fig. 3.3 Nanopatterning techniques. **a** AFM tip moving left to right above a 3 u.c. LAO/STO interface, locally changing the charge state of the surface creating a conducting wire. From [77]. **b** Atomic force microscope image of an 800 nm wide bridge. Polycrystalline/amorphous LAO, grown on the amorphous STO, has a lighter color, while the epitaxial LAO has a darker one. The 2DES is created only below the epitaxial LAO. From [76]

conducting interface in the device region, 2 unit cells of LAO are first deposited epitaxially over the entire substrate, after which the process described above is performed. By using this patterning method, conducting features as small as 500 nm have been achieved with e-beam lithography [76].

Conducting features down to just 2 nm have been realized through the direct atomic force microscope (AFM) writing technique [78]. A sub-critical-thickness (3 u.c.) LAO thin film is deposited on the entire substrate, which can be locally and reversibly switched between a conducting and insulating state by applying a positive or negative voltage to the AFM tip, respectively (see Fig. 3.3a). The most widely recognized mechanism of formation for this metastable conductive state is the local modification of the surface charge [79] through voltage-mediated addition and removal of water in the form of OH and H^+ [80].

3.5 Other Paths of Exploration

3.5.1 Spintronics

Spintronics is an alternative information scheme which uses the spin of the carriers, rather than their charge. Devices will require the injection, transport, modulation and detection of spin currents. The LAO/STO could be an interesting platform to test spintronics ideas, particularly owing to the possibility to modulate the Rashba effect (acting on the spin current) in this system of relatively large mobility. A first step towards such possibilities has been indirectly demonstrated with the spin injection at the LAO/STO interface from a conventional ferromagnetic metal electrode [81, 82]. Again, the behavior of the 2DES at the LAO/STO interface seems to be slightly different from what is observed in Nb-doped $SrTiO_3$ [83, 84], but the field of research is still at its infancy for $SrTiO_3$-based systems and further studies will be needed to get rid of the measurement problems and artifacts related to the tunnel barriers. In particular, so-called "non-local" measurements still need to be realized in order to directly measure the spin diffusion length.

Another interesting use of the LAO/STO system in spintronics is related to the spin to charge current conversions. Due to the Rashba interaction, spin and momentum are coupled, meaning that a charge current can imply a spin accumulation, and reciprocally. This spin accumulation can relax in a nearby material and hence produces a spin current. If a ferromagnet is placed in contact, the LAO/STO 2DES will produce a spin-torque on its magnetization, as it was first observed in 2014 [40]. The reciprocal effect (spin pumping from a ferromagnet into the 2DES, creating a charge current) has been observed recently, and very interestingly, a gate voltage can strongly modulate the amplitude (reaching values larger than what can be found in metal multilayers) and even the sign of the effect can be changed [85].

The LAO/STO system is hence showing very interesting properties in the framework of the spintronics, and its study will surely lead to other remarkable observations.

3.5.2 Diode Effects, Circuits and Sensors

Since the first decade of the 2000s, people speak about oxitronics, that is electronic circuits made of oxide systems, taking advantage of the very diverse behaviors of oxide systems.

A radically new approach has been proposed at the very beginning: the conducting circuit could be written by atomic force microscopy on an insulating three-unit-cells-thick LAO/STO interface as described in Sect. 3.4 [86]. Of course this lithography technique is not viable for consumer products, but could be useful for very peculiar applications [70, 77]. It also permits to study clean circuits of variable geometries to investigate size or quantum effects [87], which are interesting from a fundamental point of view.

A more traditional approach has been used to design circuits, starting from diodes [88] to complete oscillator circuits [89] using field effect transistors [90]. Diodes with extremely large blocking voltage can be realized, as well as very large capacitances [35] thanks to the particularities of the LAO/STO system: The 2DES can be completely expelled from below a gate electrode, changing dramatically the effective geometry of the system. These electronic components and circuits are operating at room temperature and above. They might find application in peculiar niche.

Finally, this interface also displays interesting properties as sensor, either of light [91] or of adsorbates, and hence indirectly gas [72].

References

1. S. Thiel, G. Hammerl, A. Schmehl, C. Schneider, J. Mannhart, Tunable quasi-two-dimensional electron gases in oxide heterostructures. Science **313**(5795), 1942–1945 (2006)
2. A.D. Caviglia, S. Gariglio, N. Reyren, D. Jaccard, T. Schneider, M. Gabay, S. Thiel, G. Hammerl, J. Mannhart, J.M. Triscone, Electric field control of the $LaAlO_3/SrTiO_3$ interface ground state. Nature **456**(7222), 624–627 (2008). https://doi.org/10.1038/nature07576
3. C. Bell, S. Harashima, Y. Kozuka, M. Kim, B.G. Kim, Y. Hikita, H. Hwang, Dominant mobility modulation by the electric field effect at the $LaAlO_3/SrTiO_3$ interface. Phy. Rev. Lett. **103**(22), 226802 (2009)
4. A.D. Caviglia, M. Gabay, S. Gariglio, N. Reyren, C. Cancellieri, J.M. Triscone, Tunable rashba spin-orbit interaction at oxide interfaces. Phy. Rev. Lett. **104**(12), 126803 (2010). https://doi.org/10.1103/PhysRevLett.104.126803, http://link.aps.org/doi/10.1103/PhysRevLett.104.126803
5. M. Ben Shalom, M. Sachs, D. Rakhmilevitch, A. Palevski, Y. Dagan, Tuning spin-orbit coupling and superconductivity at the $LaAlO_3/SrTiO_3$ interface: a magnetotransport study. Phy. Rev. Lett. **104**(12), 126802 (2010). https://doi.org/10.1103/PhysRevLett.104.126802, http://link.aps.org/doi/10.1103/PhysRevLett.104.126802
6. A. Fête, S. Gariglio, A. Caviglia, J.M. Triscone, M. Gabaym, Rashba induced magnetoconductance oscillations in the $LaAlO_3/SrTiO_3$ heterostructure. Phy. Rev. B **86**(20), 201105 (2012)
7. A. Caviglia, S. Gariglio, C. Cancellieri, B. Sacepe, A. Fete, N. Reyren, M. Gabay, A. Morpurgo, J.M. Triscone, Two-dimensional quantum oscillations of the conductance at $LaAlO_3/SrTiO_3$ interfaces. Phy. Rev. Lett. **105**(23), 236802 (2010)
8. A. Fête, S. Gariglio, C. Berthod, D. Li, D. Stornaiuolo, M. Gabay, J. Triscone, Large modulation of the shubnikov? de haas oscillations by the rashba interaction at the $LaAlO_3/SrTiO_3$ interface. N. J. Phy. **16**(11), 112002 (2014)
9. M. Diez, A. Monteiro, G. Mattoni, E. Cobanera, T. Hyart, E. Mulazimoglu, N. Bovenzi, C. Beenakker, A. Caviglia, Giant negative magnetoresistance driven by spin-orbit coupling at the $LaAlO_3/SrTiO_3$ interface. Phy. Rev. Lett. **115**(1), 016803 (2015)
10. L. Li, C. Richter, S. Paetel, T. Kopp, J. Mannhart, R. Ashoori, Very large capacitance enhancement in a two-dimensional electron system. Science **332**(6031), 825–828 (2011)
11. M. Rössle, K.W. Kim, A. Dubroka, P. Marsik, C.N. Wang, R. Jany, C. Richter, J. Mannhart, C. Schneider, A. Frano et al., Electric-field-induced polar order and localization of the confined electrons in $LaAlO_3/SrTiO_3$ heterostructures. Phy. Rev. Lett.**110**(13), 136805 (2013)
12. C. Cantoni, J. Gazquez, F. Miletto Granozio, M.P. Oxley, M. Varela, A.R. Lupini, S.J. Pennycook, C. Aruta, U.S. di Uccio, P. Perna et al., Electron transfer and ionic displacements at the origin of the 2d electron gas at the LAO/STO interface: direct measurements with atomic-column spatial resolution. Adv. Mater. **24**(29), 3952–3957 (2012)
13. C. Jia, S. Mi, M. Faley, U. Poppe, J. Schubert, K. Urban, Oxygen octahedron reconstruction in the $SrTiO_3/LaAlO_3$ heterointerfaces investigated using aberration-corrected ultrahigh-resolution transmission electron microscopy. Phy. Rev. B **79**(8), 081405 (2009)
14. M. Honig, J.A. Sulpizio, J. Drori, A. Joshua, E. Zeldov, S. Ilani, Local electrostatic imaging of striped domain order in $LaAlO_3/SrTiO_3$. Nat. Mater. **12**(12), 1112–1118 (2013)
15. B. Kalisky, E.M. Spanton, H. Noad, J.R. Kirtley, K.C. Nowack, C. Bell, H.K. Sato, M. Hosoda, Y. Xie, Y. Hikita et al., Locally enhanced conductivity due to the tetragonal domain structure in $LaAlO_3/SrTiO_3$ heterointerfaces. Nat. Mater. **12**(12), 1091–1095 (2013)
16. I. Pallecchi, M. Codda, E. Galleani d'Agliano, D. Marré, A.D. Caviglia, N. Reyren, S. Gariglio, J.M. Triscone, Seebeck effect in the conducting $LaAlO_3/SrTiO_3$ interface. Phy. Rev. B **81**(8), 085414 (2010). https://doi.org/10.1103/PhysRevB.81.085414
17. I. Pallecchi, F. Telesio, D. Li, A. Fête, S. Gariglio, J.M. Triscone, A. Filippetti, P. Delugas, V. Fiorentini, D. Marré, Giant oscillating thermopower at oxide interfaces. Nat. Commun. **6** (2015)

18. J.S. Kim, S.S.A. Seo, M.F. Chisholm, R. Kremer, H.U. Habermeier, B. Keimer, H.N. Lee, Nonlinear hall effect and multichannel conduction in LaAlO$_3$/SrTiO$_3$ superlattices. Phy. Rev. B **82**(20), 201,407 (2010)
19. R. Pentcheva, M. Huijben, K. Otte, W.E. Pickett, J. Kleibeuker, J. Huijben, H. Boschker, D. Kockmann, W. Siemons, G. Koster, et al., Parallel electron-hole bilayer conductivity from electronic interface reconstruction. Phy. Rev. Lett. **104**(16), 166804 (2010)
20. M.B. Shalom, A. Ron, A. Palevski, Y. Dagan, Shubnikov–de haas oscillations in SrTiO$_3$/LaAlO$_3$ interface. Phy. Rev. Lett. **105**(20), 206401 (2010)
21. S. Lerer, M. Ben Shalom, G. Deutscher, Y. Dagan, Low-temperature dependence of the thermomagnetic transport properties of the SrTiO$_3$/LaAlO$_3$ interface. Phy. Rev. B **84**(7), 075423 (2011). https://doi.org/10.1103/PhysRevB.84.075423, http://link.aps.org/doi/10.1103/PhysRevB.84.075423
22. J. Biscaras, N. Bergeal, S. Hurand, C. Grossetête, A. Rastogi, R. Budhani, D. LeBoeuf, C. Proust, J. Lesueur, Two-dimensional superconducting phase in LaAlO$_3$/SrTiO$_3$ heterostructures induced by high-mobility carrier doping. Phy. Rev. Lett. **108**(24), 247004 (2012)
23. P. Brinks, W. Siemons, J. Kleibeuker, G. Koster, G. Rijnders, M. Huijben, Anisotropic electrical transport properties of a two-dimensional electron gas at SrTiO$_3$-LaAlO$_3$ interfaces. Appl. Phy. Lett. **98**(24), 242904 (2011)
24. A. Joshua, S. Pecker, J. Ruhman, E. Altman, S. Ilani, A universal critical density underlying the physics of electrons at the LaAlO$_3$/SrTiO$_3$ interface. Nat. Commun. **3**, 1129 (2012)
25. M.B. Shalom, C. Tai, Y. Lereah, M. Sachs, E. Levy, D. Rakhmilevitch, A. Palevski, Y. Dagan, Anisotropic magnetotransport at the SrTiO$_3$/LaAlO$_3$ interface. Physical Review B **80**(14), 140403 (2009)
26. A. Annadi, Z. Huang, K. Gopinadhan, X.R. Wang, A. Srivastava, Z. Liu, H.H. Ma, T. Sarkar, T. Venkatesan et al., Fourfold oscillation in anisotropic magnetoresistance and planar hall effect at the LaAlO$_3$/SrTiO$_3$ heterointerfaces: effect of carrier confinement and electric field on magnetic interactions. Phys. Rev. B **87**(20), 201102 (2013)
27. Y. Liao, T. Kopp, C. Richter, A. Rosch, J. Mannhart, Metal-insulator transition of the LaAlO$_3$/SrTiO$_3$ interface electron system. Phy. Rev. B **83**(7), 075402 (2011)
28. C. Cancellieri, M.L. Reinle-Schmitt, M. Kobayashi, V.N. Strocov, P. Willmott, D. Fontaine, P. Ghosez, A. Filippetti, P. Delugas, V. Fiorentini, Doping-dependent band structure of LaAlO$_3$/SrTiO$_3$ interfaces by soft x-ray polarization-controlled resonant angle-resolved photoemission. Phy. Rev. B **89**(12), 121412 (2014)
29. J. Ruhman, A. Joshua, S. Ilani, E. Altman, Competition between kondo screening and magnetism at the LaAlO$_3$/SrTiO$_3$ interface. Phy. Rev. B **90**(12), 125123 (2014)
30. M.H. Fischer, S. Raghu, E.A. Kim, Spin–orbit coupling in LaAlO$_3$/SrTiO$_3$ interfaces: magnetism and orbital ordering. N. J. Phy. **15**(2), 023022 (2013)
31. Y. Kim, R.M. Lutchyn, C. Nayak, Origin and transport signatures of spin-orbit interactions in one-and two-dimensional SrTiO$_3$-based heterostructures. Phy. Rev. B **87**(24), 245,121 (2013)
32. A. Brinkman, M. Huijben, M. van Zalk, J. Huijben, U. Zeitler, J.C. Maan, W.G. van der Wiel, G. Rijnders, D.H.A. Blank, H. Hilgenkamp, Magnetic effects at the interface between nonmagnetic oxides. Nat. Mater. **6**(7), 493–496 (2007). https://doi.org/10.1038/nmat1931, http://www.nature.com/doifinder/10.1038/nmat1931
33. N. Reyren, S. Thiel, A.D. Caviglia, L.F. Kourkoutis, G. Hammerl, C. Richter, C.W. Schneider, T. Kopp, A.S. Ruetschi, D. Jaccard, M. Gabay, D.A. Muller, J.M. Triscone, J. Mannhart, Superconducting interfaces between insulating oxides. Science **317**(5842), 1196–1199 (2007). https://doi.org/10.1126/science.1146006, http://www.sciencemag.org/cgi/doi/10.1126/science.1146006
34. N. Reyren, S. Gariglio, A.D. Caviglia, D. Jaccard, T. Schneider, J.M. Triscone, Anisotropy of the superconducting transport properties of the LaAlO$_3$/SrTiO$_3$ interface. Appl. Phy. Lett. **94**(11), 112506 (2009). https://doi.org/10.1063/1.3100777, http://scitation.aip.org/content/aip/journal/apl/94/11/10.1063/1.3100777
35. L. Li, C. Richter, J. Mannhart, R.C. Ashoori, Coexistence of magnetic order and two-dimensional superconductivity at LaAlO$_3$/SrTiO$_3$ interfaces. Nat. Phy. **7**(10), 762–766 (2011). https://doi.org/10.1038/nphys2080, http://www.nature.com/doifinder/10.1038/nphys2080

36. A.P. Petrović, A. Paré, T.R. Paudel, K. Lee, S. Holmes, C.H.W. Barnes, A. David, T. Wu, E.Y. Tsymbal, C. Panagopoulos, Emergent vortices at a ferromagnetic superconducting oxide interface. N. J. Phy. **16**(10), 103012 (2014). https://doi.org/10.1088/1367-2630/16/10/103012, http://stacks.iop.org/1367-2630/16/i=10/a=103012?key=crossref.95d78e23175af9dfd7f98da58155c926
37. J.A. Bert, B. Kalisky, C. Bell, M. Kim, Y. Hikita, H.Y. Hwang, K.A. Moler, Direct imaging of the coexistence of ferromagnetism and superconductivity at the LaAlO$_3$/SrTiO$_3$ interface. Nat. Phy. **7**(10), 767–771 (2011). https://doi.org/10.1038/nphys2079, http://www.nature.com/doifinder/10.1038/nphys2079
38. D.A. Dikin, M. Mehta, C.W. Bark, C.M. Folkman, C.B. Eom, V. Chandrasekhar, Coexistence of superconductivity and ferromagnetism in two dimensions. Phy. Rev. Lett. **107**(5), 056802 (2011). https://doi.org/10.1103/PhysRevLett.107.056802, http://link.aps.org/doi/10.1103/PhysRevLett.107.056802
39. M.P. Warusawithana, C. Richter, J.A. Mundy, P. Roy, J. Ludwig, S. Paetel, T. Heeg, A.A. Pawlicki, L.F. Kourkoutis, M. Zheng, M. Lee, B. Mulcahy, W. Zander, Y. Zhu, J. Schubert, J.N. Eckstein, D.A. Muller, C.S. Hellberg, J. Mannhart, D.G. Schlom, LaAlO3 stoichiometry is key to electron liquid formation at LaAlO$_3$/SrTiO$_3$ interfaces. Nat. Commun. **4**, 2351 (2013). https://doi.org/10.1038/ncomms3351
40. K. Narayanapillai, K. Gopinadhan, X. Qiu, A. Annadi, Ariando, T. Venkatesan, H. Yang, Current-driven spin orbit field in LaAlO$_3$/SrTiO$_3$ heterostructures. Appl. Phy. Lett. **105**(16), 162405 (2014). https://doi.org/10.1063/1.4899122
41. Ariando, X. Wang, G. Baskaran, Z.Q. Liu, J. Huijben, J.B. Yi, A. Annadi, A.R. Barman, A. Rusydi, S. Dhar, Y.P. Feng, J. Ding, H. Hilgenkamp, T. Venkatesan, Electronic phase separation at the LaAlO$_3$/SrTiO$_3$ interface. Nat. Commun. **2**, 188 (2011). https://doi.org/10.1038/ncomms1192, http://www.nature.com/doifinder/10.1038/ncomms1192
42. N. Pavlenko, T. Kopp, E.Y. Tsymbal, G.A. Sawatzky, J. Mannhart, Magnetic and superconducting phases at the LaAlO$_3$/SrTiO$_3$ interface: the role of interfacial Ti 3d electrons. Phy. Rev. B **85**(2), 020407 (2012). https://doi.org/10.1103/PhysRevB.85.020407, http://link.aps.org/doi/10.1103/PhysRevB.85.020407
43. L. Weston, X.Y. Cui, S.P. Ringer, C. Stampfl, Density-functional prediction of a surface magnetic phase in LaAlO$_3$/SrTiO$_3$ heterostructures induced by Al vacancies. Phy. Rev. Lett. **113**(18), 186401 (2014). https://doi.org/10.1103/PhysRevLett.113.186401, http://link.aps.org/doi/10.1103/PhysRevLett.113.186401
44. J.S. Lee, Y.W. Xie, H.K. Sato, C. Bell, Y. Hikita, H.Y. Hwang, C.C. Kao, Titanium dxy ferromagnetism at the LaAlO$_3$/SrTiO$_3$ interface. Nature Materials **12**(8), 703–706 (2013). https://doi.org/10.1038/nmat3674, http://www.nature.com/doifinder/10.1038/nmat3674
45. M.R. Fitzsimmons, N.W. Hengartner, S. Singh, M. Zhernenkov, F.Y. Bruno, J. Santamaria, A. Brinkman, M. Huijben, H.J.A. Molegraaf, J. de la Venta, I.K. Schuller, Upper limit to magnetism in LaAlO$_3$/SrTiO$_3$ heterostructures. Phy. Rev. Lett. **107**(21), 217201 (2011). https://doi.org/10.1103/PhysRevLett.107.217201, http://link.aps.org/doi/10.1103/PhysRevLett.107.217201
46. E. Lesne, N. Reyren, D. Doennig, R. Mattana, H. Jaffrès, V. Cros, F. Petroff, F. Choueikani, P. Ohresser, R. Pentcheva, A. Barthélémy, M. Bibes, Suppression of the critical thickness threshold for conductivity at the LaAlO$_3$/SrTiO$_3$ interface. Nat. Commun. **5** (2014). https://doi.org/10.1038/ncomms5291, http://www.nature.com/doifinder/10.1038/ncomms5291
47. M. Salluzzo, S. Gariglio, D. Stornaiuolo, V. Sessi, S. Rusponi, C. Piamonteze, G.M. De Luca, M. Minola, D. Marré, A. Gadaleta, H. Brune, F. Nolting, N.B. Brookes, G. Ghiringhelli, Origin of interface magnetism in BiMnO$_3$/SrTiO$_3$ and LaAlO$_3$/SrTiO$_3$ heterostructures. Phy. Rev. Lett. **111**(8), 087204 (2013). https://doi.org/10.1103/PhysRevLett.111.087204, http://link.aps.org/doi/10.1103/PhysRevLett.111.087204
48. B. Kalisky, J.A. Bert, C. Bell, Y. Xie, H.K. Sato, M. Hosoda, Y. Hikita, H.Y. Hwang, K.A. Moler, Scanning probe manipulation of magnetism at the LaAlO$_3$/SrTiO$_3$ heterointerface. Nano Lett. **12**(8), 4055–4059 (2012). https://doi.org/10.1021/nl301451e, http://pubs.acs.org/doi/abs/10.1021/nl301451e

49. F. Bi, M. Huang, H. Lee, C.B. Eom, P. Irvin, J. Levy, LaAlO$_3$ thickness window for electronically controlled magnetism at LaAlO$_3$/SrTiO$_3$ heterointerfaces. Appl. Phy. Lett. **107**(8), 082402 (2015). https://doi.org/10.1063/1.4929430, http://scitation.aip.org/content/aip/journal/apl/107/8/10.1063/1.4929430
50. T.D. Ngo, J.W. Chang, K. Lee, S. Han, J.S. Lee, Y.H. Kim, M.H. Jung, Y.J. Doh, M.S. Choi, J. Song, J. Kim, Polarity-tunable magnetic tunnel junctions based on ferromagnetism at oxide heterointerfaces. Nat. Commun. **6**, 8035 (2015). https://doi.org/10.1038/ncomms9035, http://www.nature.com/doifinder/10.1038/ncomms9035
51. M. Honig, J.A. Sulpizio, J. Drori, A. Joshua, E. Zeldov, S. Ilani, Local electrostatic imaging of striped domain order in LaAlO$_3$/SrTiO$_3$. Nat. Mater. **12**(12), 1112–1118 (2013). https://doi.org/10.1038/nmat3810, http://www.nature.com/doifinder/10.1038/nmat3810
52. S. Banerjee, O. Erten, M. Randeria, Ferromagnetic exchange, spinorbit coupling and spiral magnetism at the LaAlO$_3$/SrTiO$_3$ interface. Nat. Phy. **9**(10), 626–630 (2013). https://doi.org/10.1038/nphys2702, http://www.nature.com/doifinder/10.1038/nphys2702
53. C.S. Koonce, M.L. Cohen, J.F. Schooley, W.R. Hosler, E.R. Pfeiffer, Superconducting transition temperatures of semiconducting SrTiO$_3$. Phy. Rev. **163**(2), 380–390 (1967). https://doi.org/10.1103/PhysRev.163.380
54. J.F. Schooley, W.R. Hosler, M.L. Cohen, Superconductivity in Semiconducting SrTiO$_3$. Physical Review Letters **12**(17), 474–475 (1964). https://doi.org/10.1103/PhysRevLett.12.474, http://link.aps.org/doi/10.1103/PhysRevLett.12.474
55. E.R. Pfeiffer, J.F. Schooley, Superconducting transition temperature of Nb-doped SrTiO$_3$. Phy. Lett. **29A**(10), 589–590 (1969)
56. Y. Kozuka, M. Kim, C. Bell, B.G. Kim, Y. Hikita, H.Y. Hwang, Two-dimensional normal-state quantum oscillations in a superconducting heterostructure. Nature **462**(7272), 487–490 (2009). https://doi.org/10.1038/nature08566
57. J. Nishimura, A. Ohtomo, A. Ohkubo, Y. Murakami, M. Kawasaki, Controlled carrier generation at a polarity-discontinued perovskite heterointerface. Jpn. J. Appl. Phy. **43**(8A), L1032–L1034 (2004). https://doi.org/10.1143/JJAP.43.L1032, http://stacks.iop.org/1347-4065/43/L1032
58. G. Binnig, A. Baratoff, H.E. Hoenig, J.G. Bednorz, Two-band Superconductivity in Nb-Doped SrTiO$_3$. Phys. Rev. Lett. **45**(15), 1352–1355 (1980)
59. C. Richter, H. Boschker, W. Dietsche, E. Fillis-Tsirakis, R. Jany, F. Loder, L.F. Kourkoutis, D.A. Muller, J.R. Kirtley, C.W. Schneider, J. Mannhart, Interface superconductor with gap behaviour like a high-temperature superconductor. Nature **502**(7472), 528–531 (2013). https://doi.org/10.1038/nature12494
60. H. Boschker, C. Richter, E. Fillis-Tsirakis, C.W. Schneider, J. Mannhart, Electronphonon coupling and the superconducting phase diagram of the LaAlO$_3$-SrTiO$_3$ interface. Sci. Rep. **5**(12309) (2015). https://doi.org/10.1038/srep12309
61. G.N. Daptary, S. Kumar, P. Kumar, A. Dogra, N. Mohanta, A. Taraphder, A. Bid, Correlated non-Gaussian phase fluctuations in LaAlO$_3$/SrTiO$_3$ heterointerfaces. Phy. Rev. B **94**(8), 085104 (2016). https://doi.org/10.1103/PhysRevB.94.085104, http://link.aps.org/doi/10.1103/PhysRevB.94.085104
62. T. Schneider, S. Weyeneth, Suppression of the Berezinskii-Kosterlitz-Thouless and quantum phase transitions in two-dimensional superconductors by finite-size effects. Phy. Rev. B **90**(6), 064501 (2014). https://doi.org/10.1103/PhysRevB.90.064501, http://link.aps.org/doi/10.1103/PhysRevB.90.064501
63. B.S. Chandrasekhar, A note on the maximum critical field of high field superconductors. Appl. Phy. Lett. **1**(1), 7–8 (1962)
64. A.M. Clogston, Upper limit for the critical field in hard superconductors. Phy. Rev. Lett. **9**(6), 266–267 (1962). https://doi.org/10.1103/PhysRevLett.9.266, http://link.aps.org/doi/10.1103/PhysRevLett.9.266
65. K. Michaeli, A.C. Potter, P.A. Lee, Superconducting and ferromagnetic phases in LaAlO$_3$/SrTiO$_3$ oxide interface structures: possibility of finite momentum pairing. Phy. Rev. Lett. **108**(11), 117003 (2012). https://doi.org/10.1103/PhysRevLett.108.117003, http://link.aps.org/doi/10.1103/PhysRevLett.108.117003

66. S. Gariglio, M. Gabay, J.M. Triscone, Research update: conductivity and beyond at the LaAlO$_3$/SrTiO$_3$ interface. APL Mater. **4**(6), 060701 (2016). https://doi.org/10.1063/1.4953822
67. S. Hurand, A. Jouan, C. Feuillet-Palma, G. Singh, J. Biscaras, E. Lesne, N. Reyren, A. Barthélémy, M. Bibes, J.E. Villegas, C. Ulysse, X. Lafosse, M. Pannetier-Lecoeur, S. Caprara, M. Grilli, J. Lesueur, N. Bergeal, Field-effect control of superconductivity and Rashba spin-orbit coupling in top-gated LaAlO$_3$/SrTiO$_3$ devices. Sci. Rep. **5**(12751) (2015). https://doi.org/10.1038/srep12751
68. P.D. Eerkes, W.G. van der Wiel, H. Hilgenkamp, Modulation of conductance and superconductivity by top-gating in LaAlO$_3$/SrTiO$_3$ 2-dimensional electron systems. Appl. Phy. Lett. **103**(20), 201603 (2013). https://doi.org/10.1063/1.4829555, http://scitation.aip.org/content/aip/journal/apl/103/20/10.1063/1.4829555
69. M. Hosoda, Y. Hikita, H.Y. Hwang, C. Bell: Transistor operation and mobility enhancement in top-gated LaAlO$_3$/SrTiO$_3$ heterostructures. Appl. Phy. Lett. **103**(10), 103507 (2013). https://doi.org/10.1063/1.4820449, http://scitation.aip.org/content/aip/journal/apl/103/10/10.1063/1.4820449
70. G. Cheng, P.F. Siles, F. Bi, C. Cen, D.F. Bogorin, C.W. Bark, C.M. Folkman, J.W. Park, C.B. Eom, G. Medeiros-Ribeiro, J. Levy, Sketched oxide single-electron transistor. Nat. Nanotechnol. **6**(6), 343–347 (2011). https://doi.org/10.1038/nnano.2011.56, http://www.nature.com/doifinder/10.1038/nnano.2011.56
71. D. Stornaiuolo, S. Gariglio, A. Fête, M. Gabay, D. Li, D. Massarotti, J.M. Triscone, Weak localization and spin-orbit interaction in side-gate field effect devices at the LaAlO$_3$/SrTiO$_3$ interface. Phy. Rev. B **90**(23), 235426 (2014). https://doi.org/10.1103/PhysRevB.90.235426, http://link.aps.org/doi/10.1103/PhysRevB.90.235426
72. Y. Xie, Y. Hikita, C. Bell, H.Y. Hwang, Control of electronic conduction at an oxide heterointerface using surface polar adsorbates. Nat. Commun. **2**, 494 (2011). https://doi.org/10.1038/ncomms1501
73. J. Biscaras, S. Hurand, C. Feuillet-Palma, A. Rastogi, R.C. Budhani, N. Reyren, E. Lesne, J. Lesueur, N. Bergeal, Limit of the electrostatic doping in two-dimensional electron gases of LaXO$_3$(X = Al, Ti)/SrTiO$_3$. Sci. Rep. **4**, 6788 (2014). https://doi.org/10.1038/srep06788
74. Z. Liu, C. Li, W. Lü, X. Huang, Z. Huang, S. Zeng, X. Qiu, L. Huang, A. Annadi, J. Chen et al., Origin of the two-dimensional electron gas at laalo$_3$/srtio$_3$ interfaces: the role of oxygen vacancies and electronic reconstruction. Phy. Rev. X **3**(2), 021010 (2013)
75. C.W. Schneider, S. Thiel, G. Hammerl, C. Richter, J. Mannhart, Microlithography of electron gases formed at interfaces in oxide heterostructures. Appl. Phy. Lett. **89**(12), 122101–122101 (2006)
76. D. Stornaiuolo, S. Gariglio, N. Couto, A. Fete, A. Caviglia, G. Seyfarth, D. Jaccard, A. Morpurgo, J.M. Triscone, In-plane electronic confinement in superconducting LaAlO$_3$/SrTiO$_3$ nanostructures. Appl. Phy. Lett. **101**(22), 222601 (2012)
77. C. Cen, S. Thiel, J. Mannhart, J. Levy, Oxide Nanoelectronics on Demand. Science **323**(5917), 1026–1030 (2009). https://doi.org/10.1126/science.1168294
78. C. Cen, S. Thiel, G. Hammerl, C. Schneider, K. Andersen, C. Hellberg, J. Mannhart, J. Levy, Nanoscale control of an interfacial metal-insulator transition at room temperature. Nat. Mater. **7**(4), 298–302 (2008)
79. Y. Xie, C. Bell, Y. Hikita, H.Y. Hwang, Tuning the electron gas at an oxide heterointerface via free surface charges. Adv. Mater. **23**(15), 1744–1747 (2011)
80. F. Bi, D.F. Bogorin, C. Cen, C.W. Bark, J.W. Park, C.B. Eom, J. Levy, "water-cycle" mechanism for writing and erasing nanostructures at the LaAlO$_3$/SrTiO$_3$ interface. Appl. Phy. Lett. **97**, 173110 (2010)
81. N. Reyren, M. Bibes, E. Lesne, J.M. George, C. Deranlot, S. Collin, A. Barthélémy, H. Jaffrès, Gate-controlled spin injection at LaAlO$_3$/SrTiO$_3$ interfaces. Phy. Rev. Lett. **108**(18), 186802 (2012). https://doi.org/10.1103/PhysRevLett.108.186802, http://link.aps.org/doi/10.1103/PhysRevLett.108.186802

82. A.G. Swartz, S. Harashima, Y. Xie, D. Lu, B. Kim, C. Bell, Y. Hikita, H.Y. Hwang, Spin-dependent transport across Co/LaAlO$_3$/SrTiO$_3$ heterojunctions. Appl. Phy. Lett. **105**(3), 032406 (2014). https://doi.org/10.1063/1.4891174, http://scitation.aip.org/content/aip/journal/apl/105/3/10.1063/1.4891174
83. W. Han, X. Jiang, A. Kajdos, S.h. Yang, S. Stemmer, S.S.P. Parkin, Spin injection and detection in lanthanum- and niobium-doped SrTiO$_3$ using the Hanle technique. Nat. Commun. **4**, 1–6 (2013). https://doi.org/10.1038/ncomms3134
84. A.M. Kamerbeek, E.K. de Vries, A. Dankert, S.P. Dash, B.J. van Wees, T. Banerjee, Electric field effects on spin accumulation in Nb-doped SrTiO$_3$ using tunable spin injection contacts at room temperature. Appl. Phy. Lett. **104**(21), 212106 (2014). https://doi.org/10.1063/1.4880895, http://scitation.aip.org/content/aip/journal/apl/104/21/10.1063/1.4880895
85. E. Lesne, Y. Fu, S. Oyarzun, J.C. Rojas-Sanchez, D.C. Vaz, H. Naganuma, G. Sicoli, J.P. Attané, M. Jamet, E. Jacquet, J.M. George, A. Barthélémy, H. Jaffrès, A. Fert, M. Bibes, L. Vila, Highly efficient and tunable spin-to-charge conversion through Rashba coupling at oxide interfaces. Nat. Mater. **15**, 1261 (2016). https://doi.org/10.1038/NMAT4726
86. C. Cen, S. Thiel, G. Hammerl, C.W. Schneider, K.E. Andersen, C.S. Hellberg, J. Mannhart, J. Levy, Nanoscale control of an interfacial metalinsulator transition at room temperature. Nat. Mater. **7**(4), 298–302 (2008). https://doi.org/10.1038/nmat2136, http://www.nature.com/doifinder/10.1038/nmat2136
87. D. Stornaiuolo, S. Gariglio, N.J.G. Couto, A. Fete, A.D. Caviglia, G. Seyfarth, D. Jaccard, A.F. Morpurgo, J.M. Triscone, In-plane electronic confinement in superconducting LaAlO$_3$/SrTiO$_3$ nanostructures. Appl. Phy. Lett. **101**(22), 222601 (2012). https://doi.org/10.1063/1.4768936, http://scitation.aip.org/content/aip/journal/apl/101/22/10.1063/1.4768936
88. B. Forg, C. Richter, J. Mannhart, Field-effect devices utilizing LaAlO$_3$/SrTiO$_3$ interfaces. Appl. Phys. Lett. **100**(5), 053506 (2012). https://doi.org/10.1063/1.3682102, http://scitation.aip.org/content/aip/journal/apl/100/5/10.1063/1.3682102
89. R. Jany, C. Richter, C. Woltmann, G. Pfanzelt, B. Förg, M. Rommel, T. Reindl, U. Waizmann, J. Weis, J.A. Mundy, D.A. Muller, H. Boschker, J. Mannhart, Monolithically integrated circuits from functional oxides. Adv. Mater. Interfaces **1**(1), 1300031 (2014). https://doi.org/10.1002/admi.201300031, http://doi.wiley.com/10.1002/admi.201300031
90. C. Woltmann, T. Harada, H. Boschker, V. Srot, P.A. van Aken, H. Klauk, J. Mannhart, Field-effect transistors with submicrometer gate lengths fabricated from LaAlO$_3$-SrTiO$_3$-based heterostructures. Phy. Rev. Appl. **4**(6), 064003 (2015). https://doi.org/10.1103/PhysRevApplied.4.064003, http://link.aps.org/doi/10.1103/PhysRevApplied.4.064003
91. A. Tebano, E. Fabbri, D. Pergolesi, G. Balestrino, E. Traversa, Room-temperature in SrTiO$_3$/LaAlO$_3$ heterostructures. ACS Nano **6**(2), 1278–1283 (2012). https://doi.org/10.1021/nn203991q

Chapter 4
ARPES Studies of Two-Dimensional Electron Gases at Transition Metal Oxide Surfaces

Siobhan McKeown Walker, Flavio Y. Bruno and Felix Baumberger

Abstract High mobility two-dimensional electron liquids (2DELs) underpin today's silicon based devices and are of fundamental importance for the emerging field of oxide electronics. Such 2DELs are usually created by engineering band offsets and charge transfer at heterointerfaces. However, in 2011 it was shown that highly itinerant 2DELs can also be induced at bare surfaces of different transition metal oxides where they are far more accessible to high resolution angle resolved photoemission (ARPES) experiments. Here we review work from this nascent field which has led to a systematic understanding of the subband structure arising from quantum confinement of highly anisotropic transition metal d-states along different crystallographic directions. We further discuss the role of different surface preparations and the origin of surface 2DELs, the understanding of which has permitted control over 2DEL carrier densities. Finally, we discuss signatures of strong many-body interactions and how spectroscopic data from surface 2DELs may be related to the transport properties of interface 2DELs in the same host materials.

4.1 Introduction

Oxide surfaces and interfaces can host electronic states that differ from those in the bulk. This offers new possibilities for electronic structure design and has motivated an increasing number of studies investigating epitaxial heterostructures. The ABO_3 perovskite transition metal oxides (TMOs) have received much attention in this

---

S. McKeown Walker (✉) · F. Y. Bruno · F. Baumberger
University of Geneva, 24 Quai Ernest-Ansermet, 1211 Geneva, Switzerland
e-mail: siobhan.mckeown@unige.ch

F. Y. Bruno
e-mail: flavio.bruno@unige.ch

F. Baumberger
e-mail: felix.baumberger@unige.ch

F. Baumberger
Swiss Light Source, Paul Scherrer Institut, 5232 Villigen, Switzerland

© Springer International Publishing AG, part of Springer Nature 2018
C. Cancellieri and V. Strocov (eds.), *Spectroscopy of Complex Oxide Interfaces*, Springer Series in Materials Science 266,
https://doi.org/10.1007/978-3-319-74989-1_4

context because their quasi-cubic structures and compatible lattice constants make them well suited to heteroepitaxy growth [1]. Moreover, they show diverse bulk properties including ferro- and antiferromagnetism, ferroelectricity or superconductivity. These phases are largely controlled by the occupation of the transition metal d-shell and by subtle changes in bond angles, rendering ABO_3 TMOs suitable for electronic structure engineering in heterostructures by exploiting interfacial charge transfer, strain and octahedral tilting patterns [1–3].

Of particular interest are high-mobility two-dimensional electron liquids (2DELs) in ABO_3 perovskites. These systems have the potential to underpin a new generation of oxide electronics by exploiting not only the various phases of the parent oxide as carrier densities are tuned, but also phases and properties unique to the oxide surface or interface 2DEL [2, 4]. Determining the intricacies of the electronic band structure of such TMO 2DELs is an important step towards understanding the underlying physics of these systems and facilitates the engineering of desirable properties. However, the intrinsically buried nature of interface 2DELs poses substantial experimental challenges. High-resolution angle resolved photoemission spectroscopy (ARPES) using UV excitation, a standard technique for band structure determination, has insufficient probing depth to study important systems such as the $LaAlO_3/SrTiO_3$ (LAO/STO) interface 2DEL, despite them being buried beneath only a few unit cells. This restricts photoemission studies of interface 2DELs to the soft or hard X-ray regime where the effective resolution in energy and momentum is reduced. Other spectroscopic techniques that were successfully applied to oxide interfaces such as X-ray absorption (XAS) [5] or resonant inelastic X-ray scattering (RIXS) [6, 7] give valuable information on orbital symmetries and collective excitations but do not offer direct momentum space resolution. Microscopic electronic structure information from interfaces has also been deduced from quantum oscillation data. This technique is exceptionally precise but requires very high mobilities, which are hard to achieve in correlated electron systems, and the data is often difficult to interpret [8–11]. These challenges have motivated an alternative approach to the creation and spectroscopic investigation of oxide 2DELs. Recognizing that the fundamental electronic properties of 2DELs are defined by their host material and the electrostatic boundary conditions, in 2011 Meevasana et al. [12] and Santander-Syro et al. [13] reported that a 2DEL showing hallmarks of the band structure predicted for the LAO/STO interface can be created on the bare (001) surface of $SrTiO_3$ (STO) where it is accessible to high-resolution ARPES experiments. This approach has subsequently been extended to different TMO host materials and surface orientations revealing the fundamental electronic properties of oxide 2DELs and a common framework for describing system-to-system variation. In this chapter, we review the present status of this emerging field.

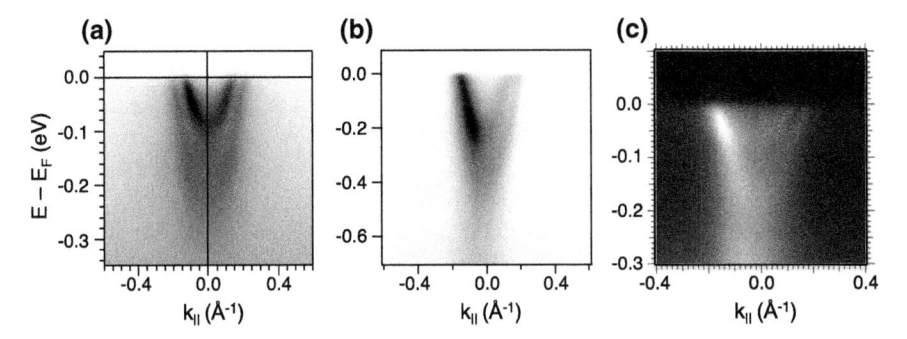

Fig. 4.1 Energy-momentum dispersions of surface 2DELs measured by VUV ARPES on the (001) surfaces of **a** SrTiO$_3$, **b** KTaO$_3$ and **c** anatase TiO$_2$. High intensity (black, black and white in (**a**), (**b**) and (**c**) respectively) delineates the electronic states of the charge accumulation layer. Adapted from Santander-Syro et al. [13], King et al. [15] and Rödel et al. [18] respectively

4.1.1 Metallic Subbands at the Surface of an Insulator

We will focus the discussion on ARPES studies of 2DELs with d-orbital character in the transition metal oxides SrTiO$_3$ [12, 13], KTaO$_3$ (KTO) [14–16] and anatase TiO$_2$ [17, 18]. We note that electron accumulation layers have also been observed on the surface of the transparent conducting oxides CdO [19, 20] and In$_2$O$_3$ [21] but these 2DELs derive from free-electron like s states and will not be discussed here. In stoichiometric form, STO, KTO and anatase TiO$_2$ are band insulators with a d^0 configuration of the transition metal ion. Importantly, all three materials are susceptible to chemical doping [22–27] which introduces electrons into the conduction band minimum resulting in a three dimensional bulk metallic state. Using appropriate surface preparations, electron accumulation layers that are independent of the residual bulk doping have been reported in all three of these TMOs. As shown in Fig. 4.1, these charge accumulation layers all show multiple subbands, which is a key-signature of quantum confinement and is adopted as the finger-print of a 2DEL in ARPES measurements throughout this review. In the following we will discuss the subband structures of 2DELs in STO, KTO and anatase TiO$_2$ in detail, exploring not only material dependent electronic properties, but also the influence of the crystallographic orientation of the surface on 2DEL characteristics. We will further discuss the origin of these metallic states on the surfaces of insulating TMOs and briefly review different approaches for calculating their band structure. We will also summarize very recent ARPES studies that provide insight into the nature of many-body interactions in oxide 2DELs.

4.2 Origin of Surface 2DELs in TMOs

A 2DEL arises as the conduction band minimum of an insulation or semiconducting crystal is dragged below the chemical potential by an electrostatic potential over a narrow region. Understanding the origin of the corresponding electric field in 2DEL systems is an important step towards achieving carrier density control, which in-turn underlies device functionality. While in conventional semiconductors it is well established that the potential gradient arises from work function mis-match, the origin of both the attractive confining potential and the excess charge carriers in TMO surface and interface systems is more ambiguous. For example the origin of the native charge carriers at the LAO/STO interface is still actively debated and consequently systematic control of their density has remained elusive.

The first two publications that reported the STO (001) surface 2DEL, suggested that the origin of both the electrostatic band bending and charge carriers may be surface oxygen vacancies (OVs) [12, 13]. In this scenario two excess electrons are released as a positively charged OV is created. This positive charge at the STO surface must be screened by the excess electrons which can form either localized states near the vacancy, or an itinerant accumulation layer which manifests as the observed 2DEL. Additionally the authors of [12] observed that the 2DEL bandwidth and density increase as the surface is irradiated with synchrotron light and remains constant in UHV conditions when not irradiated, leading to the hypothesis that the STO surface 2DEL originates from light-induced oxygen vacancies. In the following we will describe the experimental evidence for this scenario.

4.2.1 UV Induced Oxygen Vacancies

Both Santander-Syro et al. [13] and Meevasana et al. [12] saw evidence for band bending at the (001) STO surface. They measured angle integrated energy distribution curves (EDCs), similar to those in Fig. 4.2a which shows the $O2p$ valence band (VB) whose maximum is around 4 eV below the Fermi level before significant irradiation (blue), after long UV irradiation (red) and at intermediate times (grey). Its can be seen that the VB appears to shift to higher binding energies as the surface is irradiated. Together with small core level shifts observed by X-ray photoemission spectroscopy (XPS) [28, 29] this suggests the presence of downward band-bending at the surface induced by the UV radiation [12, 28, 30, 31]. This band-bending appears concomitant with the 2DEL peak at the Fermi level (see inset) and its magnitude, which can be broadly quantified by the shift of the VB leading edge mid-point, should be related to the 2DEL bandwidth. However, the exact magnitude of the surface band bending is difficult to extract from such spectra since they represent an average of the energy shift in each unit cell over the photoemission probing depth, and the form of the VB also evolves as the surface is irradiated.

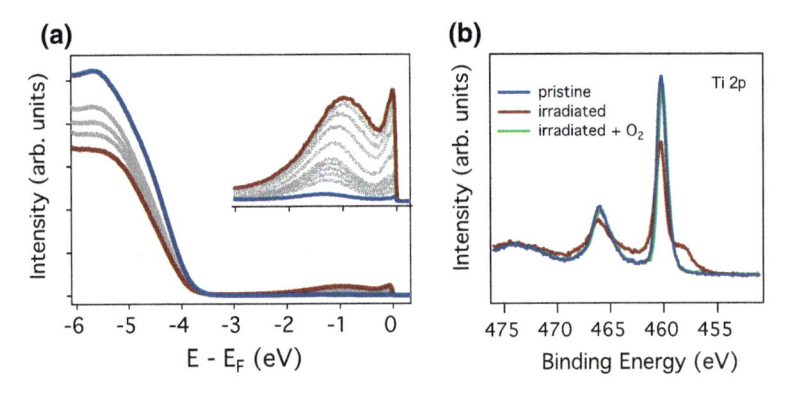

4 ARPES Studies of Two-Dimensional Electron Gases …

Fig. 4.2 **a** Angle integrated energy distribution curves showing the valence band, in-gap state and 2DEL evolution on the fractured surface of La:STO (001) after negligible (blue) or prolonged (red) UV irradiation and for intermediate irradiation times (gray) (unpublished). **b** XPS showing the Ti 2p core level under the same conditions and after exposure to 0.5 Langmuir of O_2 (green), adapted from McKeown Walker et al. [30]

In addition [12, 13] observed a non-dispersive in-gap (IG) state approximately 1.3 eV below the Fermi level that grows in intensity with increasing irradiation time (see inset of Fig. 4.2a). Previous photoemission measurements on reduced STO, Nb:STO and La:STO also observed this IG state at the STO surface and associated it with oxygen vacancy defects states [32, 33]. Further evidence that electrons localized on or near oxygen vacancy sites would form such an IG state was provided by DFT calculations of STO with oxygen deficient surfaces [34].

McKeown Walker et al. [30] demonstrated that, as shown in Fig. 4.3a–c, introducing extremely low doses of molecular oxygen into the UHV chamber eliminates the surface 2DEL and that subsequent irradiation with UV light causes the 2DEL density to recover. This explicitly demonstrated that the 2DEL is an accumulation layer of electrons at the surface of STO screening positive charge resulting from light induced oxygen vacancies. This experiment also confirmed that the IG state is associated with light-induced OVs providing evidence that these defects at the STO surface result in both itinerant and localized electronic states. McKeown Walker et al. [30] also demonstrate that the efficiency with which OV are created depends on the photon energy, suggesting that the dominant mechanism by which OV are created is inter-atomic core-hole Auger decay [35, 36]. Using this knowledge McKeown Walker et al. succeeded in controlling the density of the STO 2DEL by either measuring at a photon energy below the threshold for efficient OV creation or by measuring at higher photon energies while maintaining a partial pressure of oxygen during the measurement.

The bandwidth and density of the STO surface 2DEL do not grow indefinitely under synchrotron radiation. After a finite irradiation period the band width saturates. However, as reported by Dudy et al. [28] the intensity of both the IG state and 2DEL do not saturate over the same time scale. These authors suggest that the dynamic

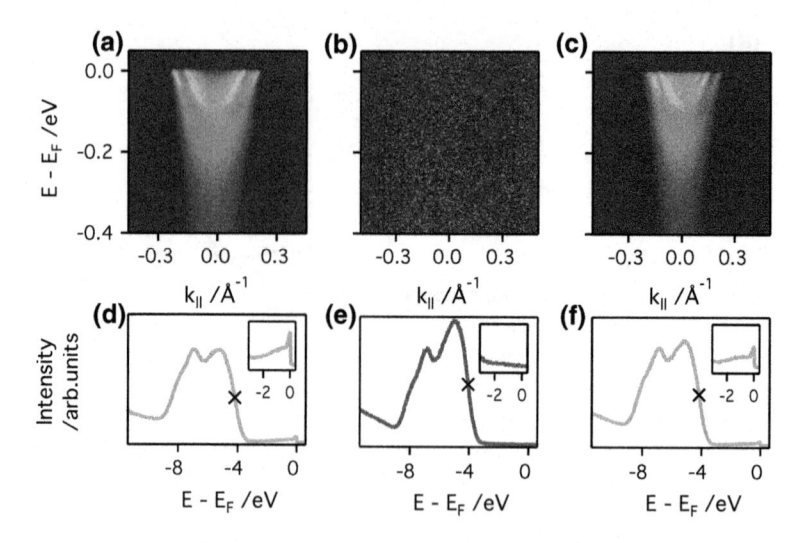

Fig. 4.3 Creation and annihilation of the 2DEL at the STO (001) surface. **a** Dispersion plot of the high density STO (001) surface 2DEL after initial irradiation with 52 eV photons. **b, c** The 2DEL disappears after in-situ exposure to 0.5 Langmuir of O_2 and reappears following further irradiation with 52 eV photons. **d–f** Angle integrated valence band spectra corresponding to the states in (**a–c**). The magnified insets show the intensity at the Fermi level. The valence band leading edge midpoint (VB LEM) is marked by a cross. All data were measured in the second Brillouin zone with 28 eV, s-polarized light. From McKeown Walker et al. [30]

equilibrium between OV creation and annihilation at finite oxygen partial pressure leads to OV clusters and electronic phase separation. Alternative explanations include the migration of ions within the lattice due to high electric fields at the surface, and that the ratio of localized and itinerant electrons donated by an oxygen vacancy evolves as a function of OV density due to a changes in the balance of correlations [37].

Qualitatively the same UV sensitive behaviour of the two-dimensional electron liquid density, VB and core level shifts and IG state intensity has been observed for various low-index surfaces of STO [31, 38, 39] and for the surface of anatase TiO_2 [18]. The carrier density of states observed by ARPES at the surface of anatase TiO_2 could be controlled by tuning the dynamic equilibrium between OV creation by UV light and re-oxidation due to finite oxygen partial pressure in the chamber [17]. Indeed the role of oxygen vacancies in TiO_2 thin films is even more dramatic with excessive irradiation leading to a local destruction of the 2DEL state [18]. Implicit in these observations is that reconstructions of the crystal surfaces [29, 40, 41] do not dominate surface charge accumulation in STO or TiO_2. This is true even in the case of (111) and (110) surfaces of STO which are polar. On the other hand, for the 2DEL at the strongly polar (001) surface of $KTaO_3$ only a weak evolution of the 2DEL bandwidth is seen as the surface is irradiated [15, 16], suggesting that defects intrinsic to the cleaved surface may induce a 2DEL in this case.

4.3 2DEL Subband Structure at the (001), (110) and (111) Surfaces of SrTiO$_3$

In this section we will describe the subband structures of STO 2DELs observed by ARPES in more detail. We will demonstrate that the characteristic features of the electronic structure can be understood on a qualitative level in terms of quantum confinement thereby justifying the identification of these states as two dimensional electron liquids. We will also discuss the role of crystallographic orientation of the surface in modulating the effects of quantum confinement on the 2DEL band structure.

4.3.1 SrTiO$_3$ (001)

Figure 4.4a shows the energy-momentum subband dispersion of the 2DEL measured on the fractured (001) surface of STO. This data, reproduced from [30], resolves five subbands, as shown schematically in Fig. 4.4b. Three subbands (L1–L3) are highly dispersive indicating a light effective mass for some of the carriers while the two shallowest subbands (H1–H2) resolved in the experiment are much less dispersive. From parabolic fits of the overall dispersion of individual subbands we estimate effective masses of $\sim 0.6\, m_e$ for L1–L3 and 9–15 m_e for (H1–H2) [42]. These values are comparable to DFT bulk band masses of the STO conduction band [43, 44] for L1–L3 while they are significantly higher than the heaviest bulk masses for H1–H2. King et al. [42] showed that this disparity between the light and heavy subbands arises due the electron-phonon interaction. For the light subbands, which have a bandwidth exceeding the Debye frequency, the dominant effect of electron-phonon interaction is an additional low-energy renormalization of the dispersion clearly discernible in the form of a kink in the subband dispersion, indicated by an arrow in Fig. 4.4. On the other hand, the shallow heavy bands are close to the anti-adiabatic limit and are thus subject to an overall renormalization of the entire occupied bandwidth, which causes the increased overall effective masses.

We have already discussed that these states are induced at the crystal surface and therefore cannot be considered bulk bands. This is also evident from the observation of 5 clearly non-degenerate states at the Γ point while the conduction band minimum of bulk STO is formed by only three approximately degenerate bands. The formation of multiple subbands, as well as the higher relative energy of the heavy subbands, are hallmarks of quantum confinement of the 3D bulk conduction band near the surface. Therefore we associate the indices 1, 2 and 3 of L1–L3 and H1, H2 with the principal quantum numbers of individual quantum well states. The three light subbands all correspond to circular Fermi surface sheets while the heavy subbands form the long axes of cigar-shaped Fermi surface sheets, as sketched in the inset of Fig. 4.4b [13, 15]. By considering the shape of the Fermi surface sheets, the spatial anisotropy of the t_{2g} orbitals that form the conduction band of STO and the polarization dependence of the photoemission matrix elements for t_{2g} states, it was shown that the light subbands are formed by electrons in d_{xy} orbitals, while the

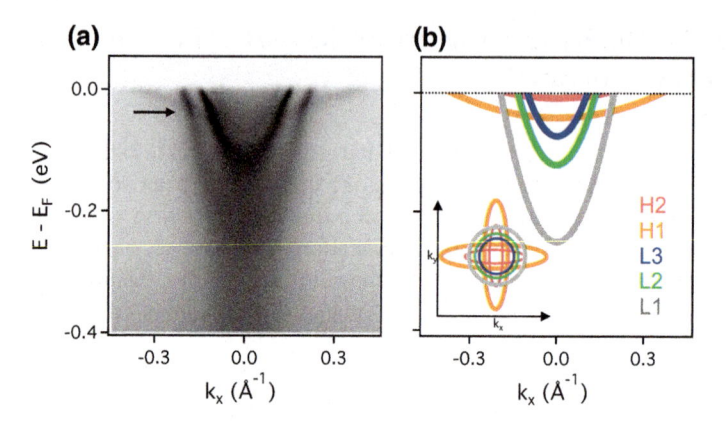

Fig. 4.4 Electronic structure of the 2DEL at the (001) surface of SrTiO$_3$. **a** Energy momentum dispersion plot measured in the second Brillouin zone. Data shown are the sum of measurements taken with s- and p-polarized light at 52 eV photon energy at 10 K. The arrow indicates the kink in the dispersion of the L1, L2 and L3 subbands due to electron-phonon interaction. **b** Schematic of the subbands resolved in the ARPES measurements of (**a**). The three light bands (L1, L2, L3) and the two heavy bands (H1 and H2) are sketched in grey, green, blue, orange and coral respectively. L1, L2 and L3 have predominantly d_{xy}-orbital character and H1, H2 have predominantly $d_{xz/yz}$-orbital character. The inset shows a sketch of the corresponding Fermi surface using the same colour scheme to distinguish between the Fermi surface sheets. Adapted from McKeown Walker et al. [30]

heavy bands have $d_{xz/yz}$ orbital character [13, 42, 45]. The surface 2DEL has thus the same orbital ordering found for the LAO/STO interface 2DEL. As described in Sect. 4.3.1.2 this is a natural consequence of quantum confinement along the [46] surface normal.

Quantum well states are by definition two-dimensional and do not disperse along the confinement direction. Therefore, measuring the subband dispersion along k_z is a direct test for the presence of quantum confinement. Experimentally this is achieved by varying the photon energy of the exciting radiation in order to probe the band dispersion perpendicular to the crystal surface. One such measurement for the STO (001) 2DEL from [47] is shown in Fig. 4.5. The top panel of the data-cube shows the Fermi surface in the $k_x k_z$ plane over a full Brillouin zone. The Fermi wave vectors for the first two light subbands are clearly resolved over an extended range of k_z and are found to be constant within the accuracy of the experiment. This non-dispersive behaviour along the k_z axis confirms the two-dimensional nature of the d_{xy} subbands already inferred in [12, 13]. For the heavy $d_{xz/yz}$ subbands the situation is less clear. These bands have generally lower spectral weight which is strongly suppressed away from the bulk Γ points making it more difficult to trace their dispersion over an extended range in k_z in order to unambiguously determine their dimensionality. Santander-Syro et al. [13] reported that the first heavy subband H1 is non-dispersive along k_z and thus two-dimensional, while Plumb et al. [29] described H1 as largely three-dimensional. The dimensionality of H2 has not been investigated in the literature to date. An independent argument for a strict two-dimensionality

4 ARPES Studies of Two-Dimensional Electron Gases ...

Fig. 4.5 Two dimensionality of the STO (001) 2DEL. Top Panel: Constant energy map at E_F in the $k_x k_z$ plane. The k_z range shown is approximately one Brillouin zone perpendicular to the sample surface. The non-dispersive nature of the Fermi wave vectors along k_z is indicative of the two dimensional nature of the state. Front Panel: Energy-momentum subband dispersion of the 2DEL. Adapted from Wang et al. [47]

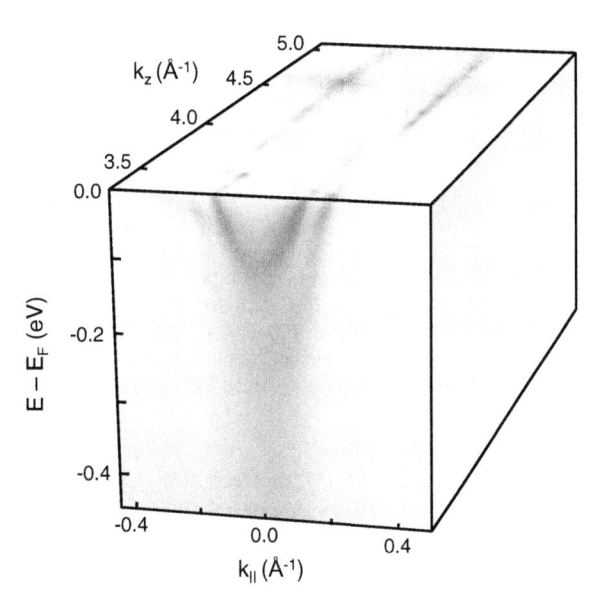

of H1 and H2 comes from their binding energies. As shown in Sect. 4.3.1.2 the degeneracy lifting at the Γ point between L1 and H1 as well as between H1 and H2 follow naturally from quantum confinement of the bulk conduction band while alternative interpretations are not evident.

The spectral weight distribution of 2D states along k_z is related to the Fourier transform of the real space wave function convolved with the k_z distribution of the photoelectron wave function. Consequently, the spectral weight distribution of 2D states along k_z encodes information about the real space extent of their wavefunctions. Therefore, from the strong intensity of L1 over a full Brillouin zone in k_z we can infer that this state is mostly confined to a single unit cell along the surface normal, while the limited and periodic intensity distribution of H1 indicates a more spatially extended wave function. This is in qualitative agreement with the tight-binding supercell model presented in Sect. 4.3.1.2, even if the complicated oscillations of the photoemission matrix elements have so far prohibited explicit determination of the spatial extent of the wavefunctions of individual quantum well states.

The STO (001) surface 2DEL with a saturated bandwidth of ∼250 meV described in this section has a large carrier density. Evaluating the Luttinger volume of the Fermi surfaces of L1–L3 and H1 for the 2DEL shown in Fig. 4.4 gives $n_{2D} \sim 2 \times 10^{14}$ cm^{-2}. There is, however, a significant uncertainty in this value arising from higher-order subbands with small volumes and low spectral weight that are not included in this estimate. Moreover, many measurements reported in the literature resolved fewer subbands and correspondingly quote lower values for n_{2D} for the same occupied bandwidth of L1. Despite this uncertainty, the saturated carrier density of the STO (001) surface 2DEL is clearly higher than the sheet carrier densities

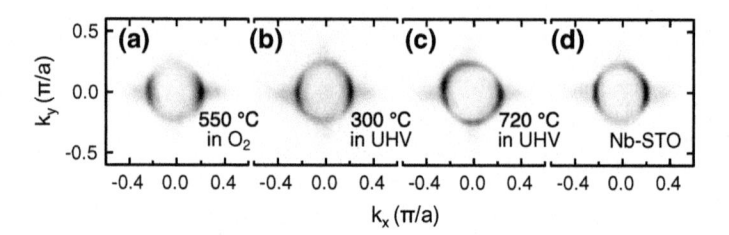

Fig. 4.6 The Fermi surface of the STO (001) 2DEL in substrates of stoichiometric STO annealed at **a** 550 °C in 100 mbar O_2 for 2 h, **b** 300 °C in UHV for 15 h **c** 720 °C in UHV for 1 h and **d** Nb:STO annealed at 550 °C in 100 mbar O_2 for 2 h. The form of the Fermi surface is insensitive to the annealing conditions and bulk doping and is the same as observed on cleaved STO surfaces. This demonstrates the universality of the STO (001) surface 2DEL. From Plumb et al. [29]

reported for the LAO/STO interface 2DEL of typically $n_{2D} = 0.5 - 5 \times 10^{13}$ cm^{-2} [48] and approaches the value of 0.5 electrons per unit cell (3.3×10^{14} cm^{-2}) found in the ideal polar catastrophe scenario for the LAO/STO interface.

4.3.1.1 Universality of TMO Surface 2DEL Band Structure

The first publications of the field [12, 13] showed that a 2DEL with virtually identical band structure and density is observed on the bare fractured surfaces of La:STO, stoichiometric STO and reduced STO single crystals with bulk carrier densities up to $n_{3D} \sim 1 \times 10^{20}$ cm^{-3}. Subsequently, it was shown that the 2DEL can also be induced on TiO_2 terminated wafers obtained by mechanical polishing, ex situ etching and different in-situ surface preparations [29, 40, 47, 49]. Figure 4.6 shows the Fermi surface of the STO (001) 2DEL observed on TiO_2-terminated wafers following different in situ annealing procedures. The circular Fermi surface of the first light subband and two cigar-like Fermi surface sheets can be seen in all cases, just as found on the fractured STO (001) surface. The characteristic features of the electronic structure and in particular the presence of multiple orbitally-ordered subbands, do not change depending on the annealing procedure. This indicates a remarkable universality of the STO (001) 2DEL subband structure. It is insensitive to the bulk doping level of the single crystal, the origin of the residual bulk doping, the marked differences in macroscopic surface roughness between fractured and polished surfaces and the various terminations and reconstructions yielded by different annealing procedures.

4.3.1.2 Bandstructure Modelling

Santander-Syro et al. [13] first pointed out that the energetic ordering of the subband ladder in the STO (001) surface 2DEL can be understood qualitatively with a simple model of quantum confinement of strongly anisotropic bands with t_{2g} orbital character. As illustrated in Fig. 4.7b, each t_{2g} orbital is associated with electron motion with

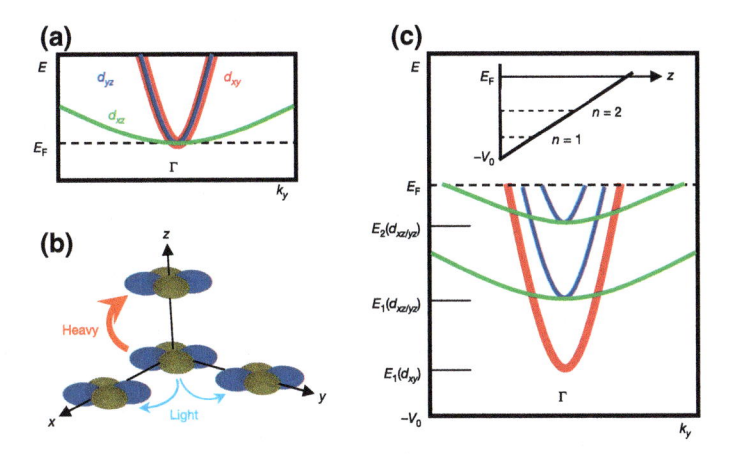

Fig. 4.7 Modelling the effects of quantum confinement in a wedge-shaped potential well. **a** Simplified bulk band structure of STO along the k_y axis. The colours indicate the orbital character as indicated. **b** Cartoon of the allowed motion of an electron between d_{xy} orbitals in a cubic structure. For a band formed of such electrons, the direction in which the effective band mass would be "heavy" or "light" is indicated. This is determined by the small or large overlap of the orbital in that direction, respectively. **c** The subband structure expected using the approximate solution of a particle of mass m_z in a wedge potential to define the subband confinement energies at the Γ point. The inset shows the wedge potential profile as a function of z, the direction perpendicular to the crystal surface. From Santander-Syro et al. [13]

a light effective mass along two crystallographic axes and a heavy mass along the third direction. It follows that 2DEL subbands with a heavy in-plane effective mass derive from orbitals with light out-of-plane mass m_z while light subbands can have either a light or heavy out-of-plane mass. If bands with these orbital characteristics are subjected to a simple wedge potential as sketched in the inset of Fig. 4.7c, they experience an energy shift proportional to $m_z^{-1/3}$. This shift relative to the bottom of the potential well will be smallest for the d_{xy} band which has light effective mass in the surface plane and a heavy mass along the confinement direction. Thus the lowest order d_{xy} band will sit near the bottom of the wedge potential. Conversely the bands with out-of-plane $d_{xz/yz}$ orbital character will be pushed to higher energy. This is shown in Fig. 4.7c and qualitatively reproduces experimental results from ARPES at the STO surface and from X-ray linear dichroism (XLD) of the LAO/STO interface [5]. It is clear however that this simple wedge model cannot accurately predict the relative subband energies seen in ARPES experiments.

More realistic confinement potentials can be obtained from a self-consistent solution of the Poisson and Schrödinger equations [12, 50]. Using an accurate tight-binding parametrization of the bulk conduction band and including surface band bending as an on-site potential term in a large supercell extending several tens of unit cells perpendicular to the surface, these calculations reproduce experimental band structures in various quantum confined systems to a high degree of accuracy [15, 42, 51–55]. The subband structure for such a calculation based on an ab initio DFT calculation of bulk cubic STO including spin-orbit coupling, is shown in Fig. 4.8a. The electrostatic boundary conditions are chosen such that the total calculated bandwidth matches the saturated bandwidth of the 2DEL. The resulting confinement energies of the subbands are in good agreement with ARPES data as demonstrated in Fig. 4.9a. Such calculations can be used to estimate the finite extent of the 2DEL along the confinement direction by projecting the eigenstates onto atomic layers as shown in Fig. 4.8 [42]. The confinement energy shifts are directly related to the spatial extent of the wavefunctions perpendicular to the surface, giving the intuitive result that the L1 d_{xy} band is very spatially confined, as indicated by the high intensity of the band in Fig. 4.8b. The wavefunctions of higher order d_{xy} bands are peaked successively further below the surface and the high intensity of the H1 subband in Fig. 4.8d shows that the wavefunction of the first $d_{xz/yz}$ band is centred 3–5 unit cells below the surface. The distribution of intensity for the heavy band shows that it is also much more spatially extended than the d_{xy} bands. This subband-specific spatial profile of the wavefunctions reflects the form of the potential well and is consistent with the matrix element structure seen in Fig. 4.5 and discussed in Sect. 4.3.1. From this it is clear that the spatial extent of the 2DEL as a whole is not a single well defined quantity. While the total charge density will always peak near the surface/interface, it can have very long tails extending deep into the bulk. These tails are often not seen in spectroscopic experiments whose signal strengths are typically proportional to the charge density. However, they might have a strong influence on transport properties which are often dominated by the most mobile carriers that might reside far from the interface where scattering is generally lower. We also note that the shape of the confinement potential and total carrier density distribution in oxide 2DELs is a strong function of carrier density. This is particularly true for STO since the strong suppression of its dielectric constant in an electric field progressively enhances the confinement at high carrier density. These considerations might explain why transport measurements of the LAO/STO (001) 2DEL often find thicknesses >10 nm [56, 57], while spectroscopic studies and DFT calculations typically report a confinement of the 2DEL within <2 nm for both surface and interface systems [42, 50, 58–60].

Full ab-initio electronic structure calculations of STO surface 2DELs have been presented in [62, 63] and qualitatively agree with the results of tight-binding supercell calculations. While such calculations are suitable for studying the behaviour of oxygen vacancies, direct comparison of the resulting band structure with experiment is hindered by limitations in the size of the supercell and the ordered nature of vacancy arrangements in density functional calculations that impose translational invariance in the surface plane.

4.3.1.3 Rashba Spin Orbit Coupling

The confinement potential associated with a surface 2DEL inherently breaks inversion symmetry, which lifts the constraint of spin degeneracy from the band structure. Spin splitting may result due to the coupling of an electron's motion to its spin via the effective in-plane magnetic field resulting from a Lorentz transformation of the

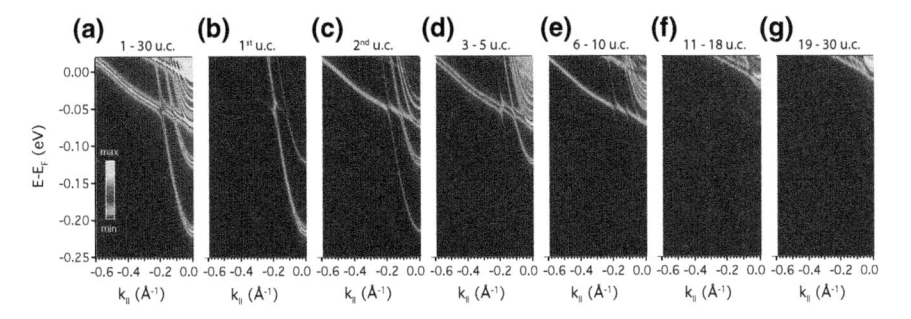

Fig. 4.8 Layer-projected tight binding superecell calculations of the electronic structure of the STO (001) surface 2DEL integrated over **a** the full 30 unit cell supercell, and **b–e** individual or few unit cell regions as labelled in the figure. High intensity (yellow) corresponds to a high wavefuncion weight. This calculation is based on and ab initio DFT parametrization of bulk STO, including a field dependent dielectric constant as proposed in [50, 61] and implements electrostatic boundary conditions that reproduce the overall bandwidth of the 2DEL. From King et al. [42]

symmetry-breaking electric field at the surface. This is known as the Rashba spin-orbit interaction [64]. The magnitude of spin-splitting in a Rashba system is linearly proportional to the strength of the electric field. This behaviour is observed in conventional semiconductor 2DELs [65] and is a prerequisite for many applications in spintronics such as the spin-field effect transistors (spin-FET) [66]. Rashba spin-splitting is also expected to scale with the strength of the atomic spin-orbit interaction (SOI) in the host material. Therefore considering the light atomic masses of the constituent elements of STO and consequently small atomic SOI, a small Rashba effect might naïvely be expected in STO based 2DELs. Surprisingly though, a substantial gate-tunable spin splitting of 2–10 meV has been deduced from transport experiments for both the LAO/STO interface 2DEL [8, 67–69] and electrolyte-gated STO [70]. The spin-splitting was found to have a strongly non-linear dependence on gate-field and weak antilocalization measurements suggest that it is proportional to k^3 rather than being k-linear as expected in the simplest models of Rashba spin-splitting [70].

Using band structure calculations based on relativistic DFT shown in Fig. 4.9a, b, Zhong et al. [71] demonstrated that these behaviours can be understood as the signatures of an unconventional Rashba spin-splitting arising from the multi-orbital nature of 2DELs in STO. Figure 4.9a, b shows the band structure for a single LAO/STO interface. In the region of the avoided crossings of light and heavy subbands the spin splitting is dramatically enhanced and clearly deviates from a k-linear form. A strikingly similar spin structure of the STO surface 2DEL can be seen in the tight binding supercell (TBSC) calculations of [42, 54] as shown in Fig. 4.9c, d and has been found by several other theoretical studies [42, 46, 72–76]. The authors of [42] related this enhancement to the finite orbital angular momentum that arises at the crossings of bands of different orbital character and augments the spin-orbit interaction at these particular locations in momentum-space. In this case the gate voltage used in transport experiments not only directly tunes the Rashba coefficient in STO 2DELs, but indirectly tunes the spin splitting by controlling the

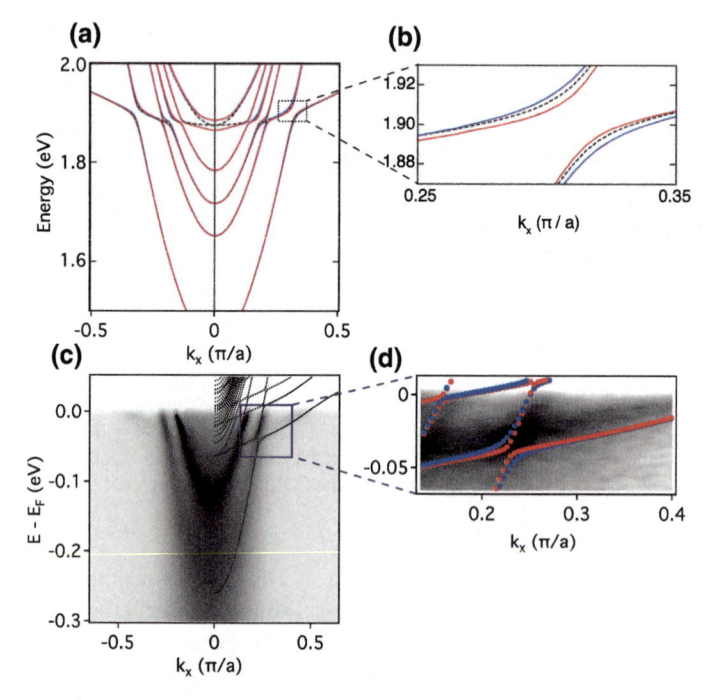

Fig. 4.9 Unconventionl Rashba-like spin splitting of STO (001) 2DEL subband structures. **a** DFT band structure of a single *n*-type vacuum/LAO/STO interface. **b** The boxed region in (**a**) is enlarged which reveals that the bands are not spin degenerate, and that spin-splitting is enhanced at the avoided crossings of light and heavy subbands. **c** Band structure of the STO surface found from tight-binding supercell calculations based on ab initio DFT electronic structure of bulk STO (black lines), overlayed on ARPES data (grascale image) of the STO (001) surface 2DEL. This shows good agreement with the experimental confinement energies and many similarities with the interface calculation in (**a**). **d** The boxed region in (**c**) is enlarged and shows a similar spin-splitting as seen in (**b**) for the LAO/STO interface. In **a**, **b** and **d** opposite in-plane spin channels are coloured red and blue. Adapted from Zhong et al. [71] and McKeown Walker et al. [54]

band filling and band structure. Indeed this indirect electrostatic tuning may be more important in such a system since the spin splitting is enhanced by approximately an order of magnitude at the avoided crossings of the $d_{xz/yz}$ and d_{xy} bands. However, this unconventional spin-splitting never exceeds ∼10 meV and thus remains below the resolution of high-resolution ARPES measurements such those of Fig. 4.4.

In order to gain direct spectroscopic insight into the spin structure of the STO (001) surface 2DEL Santander-Syro et al. [49] and McKeown Walker et al. [54] performed spin and angle resolved photoemission spectroscopy (SARPES) experiments. However, these authors reported conflicting results. McKeown Walker et al. measured no significant spin polarization of the photocurrent above the ∼5% noise level of their experiment. Considering the complex subband structure and poor experimental resolution in SARPES, they reason that such a negligible photocurrent polarization is fully consistent with the spin splitting shown in Fig. 4.9. However, direct experimental confirmation of this unconventional Rashba effect remains elusive. Conversely Santander-Syro et al. measured a large polarization of the photocurrent which prompted them to propose an entirely new interpretation of the universal subband structure measured at the surface of STO (001) crystals. They propose that bands L1 and L2 are the fully spin polarized components of a single d_{xy} subband. In a Rashba system the spin states must be degenerate at the Brillouin zone centre. However, it is well established that L1 and L2 of the STO surface 2DEL are not degenerate. Santander-Syro et al. speculate that the presence of ferromagnetic domains, which generate a large Zeeman-like term lifting the Γ point degeneracy, could account for this. Indeed some DFT calculations for ordered oxygen vacancy arrangements at the STO (001) surface show magnetic solutions of comparable energy to the paramagnetic case [62, 63]. However, in these calculations the remnant in-plane spin component due to the Rashba effect is an order of magnitude smaller than that measured by Santander-Syro et al. To date, the origin of the discrepancy between these two SARPES experiments is unclear and the details of the spin structure remain elusive.

4.3.2 SrTiO₃ (111) and (110) surface 2DELs

Two-dimensional electron liquids in ABO_3 transition metal oxides oxides are by no means restricted to the (001) plane. 2DELs have been successfully engineered at the bare (111) and (110) surfaces of SrTiO₃ [31, 38, 39] and at interfaces with these orientations [77–79]. These studies are motivated, in part, by theoretical predictions of novel ferromagnetic and ferroelectric states and topological phases in (111) bilayers of cubic perovskites [80, 81] and the intrinsic in-plane anisotropy of the (110) plane. In the following we will discuss the overall electronic structure of (111) and (110) orientated surface 2DELs on STO and relate it to the framework for quantum confinement developed in Sect. 4.3.1.2 and its interplay with the different symmetries of these surfaces.

Figure 4.10a shows the Fermi surface of the STO (111) surface. Three intersecting elliptical Fermi surface sheets with an overall six-fold symmetry can be seen. Each of these can be associated with the projection of a single t_{2g} component of the conduction band onto the (111) plane. As seen in Fig. 4.10b, which shows the bands dispersing along the long axis of one ellipse, within the accuracy of the experiment these electron like bands are degenerate at the Γ point. This is a natural consequence of the 120° rotational equivalence of the t_{2g} orbitals in the (111) plane, which causes all three bands to have the same effective mass along the confinement direction. When applied to the STO (110) plane, these arguments predict degenerate bands of $d_{xz/yz}$ character, due to the "semi-heavy" hopping for these orbitals, and a d_{xy} band with weaker confinement. This is indeed the case, as seen in Fig. 4.11a, b which shows the Fermi surface and a band dispersion from the STO (110) 2DEL [39]. Thus the STO (110) 2DEL is orbitally polarized, although the relatively small variation

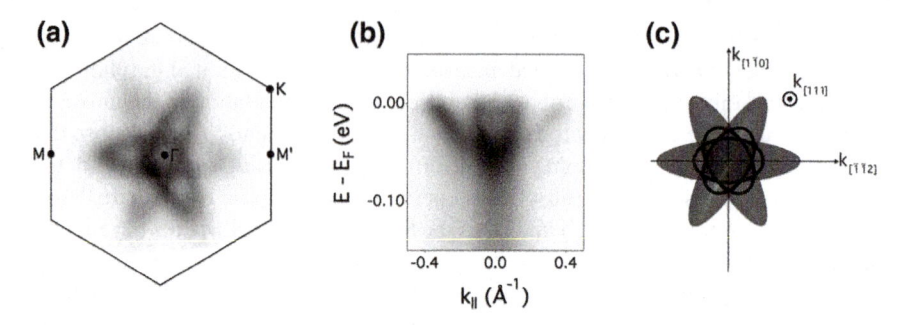

Fig. 4.10 Electronic structure of the STO (111) surface 2DEL. **a** Fermi surface showing three elliptical Fermi surface sheets. The black hexagon indicates the surface Brilluion zone and the Black dot indicates the Γ point. **b** Dispersion along the Γ-M' direction. Spectral weight at the Fermi level is from a second subband. **c** Projection onto the (111) plane (red shape) and cut at the Γ point (black lines) of a model STO bulk Fermi surface. The form of the STO (111) 2DEL Fermi surface closely ressembles the projection. Adapted from McKeown Walker et al. [31]

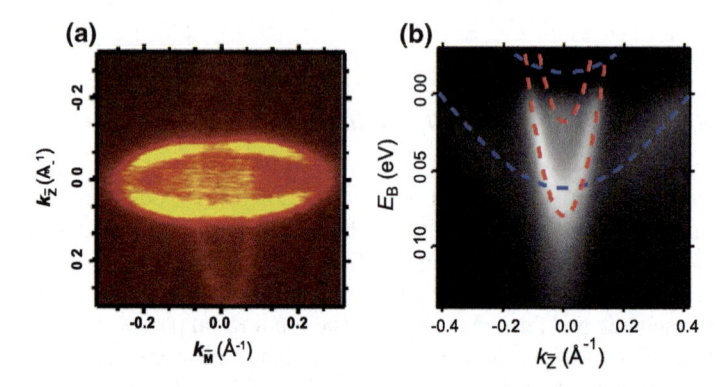

Fig. 4.11 Electroninc structure of the STO (110) surface 2DEL. **a** Fermi surface showing two intersecting elliptical Fermi surface sheets. The $d_{xz/yz}$ bands are degenerate and have higher intensity here due to the polarization of the exciting radiation. **b** Dispersion along the $k_{\bar{z}}$ direction. The dashed red (blue) lines indicate the dispersion of the $d_{xz/yz}(d_{xy})$ states and possible higher order subbands. Adapted from Wang et al. [39]

between the band masses for the $d_{xz/yz}$ and d_{xy} bands along the [55] direction leads to a much less dramatic orbital reconstruction than found for the STO (001) 2DEL.

Notably the two dimensional carrier densities of the fully saturated 2DELs on all three low index surfaces of STO are comparable to each other, with $n_{2D} \sim 2 \times 10^{14}$ cm^{-2}. Additionally, models reproducing the saturated band structures of both the (111) and (001) 2DELs find very similar confinement fields at the surface [31, 42]. This suggests a common origin for the saturation of the 2DEL bandwidth controlled by a physical limit on the electric field strength at the surface. One possible mechanism for such a limit could be the onset of diffusion of charged oxygen vacancies as the electric field increases. The common carrier density implies that 2DELs on different low index surfaces will have different bandwidths and spatial extents. This is evident from the dispersion plots of Figs. 4.4, 4.10 and 4.11 and photon energy dependent measurements that demonstrate the extended nature of the electron density of (110) and (111) 2DELs, in analogy to the heavy $d_{xz/yz}$ band of the STO (001) 2DEL.

Both the STO (111) and (110) surfaces show dramatic in plane subband mass enhancements with respect to the dispersion of bulk carriers along the same direction. As shown in [31, 39] this is an intriguing effect of quantum confinement, rather than a manifestation of many-body interactions. Intuitively this mass enhancement can be understood as the result of the projection of the three dimensional bulk Fermi surface onto a 2D plane. As illustrated in Fig. 4.10c for the (111) plane, the contours of the projection (red) are elongated with respect to those of the bulk dispersion in that plane (black lines) and thus correspond to higher effective masses. An alternative way of thinking is that the in-plane subband mass is enhanced as a result of the zig-zag hopping of carriers moving in the surface plane, which, in conjunction with surface band banding leads to reduced hopping elements and thus enhanced effective masses in-plane.

Along both the [111] and [110] directions $SrTiO_3$ can be viewed as a stack of charged planes. Therefore the charge accumulation mechanism at (111) or (110) surfaces might be expected to be different from what is observed for the neutrally stacked [001] direction. For example, surface reconstructions that compensate the polarity of the surface could induce spontaneous charge accumulation. However the origin of the (111) and (110) surface 2DELs has very much the same phenomenology as the (001) surface, as discussed in Sect. 4.2.1. Additionally, there is no evidence from ARPES that surface reconstructions influence the (111) and (110) surface 2DEL band structures. For the case of the STO (110) 2DEL where the surface is known to have a 4×1 reconstruction, this has been attributed to the protective nature of the insulating over-layer of titania formed by this reconstruction. Wang et al. proposed that light-induced oxygen vacancies migrate below the overlayer and dope electrons which are not perturbed by the potential of the surface reconstruction. For the case of the STO (111) 2DEL, where details of the surface termination are not known, the insensitivity of the states to possible reconstructions was attributed to the wavefunctions of the 2DEL being centred far below the surface. The envelope wavefunction solutions of self-consistent tight binding supercell calculations [31] show the wavefunctions peak ≈6 Ti layers below the surface. This is in line with the observation that back-folded bands have been observed for the 2DELs at the reconstructed STO (001) and anatase TiO_2 (001) surfaces, where the d_{xy} subbands are more tightly confined at the surface [82]. It may also provide these systems with some degree of insensitivity to surface or interface impurities.

4.4 Surface 2DELs in Other Transition Metal Oxides

The diverse bulk properties of the large number of transition metal oxides suitable for heteroepitaxy hold much potential for both fundamental studies and applications.

Inducing 2DELs in insulating oxides other than STO is an important step towards unlocking this potential. However, so far little is known about the prerequisites on host materials and interface properties required to this end. To date, besides on SrTiO$_3$, highly itinerant surface 2DELs have been observed on KTaO$_3$ [14–16] and anatase TiO$_2$ [18, 82] which both have an empty d-shell in stoichiometric form, while attempts to induce a 2DEL in rutile TiO$_2$ and the Mott insulator LaTiO$_3$ were unsuccessful [18, 83]. In the case of rutile TiO$_2$, this might be related to the strong tendency of bulk samples to localize excess carriers. Hole doped bulk LaTiO$_3$ on the other hand, is known to host itinerant carriers [3] suggesting that its fundamental material properties should not prohibit the formation of 2DELs. This highlights one of the key challenges of this field. While it is clear that the creation of a 2DEL requires chemical doping or charge transfer across an interface, it is often hard to predict whether carrier doping of insulating oxides induces metallicity.

4.4.1 KTaO$_3$

KTaO$_3$ shares important properties with SrTiO$_3$. Both are ABO$_3$ perovskites with empty d-shell and are close to a ferroelectric instability. However, unlike in STO, the bulk truncated (001) surface of KTO is strongly polar. Additionally, in KTO the effective masses of the conduction band are lighter and the spin-orbit interaction in the Ta $5d$ shell is more than an order of magnitude larger than in the Ti $3d$ states. This suggests that KTO might be a suitable material on which to engineer 2DELs with high mobility and large tunable Rashba splitting. However, the lack of established surface preparation recipes resulting in well-ordered surfaces with single termination and the poor stability of the KTO (001) surface at high temperature have thus far prohibited the growth of heteroepitaxial interfaces with the high quality that is routinely achieved with STO substrates. Indeed, KTO based 2DELs were first induced with a parylene gate dielectric [84] and by electrolyte gating [85], while the first oxide interface inducing a 2DEL in KTO was only reported recently [86]. Notably, these studies found superconductivity at high carrier density [85] and reported spin-precession lengths that are significantly shorter than in InGaAs and tunable over a very wide range varying from 20 to 60 nm with gate voltage [84] suggesting much potential of KTO for spintronic devices. Yet, the Rashba effect which causes the spin-precession and even the overall band structure of KTO based 2DELs remain poorly understood.

King et al. reported a 2DEL on the bare (001) surface of KTO and studied its band structure with a combination of ARPES and tight-binding supercell calculations [15]. The experimental data showed an occupied band width of ∼400 meV and resolved two isotropic light subbands with effective masses of ∼0.3 m_e and a shallower subband with $m^* \sim 2 - 3\,m_e$ contributing an elliptical Fermi surface. These results were confirmed by Santander-Syro et al. [16] and could largely be reproduced by band structure calculations although the agreement is not as good as in STO (001). In particular, finer details, such as hybridization gaps between the subbands and the theoretically predicted Rashba splitting could not be resolved experimentally. From the line width, King et al. deduced a upper limit for the Rashba splitting of ~0.02 Å^{-1} [15], which is consistent with the spin precession lengths reported in [84] but still more than an order of magnitude lower than in other surface systems containing heavy atoms, such as the L-gap surface state on Au(111) [87] or the surface 2DEL on the topological insulator Bi_2Se_3 [88].

The authors of [15] attributed the modest Rashba splitting in KTO to the particular orbital character of the bulk states from which the 2DEL derives. The strong spin-orbit interaction in KTO restores the orbital angular momentum, which is largely quenched in the $3d$ counterpart STO. Its conduction band is thus more appropriately described by total angular momentum states rather than the crystal field eigenstates commonly used in STO. The conduction band edge at the Γ point of KTO is formed by a quartet of $J_{\text{eff}} = 3/2$ states, which are a linear combination of all three t_{2g} orbitals. The $J_{\text{eff}} = 1/2$ doublet is split off by the spin-orbit gap $\Delta_{SO} \approx 400$ meV. This splitting is larger than the occupied band width of KTO based 2DELs reported in the literature [15, 16], which therefore derive from the $J_{\text{eff}} = 3/2$ states. To a first approximation, the 2DEL in KTO has thus the same orbital composition as 2D hole gases in typical III-V semiconductors such as GaAs which suppresses the k-linear Rashba term [89]. It should be noted however that, due to the same physics that produces orbital ordering in the STO (001) 2DEL, the strong quantum confinement in KTO (001) modifies the orbital character and re-introduces a significant orbital polarization as pointed out in [16].

4.4.2 Anatase TiO_2

TiO_2 crystallizes in the rutile and anatase structures and is one of the most intensely studied TMOs due to its diverse applications in heterogeneous catalysis, photocatalysis, gas-sensing or photovoltaics and its use as transparent conductive coating or simply as biocompatible white pigment. While the surface science of TiO_2 is intensely studied [90], TiO_2 has so far received less attention as a host material for oxide 2DELs and only a few studies reported conductive interfaces with other TMOs [91, 92]. Similar to STO, TiO_2 is susceptible to the creation of light-induced oxygen vacancies [17, 36], which was exploited by Moser et al. [17] to dope anatase bulk single crystals and PLD-grown thin films inducing quasi-3D electronic states which showed strong signatures of electron-phonon interaction in the ARPES spectra. Subsequently, Rödel et al. showed that under appropriate conditions fully 2D quantum confined states with the characteristic subband ladder of a 2DEL can be induced on the (001) and (101) surface of anatase TiO_2 [18]. Intriguingly though, the same approach did not induce itinerant carriers in rutile TiO_2 [18].

Unlike in STO, the conduction band minimum of anatase TiO_2 is of pure d_{xy} orbital character with the other t_{2g} orbitals split off by ~0.5 eV. This results in a particularly simple electronic structure of anatase TiO_2 based 2DELs with isotropic Fermi surfaces of all subbands on the (001) surface and concentric elliptical Fermi surfaces on the (101) surface [18]. The d_{xy} orbital character of anatase TiO_2 based

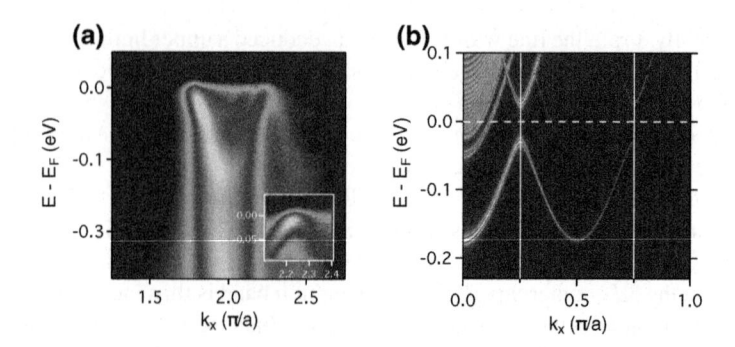

Fig. 4.12 **a** 2DEL subband dispersion at the surface of anatase TiO$_2$ (001) thin film terminated by a (1 × 4) surface reconstruction. Backfolding of the $n = 1$ quantum well state at the superlattice zone boundary is highlighted in the inset where the opening of a superlattice band gap is observed. **b** Tight binding supercell calculation of the spectral weight distribution for a 4 × 30 supercell with an in-plane potential modulation of ∼100 meV, estimated from ab-initio calculations of core-levels, imposed in the topmost unit cell. For details see Wang et al. [82]

2DELs also leads to a particularly strong confinement of the carriers in a narrow layer below the surface. This was exploited by Wang et al. to demonstrate a periodic lateral modulation of the 2DEL by the (1 × 4) surface reconstruction of in-situ grown anatase TiO$_2$ thin films with (001) orientation [82]. Tuning the Fermi wave vector of the first subband to coincide with the superlattice Brillouin zone boundary corresponding to the surface reconstruction, the authors of [82] found a sizeable superlattice band gap at the Fermi level, as shown in Fig. 4.12. This suggests a new route towards electronic structure engineering in oxide 2DELs by exploiting the ubiquitous surface reconstructions of TMOs.

4.5 Many-Body Interactions in TMO 2DELs

The thermodynamic and transport properties of transition metal oxides are often dominated by many-body interactions. Prominent examples include high temperature superconductivity in cuprates, colossal magnetoresistance in manganites or the ubiquitous metal-insulator transitions in ultrathin TMO films [93, 94]. Unlocking the full potential of TMO 2DELs will require an improved understanding of these interactions as the interfacial carrier density is tuned, which is a formidable task. Here, we briefly review the first microscopic measurements of many-body interactions in oxide surface 2DELs using ARPES [17, 40, 42, 47]. These studies all focus on electron-phonon interaction (EPI) in anatase TiO$_2$ and SrTiO$_3$ as the density of itinerant carriers is tuned by controlling the oxygen vacancy concentration using the methods described in Sect. 4.2.1.

Electron-phonon interaction in STO dominates the mobility of interface 2DELs at elevated temperatures [95], contributes to the large thermoelectric coefficient of depleted 2DELs [96] and has been invoked as the pairing glue for superconductivity [97, 98]. Yet, until recently little was known about its nature and strength. Due to its strongly ionic character, lightly doped STO was often considered a model system for the nonlocal Fröhlich interaction describing the dielectric screening of an excess charge by longitudinal optical (LO) phonons [99]. In this model, EPI is strongly peaked at small momentum transfer **Q** and dominated by coupling to the highest LO branch [99]. On the other hand, STO is well known for its large static dielectric constant implying soft transverse modes, which can eventually condense into a ferroelectric state [100, 101]. A recent theoretical study found that coupling to such a soft mode near a quantum critical point reproduces the superconducting dome of STO [97].

Early experimental studies of EPI in STO focused on doped bulk samples. Van Mechelen et al. [44] and Devreese et al. [102] showed that a pronounced mid-infrared peak in optical spectra can be reproduced quantitatively by the Fröhlich model with a moderate coupling constant of $\alpha \approx 2$ deduced from the independently determined static and high-frequency dielectric constants respectively. In this picture, the system remains fully itinerant despite the relatively large mass enhancement from EPI of $m^*/m_{band} \sim 2 - 3$ and forms a liquid of large polarons. On the other hand ARPES, which gives more direct insight into EPI, provided conflicting results. Chang et al. [103] reported signatures consistent with coupling to the highest LO phonon branch with frequency $\Omega_{LO,4} \approx 100$ meV, as observed by van Mechelen et al. [44], while Meevasana et al. [104] reported a perturbative EPI with much stronger coupling to a soft LO mode than expected in the Fröhlich model. While this discrepancy was never fully resolved, we speculate that it arises at least partially from accidental surface doping in the latter study. The ARPES study by Moser et al. of quasi-3D carriers in anatase TiO_2 created by photo-induced oxygen vacancies reported replica bands characteristic of Fröhlich polarons at low carrier density and a progressive screening of EPI with increasing density [17] which has similarities to what is observed in STO (001) as will be described in detail in this section.

The marked influence of the carrier concentration on the spectral function is evident from the data on the STO (001) surface 2DEL of Wang et al. [47] reproduced in Fig. 4.13. Using in-situ prepared surfaces, these authors could reduce the vacancy formation rate permitting a systematic study of the spectral function for carrier densities spanning nearly an order of magnitude from $\sim 3 \times 10^{13}$ to 2×10^{14} cm^{-2}. At the lowest densities corresponding to an occupied quasiparticle bandwidth of ~ 20 meV, the spectra of [47] show dispersive replica bands shifted by multiples of 100 meV to higher energy. This implies preferential coupling with small momentum transfer ($\mathbf{Q} \ll \pi/a$) to the LO_4 mode of STO, which is the hallmark of Fröhlich polarons, quasiparticles formed by an excess electron dressed by a polarization cloud extending over several lattice sites that follows the charge as it propagates through the crystal [17, 40, 99]. Figure 4.13g shows that the spectral weight at this density is clearly dominated by excitations involving one or more phonons (blue peaks in Fig. 4.13g). The quasiparticle residue, which gives the relative spectral weight of the coherent quasiparticle (yellow in Fig. 4.13g), was observed to be $Z \approx 0.2$ corresponding to a coupling constant $\alpha \approx 2.8$ in the Fröhlich model [47, 105], in fair

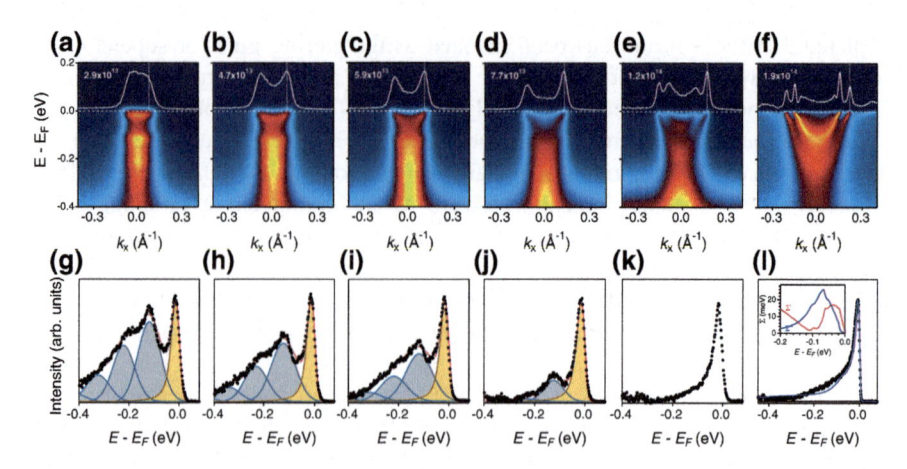

Fig. 4.13 Evolution of the spectral function in the STO (001) surface 2DEL with increasing carrier density. **a–f** Raw dispersion plots and MDCs extracted at the Fermi level. Carrier densities are indicated in units of cm^{-2}. **g–l** EDCs at the Fermi wave vector k_F indicated in the image plots. Fits in **g–j** use a Franck-Condon model with a single phonon mode of 100 meV. The thin blue line in (**l**) is a spectral function calculated for the conventional Eliashberg self-energy in the high-density limit reported in [42] and shown in the inset. Adapted from Wang et al. [47]

agreement with the analysis of optical spectra in lightly doped bulk STO [44, 102]. Comparing the Luttinger volume of the surface 2DEL with sheet carrier densities at the LAO/STO interface deduced from the Hall coefficient, the authors of [47] further concluded that interface superconductivity likely derives from a liquid of large polarons or possibly bipolarons.

As the density is increased, spectral weight is gradually transferred from the replicas to the quasiparticle band implying a decreasing effective coupling constant α. Concomitant, the mass enhancement m^*/m_{band} decreases from 2.4 to 1.7 m_e, following the trend $m^*/m_{\text{band}} = 1/(1 - \alpha/6)$ expected for Fröhlich polarons at intermediate coupling strengths [99]. Using exact diagonalization, the authors of [47] showed that the observed evolution of EPI can be traced back to a gradual transition from dielectric screening at low density to dominantly electronic screening at high density qualitatively consistent with earlier ARPES measurements on quasi-3D states in oxygen deficient anatase TiO_2 [17]. Superconducting susceptibilities calculated within the same approach further showed that the dominant pairing channel has s-wave symmetry and indicated that a competition between the opposite trends of density of states and effective coupling strength underlies the dome shaped superconductivity observed at the LAO/STO interface [106].

It is worth noting that the effect of increasing carrier density in the STO (001) 2DEL is not limited to progressive screening of the long-range Fröhlich interaction. At the saturation density of the surface 2DEL, the LO_4 mode is almost completely screened as is evident from the absence of any spectral signatures at 100 meV in Fig. 4.13f. However, the coupling to lower frequency modes increases far beyond the predictions of the Fröhlich model [42, 104, 107] pointing at a remarkable complexity of EPI in oxide 2DELs. We also point out that a significant coupling to the soft ferroelectric mode predicted in [97] cannot be excluded from published ARPES data on surface or interface 2DELs in STO due to the difficulty of quantifying the quasiparticle dispersion at very low energy.

Electron-phonon interaction at the LAO/STO interface has recently been studied by tunneling spectroscopy and soft X-ray ARPES. Investigating tunnel junctions to interface 2DELs with ∼30 meV occupied bandwidth, Boschker et al. [108] found inelastic tunneling attributed to EPI with dominant coupling to the LO_4 mode and progressively weaker contributions from the softer LO modes of STO. Varying the chemical potential over ∼5 meV using a back gate did not change the coupling strength significantly. While tunneling does not give absolute coupling strengths as they can be deduced from ARPES specta, the dominant contribution from the LO_4 mode is in agreement with the ARPES results on the STO (001) surface 2DEL for similar bandwidths [47]. The results on the surface 2DEL of [39] were further confirmed by a soft X-ray ARPES study of an LAO/STO interface with $n_{2D} \approx 8 \times 10^{13}$ cm^{-2} [109], reporting a replica band and $Z \approx 0.4$ in excellent agreement with the value reported in [39] for the same density. This strongly suggests that the nature and strength of EPI is similar for interface and surface 2DELs of the same density. We note, however, that manifestations of EPI in transport properties might differ between these two systems due to the different nature and density of defects.

4.6 Discussion

The study of 2DELs at the surface of 3D transition metal oxides is clearly motivated by the importance of interface 2DELs for oxide electronics. However, since the first discovery of surface 2DELs on STO (001) in 2011 [12, 13], their investigation by ARPES has, to some extent, evolved into its own sub-field with a number of interesting results from several groups as summarized in this chapter. These include the comprehensive characterization of the subband masses, the subband ordering and the resulting orbital polarization for different surface planes and the rationalization of these effects in terms of quantum confinement [12, 13, 15, 31, 39, 42]; the identification of light-induced oxygen vacancies as the microscopic origin of the 2DEL carriers which has permitted tuning of the carrier density over a wide range [12, 30, 31]; and the observation of a complex evolution of electron-phonon interaction with carrier density giving rare microscopic insight into the many-body interactions governing important properties of oxide 2DELs [40, 42, 47]. Most of these studies have focused on different surface orientations of STO but 2DELs have also been induced and studied at low-index surfaces of $KTaO_3$ and anatase TiO_2 [15, 16, 18, 82].

While these results are interesting in their own right, their relation with the properties of interface 2DELs is not always clear. This being said, for the intensely studied STO (001) 2DEL a number of experiments suggest that some important properties are

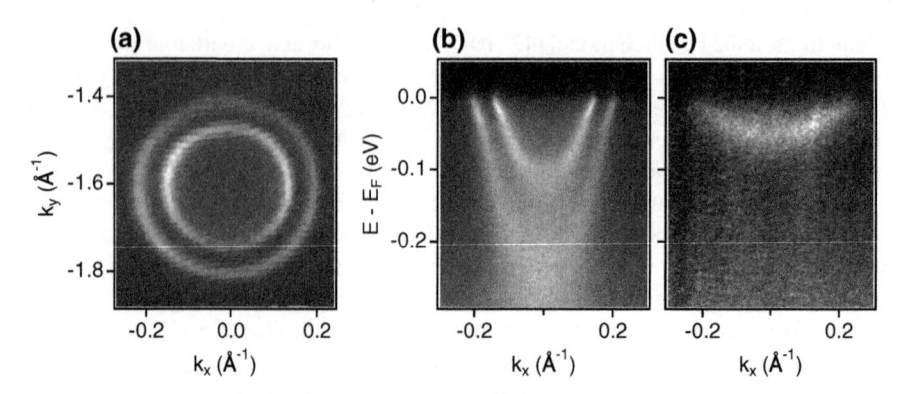

Fig. 4.14 Electronic structure of the 2DEL at the interface of STO (001) and aluminium oxide formed when 2 Å Al is deposited on the bare TiO$_2$ terminated (001) surface of STO at room temperature. A redox reaction at the interface occurs and the surface of STO becomes reduced and doped with itinerant electrons which have the Fermi surface shown in (**a**). The Luttinger volume of these two concentric circular Fermi surface sheets is as found for the 2DEL at the bare (001) surface. The electron dispersions along the k_x axis are shown in (**b**) and (**c**) for orthogonal polarizations of the exciting radiation revealing the light d_{xy} and heavy $d_{xz/yz}$ bands respectively. This orbital polarization is the same as found for the 2DEL at the bare (001) surface. Adapted from Rödel et al. [45]

universal in the sense that they are largely determined by the host material and crystallographic orientation, rather than by structural details or the origin of the charges. For instance, orbital polarization is evident in the surface 2DEL [13, 15, 42] and has also been measured directly for the LAO/STO interface using X-ray linear dichroism [5]. The Rashba spin-splitting of the surface 2DEL discussed in [42, 54] is also in fair agreement with weak antilocalization and quantum oscillation measurements and calculations of interface 2DELs [67–69, 110]. Moreover, the nature and strength of electron-phonon interaction at both the bare STO (001) surface and the LAO/STO (001) interface were found to be in good agreement [47, 108, 109]. The universality of electronic properties is further supported by a recent experiment demonstrating that room temperature deposition of ≈2 Å Al on a TiO$_2$ terminated STO (001) wafer surface donates electrons and induces a 2DEL in STO [45] by strongly reducing the surface. As seen in Fig. 4.14 the subband structure of this 2DEL bears all the hall marks of the STO surface 2DEL, while in fact it exists at the interface of STO and aluminium oxide.

On the other hand, some basic 2DEL properties such as the carrier density and occupied bandwidth remain controversial. For the surface 2DELs discussed in this chapter, the latter is evident from the ARPES data while the total carrier concentration and density of states at the Fermi level are difficult to determine experimentally since the closely spaced shallow subbands predicted by band structure calculations have so far eluded detection. In interface systems, both carrier density and bandwidth are difficult to quantify. The Hall effect, commonly used as a measure of the carrier density, is strongly non-linear for the LAO/STO interface and its quantitative relation to the itinerant carrier density is model dependent [111]. Furthermore estimates of the occupied bandwidth based on quantum oscillation and spectroscopic measurements range from ≈5 meV [10] up to ≈200 meV [6, 108] for LAO/STO systems with similar Hall densities. Thus so far it has proved difficult engineer these fundamental properties through control of the growth conditions. A direct comparison of results from ARPES and transport experiments on a single STO 2DEL sample could provide more insight and help to establish the extent to which electronic properties can be considered universal. However, this remains a challenging task as the surface 2DEL is not accessible to standard magneto-transport experiments while ARPES experiments on interfaces are limited to the soft X-ray regime where the effective resolution has so far precluded results that resolve the full subband structure. In this regard, the aluminium oxide/STO interface studied by ARPES in [45] provides an interesting opportunity to overcome these difficulties as it is much more amenable to magneto-transport experiments than the bare STO surface due to the protective nature of the aluminium oxide over-layer.

Another aspect that has perhaps not received the attention it deserves is the relation between sample environment and 2DEL properties. In Sect. 4.2.1 we described the exceptional sensitivity of the cleaved STO (001) surface to light induced oxygen vacancy formation as well as to the residual oxygen partial pressure. Adsorbates on the LAO surface and irradiation even with visible light were also found to strongly affect the LAO/STO interface 2DEL [112–114]. Yet, the understanding of the effects of different sample preparations and environments including electron and photon beams, as they are used in many experiments, is only in its infancy and much work remains to be done to improve the consistency of experiments and ultimately the stability of devices. On the other hand, the sensitivity of emergent properties in oxides to defects, including those introduced by experimental probes, offers entirely new perspectives for tailoring properties on the nanoscale [12, 115, 116].

Acknowledgements The authors acknowledge financial support through the University of Geneva and the Swiss National Science Foundation and would like to thank C. Bernhard, R. Claessen, A. Fête, S. Gariglio, K. Held, P. D. C. King, F. Lechermann, L. D. Marks, W. Meevasana, N. C. Plumb, M. Radovic, V. Strocov, A. Tamai, J. M. Triscone, D. van der Marel and Z. Wang for discussions.

References

1. P. Zubko, S. Gariglio, P. Gabay, M. Ghosez, J.-M. Triscone, Interface physics in complex oxide heterostructures. Ann. Rev. Condens. Matter Phys. **2**, 141–165 (2011)
2. J. Mannhart, D.G. Schlom, Oxide interfaces-an opportunity for electronics. Science **327**, 1607–1611 (2010)
3. Y. Tokura, Y. Taguchi, Y. Okada, Y. Fujishima, T. Arima, K. Kumagai, Y. Iye, Filling dependence of electronic properties on the verge of metal-Mott-insulator transition in $Sr_{1-x}La_xTiO_3$. Phys. Rev. Lett. **70**, 2126 (1993)
4. R. Jany, C. Richter, C. Woltmann, G. Pfanzelt, B. Förg, M. Rommel, T. Reindl, U. Waizmann, J. Weis, J.A. Mundy, D.A. Muller, H. Boschker, J. Mannhart, Monolithically integrated circuits

from functional oxides. Adv. Mater. Interfaces **1**, 1300031 (2014)
5. M. Salluzzo, J.C. Cezar, N.B. Brookes, V. Bisogni, G.M. De Luca, C. Richter, S. Thiel, J. Mannhart, M. Huijben, A. Brinkman, G. Rijnders, G. Ghiringhelli, Orbital reconstruction and the two-dimensional electron gas at the $LaAlO_3/SrTiO_3$ interface. Phys. Rev. Lett. **102**, 166804 (2009)
6. G. Berner, S. Glawion, J. Walde, F. Pfaff, H. Hollmark, L.-C. Duda, S. Paetel, C. Richter, J. Mannhart, M. Sing, R. Claessen, $LaAlO_3/SrTiO_3$ oxide heterostructures studied by resonant inelastic X-ray scattering. Phys. Rev. B **82**, 241405 (2010)
7. K.-J. Zhou, M. Radovic, J. Schlappa, V. Strocov, R. Frison, J. Mesot, L. Patthey, T. Schmitt, Localized and delocalized Ti $3d$ carriers in $LaAlO_3/SrTiO_3$ superlattices revealed by resonant inelastic X-ray scattering. Phys. Rev. B **83**, 201402(R) (2011)
8. A. Fête, S. Gariglio, A.D. Caviglia, J.-M. Triscone, M. Gabay, Rashba induced magneto-conductance oscillations in the $LaAlO_3$-$SrTiO_3$ heterostructure. Phys. Rev. B **86**, 201105(R) (2012)
9. M. Kim, C. Bell, Y. Kozuka, M. Kurita, Y. Hikita, H.Y. Hwang, Fermi surface and superconductivity in low-density high-mobility δ-doped $SrTiO_3$. Phys. Rev. Lett. **107**, 106801 (2011)
10. A. McCollam, S. Wenderich, M.K. Kruize, V.K. Guduru, H.J.A. Molegraaf, M. Huijben, G. Koster, D.H.A. Blank, G. Rijnders, A. Brinkman, H. Hilgenkamp, U. Zeitler, J.C. Maan, Quantum oscillations and subband properties of the two-dimensional electron gas at the $LaAlO_3/SrTiO_3$ interface. APL Mater. **2**, 022102 (2014)
11. P. Moetakef, D.G. Ouellette, J.R. Williams, S. James Allen, L. Balents, D. Goldhaber-Gordon, S. Stemmer, Quantum oscillations from a two-dimensional electron gas at a Mott/band insulator interface. Appl. Phys. Lett. **101**, 151604 (2012)
12. W. Meevasana, P.D.C. King, R.H. He, S.-K. Mo, M. Hashimoto, A. Tamai, P. Songsiriritthigul, F. Baumberger, Z.-X. Shen, Creation and control of a two-dimensional electron liquid at the bare $SrTiO_3$ surface. Nat. Mater. **10**, 114–118 (2011)
13. A.F. Santander-Syro, O. Copie, T. Kondo, F. Fortuna, S. Pailhès, R. Weht, X.G. Qiu, F. Bertran, A. Nicolaou, A. Taleb-Ibrahimi, P. Le Fèvre, G. Herranz, M. Bibes, N. Reyren, Y. Apertet, P. Lecoeur, A. Barthélémy, M.J. Rozenberg, Two-dimensional electron gas with universal subbands at the surface of $SrTiO_3$. Nature **469**, 189–193 (2011)
14. C. Bareille, F. Fortuna, T.C. Rödel, F. Bertran, M. Gabay, O.H. Cubelos, A. Taleb-Ibrahimi, P. Le Fèvre, M. Bibes, A. Barthélémy, T. Maroutian, P. Lecoeur, M.J. Rozenberg, A.F. Santander-Syro, Two-dimensional electron gas with six-fold symmetry at the (111) surface of KTaO3. Sci. Rep. **4**, 3586 (2014)
15. P.D.C. King, R.H. He, T. Eknapakul, P. Buaphet, S.-K. Mo, Y. Kaneko, S. Harashima, Y. Hikita, M.S. Bahramy, C. Bell, Z. Hussain, Y. Tokura, Z.-X. Shen, H.Y. Hwang, F. Baumberger, W. Meevasana, Subband structure of a two-dimensional electron gas formed at the polar surface of the strong spin-orbit perovskite $KTaO_3$. Phys. Rev. Lett. **108**, 117602 (2012)
16. A.F. Santander-Syro, C. Bareille, F. Fortuna, O. Copie, M. Gabay, F. Bertran, A. Taleb-Ibrahimi, P. Le Fèvre, G. Herranz, N. Reyren, M. Bibes, A. Barthélémy, P. Lecoeur, J. Guevara, M.J. Rozenberg, Orbital symmetry reconstruction and strong mass renormalization in the two-dimensional electron gas at the surface of $KTaO_3$. Phys. Rev. B **86**, 121107(R) (2012)
17. S. Moser, L. Moreschini, J. Jaćimović, O.S. Barišić, H. Berger, A. Magrez, Y.J. Chang, K.S. Kim, A. Bostwick, E. Rotenberg, L. Forró, M. Grioni, Tunable polaronic conduction in anatase TiO_2. Phys. Rev. Lett. **110**, 196403 (2013)
18. T.C. Rödel, F. Fortuna, F. Bertran, M. Gabay, M.J. Rozenberg, A.F. Santander-Syro, P. Le Fèvre, Engineering two-dimensional electron gases at the (001) and (101) surfaces of TiO_2 anatase using light. Phys. Rev. B **92**, 041106(R) (2015)
19. P.D.C. King, T.D. Veal, C.F. McConville, J. Zúñiga Pérez, V. Muñoz Sanjosé, M. Hopkinson, E.D.L. Rienks, M.F. Jensen, P. Hofmann, Surface band-gap narrowing in quantized electron accumulation layers. Phys. Rev. Lett. **104**, 256803 (2010)
20. L.F.J. Piper, L. Colakerol, P.D.C. King, A. Schleife, J. Zúñiga Pérez, P.A. Glans, T. Learmonth, A. Federov, T.D. Veal, F. Fuchs, V. Muñoz Sanjosé, F. Bechstedt, C.F. McConville, K.E. Smith, Observation of quantized subband states and evidence for surface electron accumulation in CdO from angle-resolved photoemission spectroscopy. Phys. Rev. B **78**, 165127 (2008)

21. K.H.L. Zhang, R.G. Egdell, F. Offi, S. Iacobucci, L. Petaccia, S. Gorovikov, P.D.C. King, Microscopic origin of electron accumulation in In_2O_3. Phys. Rev. Lett. **110**, 056803 (2013)
22. L. Forro, O. Chauvet, D. Emin, L. Zuppiroli, H. Berger, F. Lévy, High mobility ntype charge carriers in large single crystals of anatase (TiO2). J. Appl. Phys. **75**, 633–635 (1994)
23. J. Jaćimović, R. Gaál, A. Magrez, J. Piatek, L. Forró, L.S. Nakao, Y. Hirose, T. Hasegawa, Low temperature resistivity, thermoelectricity, and power factor of Nb doped anatase TiO_2. Appl. Phys. Lett. **102**, 013901 (2013)
24. J. Schooley, W. Hosler, E. Ambler, J. Becker, M. Cohen, C. Koonce, Dependence of the superconducting transition temperature on carrier concentration in semiconducting $SrTiO_3$. Phys. Rev. Lett. **14**, 305–307 (1965)
25. A. Spinelli, M.A. Torija, C. Liu, C. Jan, C. Leighton, Electronic transport in doped $SrTiO_3$: conduction mechanisms and potential applications. Phys. Rev. B **81**, 155110 (2010)
26. H. Uwe, J. Kinoshita, K. Yoshihiro, C. Yamanouchi, T. Sakudo, Evidence for light and heavy conduction electrons at the zone center in $KTaO_3$. Phys. Rev. B **19**, 3041–3044 (1979)
27. S.H. Wemple, Some transport properties of oxygen-deficient single-crystal potassium tantalate ($KTaO_3$). Phys. Rev. **137**, A1575 (1965)
28. L. Dudy, M. Sing, P. Scheiderer, J.D. Denlinger, P. Schütz, J. Gabel, M. Buchwald, C. Schlueter, T.-L. Lee, R. Claessen, In Situ control of separate electronic phases on SrTiO3 surfaces by oxygen dosing. Adv. Mater. **28**, 7443–7449 (2016)
29. N.C. Plumb, M. Salluzzo, E. Razzoli, M. Månsson, M. Falub, J. Krempasky, C.E. Matt, J. Chang, M. Schulte, J. Braun, H. Ebert, J. Minár, B. Delley, K.-J. Zhou, T. Schmitt, M. Shi, J. Mesot, L. Patthey, M. Radović, Mixed dimensionality of confined conducting electrons in the surface region of $SrTiO_3$. Phys. Rev. Lett. **113**, 086801 (2014)
30. S. McKeown Walker, F.Y. Bruno, Z. Wang, A. de la Torre, S. Riccó, A. Tamai, T.K. Kim, M. Hoesch, M. Shi, M.S. Bahramy, P.D.C. King, F. Baumberger, Carrier-density control of the $SrTiO_3$ (001) surface 2D electron gas studied by ARPES. Adv. Mater. **27**, 3894–3899 (2015)
31. S. McKeown Walker, A. de la Torre, F.Y. Bruno, A. Tamai, T.K. Kim, M. Hoesch, M. Shi, M.S. Bahramy, P.D.C. King F. Baumberger, Control of a two-dimensional electron gas on SrTiO3(111) by atomic oxygen. Phys. Rev. Lett. **113**, 177601 (2014)
32. Y. Aiura, I. Hase, H. Bando, T. Yasue, T. Saitoh, D.S. Dessau, Photoemission study of the metallic state of lightly electron-doped $SrTiO_3$. Surf. Sci. **515**, 61–74 (2002)
33. V.E. Henrich, G. Dresselhaus, H.J. Zeiger, Surface defects and the electronic structure of SrTiO3 surfaces. Phys. Rev. B **17**, 4908–4921 (1978)
34. H.O. Jeschke, J. Shen, R. Valenti, Localized versus itinerant states created by multiple oxygen vacancies in SrTiO3. New J. Phys. **17**, 023034 (2015)
35. M.L. Knotek, Stimulated desorption from surfaces. Phys. Today **37**, 24 (1984)
36. M.L. Knotek, P.J. Feibelman, Ion desorption by core-hole auger decay. Phys. Rev. Lett. **40**, 964–967 (1978)
37. C. Lin, A.A. Demkov, Electron correlation in oxygen vacancy in $SrTiO_3$. Phys. Rev. Lett. **111**, 217601 (2013)
38. T.C. Rödel, C. Bareille, F. Fortuna, C. Baumier, F. Bertran, P. Le Fèvre, M. Gabay, O. Hijano Cubelos, M.J. Rozenberg, T. Maroutian, P. Lecoeur, A.F. Santander-Syro, Orientational tuning of the fermi sea of confined electrons at the $SrTiO_3$. Phys. Rev. Appl. **1**, 051002 (2014)
39. Z. Wang, Z. Zhong, X. Hao, S. Gerhold, B. Stöger, M. Schmid, J. Sánchez-Barriga, A. Varykhalov, C. Franchini, K. Held, U. Diebold, Anisotropic two-dimensional electron gas at SrTiO3(110). Proc. Natl. Acad. Sci. U.S. Am. **111**, 3933–3937 (2014)
40. C. Chen, J. Avila, E. Frantzeskakis, A. Levy, M.C. Asensio, Observation of a two-dimensional liquid of Fröhlich polarons at the bare SrTiO3 surface. Nat. Commun. **6**, 8585 (2015)
41. P. Delugas, V. Fiorentini, A. Mattoni, A. Filippetti, Intrinsic origin of two-dimensional electron gas at the (001) surface of $SrTiO_3$. Phys. Rev. B **91**, 115315 (2015)
42. P.D.C. King, S. McKeown Walker, A. Tamai, A. de la Torre, T. Eknapakul, P. Buaphet, S.-K. Mo, W. Meevasana, M.S. Bahramy, F. Baumberger, Quasiparticle dynamics and spin-orbital texture of the $SrTiO_3$ two-dimensional electron gas. Nat. Commun. **5**, 3414 (2014)

43. D. van der Marel, J.L.M. van Mechelen, I.I. Mazin, Common Fermi-liquid origin of T^2 resistivity and superconductivity in n-type SrTiO$_3$. Phys. Rev. B **84**, 205111 (2011)
44. J.L.M. van Mechelen, D. van der Marel, C. Grimaldi, A. Kuzmenko, N. Armitage, N. Reyren, H. Hagemann, I. Mazin, Electron-phonon interaction and charge carrier mass enhancement in SrTiO$_3$. Phys. Rev. Lett. **100**, 226403 (2008)
45. T.C. Rödel, F. Fortuna, S. Sengupta, E. Frantzeskakis, P.L. Fèvre, F. Bertran, B. Mercey, S. Matzen, G. Agnus, T. Maroutian, P. Lecoeur, A.F. Santander-Syro, Universal fabrication of 2D electron systems in functional oxides. Adv. Mater. **28**, 1976–1980 (2016)
46. L.W. van Heeringen, G.A. De Wijs, A. McCollam, J.C. Maan, A. Fasolino, $k \cdot p$ subband structure of the LaAlO$_3$/SrTiO$_3$ interface. Phys. Rev. B **88**, 205104 (2013)
47. Z. Wang, S. McKeown Walker, A. Tamai, Y. Wang, Z. Ristic, F.Y. Bruno, A. de la Torre, S. Riccò, N.C. Plumb, M. Shi, P. Hlawenka, J. Sánchez-Barriga, A. Varykhalov, T.K. Kim, M. Hoesch, P.D.C. King, W. Meevasana, U. Diebold, J. Mesot, B. Moritz, T.P. Devereaux, M. Radovic, F. Baumberger, Tailoring the nature and strength of electron-phonon interactions in the SrTiO$_3$(001) 2D electron liquid. Nat. Mater. **15**, 835–839 (2016)
48. A. Fête, C. Cancellieri, D. Li, D. Stornaiuolo, A.D. Caviglia, S. Gariglio, J.-M. Triscone, Growth-induced electron mobility enhancement at the LaAlO$_3$/SrTiO$_3$ interface. Appl. Phys. Lett. **106**, 051604 (2015)
49. A.F. Santander-Syro, F. Fortuna, C. Bareille, T.C. Rödel, G. Landolt, N.C. Plumb, J.H. Dil, M. Radović, Giant spin splitting of the two-dimensional electron gas at the surface of SrTiO$_3$. Nat. Mater. **13**, 1085–1090 (2014)
50. M. Stengel, First-principles modeling of electrostatically doped perovskite systems. Phys. Rev. Lett. **106**, 136803 (2011)
51. M.S. Bahramy, P.D.C. King, A. de la Torre, J. Chang, M. Shi, L. Patthey, G. Balakrishnan, P. Hofmann, R. Arita, N. Nagaosa, F. Baumberger, Emergent quantum confinement at topological insulator surfaces. Nat. Commun. **3**, 1159 (2012)
52. S. Gariglio, A. Fête, J.-M. Triscone, Electron confinement at the LaAlO$_3$/SrTiO$_3$ interface. J. Phys. Condens. Matter **27**, 283201 (2015)
53. P.D.C. King, T. Veal, C. McConville, Nonparabolic coupled Poisson-Schrödinger solutions for quantized electron accumulation layers: band bending, charge profile, and subbands at InN surfaces. Phys. Rev. B **77**, 125305 (2008)
54. S. McKeown Walker, S. Riccò, F.Y. Bruno, A. de la Torre, A. Tamai, E. Golias, A. Varykhalov, D. Marchenko, M. Hoesch, M.S. Bahramy, P.D.C. King, J. Sánchez-Barriga, F. Baumberger, Absence of giant spin splitting in the two-dimensional electron liquid at the surface of SrTiO$_3$ (001). Phys. Rev. B **93**, 245143 (2016)
55. R. Yukawa, K. Ozawa, S. Yamamoto, R.-Y. Liu, I. Matsuda, Anisotropic effective mass approximation model to calculate multiple subband structures at wide-gap semiconductor surfaces: application to accumulation layers of SrTiO$_3$ and ZnO. Surf. Sci. **641**, 224–230 (2015)
56. M. Basletic, J.-L. Maurice, C. Carrétéro, G. Herranz, O. Copie, M. Bibes, E. Jacquet, K. Bouzehouane, S. Fusil, A. Barthélémy, Mapping the spatial distribution of charge carriers in LaAlO$_3$/SrTiO$_3$ heterostructures. Nat. Mater. **7**, 621–625 (2008)
57. N. Reyren, S. Gariglio, A.D. Caviglia, D. Jaccard, T. Schneider, J.-M. Triscone, Anisotropy of the superconducting transport properties of the LaAlO$_3$/SrTiO$_3$ interface. Appl. Phys. Lett. **94**, 112506 (2009)
58. C. Cancellieri, M.L. Reinle-Schmitt, M. Kobayashi, V.N. Strocov, T. Schmitt, P.R. Willmott, S. Gariglio, J.-M. Triscone, Interface fermi states of LaAlO$_3$/SrTiO$_3$ and related heterostructures. Phys. Rev. Lett. **110**, 137601 (2013)
59. A. Dubroka, M. Rössle, K.W. Kim, V.K. Malik, L. Schultz, S. Thiel, C.W. Schneider, J. Mannhart, G. Herranz, O. Copie, M. Bibes, A. Barthélémy, C. Bernhard, Dynamical response and confinement of the electrons at the LaAlO$_3$/SrTiO$_3$ interface. Phys. Rev. Lett. **104**, 156807 (2010)
60. M. Sing, G. Berner, K. Goß, A. Müller, A. Ruff, A. Wetscherek, S. Thiel, J. Mannhart, S.A. Pauli, C.W. Schneider, P.R. Willmott, M. Gorgoi, F. Schäfers, R. Claessen, Profiling

the interface electron gas of LaAlO$_3$/SrTiO$_3$ heterostructures with hard X-ray photoelectron spectroscopy. Phys. Rev. Lett. **102**, 176805 (2009)
61. O. Copie, V. Garcia, C. Bödefeld, C. Carrétéro, M. Bibes, G. Herranz, E. Jacquet, J.-L. Maurice, B. Vinter, S. Fusil, K. Bouzehouane, H. Jaffrès, A. Barthélémy, Towards two-dimensional metallic behavior at LaAlO$_3$/SrTiO$_3$ interfaces. Phys. Rev. Lett. **102**, 216804 (2009)
62. M. Altmeyer, H.O. Jeschke, O. Hijano-Cubelos, C. Martins, F. Lechermann, K. Koepernik, A.F. Santander-Syro, M.J. Rozenberg, R. Valentí, M. Gabay, Magnetism, spin texture, and in-gap states: atomic specialization at the surface of oxygen-deficient SrTiO$_3$. Phys. Rev. Lett. **116**, 157203 (2016)
63. A.C. Garcia-Castro, M.G. Vergniory, E. Bousquet, A.H. Romero, Spin texture induced by oxygen vacancies in strontium perovskite (001) surfaces: a theoretical comparison between SrTiO$_3$ and SrHfO$_3$. Phys. Rev. B **93**, 045405 (2016)
64. Y.A. Bychkov, E.I. Rashba, Properties of a 2D electron gas with lifted spectral degeneracy. JETP Lett. **39**, 78–81 (1984)
65. J. Nitta, T. Akazaki, H. Takayanagi, T. Enoki, Gate control of spin-orbit interaction in an inverted In$_{0.53}$Ga$_{0.47}$As/In$_{0.48}$Al$_{0.48}$As heterostructure. Phys. Rev. Lett. **78**, 1335–1338 (1997)
66. H.C. Koo, J.H. Kwon, J. Eom, J. Chang, S.H. Han, M. Johnson, Control of spin precession in a spin-injected field effect transistor. Science **325**, 1515–1518 (2009)
67. A.D. Caviglia, M. Gabay, S. Gariglio, N. Reyren, C. Cancellieri, J.-M. Triscone, Tunable Rashba spin-orbit interaction at oxide interfaces. Phys. Rev. Lett. **104**, 126803 (2010)
68. S. Hurand, A. Jouan, C. Feuillet-Palma, G. Singh, J. Biscaras, E. Lesne, N. Reyren, A. Barthélémy, M. Bibes, J.E. Villegas, C. Ulysse, X. Lafosse, M. Pannetier-Lecoeur, S. Caprara, M. Grilli, J. Lesueur, N. Bergeal, Field-effect control of superconductivity and Rashba spin-orbit coupling in top-gated LaAlO$_3$/SrTiO$_3$ devices. Sci. Rep. **5**, 12751 (2015)
69. H. Liang, L. Cheng, L. Wei, Z. Luo, G. Yu, C. Zeng, Z. Zhang, Nonmonotonically tunable Rashba spin-orbit coupling by multiple-band filling control in SrTiO$_3$-based interfacial d-electron gases. Phys. Rev. B **92**, 075309 (2015)
70. H. Nakamura, T. Koga, T. Kimura, Experimental evidence of cubic Rashba effect in an inversion-symmetric oxide. Phys. Rev. Lett. **108**, 206601 (2012)
71. Z. Zhong, A. Tóth, K. Held, Theory of spin-orbit coupling at LaAlO$_3$/SrTiO$_3$ interfaces and SrTiO$_3$ surfaces. Phys. Rev. B **87**, 161102 (2013)
72. G. Khalsa, B. Lee, A.H. MacDonald, Theory of t_{2g} electron-gas Rashba interactions. Phys. Rev. B **88**, 041302 (2013)
73. P. Kim, K.T. Kang, G. Go, J.H. Han, Nature of orbital and spin Rashba coupling in the surface bands of SrTiO$_3$ and KTaO$_3$. Phys. Rev. B **90**, 205423 (2014)
74. Y. Kim, R.M. Lutchyn, C. Nayak, Origin and transport signatures of spin-orbit interactions in one- and two-dimensional SrTiO$_3$-based heterostructures. Phys. Rev. B **87**, 245121 (2013)
75. K.V. Shanavas, Theoretical study of the cubic Rashba effect at the SrTiO$_3$(001) surfaces. Phys. Rev. B **93**, 045108 (2016)
76. J. Zhou, W.Y. Shan, D. Xiao, Spin responses and effective Hamiltonian for the two-dimensional electron gas at the oxide interface LaAlO$_3$/ SrTiO$_3$. Phys. Rev. B **91**, 241302(R) (2015)
77. A. Annadi, Q. Zhang, X. Renshaw Wang, N. Tuzla, K. Gopinadhan, W.M. Lü, A. Roy Barman, Z.Q. Liu, A. Srivastava, S. Saha, Y.L. Zhao, S.W. Zeng, S. Dhar, E. Olsson, B. Gu, S. Yunoki, S. Maekawa, H. Hilgenkamp, T. Venkatesan, Ariando, Anisotropic two-dimensional electron gas at the LaAlO$_3$/SrTiO$_3$ (110) interface. Nat. Commun. **4**, 1838 (2013)
78. G. Herranz, F. Sánchez, N. Dix, M. Scigaj, J. Fontcuberta, High mobility conduction at (110) and (111) LaAlO$_3$/SrTiO$_3$ interfaces. Sci. Rep. **2**, 758 (2012)
79. S. Raghavan, J.Y. Zhang, S. Stemmer, Two-dimensional electron liquid at the (111) SmTiO$_3$/SrTiO$_3$ interface. Appl. Phys. Lett. **106**, 132104 (2015)
80. W.E. Doennig, D. Pickett, R. Pentcheva, Massive symmetry breaking in LaAlO$_3$/SrTiO$_3$ (111) quantum wells: a three-orbital strongly correlated generalization of graphene. Phys. Rev. Lett. **111**, 126804 (2013)

81. D. Xiao, W. Zhu, Y. Ran, N. Nagaosa, S. Okamoto, Interface engineering of quantum hall effects in digital transition metal oxide heterostructures. Nat. Commun. **2**, 596 (2011)
82. Z. Wang, Z. Zhong, S. McKeown Walker, Z. Ristic, J.-Z. Ma, F.Y. Bruno, S. Riccò, G. Sangiovanni, G. Eres, N.C. Plumb, L. Patthey, M. Shi, J. Mesot, F. Baumberger, M. Radović, Atomically precise lateral modulation of a two-dimensional electron liquid in anatase TiO_2 thin films. Nano Lett. **17**(4), 2561–2567 (2017)
83. S. McKeown Walker, Two dimensional electron liquids at oxide surfaces studied by angle resolved photoemission spectroscopy. Ph.D. thesis, University of Geneva, 24 Quai Ernest-Ansermet, Geneva, CH-1211, Switzerland (2016)
84. H. Nakamura, T. Kimura, Electric field tuning of spin-orbit coupling in $KTaO_3$ field-effect transistors. Phys. Rev. B **80**, 121308(R) (2009)
85. K. Ueno, S. Nakamura, H. Shimotani, H.T. Yuan, N. Kimura, T. Nojima, H. Aoki, Y. Iwasa, M. Kawasaki, Discovery of superconductivity in $KTaO_3$ by electrostatic carrier doping. Nat. Nanotechnol. **6**, 408–412 (2011)
86. K. Zou, S. Ismail-Beigi, K. Kisslinger, X. Shen, D. Su, F.J. Walker, C.H. Ahn, $LaTiO_3/KTaO_3$ interfaces: a new two-dimensional electron gas system. APL Mater. **3**, 036104 (2015)
87. S. LaShell, B.A. McDougall, E. Jensen, Spin Splitting of an Au(111) surface state band observed with angle resolved photoelectron spectroscopy. Phys. Rev. Lett. **77**, 3419–3422 (1996)
88. P.D.C. King, R. Hatch, M. Bianchi, R. Ovsyannikov, C. Lupulescu, G. Landolt, B. Slomski, J.H. Dil, D. Guan, J. Mi, E. Rienks, J. Fink, A. Lindblad, S. Svensson, S. Bao, G. Balakrishnan, B. Iversen, J. Osterwalder, W. Eberhardt, F. Baumberger, P. Hofmann, Large tunable Rashba spin splitting of a two-dimensional electron gas in Bi2Se3. Phys. Rev. Lett. **107**, 96802 (2011)
89. R. Winkler, *Spin-orbit Coupling Effects in Two-Dimensional Electron and Hole Systems* (Springer, Berlin, 2003)
90. U. Diebold, The surface science of titanium dioxide. Surf. Sci. Rep. **48**, 53–229 (2003)
91. M. Minohara, T. Tachikawa, Y. Nakanishi, Y. Hikita, L.F. Kourkoutis, J.-S. Lee, C.-C. Kao, M. Yoshita, H. Akiyama, C. Bell, H.Y. Hwang, Atomically engineered metal-insulator transition at the $TiO_2/LaAlO_3$ heterointerface. Nano Lett. **14**, 6743–6746 (2014)
92. T. Sarkar, K. Gopinadhan, J. Zhou, S. Saha, J.M.D. Coey, Y.P. Feng, T.Venkatesan Ariando, Electron transport at the TiO_2 surfaces of rutile, anatase, and strontium titanate: the influence of orbital corrugation. ACS Appl. Mater. Interfaces **7**(44), 24616 (2015)
93. P.D.C. King, H.I. Wei, Y.F. Nie, M. Uchida, C. Adamo, S. Zhu, X. He, I. Božović, D.G. Schlom, K.M. Shen, Atomic-scale control of competing electronic phases in ultrathin $LaNiO_3$. Nat. Nanotechnol. **9**, 443–447 (2014)
94. K. Yoshimatsu, T. Okabe, H. Kumigashira, S. Okamoto, S. Aizaki, A. Fujimori, M. Oshima, Dimensional-crossover-driven metal-insulator transition in srvo3 ultrathin films. Phys. Rev. Lett. **104**, 147601 (2010)
95. E. Mikheev, B. Himmetoglu, A.P. Kajdos, P. Moetakef, T.A. Cain, C.G. Van De Walle, S. Stemmer, Limitations to the room temperature mobility of two- and three-dimensional electron liquids in $SrTiO_3$. Appl. Phys. Lett. **106**, 062102 (2015)
96. I. Pallecchi, F. Telesio, D. Li, A. Fête, S. Gariglio, J.-M. Triscone, A. Filippetti, P. Delugas, V. Fiorentini, D. Marré, Giant oscillating thermopower at oxide interfaces. Nat. Commun. **6**, 6678 (2015)
97. J.M. Edge, Y. Kedem, U. Aschauer, N.A. Spaldin, A.V. Balatsky, Quantum critical origin of the superconducting dome in $SrTiO_3$. Phys. Rev. Lett. **115**, 247002 (2015)
98. L.P. Gor'kov, Phonon mechanism in the most dilute superconductor: n-type $SrTiO_3$. Proc. Natl. Acad. Sci. U. S. Am. **113**, 4646–4651 (2015)
99. J.T. Devreese, A.S. Alexandrov, Fröhlich polaron and bipolaron: recent developments. Rep. Prog. Phys. **72**, 066501 (2009)
100. M. Itoh, R. Wang, Quantum ferroelectricity in $SrTiO_3$ induced by oxygen isotope exchange. Appl. Phys. Lett. **76**, 221 (2000)
101. S.E. Rowley, L.J. Spalek, R.P. Smith, M.P.M. Dean, M. Itoh, J.F. Scott, G.G. Lonzarich, S.S. Saxena, Ferroelectric quantum criticality. Nat. Phys. **10**, 367–372 (2014)

102. J.T. Devreese, S.N. Klimin, J.L.M. van Mechelen, D. van der Marel, Many-body large polaron optical conductivity in Nb:SrTiO$_3$. Phys. Rev. B **81**, 125119 (2010)
103. Y.J. Chang, A. Bostwick, Y.S. Kim, K. Horn, E. Rotenberg, Structure and correlation effects in semiconducting SrTiO$_3$. Phys. Rev. B **81**, 235109 (2010)
104. W. Meevasana, X.J. Zhou, B. Moritz, C.-C. Chen, R.H. He, S.-I. Fujimori, D.H. Lu, S.-K. Mo, R.G. Moore, F. Baumberger, T.P. Devereaux, D. van der Marel, N. Nagaosa, J. Zaanen, Z.-X. Shen, Strong energy-momentum dispersion of phonon-dressed carriers in the lightly doped band insulator SrTiO$_3$. New J. Phys. **12**, 023004 (2010)
105. A.S. Mishchenko, N.V. Prokofev, A. Sakamoto, B.V. Svistunov, Diagrammatic quantum Monte Carlo study of the Fröhlich polaron A. Phys. Rev. B **62**, 6317–6336 (2000)
106. A.D. Caviglia, S. Gariglio, N. Reyren, D. Jaccard, T. Schneider, M. Gabay, S. Thiel, G. Hammerl, J. Mannhart, J.-M. Triscone, Electric field control of the LaAlO$_3$/SrTiO$_3$ interface ground state. Nature **456**, 624–627 (2008)
107. G. Verbist, F.M. Peeters, J.T. Devreese, Extended stability region for large bipolarons through interaction with multiple phonon branches. Ferroelectrics **130**, 27–34 (1992)
108. H. Boschker, C. Richter, E. Fillis-Tsirakis, C.W. Schneider, J. Mannhart, Electron–phonon coupling and the superconducting phase diagram of the LaAlO$_3$–SrTiO$_3$ interface. Sci. Rep. **5**, 12309 (2015)
109. C. Cancellieri, A.S. Mishchenko, U. Aschauer, A. Filippetti, C. Faber, O.S. Barisic, V.A. Rogalev, T. Schmitt, N. Nagaosa, V.N. Strocov, Polaronic metal state at the LaAlO$_3$/SrTiO$_3$ interface. Nat. Commun. **7**, 10386 (2016)
110. A. Fête, S. Gariglio, C. Berthod, D. Li, D. Stornaiuolo, M. Gabay, J.-M. Triscone, Large modulation of the Shubnikov-de Haas oscillations by the Rashba interaction at the LaAlO$_3$/SrTiO$_3$ interface. New J. Phys. **16**, 112002 (2014)
111. A. Fête, Magnetotransport experiments at the LaAlO$_3$/SrTiO$_3$ interface. Ph.D. thesis, University of Geneva, 24 Quai Ernest-Ansermet, Geneva, CH-1211, Switzerland (2014)
112. K.A. Brown, S. He, D.J. Eichelsdoerfer, M. Huang, I. Levy, H. Lee, S. Ryu, P. Irvin, J. Mendez-Arroyo, C.-B. Eom, C.A. Mirkin, J. Levy, Giant conductivity switching of LaAlO3/SrTiO3 heterointerfaces governed by surface protonation. Nat. Commun. **7**, 10681 (2016)
113. P. Scheiderer, F. Pfaff, J. Gabel, M. Kamp, M. Sing, R. Claessen, Surface-interface coupling in an oxide heterostructure: impact of adsorbates on LaAlO$_3$/SrTiO$_3$. Phys. Rev. B **92**, 195422 (2015)
114. A. Tebano, E. Fabbri, D. Pergolesi, G. Balestrino, E. Traversa, Room-temperature giant persistent photoconductivity in SrTiO$_3$/LaAlO$_3$ heterostructures. ACS Nano **6**, 1278–1283 (2012)
115. C. Cen, S. Thiel, G. Hammerl, C.W. Schneider, K.E. Andersen, C.S. Hellberg, J. Mannhart, J. Levy, Nanoscale control of an interfacial metal-insulator transition at room temperature. Nat. Mater. **7**, 298–302 (2008)
116. G. Cheng, M. Tomczyk, S. Lu, J.P. Veazey, M. Huang, P. Irvin, S. Ryu, H. Lee, C.-B. Eom, C.S. Hellberg, J. Levy, Electron pairing without superconductivity. Nature **521**, 196–199 (2015)

Chapter 5
Photoelectron Spectroscopy of Transition-Metal Oxide Interfaces

M. Sing and R. Claessen

Abstract Transition metal oxides exhibit a plethora of intrinsic functionalities like superconductivity, magnetism or multiferroicity. To put these to practical use requires the integration of suited oxide materials within thin film structures where the active regions with switchable and tunable physical properties often are the very interfaces. Fundamental knowledge on the chemical and electronic interface structure is key to design target properties for working devices. Here we will show that photoelectron spectroscopy is a powerful tool to obtain such kind of information if high enough photon energies in the soft and hard X-ray regime are employed to enhance the probing depth and hence get access to the electronic structure of buried layers and interfaces.

5.1 Introduction

As was pointed out in the introductory Chap. 1, the big challenge posed to photoelectron spectroscopy of transition-metal oxide interfaces is two-fold: first, to access the interface, which is buried under a typically several nanometers thick overlayer, at all, given the notoriously low information depth of photoemission; and second, to extract information on only the interface region, although photoemission integrates electrons basically from all depths, the signal being damped exponentially with increasing distance from the surface.

One can overcome the first quite simply going to higher photon, i.e., photoelectron kinetic energies, thereby making use of the longer inelastic mean free path of electrons in solids according to what is called the "universal curve" [1]. To cope with the second, in general, is less straightforward. Hence, a major part of this chapter will

---

M. Sing (✉) · R. Claessen
Physikalisches Institut and Röntgen Center for Complex Material Systems (RCCM),
Universität Würzburg, Am Hubland, 97074 Würzburg, Germany
e-mail: sing@physik.uni-wuerzburg.de

R. Claessen
e-mail: claessen@physik.uni-wuerzburg.de

deal with the question of how interface contrast can be achieved in a specific case and what kind of information about the electronic and chemical interface properties can be deduced from the various variants of photoelectron spectroscopy when applied to buried transition-metal oxide interfaces. Here we will confine ourselves to angle-integrated photoemission with hard and soft X-rays. Soft X-ray based angle-resolved work is reviewed in Chap. 6.

5.2 Depth Profiling

Historically, the first photoelectron spectra were recorded to examine core-levels because even with limited energy resolution ($\sim 1 - 2$ eV) and moderate photon densities as provided by lab X-ray sources valuable information can be obtained, e.g., on the chemical surface composition or the valence state of an atomic species in the sample [2]. Suitable core-levels such as O $1s$ or TM $2p$ (TM: transition metal) are accessible with standard emission lines like Al K_α ($h\nu = 1486.6$ eV) that warrant high enough probing depth to ensure interface sensitivity. In addition, compared to valence orbitals such as O $2p$ or TM $3d$ their photoabsorption cross sections are large at these photon energies and they exhibit completely filled shells, resulting in high count rates and superior statistics. Although core electrons are tightly bound to the atomic nuclei and occupy states which largely retain their atomic character also in a solid, core electron photoemission spectra contain distinct information also on the valence states [2]. For instance, the relative shift of binding energies of a core level of a certain element, the so-called chemical shift, reflects its oxidation state, i.e., the occupancy of the valence shell or the chemical environment. An asymmetric tail to higher binding energies may reflect the coupling to conduction band excitations, pointing to a metallic sample. Additional lines in core level spectra may originate from multiplet excitations due to the coupling of the core hole to the valence shell or from charge transfer from neighboring ligand atoms. Finally, at interfaces and surfaces the redistribution of valence charge can lead to band bending which again causes a spatially dependent shift of core level binding energies within the space charge region.

The observation of a conducting interface in transport measurements of the prototypical LaAlO$_3$/SrTiO$_3$ heterostructure and its tentative explanation as induced by electronic reconstruction (cf. Chap. 2) immediately raised questions regarding the orbital character of the states hosting the mobile electrons, the extension of the conducting region and the charge carrier density. From the above it appears quite natural to look at core level spectra by hard X-ray photoelectron spectroscopy (HAXPES) to gain insight into the microscopic nature of the conducting electron system in LaAlO$_3$/SrTiO$_3$.

Figure 5.1b and c display spectra of the spin-orbit-split Ti $2p$ core level for two samples, recorded with a photon energy of 3 keV at various photoelectron emission angles θ with respect to the surface normal. The interesting feature, containing information on the valence electronic structure, is the tiny spectral weight at the lower

5 Photoelectron Spectroscopy of Transition-Metal Oxide Interfaces

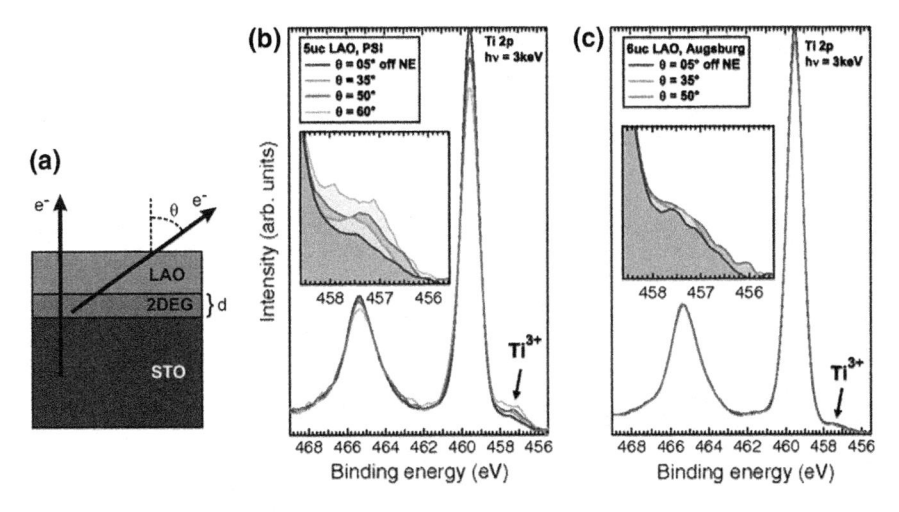

Fig. 5.1 **a** Sketch of the measuring geometry for angle-dependent hard X-ray photoemission experiments on a LaAlO$_3$/SrTiO$_3$ heterostructure with an interface two-dimensional electron gas (2DEG). **b, c**: Ti $2p$ spectra of two different LaAlO$_3$/SrTiO$_3$ heterostructures with varying emission angle θ, measured with respect to normal emission (NE). (from [3])

binding energy side of the main line. It stems from the emission of photoelectrons in the Ti $2p$ shell as well, but of Ti ions with an extra electron in the otherwise empty $3d$ shell of undoped SrTiO$_3$. The additional screening of the core potential by such an extra electron causes a chemical shift of the core level binding energies to lower values. The tiny spectral weight hence is due to Ti ions in the 3+ oxidation state (instead of 4+ as in undoped SrTiO$_3$) and therefore evidence for the two-dimensional electron system at the interface. Its intensity is a measure of the density of Ti $3d$ electrons, i.e., the charge carrier density of the interfacial electron system.

From the spectra in Fig. 5.1 it is also seen that the Ti^{3+} emission increases with photoelectron emission angle θ. Since according to the so-called universal curve the inelastic mean free path of electrons in solids is limited to only several nanometers even at kinetic energies of about 3 keV (cf. Chap. 1), the information depth of photoemission can be effectively changed if the detection angle of photoelectrons is varied from normal emission (most bulk sensitive) towards grazing emission (most surface sensitive) according to: $\lambda_{eff} = \lambda_{IMFP} \cdot \cos\theta$ where λ_{eff} is the inelastic mean free path for a given electron kinetic energy (see Fig. 5.1a). Therefore, the angle dependence seen in the spectra of Fig. 5.1 means that the vertical extension of the conducting region at the interface is of the order of the inelastic mean free path. Was the extension significantly larger, no angle dependence would be observed.

The quantitative angle dependence of the Ti^{3+}/Ti^{4+} ratio can be used to infer estimates for the sheet carrier density and the thickness of the two-dimensional electron system. To this end a model for the vertical charge distribution in the conducting layer has to be assumed, in the most simple case just a homogeneous electron distribution of thickness d. This situation is sketched in 5.1a. To simulate the angle dependence

Fig. 5.2 Schematic band diagram of a LaAlO$_3$/SrTiO$_3$ heterostructure. Drawn are the band edges with a finite slope across the film for the standard electronic reconstruction scenario and flat bands as derived from photoemission. The valence band edges are traced by film core levels. For both situations simulated photoemission spectra are depicted, taking into account relative energetic shifts and exponential damping of the photoemission spectra of each layer. (from [5])

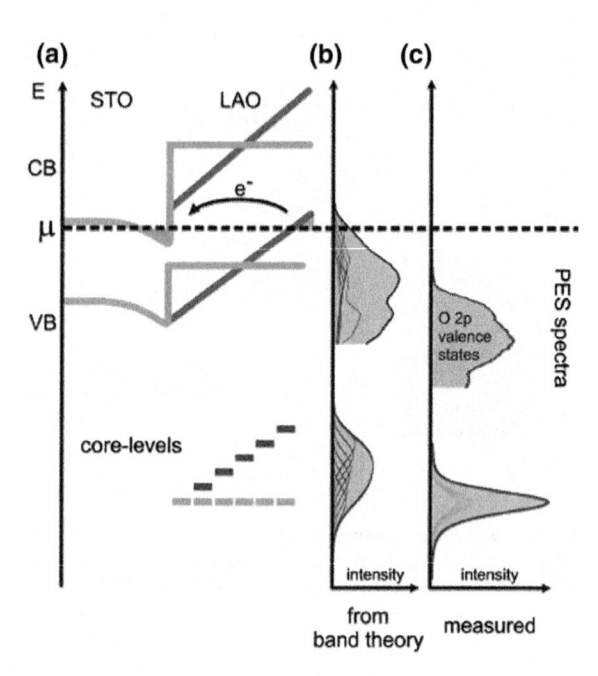

of the Ti^{3+}/Ti^{4+} ratio one only needs to integrate the contributions to the Ti^{3+} and Ti^{4+} emission from the respective Ti ions according to their vertical distribution, taking the exponential damping factor $e^{-z/\lambda_{eff}}$ for photoelectrons emitted at a depth z below the interface into account [3]. The thickness-dependent sheet carrier densities inferred from such an analyis are about an order of magnitude lower than expected in an ideal electronic reconstruction scenario. Furthermore, they do not exhibit a sharp step at the critical thickness as is seen in Hall measurements [4] but rather display a continuous increase with growing LaAlO$_3$ thickness. Finally, the conducting region turns out to comprise only a few unit cells at room temperature.

Indeed, the most convincing proof of the standard electronic reconstruction scenario probably would be the direct observation of the built-in potential which should be detectable in photoemission as is sketched in Fig. 5.2 [5]. The potential gradient across the film will shift the LaAlO$_3$ valence band edge towards and above the chemical potential. As a result, spectral weight should appear in photoemission at and near the chemical potential when holes are doped into the O $2p$-derived states at the very surface due to the redistribution of electrons to the interface. Similarly, the LaAlO$_3$-related core levels will track the slope of the valence band edge. This in turn will lead to asymmetrically broadened line shapes with shifted energies relative to the case of flat bands.

Looking first at the valence band photoemission spectrum of a metallic sample in Fig. 5.3a, a clear gap is observed between the valence band edge and the chemical potential. The gap size reflects the pristine band gap of SrTiO$_3$ modulo a band offset (cf. Sect. 5.3). Obviously, the expectation regarding the shift of the LaAlO$_3$ valence

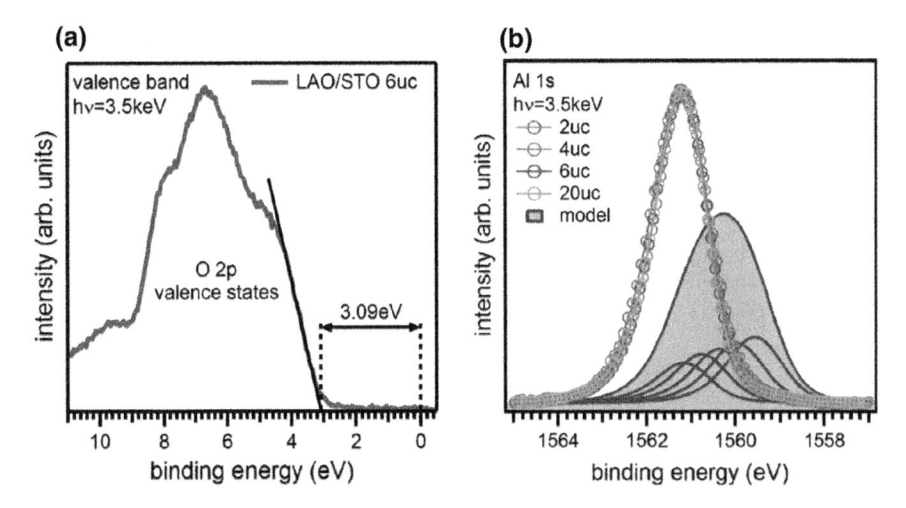

Fig. 5.3 a Valence band spectrum exhibiting a finite gap between valence band edge and chemical potential. **b** Al $1s$ core-level spectra for various thicknesses of the LaAlO$_3$ film. No changes in line shape and energy are observed. A model spectrum as expected in the presence of an internal field is also shown. (from [5])

band edge is not matched by experiment. The same is true if one analyses the Al $1s$ core level spectra depicted in Fig. 5.3b. The line shapes and energies are unaltered for samples with various film thicknesses, ranging from subcritical to far above the threshold for metallicity. If the standard electronic reconstruction scenario holds, all samples should exhibit a different potential gradient, being reflected in distinctly different core level spectra. A simulated spectrum for a sample with a 6 unit cells thick film is shown for comparison. In conclusion, a built-in potential is absent in photoemission on LaAlO$_3$/SrTiO$_3$. Interestingly, on the related LaCrO$_3$/SrTiO$_3$ heterostructure with a polar discontinuity both core level and valence band photoemission clearly indicate a built-in potential, confirming that photoemission indeed is capable to detect it if present [6].

5.3 Band Bending and Offset

In the strive towards exploiting oxide heterointerfaces in future electronics[1] further important parameters to know and finally control beside charge carrier concentration and extension of the conducting region are the band offset at the interface as well as the bending in the space charge region. This information can also readily be extracted from HAXPES of core-levels and valence band.

---

[1] Confer the famous saying "Often it may be said that *the interface is the device.*" by H. Kroemer in his Nobelprize lecture. Clearly, at this time — in 2000 — he referred to the non-oxide semiconductor heterostructures that have been revolutionizing our daily life since the 70s to date.

Fig. 5.4 Decomposition of the valence band spectrum of a LaAlO$_3$/SrTiO$_3$ heterostructure into the contributions of LaAlO$_3$ and SrTiO$_3$. The band offset can directly by read off from the relative shift of the O 2p derived valence band edges of the LaAlO$_3$ and SrTiO$_3$ spectra. (from [5])

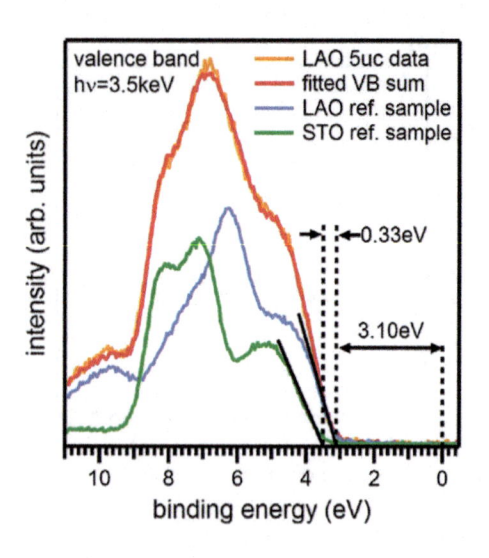

The method by Kraut et al. [7] represents the most simple approach to determine the band offset and was first applied by Segal and coworkers to LaAlO$_3$ films grown by molecular beam epitaxy on SrTiO$_3$ [8] substrates. The band offset is determined from the energy difference of two core-levels that are specific for substrate and film of a heterostructure. This is compared with the energy difference of the same two core levels in the bulk constituents where the energies are referred to the respective valence band edges. If the band offset is zero the energy differences of these core-levels will equal each other. From the analysis of the La $4d_{5/2}$-Sr $3d_{5/2}$ pair of core levels Segal and coworkers find that the valence band maximum of the LaAlO$_3$ film lies 0.35 eV below that of SrTiO$_3$. In contrast, other studies on films grown by pulsed laser deposition find it above with a certain scatter of the reported band offsets [3, 9–11]. While a systematic investigation of the band offsets in dependence of growth conditions and sample properties is lacking, it is clear that variations in the charge distribution due to, e.g., oxygen vacancies or cation intermixing and off-stoichiometries will cause different band alignments and bendings at the interface.

More laborious ways to extract information about the band alignment use some kind of data fitting with or without additional model assumptions. As an example, it is illustrated in Fig. 5.4 for a LaAlO$_3$/SrTiO$_3$ heterostructure with a 5 unit cell thick film how the valence band offset can be deduced from a decomposition of the valence band spectrum into the individual contributions from LaAlO$_3$ and SrTiO$_3$ obtained from photoemission measurements recorded under the same conditions on reference samples. The fitting parameters of the superposition of the reference spectra to the LaAlO$_3$/SrTiO$_3$ valence band spectrum are their relative energetic shift and intensities [5]. The linear extrapolation of the leading edges of the O 2p derived valence states to zero baseline as indicated in the figure can be taken as measure of the valence band offset. Sign and modulus agree well with the results from the method by Kraut et al. [5].

5 Photoelectron Spectroscopy of Transition-Metal Oxide Interfaces 93

Fig. 5.5 Illustration of the analysis of band-bending effects for the Sr $3d$ core-level spectrum in γ-Al$_2$O$_3$/SrTiO$_3$. The emission from each SrO layer is exponentially damped and shifted in energy according to the bending potential $E_{bb}(z)$. The resulting spectrum is a superposition of the spectra within the bending zone and the bulk contribution from all other layers. (from [14])

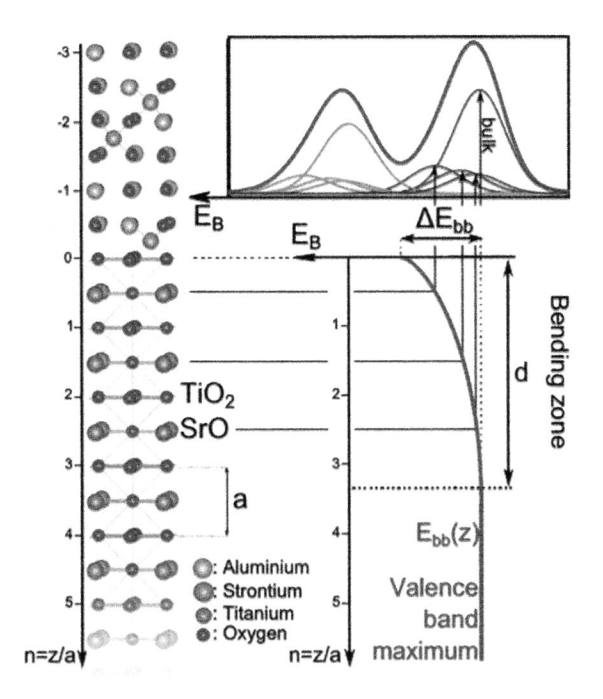

A quantitative account of the band bending in the space charge region of the SrTiO$_3$ substrate inevitably requires a fitting analysis along the lines already mentioned above. In a rigid band picture, the core-level energies will trace the bent valence band maximum causing a slight asymmetry and energy shift of the line shapes in the photoemission spectra (see Fig. 5.2). In the following we exemplify the procedure in case of the heterostructure γ-Al$_2$O$_3$ [12, 13] which is closely related to the LaAlO$_3$/SrTiO$_3$ system. It likewise exhibits a non-polar/polar heterointerface with a polar discontinuity of similar size, the film crystal structure being a defect spinel instead of a perovskite as in LaAlO$_3$/SrTiO$_3$. Although it displays a critical thickness for the formation of a conducting interface as well, conductivity is quenched by post-oxidation in contrast to LaAlO$_3$/SrTiO$_3$, indicating that oxygen vacancies at the interface act as charge reservoir (cf. Sect. 5.4). The interest in this heterointerface is because—if grown under suitable conditions—an order of magnitude higher mobilities and charge carrier concentrations compared to LaAlO$_3$/SrTiO$_3$ can be achieved.

In Fig. 5.5a cross-section of the heterostructure is sketched with the layers above and below the interface ($n = 0$) numbered by integers. The fit function is a superposition of pristine core-level line shapes (Voigt profiles), one for each layer in the bending region, that are shifted in energy according to the assumed bending potential $E_{bb}(z)$ and weighted by an exponential damping factor $e^{-z/\lambda_{eff}}$ where z denotes the depth of the layer with respect to the interface and λ_{eff} is the inelastic mean free path. A further contribution has to be added representing the signal from the bulk,

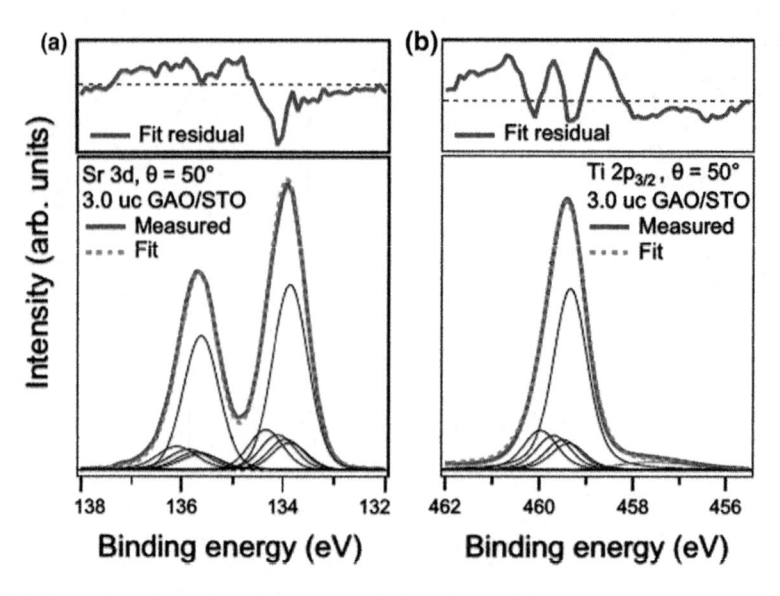

Fig. 5.6 Exemplary fits and residuals for the Sr $3d$ (a) and Ti $2p_{3/2}$ (b) core-level spectra of a γ-Al$_2$O$_3$/SrTiO$_3$ heterostructure. For details see text. (from [14])

i.e., all other layers outside the bending region, by summing up the unshifted line shapes with only the damping being taken into account. The fitting variables comprise all parameters needed to describe the bending potential and an additional one for normalization. The relative weights of all contributions are fixed by the damping factors.

Figure 5.6a, b shows typical fit results for the Sr $3d$ doublet and the Ti $2p_{3/2}$ line, respectively, assuming a bending potential within the Schottky approximation of the form $E_{bb}(z) = \Delta E_{bb}(z/d - 1)^2$ for $0 < z < d$ and 0 elsewhere. As can be judged from the spectra the band-bending effects are subtle. To obtain robust results a fitting scheme has to be implemented that minimizes χ^2 for all measured peaks at once. In the example shown the global fitting was applied to 48 peaks with the number of independent fit variables reduced by physical constraints to 47, i.e., $\sim$1 per peak [14].

From this fitting analysis the behavior of the bending potential at the interface can be deduced. Together with the results from the analysis of band offset and potential gradient and with literature values for the band gaps of γ-Al$_2$O$_3$ and SrTiO$_3$ the complete band arrangement of the heterostructure can be obtained as summarized in Fig. 5.7. The maximum depth of the bending amounts to about 600 meV with the bending zone stretching about 1.5 nm into the substrate. The valence band maximum of γ-Al$_2$O$_3$ lies about 600 meV below that of SrTiO$_3$, meaning that both are aligned at the interface.

Fig. 5.7 Band alignment and bending at the spinel/perovskite heterointerface of γ-Al$_2$O$_3$/SrTiO$_3$ as deduced from hard X-ray photoemission measurements. (from [14])

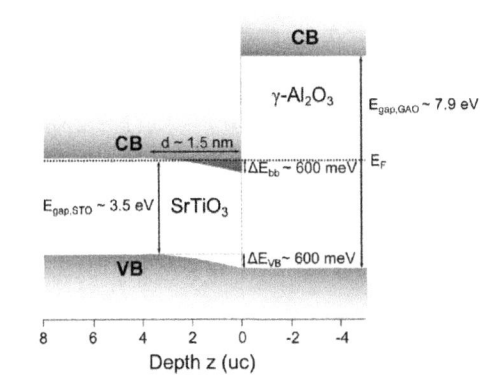

5.4 Role of Oxygen Vacancies

As already discussed in the introductory chapters of this book, the physical origin of the interfacial two-dimensional electron system in LaAlO$_3$/SrTiO$_3$ is still a matter of debate. While it is tempting to attribute its formation to an electronic reconstruction resulting from the polar discontinuity at the interface, it cannot be ruled out that chemical effects such as cation intermixing and/or off-stochiometries play a crucial or at least additional role. This concerns in particular the effect of oxygen vacancies (O$_{vac}$) which in oxides are known to act as electron donors. In fact, it has recently been established that oxygen depletion of bare surfaces of SrTiO$_3$ leads to formation of a metallic 2D surface band [15–17]. It therefore will not come as a surprise that oxygen defects also affect the properties of the two-dimensional electron system in LaAlO$_3$/SrTiO$_3$. For example, varying the oxygen pressure during PLD growth of the heterostructure, thereby adjusting its actual O-stoichiometry, will cause the sheet resistance to change by several orders of magnitude [18]. The effect of O$_{vac}$ doping can also be traced directly in the Ti 2p spectrum as depicted in Fig. 5.8. The Ti^{3+} spectral weight strongly increases with decreasing growth pressure, indicating that the lack of oxygen during the PLD process leads to formation of oxygen vacancies, which in turn causes electron doping—and hence a valence change—of an increasing fraction of Ti ions. In order to remove these vacancies, i.e. fully oxidize the heterostructure, a postoxidation procedure is applied, in which the samples are exposed to an extreme O$_2$-atmosphere of 500 mbar immediately after growth while cooling down to room temperature [19]. The effect is shown in Fig. 5.8c. Interestingly, a stable and reproducible Ti^{3+} signal is observed, despite the absence of O$_{vac}$-doping and independent of the actual O-stoichiometry before postoxidation. The residual concentration of interfacial Ti 3d electrons even for a fully oxidized heterostructure thus already hints at the existence of an *intrinsic* mechanism for the formation of a conducting interface in LaAlO$_3$/SrTiO$_3$, distinct from simple O$_{vac}$ doping. We note in passing that great care has to be taken in this high-pressure postoxidation procedure to avoid water or hydrogen surface contamination, as

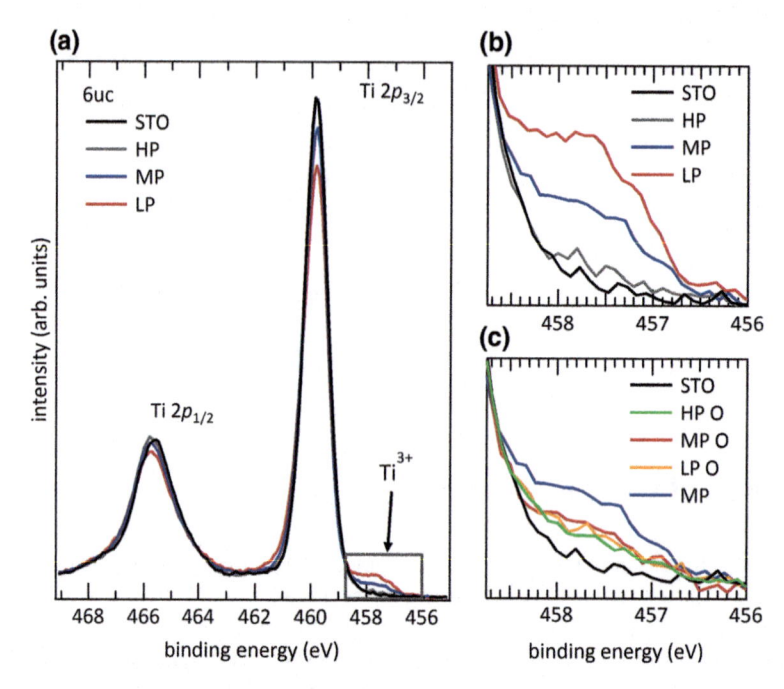

Fig. 5.8 a Ti $2p$ spectra of LaAlO$_3$/SrTiO$_3$ heterostructures grown by pulsed laser deposition (PLD) at different oxygen pressures (LP = low pressure, 10^{-5} mbar; MP = medium pressure, 10^{-3} mbar; HP = high pressure, 10^{-1} mbar) in comparison to a pristine SrTiO$_3$ substrate. **b** blow-up of the Ti^{3+} region. **c** Ti^{3+} region after in situ post-growth oxidation at 500 mbar O$_2$ (noted by "O")

surface protonation can also induce Ti $3d$-carriers in the LaAlO$_3$/SrTiO$_3$ interface [20, 21].

All electronic structure information on LaAlO$_3$/SrTiO$_3$ discussed so far has been obtained from core levels probed by HAXPES. However, the Ti $2p$ core level spectra alone cannot provide detailed information on origin and nature of the interfacial Ti $3d$ electrons beyond their bare existence. *Direct* spectroscopic access to these valence electrons is therefore highly desirable. However, HAXPES, despite its enhanced probing depth, cannot be applied for this purpose, as the optical (dipole, and at these high energies also quadrupole) transition matrix element for direct photoexcitation of the Ti $3d$ states (depicted in Fig. 5.9a) drops dramatically by several orders of magnitude when going to hard X-ray photon energies, making it much smaller than for the core levels. In combination with the buried location of the two-dimensional electron system and its low carrier density (as known from Hall effect measurements) no usable photoemission signal of the Ti $3d$ valence states can be obtained by HAXPES.

Fortunately, *resonant* photoemission with soft X-rays at the Ti L absorption edge provides an alternative mechanism to probe the relevant interface electrons (see Fig. 5.9b) [22, 23]. In this process the energy of the impinging photon is first

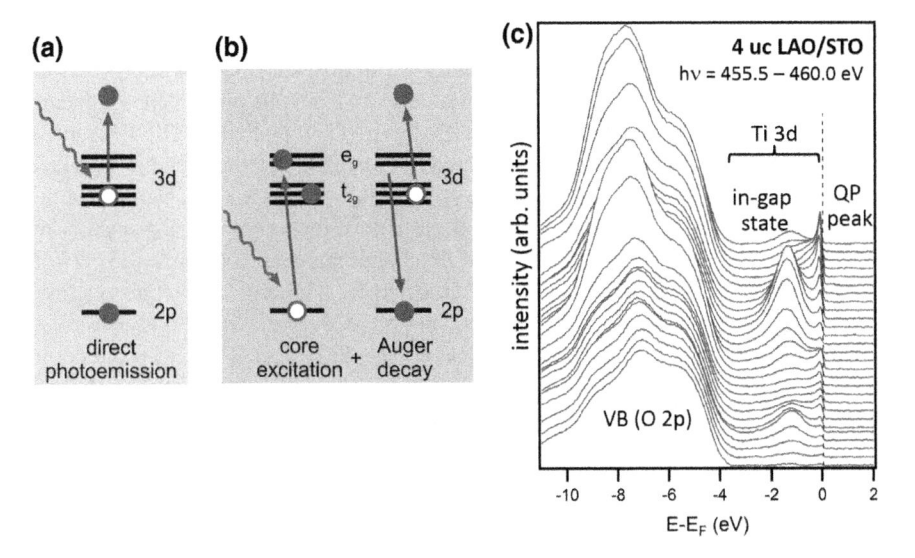

Fig. 5.9 **a** Schematic view of direct photoemission from a Ti $3d$ level and **b** a resonant photoemission process at the Ti L-edge, consisting of an initial excitation of a Ti $2p$ core electron into the $3d$ shell and a subsequent Auger decay (super-Coster-Kronig transition) with the same final state reached in the direct photoemission process. The example shown here corresponds to an e_g resonance. **c** Resonant photoemission spectra of a LaAlO$_3$/SrTiO$_3$ heterostructure around the Ti L-edge with the appearance of the Ti $3d$-derived quasiparticle (QP) and in-gap (IG) peaks

stored by exciting a $2p$ core electron into an empty $3d$ state (L-edge absorption). This intermediate excited configuration immediately decays via an Auger process, in which the $2p$ core hole is filled up again and one $3d$ electron is ejected into the vacuum. The final state reached in this way is identical to that of the direct photoemission process, but its intensity is determined by the transition matrix elements for the initial core excitation ($2p^6\ 3d^n \to 2p^5\ 3d^{n+1}$, n being the occupancy of the $3d$ shell, here: $n=1$) and the subsequent Auger decay ($2p^5\ 3d^{n+1} \to 2p^6\ 3d^{n-1}$ + photoelectron). Due to the high absorption probability at transition metal L-edges this excitation channel leads to a giant enhancement of the photoemission signal, and because the involved processes are all local, i.e. take place on the same atom, resonant photoelectron spectroscopy (ResPES) is element and orbital specific, in our case to the Ti $3d$ states.

Direct spectroscopy of the interface two-dimensional electron system in LaAlO$_3$/SrTiO$_3$ by ResPES has first been demonstrated by Drera et al. [24] at moderate energy resolution. In the meantime several dedicated soft X-ray beamlines have become available at various synchrotron radiation facilities worldwide, which allow high-resolution and high-countrate ResPES experiments of buried interfaces. As an example, Fig. 5.9c shows a series of resonant valence band spectra of a LaAlO$_3$/SrTiO$_3$ heterostructure taken in a narrow photon energy band around the Ti L-edge. Besides the mainly O $2p$-derived valence band pronounced intensity of two distinct features is observed in the gap region from the Fermi level to approximately 3.5 eV below, which display strong but different intensity variations with photon energy. From their appearance at the L-edge (and its absence away from this photon energy range) both features can unambiguously be identified as of Ti $3d$ character. The sharp peak observed directly at the Fermi energy is readily attributed to the metallic interface and is hence referred to as "quasiparticle (QP) peak". The *itinerant* character of the underlying states has been confirmed by angle-resolved photoelectron spectroscopy at the L-edge (SX-ARPES) which allows direct mapping out their k-space dispersion (see [25, 26] and the detailed discussion in Chap. 6). The broader feature centered around -1.3 eV, due to its location labeled "in-gap (IG) state", was already observed in early low-energy photoemission studies on bare $SrTiO_3$ crystals, where it could be related to the presence of oxygen vacancies [27]. As will be shown further below, this is also the case for the heterostructure. Indeed, more recent theoretical studies have demonstrated that O_{vac} as positively charged point defects are able to pull down $3d$ levels of the neighboring Ti atoms from the conduction band into the band gap, thereby creating *localized* Ti electrons [28–32].

The mobile versus localized character of the QP and IG peaks, respectively, is also reflected in their distinct photon energy dependence. Figure 5.10 displays the full ResPES signal of the gap region in Fig. 5.9c as false color map. Plotting the intensity of the QP and IG peaks (i.e. for their respective fixed binding energies) as function of photon energy yields the so-called constant initial state (CIS) spectra displayed in the center panel of Fig. 5.10. They look clearly quite different, and it is instructive to compare their spectral shape to the X-ray absorption spectra of

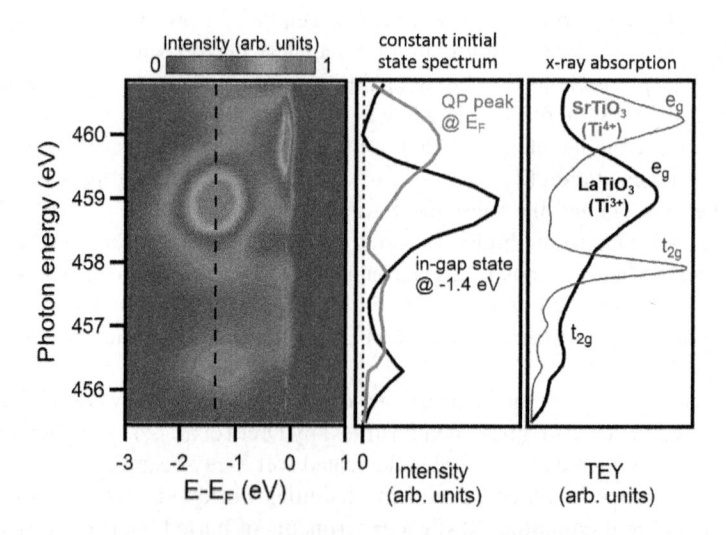

Fig. 5.10 Left: false color representation of the Ti L-edge ResPES spectra in Fig. 5.9c, plotted versus binding energy (only gap region) and photon energy. Center: constant initial state (CIS) spectra for the QP and IG peaks corresponding to the cuts along the broken lines, respectively, in the color map. Right: Ti L-edge spectra for a tetravalent ($SrTiO_3$) and a trivalent ($LaTiO_3$) titanate

trivalent and tetravalent Ti, as measured on LaTiO$_3$ (for Ti^{3+}) and SrTiO$_3$ (for Ti^{4+}), respectively (see right panel). While the CIS curve of the in-gap peak traces the Ti^{3+} absorption spectrum quite closely, in agreement with the static Ti $3d^1$ configuration of this localized state, the CIS spectrum of the quasiparticle peak is shifted towards higher photon energies by almost 1 eV and clearly contains spectral signatures of the Ti^{4+} absorption. In addition, the QP peak resonates over a wider energy range than the in-gap state. This phenomenology is known from ResPES on (conducting) $3d$ transition metals [33], indicating that it is a hallmark of itinerant $3d$ electrons.

As a common feature, interestingly, the QP and IG peaks both show their strongest enhancement at the e_g resonance of the Ti^{4+} and Ti^{3+} absorption spectra, respectively. Finally, resonance effects are observed also in the main valence band region (denoted as "VB (O $2p$)" in Fig. 5.9c). This behavior originates from the small, but finite O $2p$-Ti $3d$ hybridization of the valence band states. The enhanced resonance seen especially in the region between -6 and -9 eV is in excellent agreement with the partial Ti $3d$ density of states previously obtained from band calculations and X-ray standing wave HAXPES experiments [34].

Applying ResPES to the study of oxygen vacancies in LaAlO$_3$/SrTiO$_3$ requires a controlled way of adjusting the O$_{vac}$ concentration in the heterostructure. As already discussed for the HAXPES data in Fig. 5.8, this can be achieved by varying the oxygen pressure during PLD growth. However, this implies the preparation of separate samples for each vacancy concentration to be studied. A much simpler and systematic approach is the in situ control of the oxygen stoichiometry by the combined use of photon-induced oxygen depletion and oxygen dosing. It had already been known from previous photoemission experiments of bare SrTiO$_3$ surfaces that irradiation with photons beyond a certain threshold energy lead to a loss of oxygen [17, 35], most likely by a so-called Feibelman-Knotek process in which a core-excitation with subsequent interatomic Auger decay causes a deionization of the O^{2-} anion and hence a breaking of its ionic bond [36]. The effect is demonstrated by the Ti $2p$ HAXPES spectra in Fig. 5.11b. The series starts with a spectrum of a postoxidized (i.e., nominally O-stoichiometric) sample, which already shows some finite Ti^{3+} signal, partly due to the "intrinsic" mechanism as discussed above. With increasing X-ray exposure time the trivalent signal is seen to strongly grow, attributed to the photon-induced oxygen loss and the concomitant O$_{vac}$ doping of the interface (at work already during the measurement of the first spectrum and hence causing part of its Ti^{3+} signal). The effect can be reversed by dosing the surface with a steady low-pressure flow of molecular oxygen out of a metal capillary just in front of the sample (see Fig. 5.11a for a sketch of the experimental setup), as is clearly seen by the strongly reduced (but still finite) Ti^{3+} weight of the lowermost spectrum in Fig. 5.11b. Note that the O$_2$ flow is applied in presence of the X-ray beam, so that a large fraction of the molecules will photo-dissociate and diffuse as atomic oxygen through the LaAlO$_3$ film into the SrTiO$_3$ substrate. Similar dosing experiments have recently been conducted on bare SrTiO$_3$ surfaces and allowed a very systematic study of the metallic 2D surface states formed upon O depletion.

The HAXPES spectra in Fig. 5.11b have been taken at beamline I09 of the Diamond Light Source in the UK. This undulator beamline not only supplies a high-flux

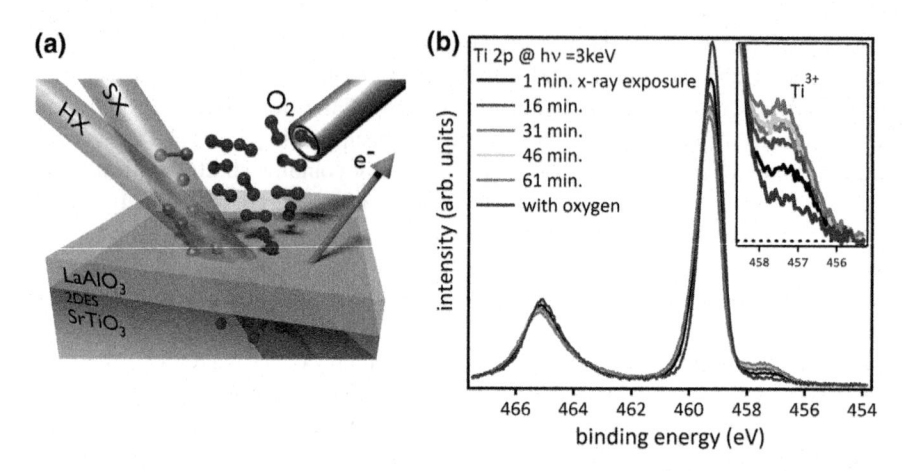

Fig. 5.11 a Schematic view of the oxygen dosing experiments conducted at beamline I09 of the Diamond Light Source (UK). Note that this beamline provides both a hard X-ray and a soft X-ray beam directed at the very same sample spot. b Ti $2p$ HAXPES spectra of a LaAlO$_3$/SrTiO$_3$ (4 unit cells) sample as function of X-ray exposure time and with oxygen dosing

hard X-ray (HX) beam, responsible for the rapid photon-induced O loss, but in a rather unique configuration provides an *additional* independent soft X-ray (SX) beam directed at the same sample spot [37]. In the following we will discuss Ti L-edge ResPES results on oxygen dosing measured with this SX beam, whereas the HX beam has been used for the actual O depletion [38]. It has to be noted that the SX photons also cause oxygen loss, however, due to its much higher flux (and energy) the HX beam proved to be much more efficient in that respect, with time scales for O depletion almost one order faster than for the SX beam.

The results of these ResPES experiments are summarized in Fig. 5.12. The photon energy was tuned either to maximally enhance the emission from the QP peak at the chemical potential (Fig. 5.12a) or the IG state at about 1.3 eV binding energy (Fig. 5.12b). The QP spectral weight increases upon irradiation with X-rays, i.e. progressive O depletion, starting from the most strongly oxidized state (blue curve) to the one with saturated O depletion (red curve), as can best be judged if measured on the QP resonance (Fig. 5.12a). Interestingly, the QP peak cannot be completely quenched under strongest O dosing. Since for the same state no IG weight can be discerned at the IG resonance photon energy—note that the remaining intensity tail reaching into the pristine gap stems from the LAO valence band—, we consider this state fully oxidized. The increase of QP intensity upon O depletion is paralleled by an increase of the IG spectral weight, and both can be fully reversed upon O dosing (cf. bottom and top graphs of Fig. 5.12b).

The qualitative picture drawn can be corroborated by a quantitative analysis as visualized in Fig. 5.12c. Here we plot the QP (pink) and IG intensities (green) versus irradiation time with the oxygen dosing switched off and on. The intensities were obtained from the areas under the PES spectra within the intervals marked in Fig. 5.12a and b. Again, the interesting observation is the persistent QP intensity even under full oxidation while the IG weight is completely suppressed. If this finite QP intensity is subtracted, the resulting curve (grey) and the IG intensity curve (green) are essentially congruent, meaning that the electrons provided by the oxygen vacancies contribute in a fixed proportion to mobile and immobile states, respectively [17].

Further, a comparison to results obtained in the same way on bare $SrTiO_3$ is elucidating. Here, a surface two-dimensional electron system is exclusively induced by oxygen vacancies and concomitant band bending [15–17, 27, 35, 39–41]. The spectra recorded at the QP resonance are displayed in Fig. 5.12d, the IG and QP intensities derived from a quantiative analysis in Fig. 5.12e. Similar to the heterostructure, QP and IG intensities increase and decrease in proportion upon O depletion and refilling, respectively. Strikingly, however, upon oxygen dosing both IG and QP weights are suppressed completely turning the surface insulating, in stark contrast to the $LaAlO_3/SrTiO_3$ heterostructure. We note in passing, that while these data have been taken on (001)-oriented surfaces and interfaces, identical results have been obtained for the (111)-orientation [38].

As a result, these O dosing experiments establish the fact that the introduction of oxygen vacancies leads to electron doping into two separate classes of Ti $3d$-derived states, namely localized and delocalized ones. This is consistent with recent theoretical studies which find the oxygen vacancies to generate Ti $3d$ states in the $SrTiO_3$ band gap. However, these states can only be singly occupied, as a combination of strong on-site Coulomb repulsion in the $3d$ shell and bonding-antibonding splitting between the Ti atoms on either side of the vacancy will push the next available local $3d$ level way above the conduction band minimum [29, 31, 32]. Thus, the "other" of the two electrons supplied by each O vacancy will populate the minimum of the (also Ti $3d$-like) conduction band and hence contribute to the metallic interface. The O dosing experiments unambiguously demonstrate that localized and mobile electrons are indeed induced in constant proportion, although the unknown photoemission cross sections for IG and QP peaks do not allow us to verify a simple 1:1 ratio as expected from these considerations. The clustering of vacancies, energetically favorable with respect to the formation of isolated O_{vac} defects, will have additional impact on the ratio between localized and mobile $3d$ electrons and also induce pronounced broadening of the IG peak [30].

It is tempting to relate the O_{vac}-induced generation of singly-occupied localized $3d$ states and, hence, formation of local spin 1/2 magnetic moments to the observation of ferromagnetic behavior in some $LaAlO_3/SrTiO_3$ samples [42, 43]. While the precise origin of magnetism has not yet been clarified, one could conceive a RKKY-like coupling between the local moments mediated by the itinerant carriers. In this picture the emergence of magnetism necessitates the presence of O vacancies which critically depends on the precise PLD growth conditions and sample history (e.g., storage conditions after PLD growth), which may explain why magnetic behavior is seen in some samples but not in others.

The most important result of the O dosing experiments is, however, the observation of a persistent quasiparticle component in fully oxidized $LaAlO_3/SrTiO_3$ heterostructures (above the critical thickness), whereas the localized states are completely

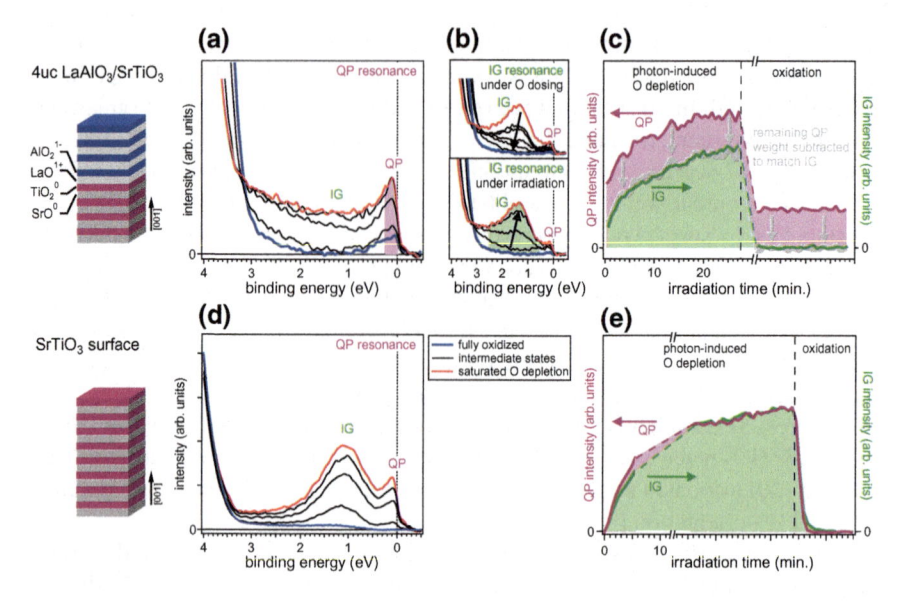

Fig. 5.12 Valence band photoemission spectra at different resonance energies for a metallic LaAlO$_3$/SrTiO$_3$ heterostructure and bare Nb-doped SrTiO$_3$ with various oxygen vacancy concentrations and quantitative behavior of the QP and IG intensities. **a** QP resonance spectra under irradiation with X-rays. **b** IG resonance spectra under irradiation only (bottom) and with simultaneous oxygen dosing (top). The blue curve corresponds to the fully oxidized state, the red to the one with saturated O depletion. The black arrows indicate the chronology of the measurements. **c** QP (pink curve) and IG (green curve) spectral weights plotted versus X-ray irradiation time with the oxygen dosing switched off and on. If the constant value for the persistent QP intensity in the fully oxidized state is subtracted (grey curve), the QP and IG intensity curves are essentially congruent. **d** Similar as (a), but for the bare (001) surface of Nb-doped SrTiO$_3$. **e** QP and IG intensities versus X-ray irradiation time for the bare surface. QP and IG features become fully suppressed upon oxidation. (from [38])

quenched. This clearly demonstrates the existence of a generic mechanism for the formation of a conductig interface in these heterostructures distinct from the simple O_{vac} doping seen on the bare SrTiO$_3$ surface. This behavior is also different from that seen in heterostructures comprising an *amorphous* LaAlO$_3$ film (*a*-LaAlO$_3$/SrTiO$_3$). While these samples also display a two-dimensional electron system beyond a critical film thickness, it results exclusively from O_{vac} doping and can be completely suppressed by full oxidation, as seen both in transport [44, 45] and photoemission (not shown here, [38]). This proves that the generic quasiparticle component in epitaxial heterostructures does not only require the presence of the LaAlO$_3$ film alone, but also its crystalline and hence polar order, which in turn identifies again the polar discontinuity at the interface as essential driving factor. As however already discussed in the previous section, the simple electronic reconstruction scenario based on that premise is in conflict with the failure of photoemission to detect any built-in potential in the LaAlO$_3$ film [5, 8, 46].

An alternative and energetically less costly scenario is the formation of electronically active defects at the very film surface [20, 21, 47–50]. They are generated at and above the critical thickness when the energy gain by transferring the released charge to the interface outweighs their formation enthalpy. As a consequence of these positively charged defects and the corresponding negative charge at the interface, the potential gradient in the film is compensated and the observed flat-band situation occurs. The elucidation of the precise nature of this generic mechanism is still an ongoing challenge (see also Chap. 2 for a discussion of the various models), which in SrTiO$_3$-based heterostructures—as discussed, e.g., in the context of possible magnetism—is often further complicated by this material's propensity to easily form oxygen vacancies. Fortunately, high-energy photoemission represents a valuable tool to identify and single out the effects of O$_{vac}$-doping.

5.5 Conclusions and Outlook

As we have demonstrated for selected examples photoelectron spectroscopy employing soft and hard X-rays is a powerful method for the investigation of functional oxide layered structures where the active region for potential applications is often buried several nanometers below the surface. It allows for the identification of electronic states and their orbital character as well as for chemical depth profiling and the characterization of the interface and film electronic structure in terms of charge carrier concentrations and band alignment. Such information is key for tailored and switchable functionalities of oxide heterostructures and superlattices. Of particular use is the capability of resonant soft X-ray photoemission to probe oxygen vacancy induced defect states. These may turn out to become technologically relevant in device concepts based on a spatial separation of donors and conducting layers to obtain higher mobilities.

Acknowledgements We gratefully acknowledge financial support by the Deutsche Forschungsgemeinschaft (FOR 1162, FOR 1346, SFB 1170) and the German Federal Ministry of Education and Research. The synchrotron experiments were conducted at BESSY II (Helmholtz-Zentrum Berlin, Germany), PETRA III (Deutsches Elektronen-Synchrotron, Hamburg, Germany), Swiss Light Source (Paul Scherrer Institut, Villigen, Switzerland) and Diamond Light Source (Harwell Science and Innovation Campus, Didcot, United Kingdom).

References

1. S. Suga, A. Sekiyama, *Photoelectron Spectroscopy - Bulk and Surface Electronic Structures*. Springer Series in Optical Sciences, vol. 176 (Springer, Berlin, 2014)
2. S. Hüfner, *Photoemission Spectroscopy* (Springer, Berlin, 1995)
3. M. Sing, G. Berner, K. Goß, A. Müller, A. Ruff, A. Wetscherek, S. Thiel, J. Mannhart, S.A. Pauli, C.W. Schneider, P.R. Willmott, M. Gorgoi, F. Schäfers, R. Claessen, Phys. Rev. Lett. **102**, 176805 (2009)

4. S. Thiel, G. Hammerl, A. Schmehl, C.W. Schneider, J. Mannhart, Science **313**, 1942 (2006)
5. G. Berner, A. Müller, F. Pfaff, J. Walde, C. Richter, J. Mannhart, S. Thiess, A. Gloskovskii, W. Drube, M. Sing, R. Claessen, Phys. Rev. B **88**, 115111 (2013)
6. S.A. Chambers, L. Qiao, T.C. Droubay, T.C. Kaspar, B.W. Arey, P.V. Sushko, Phys. Rev. Lett. **107**, 206802 (2011)
7. E.A. Kraut, R.W. Granta, J.R. Waldrop, S.P. Kowalczyk, Phys. Rev. Lett. **44**, 1620 (1980)
8. Y. Segal, J.H. Ngai, J.W. Reiner, F.J. Walker, C.H. Ahn, Phys. Rev. B **80**, 241107 (2009)
9. S. Chambers, M. Engelhard, V. Shutthanandan, Z. Zhu, T. Droubay, L. Qiao, P. Sushko, T. Feng, H. Lee, T. Gustafsson, E. Garfunkel, A. Shah, J.M. Zuo, Q. Ramasse, Surf. Sci. Rep. **65**, 317 (2010)
10. L. Qiao, T.C. Droubay, T. Varga, M.E. Bowden, V. Shutthanandan, Z. Zhu, T.C. Kaspar, S.A. Chambers, Phys. Rev. B **83**, 085408 (2011)
11. G. Drera, G. Salvinelli, A. Brinkman, M. Huijben, G. Koster, H. Hilgenkamp, G. Rijnders, D. Visentin, L. Sangaletti, Phys. Rev. B **87**, 075435 (2013)
12. Y.Z. Chen, N. Bovet, F. Trier, D.V. Christensen, F.M. Qu, N.H. Andersen, T. Kasama, W. Zhang, R. Giraud, J. Dufouleur, T.S. Jespersen, J.R. Sun, A. Smith, J. Nygård, L. Lu, B. Büchner, B.G. Shen, S. Linderoth, N. Pryds, Nat. Commun. **4**, 1371 (2013)
13. Y.Z. Chen, N. Bovet, T. Kasama, W.W. Gao, S. Yazdi, C. Ma, N. Pryds, S. Linderoth, Adv. Mater. **26**, 1462 (2013)
14. P. Schütz, F. Pfaff, P. Scheiderer, Y.Z. Chen, N. Pryds, M. Gorgoi, M. Sing, R. Claessen, Phys. Rev. B **91**, 165118 (2015)
15. W. Meevasana, P. King, R.H. He, S.K. Mo, M. Hashimoto, A. Tamai, P. Songsiriritthigul, F. Baumberger, Z.X. Shen, Nat. Mater. **10**, 114 (2011)
16. A.F. Santander-Syro, O. Copie, T. Kondo, F. Fortuna, S. Pailhès, R. Weht, X.G. Qiu, F. Bertran, A. Nicolaou, A. Taleb-Ibrahimi, P. Le Fèvre, G. Herranz, M. Bibes, N. Reyren, Y. Apertet, P. Lecoeur, A. Barthélémy, M.J. Rozenberg, Nature **469**, 189 (2011)
17. L. Dudy, M. Sing, P. Scheiderer, J.D. Denlinger, P. Schütz, J. Gabel, M. Buchwald, C. Schlueter, T.L. Lee, R. Claessen, Adv. Mater. **28**, 7443 (2016)
18. A. Brinkman, M. Huijben, M. van Zalk, J. Huijben, U. Zeitler, J.C. Maan, W.G. van der Wiel, G. Rijnders, D.H.A. Blank, H. Hilgenkamp, Nat. Mater. **6**, 493 (2007)
19. C. Cancellieri, N. Reyren, S. Gariglio, A.D. Caviglia, A. Fête, J.M. Triscone, Europhys. Lett. **91**, 17004 (2010)
20. N.C. Bristowe, P.B. Littlewood, E. Artacho, Phys. Rev. B **83**, 205405 (2011)
21. P. Scheiderer, F. Pfaff, J. Gabel, M. Kamp, M. Sing, R. Claessen, Phys. Rev. B **92**, 195422 (2015)
22. L.C. Davis, Phys. Rev. B **25**, 2912 (1982)
23. J. W. Allen, Resonant photoemission of solids with strongly correlated electrons, in *Synchrotron Radiation Research: Advances in Surface Science and Low Dimensional Science*, ed. by R.F. Bachrach (Plenum, New York, 1992)
24. G. Drera, F. Banfi, F.F. Canova, P. Borghetti, L. Sangaletti, F. Bondino, E. Magnano, J. Huijben, M. Huijben, G. Rijnders, D.H.A. Blank, H. Hilgenkamp, A. Brinkman, Appl. Phys. Lett. **98**, 052907 (2011)
25. G. Berner, M. Sing, H. Fujiwara, A. Yasui, Y. Saitoh, A. Yamasaki, Y. Nishitani, A. Sekiyama, N. Pavlenko, T. Kopp, C. Richter, J. Mannhart, S. Suga, R. Claessen, Phys. Rev. Lett. **110**, 247601 (2013)
26. C. Cancellieri, M.L. Reinle-Schmitt, M. Kobayashi, V.N. Strocov, P.R. Willmott, D. Fontaine, P. Ghosez, A. Filippetti, P. Delugas, V. Fiorentini, Phys. Rev. B **89**, 121412(R) (2014)
27. Y. Aiura, I. Hase, H. Bando, T. Yasue, T. Saitoh, D.S. Dessau, Surf. Sci. **515**, 61 (2002)
28. J. Shen, H. Lee, R. Valentí, H.O. Jeschke, Phys. Rev. B **86**, 195119 (2012)
29. C.W. Lin, A.A. Demkov, Phys. Rev. Lett. **111**, 217601 (2013)
30. H.O. Jeschke, J. Shen, R. Valentí, New J. Phys. **17**, 023034 (2015)
31. M. Altmeyer, H.O. Jeschke, O. Hijano-Cubelos, C. Martins, F. Lechermann, K. Koepernik, M.J.R.A.F. Santander-Syro, R. Valentí, M. Gabay, Phys. Rev. Lett. **116**, 157203 (2016)

32. F. Lechermann, H.O. Jeschke, A.J. Kim, S. Backes, R. Valentí, Phys. Rev. B **93**, 121103(R) (2016)
33. T. Kaurila, J. Väyrynen, M. Isokallio, J. Phys. Condens. Matter **9**, 6533 (1997)
34. S. Thiess, T.L. Lee, F. Bottind, J. Zegenhagen, Solid State Commun. **150**, 553 (2010)
35. S.M. Walker, F.Y. Bruno, Z. Wang, A. de la Torre, S. Riccó, A. Tamai, T.K. Kim, M. Hoesch, M. Shi, M.S. Bahramy, P.D.C. King, F. Baumberger, Adv. Mater. **27**, 3894 (2015)
36. M.L. Knotek, P.J. Feibelman, Phys. Rev. Lett. **40**, 964 (1978)
37. http://www.diamond.ac.uk/Beamlines/Surfaces-and-Interfaces/I09.html
38. J. Gabel, M. Zapf, P. Scheiderer, P. Schütz, L. Dudy, M. Stübinger, C. Schlueter, T.L. Lee, M. Sing, R. Claessen (2016) (Submitted for publication)
39. T. Rödel, C. Bareille, F. Fortuna, C. Baumier, F. Bertran, P. Le Fèvre, M. Gabay, O. Hijano Cubelos, M.J. Rozenberg, T. Maroutian, P. Lecoeur, A. Santander-Syro, Phys. Rev. Appl. **1**, 051002 (2014)
40. S. McKeown Walker, A. de la Torre, F.Y. Bruno, A. Tamai, T.K. Kim, M. Hoesch, M. Shi, M.S. Bahramy, P.D.C. King, F. Baumberger, Phys. Rev. Lett. **113**, 177601 (2014)
41. Z. Wang, Z. Zhong, X. Hao, S. Gerhold, B. Stöger, M. Schmid, J. Sánchez-Barriga, A. Varykhalov, C. Franchini, K. Held, U. Diebold, PNAS **111**, 3933 (2014)
42. L. Li, C. Richter, J. Mannhart, R.C. Ashoori, Nat. Phys. **7**, 762 (2011)
43. J.A. Bert, B. Kalisky, C. Bell, M. Kim, Y. Hikita, H.Y. Hwang, K.A. Moler, Nat. Phys. **7**, 767 (2011)
44. Y. Chen, N. Pryds, J.E. Kleibeuker, G. Koster, J. Sun, E. Stamate, B. Shen, G. Rijnders, S. Linderoth, Nano Lett. **11**, 3774 (2011)
45. G. Herranz, F. Sánchez, N. Dix, M. Scigaj, J. Fontcuberta, Sci. Rep. **2**, 758 (2012)
46. E. Slooten, Z. Zhong, H.J.A. Molegraaf, P.D. Eerkes, S. de Jong, F. Massee, E. van Heumen, M.K. Kruize, S. Wenderich, J.E. Kleibeuker, M. Gorgoi, H. Hilgenkamp, A. Brinkman, M. Huijben, G. Rijnders, D.H.A. Blank, G. Koster, P.J. Kelly, M.S. Golden, Phys. Rev. B **87**, 085128 (2013)
47. Z. Zhong, P.X. Xu, P.J. Kelly, Phys. Rev. B **82**, 165127 (2010)
48. Y. Li, S.N. Phattalung, S. Limpijumnong, J. Kim, J. Yu, Phys. Rev. B **84**, 245307 (2011)
49. N. Pavlenko, T. Kopp, E.Y. Tsymbal, J. Mannhart, G.A. Sawatzky, Phys. Rev. B **86**, 064431 (2012)
50. L. Yu, A. Zunger, Nat. Commun. **5**, 5118 (2014)

Chapter 6
Electrons and Polarons at Oxide Interfaces Explored by Soft-X-Ray ARPES

Vladimir N. Strocov, Claudia Cancellieri and Andrey S. Mishchenko

Abstract Soft-X-ray ARPES (SX-ARPES) with its enhanced probing depth and chemical specificity allows access to fundamental electronic structure characteristics—momentum-resolved spectral function, band structure, Fermi surface—of systems difficult and even impossible for the conventional ARPES such as three-dimensional materials, buried interfaces and impurities. After a recap of the spectroscopic abilities of SX-ARPES, we review its applications to oxide interfaces, focusing on the paradigm $LaAlO_3/SrTiO_3$ interface. Resonant SX-ARPES at the Ti L-edge accentuates photoemission response of the mobile interface electrons and exposes their d_{xy}-, d_{yz}- and d_{xz}-derived subbands forming the Fermi surface in the interface quantum well. After a recap of the electron-phonon interaction physics, we demonstrate that peak-dip-hump structure of the experimental spectral function manifests the Holstein-type large polaron nature of the interface charge carriers, explaining their fundamentally reduced mobility. Coupling of the charge carriers to polar soft phonon modes defines dramatic drop of mobility with temperature. Oxygen deficiency adds another dimension to the rich physics of $LaAlO_3/SrTiO_3$ resulting from co-existence of mobile and localized electrons introduced by oxygen vacancies. Oxygen deficiency allows tuning of the polaronic coupling and thus mobility of the charge carriers, as well as of interfacial ferromagnetism connected with various atomic configurations of the vacancies. Finally, we discuss spectroscopic evidence of phase separation at the $LaAlO_3/SrTiO_3$ interface. Concluding, we put prospects

---

V. N. Strocov (✉)
Swiss Light Source, Paul Scherrer Institute, 5232 Villigen-PSI, Switzerland
e-mail: vladimir.strocov@psi.ch

C. Cancellieri
EMPA, Swiss Federal Laboratories for Materials Science & Technology, Ueberlandstrasse 129, 8600 Duebendorf, Switzerland
e-mail: claudia.cancellieri@empa.ch

A. S. Mishchenko
RIKEN Center for Emergent Matter Science (CEMS), 2-1 Hirosawa, Wako, Saitama 351-0198, Japan
e-mail: mishchenko@riken.jp

© Springer International Publishing AG, part of Springer Nature 2018
C. Cancellieri and V. Strocov (eds.), *Spectroscopy of Complex Oxide Interfaces*, Springer Series in Materials Science 266,
https://doi.org/10.1007/978-3-319-74989-1_6

of SX-ARPES for complex heterostructures, spin-resolving experiments opening the totally unexplored field of interfacial spin structure, and in-operando field-effect experiments paving the way towards device applications of the reach physics of oxide interfaces.

6.1 Soft-X-Ray ARPES: From Bulk Materials to Interfaces and Impurities

Angle-resolved photoelectron spectroscopy (ARPES) as the unique experimental technique delivering direct information about electronic structure of crystalline solids resolved in electron energy E and momentum $\mathbf{k}$. While the fundamentals of ARPES as well as characteristic spectroscopic properties of this technique in different photon energy ranges have been discussed in Introduction to this volume, in this Chapter we will focus on spectroscopic abilities of soft-X-ray ARPES (SX-ARPES). Combining the $\mathbf{k}$ resolution with enhanced probing depth and chemical specificity, this technique is ideally suited for buried interface and impurity systems which are in the core of nowadays and future electronic and spintronic devices.

6.1.1 Virtues and Challenges of Soft-X-Ray ARPES

SX-ARPES as a spectroscopic technique exploiting the region of photon energies hv around 1 keV has been pioneered at SPRing-8 [1–3] and has recently been boosted with advent of synchrotron sources delivering high photon flux in this energy range such as the Swiss Light Source (SLS) and Diamond Light Source (DLS). SX-ARPES features a few fundamental advantages compared to the conventional VUV-ARPES with its hv region around 20–100 eV:

Probing depth. The "universal curve" of the photoelectron attenuation length λ (see, for example Powell et al. [4]) shows that VUV-ARPES is characterised by a probing depth of a few Å, which makes this technique extremely surface sensitive, and limits its applications basically to atomically clean surfaces. The increase of λ in the SX-ARPES energy range enhances bulk sensitivity of the ARPES experiment as well as enables access to systems buried behind a few surface layers, often without any surface preparation. SX-ARPES is therefore highly relevant for real-world materials of importance in modern solid-state technology.

3D momentum resolution. The photoemission (PE) final state confinement within λ results, by the Heisenberg uncertainty principle, in intrinsic broadening of the corresponding surface-perpendicular momentum $k_\perp$ defined by $\Delta k_\perp = \lambda^{-1}$. The increase of λ at higher energies results therefore in improvement of the intrinsic $k_\perp$ resolution [5, 6]. In combination with free-electron final-state dispersion, this sharpens the definition of $k_\perp$ and thus of the full 3D momentum to enable precise determination of the valence band (VB) dispersions $E(\mathbf{k})$ in 3D materials (see examples below as well as [6, 7, 8]; etc). problematic for the VUV-ARPES.

Resonant photoemission spectroscopy (ResPES). The SX-ARPES energy range goes through a number of important absorption edges including the L-edges of transition metals and M-edges of rear-earths. This allows resonant excitation of their valence d- and f-states which play the crucial role in the physics of strong electron correlations. In this way resonant SX-ARPES not only reveals the elemental composition of the VB [9, 10] but also can be used to highlight the signal from buried systems.

A few applications of SX-ARPES exploiting the above virtues of enhanced probing depth, 3D momentum definition and chemical specificity will be illustrated later.

Realization of these SX-ARPES virtues comes however with a few severe challenges:

Cross-section problem. The main challenge of SX-ARPES, until now severely impeding its practical use, is that the VB photoexcitation cross-section reduces compared to the VUV energy range typically by 2–3 orders of magnitude [11] because overlap of the rapidly oscillating high-energy final states with smooth valence states reduces essentially to the small ion core region [12]. Such a dramatic signal loss has to be compensated by high flux of incoming photons, which requires the most advanced synchrotron radiation sources and beamline instrumentation, as well as efficient photoelectron detectors.

Electron-phonon scattering. In the soft-X-ray energy range the photoelectron wavelengths become comparable with phononic displacements of atoms in the unit cell, which relaxes the $\mathbf{k}$-conserving dipole selection rules. This electron-phonon interaction effect manifests itself as momentum broadening of the spectral structures as well as transfer of the coherent spectral weight I^{coh} into an incoherent background reflecting the matrix-element weighted $\mathbf{k}$-integrated density of states (DOS) [13, 14]. The reduction of I^{coh} relative to that at zero temperature $I^{coh}_{T=0}$ can be expressed, in the first approximation, as $I^{coh} = W(T)I^{coh}_{T=0}$, where $W(T)$ is the photoemission Debye-Waller factor $W(T) = e^{-\Delta G^2 U_0^2(T)}$ with $\Delta G \propto \sqrt{h\nu}$ expressing the momentum transfer between the initial and final state and U_0 the *rms* thermal atomic displacement [15]. Cryogenic sample cooling, ultimately towards liquid-He temperatures, is therefore imperative to achieve adequate $\mathbf{k}$-resolution of SX-ARPES. More on the temperature effects can be found in Chap. 7.

Further increase of $h\nu$ into the multi-keV energy range pushes λ to more than 50 Å [4]. The corresponding hard-X-ray ARPES (HX-ARPES, see Chap. 7) has however to stand progressive reduction of the VB cross-section and loss of the coherent spectral weight with energy. In addition, recoil from high-energy photoelectrons—essentially, emission of phonons back in the lattice [16]—smears photoelectron $\mathbf{k}$ and energy [17]. Therefore, the $\mathbf{k}$-resolving abilities of HX-ARPES seem to stay with high-Z materials like W or materials with stiff covalent bonds like GaAs [15, 18, 19] where the temperature and recoil effects are small. SX-ARPES appears therefore as a winning combination of enhanced probing depth with sufficient VB cross-section and $\mathbf{k}$-resolution.

6.1.2 Experimental Technique

The cross-section problem of SX-ARPES requires synchrotron radiation beamlines delivering high photon flux. At the time of writing, the worldwide highest brilliance soft-X-ray source was the ADRESS beamline at the SLS (for details see [20]). The beamline delivers soft-X-ray radiation with variable linear and circular polarizations in $h\nu$ range from 300 to 1600 eV with an ultimate resolving power $E/\Delta E$ of 33'000 at 1 keV. By virtue of the SLS ring energy 2.4 GeV optimal for the soft-X-ray range and the beamline design optimized for high transmission, the photon flux tops up 10^{13} photons/s/0.01%BW. Apart from SX-ARPES, the ADRESS beamline hosts a Resonant X-ray Scattering (RIXS) facility delivering complementary information on charge-neutral excitations (see Chap. 11). Dedicated SX-ARPES facilities are also available at SPRing-8, DLS, PETRA-III, and more are coming soon at MAX-IV and NSLS-II.

Another factor to increase the PE signal is a grazing-incidence experimental geometry dictated by interplay of relatively large X-ray penetration depth d with relatively small photoelectron escape depth λ [21–23]. Figure 6.1a shows the normal-emission photocurrent I_{PE} as a function of X-ray grazing incidence angle α calculated near the ends of the soft-X-ray energy interval for the valence states of two paradigm materials Cu and GaAs [23]. $I_{PE}(\alpha)$ sharply increases when going to grazing α, and reaches its maximum near the critical reflection angle α_c when the electromagnetic field concentrates in the near-surface region where the photoelectrons can escape from. Importantly, with the increase of $h\nu$ the $I_{PE}(\alpha)$ maximum shifts to very grazing α around a few degrees. Optimal for the SX-ARPES experiment will be therefore a grazing-incidence geometry [24] sketched in Fig. 6.1b. Importantly, the sample is rotated to the grazing α around the horizontal axis, because in this case the X-ray footprint on the sample blows up in the vertical plane where the synchrotron beam has much smaller size than in the horizontal. The analyzer slit is oriented in the measurement plane that enables symmetry analysis of the valence states by switching the X-ray polarization between horizontal and vertical.

The SX-ARPES setup at the ADRESS beamline [24] operates at $\alpha = 20°$ (at the moment restricted by mechanical constraints) which increases $I_{PE}(\alpha)$ by a factor of ~2 compared to the conventional $\alpha = 45°$. The light footprint on the sample is about 20×74 μm. The photoelectron analyzer PHOIBOS-150 from SPECS delivers high angular resolution $\delta\vartheta$ ~0.07° FWHM which is particularly important for SX-ARPES because of the corresponding momentum uncertainty $\delta K_{//} = 0.5124\sqrt{E_k}\sin\delta\vartheta$ magnified by high kinetic energy E_k. The CARVING manipulator provides three independent angular degrees of freedom, allowing precise navigation in **k**-space. The samples are liquid-He cooled down to 10.7 K. The practical combined (beamline plus analyzer) resolving power $E/\Delta E$ of this SX-ARPES facility is up to 20'000. This is the nowadays SX-ARPES state-of-art.

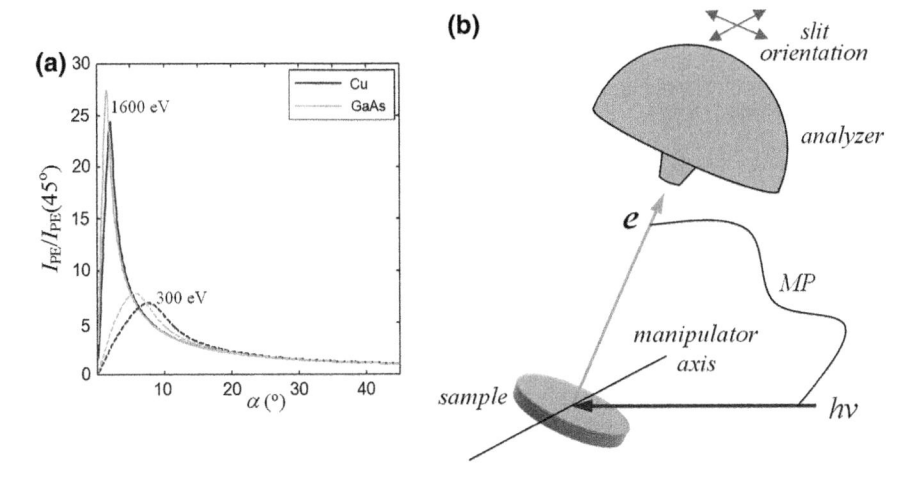

Fig. 6.1 a Dependence of the normal-emission photocurrent I_{PE} on grazing X-ray incidence angle. With increase of $h\nu$ its maximum shifts to more grazing angles (adapted from [23]); **b** Optimized geometry of the SX-ARPES experiment. The grazing X-ray incidence increases the PE signal, horizontal rotation axis reduces the light footprint on the sample, and slit oriented in the measurement plane (MP) allows symmetry analysis of the valence states [24, 25]

6.1.3 Application Examples

We will now flash through a few scientific cases achieved at the SLS which illustrate the above spectroscopic virtues of SX-ARPES. Extending the recent review [25] we will demonstrate that the recent progress in SX-ARPES instrumentation has not only overpowered the dramatic photoexcitation cross-section loss in this energy range but also pushed SX-ARPES to the most photon-hungry cases of buried interfaces and impurities.

6.1.3.1 Probing Deep: Buried Interfaces

In the Introduction to this volume we have already seen an illustration of the enhanced probing depth of SX-ARPES with experiments on GaAs(100) capped with a ~10 Å thick layer of amorphous As [26]. A recent example relevant for oxide spintronics is the spin injector heterostructure SiO_x/EuO/Si [27]. EuO is one of the most promising routes for spintronics, for entries see [28, 29]. This material, first, delivers almost 100% spin-polarized electrons inviting spin-filter applications and, second, can be integrated into the wide-spread Si technology. Furthermore, unique properties of EuO such as its colossal magnetoresistivity, metal-insulator transition, strong magneto-optic effects, tunability of the magnetic properties by doping or strain, etc. open its ways towards multi-functional devices.

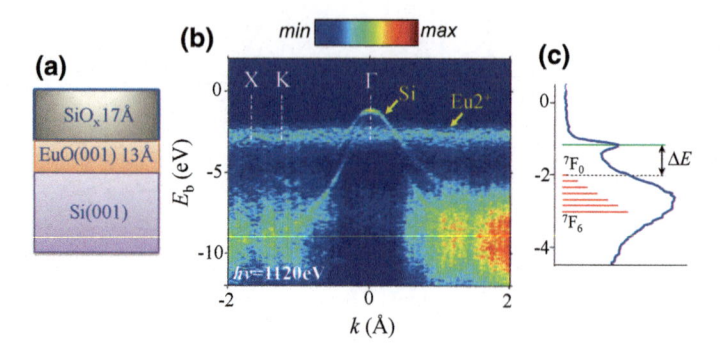

Fig. 6.2 SiOx/EuO/Si spin injector: **a** Scheme of the sample; **b** SX-ARPES image (angle-integrated component subtracted) measured at $hv = 1120$ eV, which reveals the Eu^{2+} state on top of the three-dimensional $E(\mathbf{k})$ of bulk Si along the ΓKX direction; **c** Identification of the band offset $\Delta E = 0.8$ eV which justifies the spin injecting functionality of the $SiO_x/EuO/Si$ heterostructure. Adapted from Lev et al. [27]

Our experiments on the $SiO_x/EuO/Si$ heterostructure sketched in Fig. 6.2a aimed at determination of the EuO/Si band offset, definitive for the spin injector applications. We used hv above 1 keV, allowing penetration of photoelectrons excited in the Si substrate through the SiO_x/EuO layers. The experimental ARPES intensity image in Fig. 6.2b was measured with $hv = 1120$ eV, tuning $k_\perp$ to the VB maximum of bulk Si in the Γ-point of the Brillouin zone (BZ). We see the dispersionless Eu^{2+} state in the EuO layer buried behind the 17 Å thick layer of SiO_x and, moreover, we recognize the textbook manifold of the light-hole and heavy-hole bands along the ΓKX direction of bulk Si, which is buried behind the SiO_x/EuO layer with a total thickness of 30 Å. This is an impressive example of the penetrating ability of SX-ARPES.

The energy difference between the upper edge of the Eu^{2+} multiplet and the VB maximum of Si in the Γ point, Fig. 6.2c, directly measures the EuO/Si band offset ΔE as 0.8 eV. This value is sufficiently large to make spin injection at the EuO/Si interface immune to noise and, on the other hand, it stays sufficiently small to reduce power consumption. These SX-ARPES results justify thereby the spin injecting functionality of the $SiO_x/EuO/Si$ heterostructure.

As we will see below, the SX-ARPES enhanced probing depth is essential for experiments on the paradigm TMO interface, buried LAO/STO. In this perspective we should also mention recent studies of dimensionality-tuned electronic structure of $LaNiO_3/LaAlO_3$ superlattices ([30], see Chap. 5) as well as two-dimensional electron gas in a multilayer structure formed by interfacing of the ferromagnetic Mott insulator $GdTiO_3$ with STO [31].

6.1.3.2 Momentum Resolution in 3D: Bulk Materials

We will now illustrate the use of the $k_\perp$-momentum resolution of SX-ARPES to explore 3D electronic structure of the typical transition metal dichalcogenide (TMDC) VSe$_2$. The chalcogen-metal-chalcogen layered structure of TMDCs results in their quasi-2D properties [32] but the bands derived from the out-of-plane chalcogen orbitals retain pronounced 3D character characterized by $k_\perp$ dispersion ranges of a few eV. Accurate measurements of these dispersions with VUV-ARPES are impeded by large intrinsic $k_\perp$ broadening of the low-energy final states comparable with the perpendicular extension of the BZ (see [33, 34] and references therein).

Results of SX-ARPES measurements on VSe$_2$ in an $h\nu$ range around 900 eV are compiled in Fig. 6.3 [6]. The ARPES intensity images (*b*) along two in-plane and out-of-plane directions of the BZ (*a*) clearly show the V 3d bands near E_F as well as the Se 4p ones deeper in the VB. Statistics of the experimental data is remarkable in view of the photoexcitation cross-section reduction by a factor of ~2000 for the V 3d and ~30 for the Se 4p states for our excitation energies compared to VUV–ARPES [11]. Reliable control over $k_\perp$ in the SX-ARPES experiment has allowed slicing the Fermi surface (FS) in different planes, including in- (*d*) and out-of-plane (*c*) ones. The FS topology with its characteristic symmetry pattern is extremely clear in the experimental maps. The textbook clarity of the experimental data compared to the previous VUV-ARPES results [35] comes from high $k_\perp$ definition of the high-energy final states. A detailed account of these results, in particular analysis of 3D warping of the FS to form exotic 3D charge density waves, is given in [6].

Of the recent SX-ARPES applications to bulk TMOs, precursors to our main topic of TMO interfaces, we can mention the paradigm manganite La$_{1-x}$Sr$_x$MnO$_3$ (LSMO) possessing 3D perovskite structure. Whereas previous VUV-ARPES studies suffered from insufficient $k_\perp$ resolution [36], the SX-ARPES experiments by Lev et al. [8] employing $h\nu$ around 700 eV have immediately yielded the long-sought-for canonical FS topology consisting of the alternating 3D spheroid electron pockets and cuboid hole pockets. Shadow contours of the experimental FS, reminiscent of the high-T_c cuprates [37], manifested the rhombohedral lattice distortion of LSMO. This distortion is neutral to the Jahn-Teller effect and thus polaronic coupling, but reduces the DE electron hopping along the Mn-O-Mn bonds. In interplay of the polaronic self-localization with the DE electron delocalization [38–40], this effect reduces the colossal magnetoresistance T_c. Most recently, SX-ARPES has been applied to Ce-doped CaMnO$_3$ (CMO) to reveal shadow contours of the experimental FS as signatures of the orthorhombic lattice distortion, and strong band dependent polaronic coupling (Husanu et al., unpublished).

Of other applications of SX-ARPES to bulk TMOs, we should mention a pioneering study of band structure and FS of CrO$_2$ probed through a layer of amorphous Cr$_2$O$_3$ inherently building up on its surface [41]. Further examples of the 3D electronic structure determination with SX-ARPES in various materials range from layered materials such as VSe$_2$ discussed above [6] to the conventional high-T_c superconductor MgB$_2$ with dichotomy of the Fermi states in their dimensionality [42], AlNiCo quasicrystals with anisotropic structure stabilization mechanism [43],

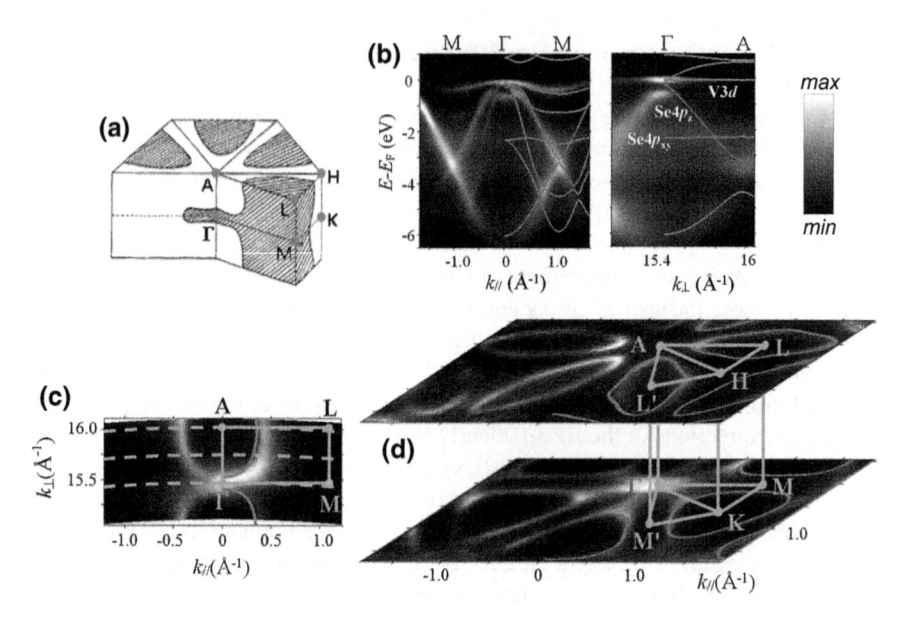

Fig. 6.3 3D electronic structure of VSe$_2$: **a** BZ with its main symmetry lines and inscribed FS; **b** ARPES intensity along the M'ΓM line measured at $h\nu = 885$ eV, and along the ΓA line under variation of $h\nu$ to scan $k_\perp$; (*c-d*) experimental FS slices in the **c** ΓALM plane (variation of $h\nu$) and **d** ΓKM and ALH planes ($h\nu = 885$ and 945 eV, respectively). The clarity of the experimental data manifests $k_\perp$ definition of the high-energy final states (adapted from Strocov et al. [6])

strongly correlated ruthenates and pnictides [44, 45], bandwidth controlled Mott materials [46], heavy fermion systems [47], etc. In the study on the 3D pnictide (Ba$_{1-x}$K$_x$) Fe$_2$As$_2$ [48] it was noted that accuracy of the electron effective mass determination is much affected by the $k_\perp$ broadening effects and therefore improves with increase of $h\nu$. Furthermore, we should particularly mention Weil semimetals where high $k_\perp$ definition of SX-ARPES was crucial to identify the 3D cones topologically connecting the surface state arcs ([49–54]. In connection with spin physics of 3D materials, we should mention bulk Rashba spin splittings in non-centrosymmetric BiTeI [55] and in the ferroelectric Rashba semiconductor α-GeTe were the splitting is coupled to the ferroelectricity [56].

6.1.3.3 Resonant Photoemission: Impurity Systems

(Ga, Mn)As with Mn concentrations of a few percent is a paradigm diluted magnetic semiconductor (DMS) where the Mn doping induces its FM properties associated with the hole carriers. Essential for physics of (Ga, Mn)As is the exact energy alignment of the Mn 3d derived impurity state (IS) and its hybridization with the GaAs host states. With Mn atoms replacing only a few percent of Ga, finding their response

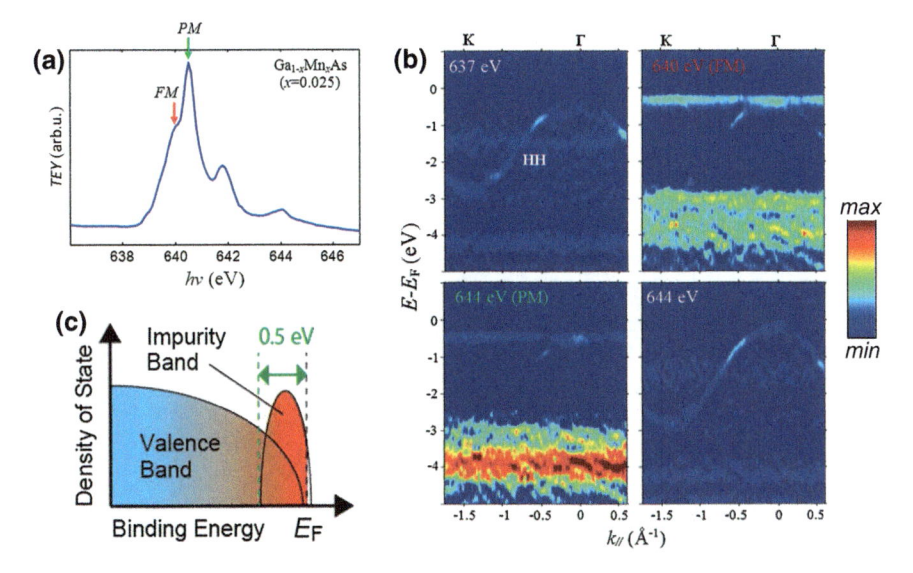

Fig. 6.4 SX-ARPES of the (Ga, Mn)As impurity system: **a** Mn L_3 XAS spectrum showing the ferromagnetic (FM) and paramagnetic (PM) components; **b** ARPES images taken across the resonance. Tuning $h\nu$ onto the FM peak unleashes the Mn impurity state in vicinity of E_F injecting the FM charge carriers; **c** Established band diagram. Adapted from Kobayashi et al. [57]

between the bulk of GaAs atoms is a "needle in a haystack" problem which solves however by taking into advantage the penetrating ability and chemical specificity of resonant SX-ARPES.

The experiments [57] were performed on (Ga, Mn)As thin films embedding Mn atoms with a concentration of 2.5% and capped by an amorphous As layer to reduce oxidation during ex-situ sample transfer [26]. Figure 6.4a shows the experimental X-ray absorption (XAS) spectrum at the Mn L_3-edge, and (*b*) the corresponding SX-ARPES images measured through the resonance. The pre-resonance image is hardly distinguishable from bare GaAs showing the light-hole band. However, tuning $h\nu$ to the first XAS peak at 640 eV, corresponding to the FM substitutional Mn ions, immediately unleashes the Mn 3*d* derived impurity state (IS) just below E_F which injects the FM charge carriers. Furthermore, simultaneous enhancement of the GaAs band identifies its hybridization with the IS. The second XAS peak at 640.5 eV, corresponding to the PM interstitial Mn atoms, leaves only an afterglow of the IS, and further increase of $h\nu$ returns us to bare band structure of the GaAs host. The ferromagnetic IS can therefore be seen only at one single excitation energy, corresponding to X-ray absorption of the FM substitutional Mn ions. In this way, chemical specificity of resonant SX-ARPES allows us to probe impurity states and their hybridization with the host states for impurity systems with atomic concentrations of a few percent (Fig. 6.5).

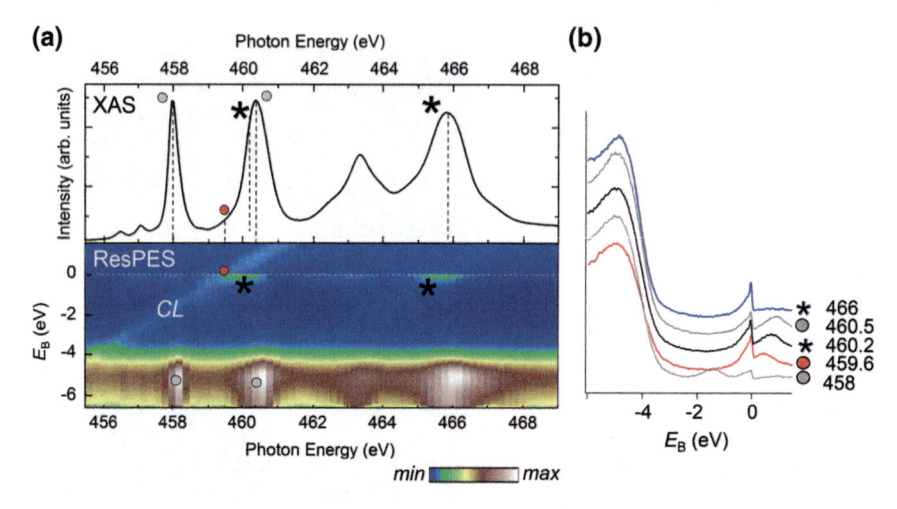

Fig. 6.5 ResPES of standard (oxygen annealed) LAO/STO: **a** XAS spectrum; **b** Map of angle-integrated ResPES intensity. *CL* marks the stray signal from the Ti 2p core level excited by second-order radiation; **c** EDCs extracted from (**b**) at selected excitation energies: the 458 and 460.5 eV are Ti^{4+} resonant, 459.6 eV gives the highest signal at E_F but overlaps with the second-order Ti 2p signal, and 460.2 and 466 eV are representative of the 2DEG. Adopted from Cancellieri et al. [58]

The band diagram of (Ga, Mn)As emerging from the results of Kobayashi et al. [57] is sketched in Fig. 6.6c. It merges the two previous models of (Ga, Mn)As [59, 60]: Whereas the FM charge carriers supplied by the IS located near E_F are characteristic of the double-exchange model, the IS hybridization with the host GaAs is characteristic of the *p-d* exchange model. The impurity character of FM charge carriers in (Ga, Mn)As explains their small mobility, impeding high-speed applications of this DMS. Recently Kobayashi et al. (unpublished) have extended this study to Be-doped InFeAs which is a potential n-type counterpart of GaMnAs for spintronics. In contrast to GaMnAs, the FM charge carriers in InFeAs were found to originate from the dispersive band states at the CB minimum. Their high mobility justifies applications of (In, Fe)As for high-speed spintronics.

An extension of the above studies to (In, Mn)As quantum dots buried in GaAs has been reported by Bouravleuv et al. [61]. Of other magnetic impurity systems, we should mention a recent study on the multiferroic compound (Ge, Mn)Te [62]. Mn doping of the host α-GeTe broke the time-reversal protected degeneracy of the Rashba bands in the Γ point. The magnitude of the corresponding Zeeman gap scaled up with Mn concentration, reflecting buildup of ferromagnetic order between the Mn atoms. Another system is V-doped topological insulator Bi$_2$Se$_3$ where ResPES has found large V weight at E_F [63] whose coexistence with the quantum anomalous Hall (QAH) effect suggests nanoscale separation of this materials into V-rich magnetic and V-deficient non-magnetic phases.

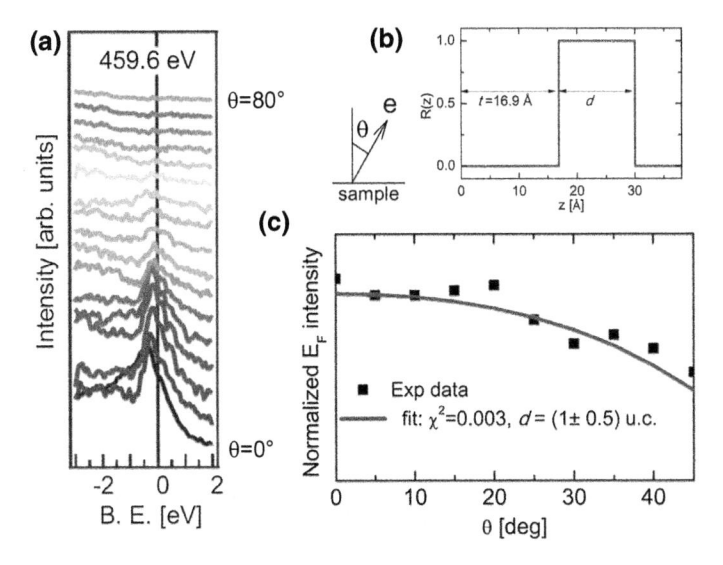

Fig. 6.6 XPS depth profiling of the 2DES: **a** Dependence of the ResPES intensity at $h\nu = 460.2$ eV and **b** its fit within the model of rectangular 2DES distribution (**c**). The fit finds the interfacial d_{xy} states within 1 u.c. in STO. Adopted from Cancellieri et al. [58]

6.2 k-Resolved Electronic Structure of LAO/STO

The interface between $LaAlO_3$ (LAO) and $SrTiO_3$ (STO) is a paradigm example of new functionalities formed at TMO interfaces of TMOs. Although these materials in their bulk form are band insulators with wide band gaps (5.6 eV for LAO and 3.2 eV for STO) their interface spontaneously forms mobile two-dimensional electron system (2DES). Its high electron mobility typical of uncorrelated electron systems co-exists with superconductivity and ferromagnetsm (for entries see the reviews [64–67]), large magnetoresistance [68] and other phenomena typical of localized correlated electrons (see Chap. 8). These properties of the 2DEG can be tuned with field effect allowing design of transistor structures with enhanced functionalities [69, 70].

The interfacial 2DES is confined within a narrow region of a few nanometres on the STO side [71–75] where the mobile electrons populate the t_{2g}-derived d_{xy}-, d_{xz}- and d_{yz}-states of Ti ions acquiring reduced valence compared to the bulk Ti^{4+}. Breaking of the 3D periodicity at the interface splits the t_{2g} bands into the d_{xy} and two degenerate d_{xz}/d_{yz} bands. Furthermore, the 2DES confinement in the interface quantum well (QW) within a narrow region of a few nanometres on the STO side further splits the d_{xy} and d_{xz}/d_{yz} bands into a ladder of subbands [72, 76, 77, 78]. This complex energy structure based on the correlated 3d orbitals, very different from conventional semiconductor heterostructures described as free particles embedded in the mean-field potential, is the source of a rich and non-trivial phenomenology. Instrumental to explore the underlying band structure of the LAO/STO interface states is SX-ARPES with its virtues of **k**-resolution, enhanced probing depth and chemical specificity. The SX-ARPES results presented below were achieved at the ADRESS beamline of the SLS.

6.2.1 Resonant Photoemission

Despite the enhanced probing depth of SX-ARPES, the buried interface 2DES can only be accessed by boosting its photoemission response using ResPES at the Ti L-edge. As we will see below, its elemental and chemical state specificity [9, 10] allows discrimination of the signal coming from the interface Ti^{3+} ions whose valence contrasts them to the Ti^{4+} ones in the STO bulk.

A typical Ti L-edge XAS spectrum of the LAO/STO heterostructure measured in total electron yield (TEY) is shown in Fig. 6.5a. Almost independent of X–ray polarization, the spectrum shows two leading peaks around 458 and 460.5 eV which correspond to electron excitation from the $2p_{3/2}$ core levels to $3d$ t_{2g} and e_g levels, respectively, of the Ti^{4+} ions. The corresponding XAS weight from the Ti^{3+} ions in the region of hv ~459.3 eV, which as can be seen in the $LaTiO_3$ XAS data [79], is hardly seen in the experimental spectrum behind the overwhelming Ti^{4+} contribution. The next pair of peaks in the XAS spectrum corresponds to electron excitation from the $2p_{3/2}$ core levels (for detailed analysis of XAS spectra of LAO/STO see Chap. 11).

The corresponding map of angle-integrated ResPES intensity in Fig. 6.5 shows four salient peaks in the VB region. Their energy position reflects the Ti^{4+} peaks in the XAS spectrum, identifying hybridization of the oxygen valence states with Ti^{4+}. Most important, at the excitation energies corresponding to the Ti^{3+} regions of the XAS spectrum the ResPES map shows clear signal of the Ti derived 2DEG near E_F. This observation arms us with the exact recipe how to experimentally access the 2DES [58, 80–82]. The corresponding EDCs should be taken away from the stray Ti $2p$ core level signal excited by second-order radiation from the beamline monochromator. We note that the present ResPES spectra, characteristic of the standard LAO/STO samples, are significantly modified by oxygen deficiency (see Sect. 6.4.1).

Before switching to **k**-resolved band structure of LAO/STO, we should mention a possibility to use ResPES measurement for XPS depth profiling of the 2DES. In the soft-X-ray energy range, measurements at room temperature suppress **k**-selectivity of the photoemission process (so-called XPS regime, [13]). In this case angle dependence of photoemission intensity $I(\theta)$ reflects attenuation of the 2DES signal in the atomic overlayers. Mathematically, this can be expressed as

$$I(\theta) = G \int_0^\infty R(z) e^{-\frac{z}{\lambda \cos\theta}} dz \qquad (6.1)$$

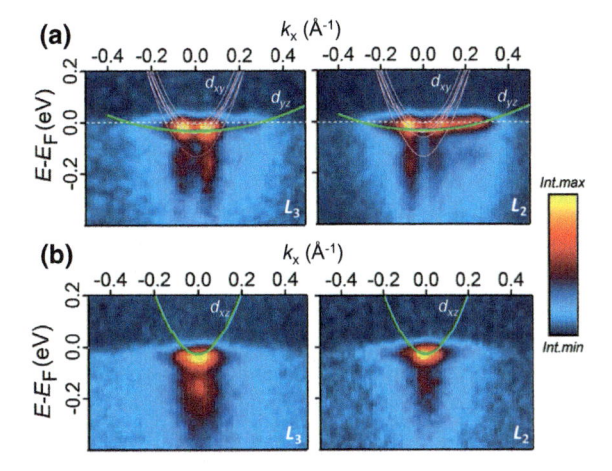

6 Electrons and Polarons at Oxide Interfaces ...   119

Fig. 6.7 Band structure of standard LAO/STO samples, measured at **a** s–polarization of X-rays, exposing the d_{xy}- and d_{yz}-bands, and **b** p-polarization, exposing the d_{xz}-bands. The DFT calculated dispersions are indicated. Renormalization of the experimental bands and tails of their spectral weight extending to lower E_b identify polaronic nature of the 2DES. The L_3- and L_2-resonance data were collected at $h\nu = 460.2$ eV and 466 eV, respectively. Adapted from Cancellieri et al. [86]

where G is a geometrical factor, $R(z)$ the spatial profile of the 2DES, and the exponent represents the photoelectron attenuation in the overlayers following the Beer-Lambert law (see, for example, [83]). Cancellieri et al. [58] have used the expression (6.1) to fit their experimental $I(\theta)$ data measured at the L_3 absorption edge, Fig. 6.6a, with $R(z)$ approximated as a rectangular shaped function with a thickness d (b). The fit (c) has revealed that the 2DES is localized within d = 1 u.c. at the STO side of the interface.

Such narrow localization of the 2DES seems to be at odds with a number of other works suggesting the 2DES extension by at least a few nm into the STO bulk [74, 84]. This controversy can be resolved based on later **k**-resolved SX-ARPES measurements (discussed in the next section). Figure 6.7a demonstrates that for excitation energy at the L_3 absorption edge the ResPES intensity is dominated by the d_{xy} band, and this band is indeed localized next to the interface. Moreover, as this region embeds the Ti^{3+} ions with maximal electron occupancy, the present results qualitatively agree with XPS depth profiling of the Ti^{3+} core level in the hard-X-ray energy range performed by Sing et al. [85]. We note that both XPS results were measured at room temperature which can also contribute to their difference to those of Copie et al. [84] and Gariglio et al. [74] measured at low temperature. In contrast to the d_{xy} bands, the d_{yz} ones extend much further into the bulk accompanied by gradual transformation of the Ti ions to the Ti^{4+} bulk valence state. These bands actually dominate the ResPES response at the L_2 absorption edge, Fig. 6.7a. Therefore, the XPS depth profiling would see there much larger extension of the 2DES.

6.2.2 Electronic Structure Fundamentals: Fermi Surface, Band Structure, Orbital Character

Band structure and FS are fundamental characteristics of the interfacial 2DES. Their **k**-resolved measurements with SX-ARPES are performed using ResPES [77, 80, 81] with the sample cooled below ~30 K in order to suppress the thermal effects reducing the coherent spectral weight (see Sect. 6.1.1). Measurements of the band dispersions, including the heavy $d_{xz/yz}$-derived bands, require challenging figures of energy resolution better than 40 meV, which for measurements of the FS can be relaxed to ~100 meV.

Typical experimental band structure of the standard LAO/STO samples along the ΓX direction of the BZ measured at two different X-ray polarizations and excitation energies is represented in Fig. 6.7 [86]. With the experimental geometry shown in Fig. 6.1b, s-polarization selects the d_{xy}- and d_{yz}-bands [77] whose wavefunction is antisymmetric relative to the ΓX symmetry line [25, 37]. The corresponding SX-ARPES data measured at the L_3- and L_2-resonances is shown in Fig. 6.7a respectively. By comparison with the overlaid $E(\mathbf{k})$ dispersions, calculated in the framework of self-interaction corrected DFT (see Chap. 8), we immediately recognize the heavy d_{yz}-band. The d_{xy}-subbands are not visible due to vanishing matrix elements, but they manifest themselves as two bright spots where they hybridize with the d_{yz}-band. Photoexcitation at the L_3-resonance delivers maximal intensity to the d_{xy}-bands and L_2-resonance to the d_{yz}-band, identifying orbital selectivity of resonance photoexcitation. Already at this point, we note strong renormalization of the d_{yz}-band compared to the DFT, and waterfalls of spectral intensity extending from the d_{xy}-spots. As we will see in Sect. 6.3.2, these spectral features manifest polaronic nature of the interface charge carriers. p-polarization of incident X-rays switches the ARPES response to the Ti d_{xz}-wavefunctions symmetric relative to the ΓX line [77]. The corresponding SX-ARPES data is shown in Fig. 6.7b for the L_3- and L_2-resonances. We note remnant signal from the d_{yz}-band manifesting slight structural distortions at the interface [87, 88] relaxing the perfect symmetry.

Typical FS of the standard LAO/STO samples collected with circular X-ray polarization at the L_3-resonance and extending over four BZs [77] is shown in Fig. 6.8a. These results are in agreement with the pioneering experiments by Berner et al. [80, 81]. The FS contours in the Γ_0 point show a circle, originating from the light d_{xy} bands, and two ellipsoids aligned along the k_x and k_y directions, originating from the heavy d_{yz} and d_{xz} bands. The intensity asymmetry relative to the horizontal ΓX line reverses with the X-ray chirality, manifesting a circular dichroism. Interestingly, the apparent FS shapes significantly vary through the four BZs. This effect is caused by different dependence of the d_{xy}, d_{xz} and d_{yz} photoemission matrix elements and thus corresponding FS sheets on (k_x, k_y). Furthermore, in line with linear dichroism of the band structure in Fig. 6.7, one can use linear X-ray polarization for orbital decomposition of the experimental FS, Fig. 6.8b: s-polarization enhances the d_{xy} and d_{xy} sheets, and p-polarization the d_{xz} one. Another interesting point is that the apparent FS contours are somewhat different between the L_2- and L_3-resonances (not shown

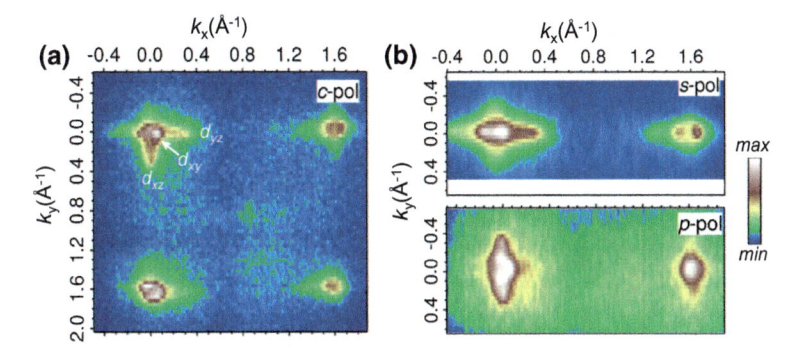

6 Electrons and Polarons at Oxide Interfaces ... 121

Fig. 6.8 FS of standard LAO/STO samples collected at circular **a** and linear **b** X-ray polarizations. s-polarization enhances the d_{xy} and d_{xy} sheets, and p-polarization the d_{xz} one. Adapted from Cancellieri et al. [77]

here for brevity) as a consequence of the orbital selectivity of resonant photoemission mentioned above. Evolution of band structure and FS of the LAO/STO interface a function of LAO overlayer thickness was studied by Plumb et al. [89].

6.2.3 Doping Effect on the Band Structure

In Chaps. 2 and 3, it is well explained how the growth conditions affect the electronic properties like conductivity, concentration n_s and mobility of the charge carriers. In particular, n_s changes consistently with the growth temperature [90] as well as with oxygen annealing [91], although in the latter case the conductivity can transform from 2D to 3D. Cancellieri et al. [77] have performed SX-ARPES studies of LAO/STO samples with different doping levels in order to correlate the band occupancy and FS shape with n_s measured in transport. Standard doped (SD) samples with $n_s = 6.5 \times 10^{13}$ e/cm^2 and low doped (LD) samples with $n_s \sim 10^{13}$ el/cm^2 were investigated. The ARPES dispersions were compared with theoretical simulations performed with the transport n_s value in order to clarify the corresponding orbital-resolved charge distributions and band occupations (a detailed theoretical analysis of the LAO/STO band structure is provided in Chap. 8).

Theoretical simulations for the SD sample in Fig. 6.9a show that for given n_s most of the charge is embedded in three bands. The two lowest have the planar d_{xy} character with a light mass $m* \approx 0.7m_0$ (m_0 is the free-electron mass) in the ΓX direction. The associated electron charge is confined entirely within the first and second TiO$_2$ layers from the interface. The third occupied band has the d_{yz} character and embeds the charge spread out-of-plane. The band is heavy along k_x ($m* \approx 9m_0$) and shifted upward in energy by ~70 meV relative to the lowest d_{xy} band. The corresponding simulated FS is consistent with the experiment, Fig. 6.9b. Luttinger count of the simulated FS demonstrates that even for the SD samples with their maximal n_s the

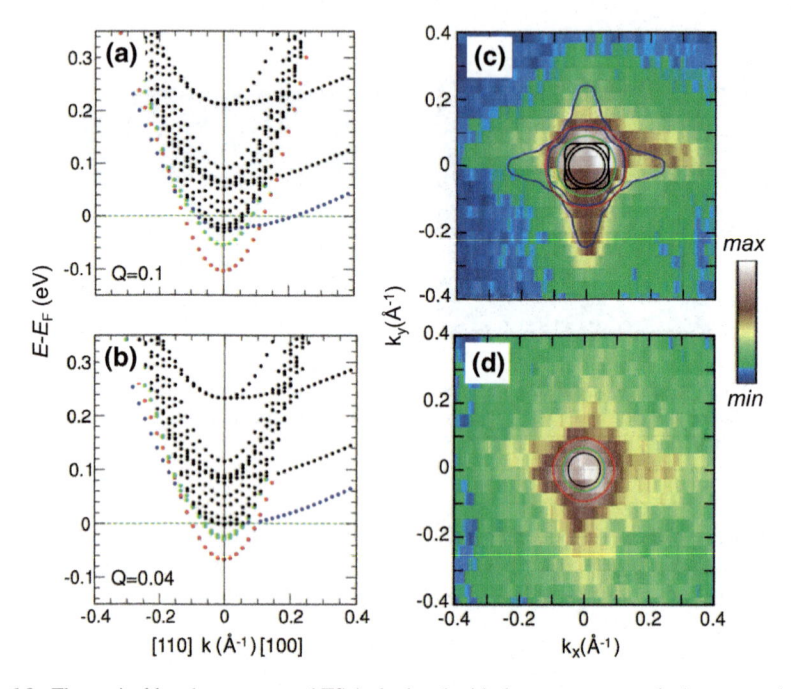

Fig. 6.9 Theoretical band structure and FS (calculated with the transport n_s value) compared with the experimental FS: **a, b** Standard and **c, d** low doped LAO/STO sample. Luttinger count of the experimental FS increases with n_s. Adapted from Cancellieri et al. [77].

ideal limit dictated by the polar catastrophe ($n_s = 3.3 \times 10^{14}$ e/cm^2 corresponding to half an electron per interface u.c.) is not actually achieved.

With decrease of doping for the LD sample, obtained by reduction of the growth temperature [90], the experimental FS reduces its Luttinger count as expressed by reduction of the d_{yz}-band Fermi vector k_F by ~0.05 Å^{-1}, which confirms coherent nature of the interfacial electron transport. The observed trend is consistent with the theoretical simulations in Fig. 6.9d incorporating the reduced n_s, where the whole band manifold shifts up in energy. The agreement, however, stays on the qualitative level because the experimental d_{yz} band stays partly occupied (although strongly reduced in its spectral weight), whereas in the calculations the d_{yz} band is entirely above E_F. Furthermore, we notice that for both SD and LD samples the experimental bands are systematically lower compared to the calculations. In other words, ARPES captures more charge carriers than detected in the transport measurements. This discrepancy hints on electronic localization and/or phase separation mechanisms at the LAO/STO interface (see Sect. 6.4.4). Under yet smaller doping the out-of-plane d_{yz} and d_{xz} orbitals will fully depopulate (Lifshitz transition) and the whole mobile charge will reside in the d_{xy} bands close to the interface with concomitant dramatic change in transport properties. Such depopulation can actually be achieved in gating experiments (Chap. 3).

6.3 Electron-Phonon Interaction and Polarons at LAO/STO

We will review now SX-ARPES results that reveal strong electron-phonon interaction (EPI) at the LAO/STO interface. This effect results in polaronic nature of the interfacial charge carriers expressed by large Holstein-type polarons. This phenomenon plays a crucial role in transport properties of LAO/STO. For related physics of bare STO interface see Chap. 4. We start with a recap of fundamentals of the polarons.

6.3.1 Basic Concepts of Polaron Physics

6.3.1.1 Qualitative Picture of Polarons

The physics of a particle interacting with its environment—a polaron—is a general problem historically started from a Fröhlich model describing an electron moving in a dielectric medium [92, 93] and disturbing neighboring lattice by long-range Coulomb forces. Very soon it was realized that similar phenomenon occurs in molecular crystals where a Holstein model [94] is applied assuming that electron is locally coupled to the host molecule deformations. Nowadays, it is already well known that depending on a "particle" and "environment" and how they interact with each other, the polaron concept applies to extreme diversity of physical phenomena [95, 96].

A simple physical picture behind the polaron phenomenon can be imagined as an electron moving in a crystal where ions displace from their equilibrium positions in response to electron-lattice interaction. This electron dragging behind a local lattice distortion—or phonon "cloud"—forms the polaron as composite charge carrier whose properties are significantly different from the bare electron, e.g. the polaron mass m^* is heavier than the bare electron mass m_0. Formation of polarons significantly modifies the properties of the system. Apart from changing m^* [97, 98], it modifies optical conductivity [99, 100], mobility [101], ARPES response [98], etc.

We restrict our following discussions to the typical situations when electrons interact with a dispersionless optical phonons of frequency ω_0. In this case the electron-phonon interaction (EPI) can by described by the Hamiltonian

$$H_{\text{int}} = \frac{1}{\sqrt{N}} \sum_{\mathbf{kq}} \Gamma(\mathbf{k}, \mathbf{q}) \left(c_{\mathbf{k}}^+ c_{\mathbf{k-q}} b_{\mathbf{q}} + c_{\mathbf{k-q}}^+ c_{\mathbf{k}} b_q^+ \right)$$

that represents the scattering of electron changing its momentum by $\mathbf{q}$ when it annihilates ($b_{\mathbf{q}}$) or creates (b_q^+) a quantum of lattice vibration (phonon) carrying momentum $\mathbf{q}$. Here $c_{\mathbf{k}}^+/c_{\mathbf{k}}$ are creation/annihilation operators of electron with momentum $\mathbf{k}$.

In general, the interaction vertex $\Gamma(\mathbf{k}, \mathbf{q})$, characterizing the EPI dependence on momenta of the particles participating in the scattering event, depends not only on phonon $\mathbf{q}$ but also on electron $\mathbf{k}$. Although the $\mathbf{k}$-dependence leads to a plenty of interesting phenomena [102] it is usually observed either in organic materials

[103, 104] or arises in case of magnetic polarons [105]. Limiting our discussion to lattice polarons in inorganic materials, we consider only a **q**-dependent vortex $\Gamma(\mathbf{q})$. However, even within this simplification there remains a variety of profoundly different EPI behaviors which are usually divided into long-range and short-range interaction.

6.3.1.2 Long-Range Versus Short-Range Electron-Phonon Interaction

The effect of EPI on physical properties depends on its long-versus short-range character. The long-range EPI with optical phonons is represented by so-called Fröhlich interaction arising from Coulomb coupling having maximal strength at zero momentum

$$\Gamma(\mathbf{q}) \sim \frac{\sqrt{\alpha}}{q^{(d-1)/2}}. \tag{6.2}$$

Here α is a dimensionless coupling constant and $d = 2$ or 3 is the dimensionality of the system [106]. Square of the potential (6.2) is the Fourier transform of the long-range three-dimensional Coulomb potential $1/r$. Thus, the EPI (6.2) implies three-dimensional interaction between the electron and polarization although the electron movement is embedded into d-dimensional space. One should note that the Fröhlich interaction does not introduce singularities into physical properties of ground state, because its contribution into a physical quantity, e.g. energy renormalization ΔE, implies the radial momentum integration

$$\Delta E \sim \int_0^\infty \frac{|\Gamma(\mathbf{q})|^2}{\omega_0 + q^2/(2m_0)} q^{d-1} dq$$

whose radial dimensionality-dependent factor q^{d-1} cancels the singularity of $\Gamma(\mathbf{q})$ at zero momentum.

The short-range EPI is represented by so-called Holstein interaction (Holstein [94] which is momentum independent

$$\Gamma(\mathbf{q}) = g \tag{6.3}$$

and arises from different mechanisms of short-range coupling to charge density distortions [107, 108]. Note that the momentum independence (6.3) implies, in contrast to the long-range one (6.2), that the interaction is local in the direct space as described by the delta-functional $\delta(\mathbf{r})$. Extending the strict definition (6.3), a terminology of Holstein-like EPI is often used for the cases where $\Gamma(\mathbf{q})$ is maximal at the Brillouin zone (BZ) boundaries [109, 110] which is typical of breathing phonon modes.

The EPI for the Holstein model is traditionally characterized by the dimensionless coupling constant

$$\lambda = \frac{2g^2}{W\omega_0} \tag{6.4}$$

defined in terms of the bare particle bandwidth W. Physically, λ is the ratio of two energies, one is the potential well due to lattice deformation and another is kinetic energy helping escape the well. Namely, the former is the polaron binding energy g^2/ω_0 at zero bandwidth $W = 0$ and the latter is half of the bare electron bandwidth $W/2$. In the hypercubic lattice of dimensionality d we have $W = 4dt$ and $m_0 = 1/2t$, where t is the electron hopping constant. If $\Gamma(\mathbf{q})$ is momentum dependent, one can introduce an effective Holstein-like dimensionless coupling constant λ by substituting $g \to \Gamma(\mathbf{q})$ and averaging the quantity (6.4) over the BZ.

The above definition of the dimensionless coupling (6.4) has several advantages. First, it is compatible with definition of λ in theory of metals [111] if one realizes that $1/W$ is roughly equal to the electronic density of states $N(E_F)$ at the Fermi energy E_F. Second, the perturbation theory for mass renormalization in metals gives very simple expression [111]

$$m^*/m_0 = 1 + \lambda \tag{6.5}$$

Note however that in the case of a single Holstein polaron there is an additional coefficient around unity in front of λ which depends on the dimensionality, lattice geometry and phonon frequency [97, 112]. Third, an important convenience of the definition (6.4) is that the critical value λ_c of the dimensionless coupling, which separates weak- and strong-coupling regimes, is around unity,

$$\lambda_c \sim 1 \tag{6.6}$$

The properties of a polaron with short-range interaction are weakly renormalized at $\lambda < \lambda_c$ and strongly modified at larger λ [97] resulting in large polaron m^* and strong lattice deformations accompanying such polaron.

The dimensionless coupling constant α of long-range coupling (6.2) is conventionally defined in terms of the dielectric properties of the crystal, and a rigorous description of this definition is rather lengthy. However, the most important information can be conveyed if we note that for the Fröhlich polaron the expressions (6.5) and (6.6) are valid if we substitute

$$\lambda \to \frac{\alpha}{6}$$

Note that in the perturbation expansion for the Fröhlich polaron $m^* = m_0/(1 - \alpha/6) \approx m_0(1 + \alpha/6)$ we use the smallness of α. The crossover between the weak- and strong-coupling regimes for the Holstein model can be a rather fast function of λ, whereas for the Fröhlich model it is always a smooth function of α. Properties of the Fröhlich polaron qualitatively change however at $\alpha_c = 6$ [98–100].

It can be shown that for Holstein coupling the electron self-energy is momentum independent for large enough E_F [113]. In such cases one can prove [114] that for the Holstein polaron

$$Z_0 \left(m^*/m_0 \right) \approx 1.$$

For the Fröhlich polaron, on the other hand, this relation is violated even for very weak EPI. Instead, the perturbation expressions for the Z-factor $Z_0 = 1 - \alpha/2$ and mass $m^*/m_0 = (1 - \alpha/6)^{-1}$ yield

$$Z_0 \left(m^*/m_0 \right) \approx 1 - \frac{\alpha}{3}.$$

The polaron properties in the strong coupling regime are considerably different from bare particle both for long- (6.2) and short-range (6.3) EPI. However, only the latter case is capable of forming so called small-radius polarons when almost the whole lattice deformation is localized at the site of the particle. The reason for such localization is the $\delta(\mathbf{r})$-character of the short-range EPI which can lead to strong on-site deformation leaving almost intact the neighboring sites. Hence, the small–radius polaron can be viewed as a particle with strong deformation at its site which is surrounded by almost unperturbed lattice. In this case the particle hopping to another site is a dramatic event associated with removal of the large lattice deformation at one site and creation of the same strong distortion at the neighbour site. Naturally, coherent movement of a small polaron is easily destroyed either by lattice imperfection or temperature induced lattice potential fluctuation [95, 96, 115]. Such phenomena are impossible in case of long-range EPI because of its $1/r$ long-range character.

One more manifestation of the EPI restricted to systems with short-range couplings is the so called "self-trapping" phenomenon [116, 117]. Self-trapping means a dramatic transformation of particle properties when system parameters are slightly changed. It occurs at $\lambda \sim \lambda_c$ when a "trapped" particle state with strong lattice deformation around it and the weakly perturbed "free" particle state may have nearly the same energy. Then, the "free" state with small m^* is effectively trapped by an admixture of the "trapped" state with large m^*, see Mishchenko et al. [118] for demonstrations of such mixing.

One has to note that the "self-trapping" phenomenon does not imply real trapping because it does not violate the translational invariance or momentum conservation. The real trapping requires in fact an attractive impurity potential breaking the translational invariance. For polarons, the corresponding lattice deformation cooperates with the attractive potential, making the impurity-assisted trapping much easier compared to bare particles [119–123]. For example, localization of a particle in a three dimensional cubic lattice requires local on-site attractive potential U to be larger than its critical value U_c roughly equal to $W/3$ [124]. For the Holstein polarons, however, U_c approaches zero with increase of λ [119–123].

6.3.1.3 Spectral Function of Polarons and Its Fingerprint in ARPES

The polarons are characterized by the electronic spectral function $A(\omega,\mathbf{k})$ that can be expressed as

$$A(\omega,\mathbf{k}) = Z_0 \delta(\omega - E(\mathbf{k})) + A_{SB}(\omega,\mathbf{k})$$

Here δ is the Dirac delta-function, Z_0 the quasiparticle (QP) weight or Z-factor, $E(\mathbf{k})$ the renormalized QP energy, and $A_{SB}(\omega,\mathbf{k})$ a high energy tail containing phonon sidebands, i.e. states where removal of electron is accompanied by excitation of one or more optical phonons. The normalization of $A(\omega,\mathbf{k})$ to unity as $\int_{-\infty}^{\infty} A(\omega,\mathbf{k})\,d\omega = 1$ allows its probabilistic interpretation [114] where an electron is removed without (or with) phonon emission with probability $Z_0 < 1$ (or $1 - Z_0 < 1$).

The shape of the high energy tail $A_{SB}(\omega,\mathbf{k})$ at zero temperature can be roughly represented as a sum of l-phonon sidebands

$$A_{SB}(\omega,\mathbf{k}) = \sum_{l=1}^{\infty} Q_l(\omega) D(\omega + l\omega_0), \quad (6.7)$$

where $D(\omega) = N^{-1} \sum_{\mathbf{k}} (E(\mathbf{k}) - \omega)$ is the normalized density of QP electronic states with dispersion $E(\mathbf{k})$ and $Q_l(\omega)$ are coefficients depending on the EPI strength, full energy width $\tilde{W}$ of the QP electronic band and its filling (for examples see Alexandrov and Ranninger [125] and Ranninger and Thibblin [126]). Energy of each l-th sideband in $A_{SB}(\omega,\mathbf{k})$ is shifted from $E(\mathbf{k})$ by $l\omega_0$ (plus, strictly speaking, a slight energy shift due to electron recoil accompanying excitation of phonon). Each sideband is broadened with $\tilde{W}$, and the individual lines in $A_{SB}(\omega,\mathbf{k})$ are resolved when $\tilde{W} \ll \omega_0$ and merge to a single broad hump when $\tilde{W} \gg \omega_0$. The above simple picture of $A_{SB}(\omega,\mathbf{k})$ formed by sidebands separated by one ω_0 breaks down in the cases of self-trapping [118] and impurity-assisted trapping [122].

So far we considered a model system including only EPI with optical phonons, where the low energy QP pole of $A(\omega,\mathbf{k})$ is a δ-function [98]. A realistic system contains many other interactions, e.g. electron-electron coupling, interaction with acoustic phonons and impurities. These interactions smear the infinitely high and narrow δ-function, but the remnant peak stays sharp in comparison with the tail of phonon sidebands. As a result, the whole $A(\omega,\mathbf{k})$ acquires a typical peak-dip-hump (PDH) structure (Fig. 6.10) where a relatively narrow QP peak is followed by broad hump at higher binding energy.

The above shape of $A_{SB}(\omega,\mathbf{k})$ is achieved only at low electron excitation energy $E(\mathbf{k}) - E_F \ll \omega_0$. This is however not the case when $E(\mathbf{k})$ is far from E_F [98, 127–129]. At excitation energy $\gg \omega_0$ the QP dispersion $E(\mathbf{k})$ undresses of the phonon cloud and becomes the bare $\varepsilon(\mathbf{k})$ dispersion. The crossover between these two regimes is characterized by the electron dispersion kink (for entries see Gunnarsson and Rosch [128]; Mishchenko et al. [130], Lanzara et al. [131], Cuk et al. [132],

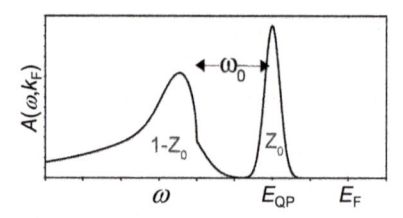

Fig. 6.10 Sketch of the polaron spectral function $A(\omega,\mathbf{k})$ at low electron excitation energy $E(\mathbf{k}) - E_F$. The characteristic peak-dip-hump (PDH) spectral shape is formed by the quasiparticle (QP) peak followed by the phonon sideband tail corresponding to a sequence of optical phonons with frequency ω_0

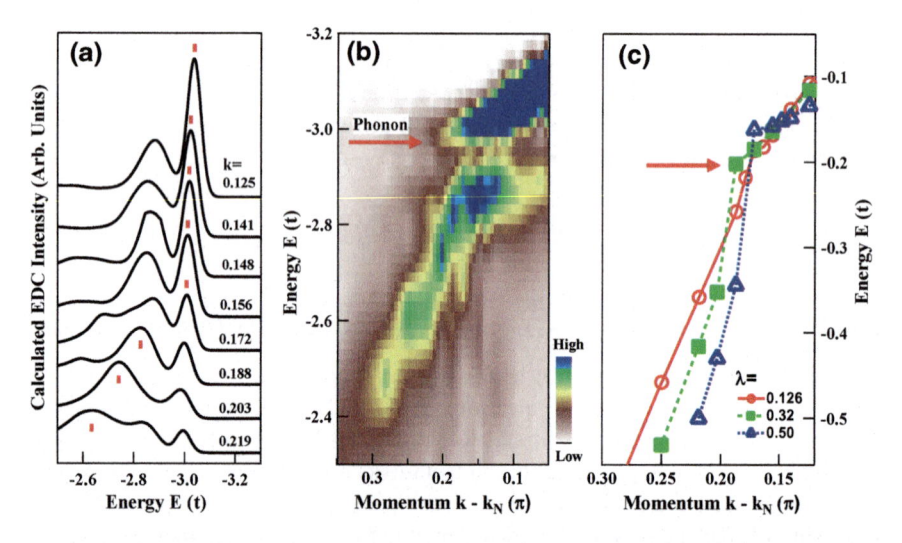

Fig. 6.11 a Typical $A(\omega,\mathbf{k})$ of high-T_c cuprates simulated within the t-J-Holstein model for $\lambda = 0.5$ with $\mathbf{k}$ varying along the nodal $(\pi/2, \pi/2) \rightarrow (0, 0)$ direction. The $A(\omega,\mathbf{k})$ maxima are marked; b Intensity plot for the same parameters; (c) Momentum dependence of the $A(\omega,\mathbf{k})$ maxima for $\lambda = 0.126$ (circles), $\lambda = 0.32$ (squares), and $\lambda = 0.5$ (triangles). The red arrow in (b) and c indicates the optical phonon energy. All energies are in units of $t = 0.4$ eV measured from the vacuum state of the t-J model (a, b) and from the top of the polaron band. Reproduced from Mishchenko et al. [130]

Devereaux et al. 133]. A typical picture of this crossover simulated for high-T_c cuprates [130] is illustrated in Fig. 6.11.

Naively, the phonon structure of the hump is often imagined similarly to the simplest situation of vibrational excitations in molecules such as gaseous hydrogen [134] where the QP peak is followed by a series of sharp lines at multiples of ω_0. This case corresponds to the situation of the independent oscillator model (IOM) [114] whose spectral function at zero temperature $T = 0$ is described by equidistant lines with weights given by Poisson distribution

$$A_{\text{SB}}(\omega) = e^{-g} \sum_{l=0}^{\infty} \frac{g^l}{l!} \delta(\omega - l\omega_0), \tag{6.8}$$

where the $l = 0$ term is the QP peak and $l > 0$ ones are satellite lines shifted by l energies ω_0. However, this Poisson distribution limit of the general $A_{\text{SB}}(\omega,\mathbf{k})$ form (6.7) can hardly be observed in condensed matter physics because the IOM is valid, first of all, only if the full electronic bandwidth $\tilde{W}$ is much smaller than ω_0 which is a quite rare case for realistic systems. Furthermore, one needs very low temperature $T \ll \omega_0$ to see the purely Poisson shape of the hump. Indeed, for finite temperature T the IOM spectral function becomes

$$A_{\text{SB}}(\omega) = e^{-g(2N+1)} \sum_{l=0}^{\infty} I_l \left\{ 2g\left[N(N+1)\right]^{1/2} \right\} e^{\left(\frac{\beta\omega_0}{2}\right)} \delta(\omega - l\omega_0),$$

where $N = \left(e^{-\frac{\beta\omega_0}{2}} - 1\right)^{-1}$ is the Bose distribution function, I_l the modified Bessel function, and $\beta = 1/T$ [114]. This function reduces to the Poisson distribution (6.8) only for $T/\omega_0 \to 0$ and otherwise has no resemblance of it. Note that in case of finite T there appear satellites with negative l values (negative energy loss). The conclusion is that the Poisson distribution of the phonon sidebands can therefore only be awaited when all of the following conditions apply: (i) dispersionless optical phonons ω_0; (ii) excitation energy in vicinity of E_F so that $E(\mathbf{k})$-$E_F \ll \omega_0$; (iii) isolated molecule vibrations with vanishing electronic bandwidth $\tilde{W} \ll \omega_0$; (iv) very low temperatures $T \ll \omega_0$. The whole list of these assumptions hardly applies to most of the cases in condensed matter physics.

Although the polarons dramatically influence physical properties of solid-state systems, retrieval of fundamental characteristics of the underlying EPI from various spectroscopic data is not straightforward. For example, m^* can in principle be determined from measurements of optical conductivity $\sigma(\omega)$ using the sum rule [135, 136]

$$\frac{1}{m_0} - \frac{1}{m^*} = \frac{2}{\pi e^2 n_0} \int_{\omega_0}^{\omega_m} \sigma(\omega)\, d\omega,$$

where e is the electron charge, n_0 the carriers concentration, and ω_m the maximal frequency which can be attributed to intra-band conductivity. However, the accuracy of such estimate of m^* depends on how trustworthy is the estimate of m_0 provided by band structure calculations.

The unique experimental probe which can capture the polaron physics is ARPES which directly measures the (matrix element weighted) electronic spectral function $A(\omega,\mathbf{k})$ most fully characterizing the EPI. Continuous progress of the ARPES technique reveals more and more cases where the polaronic $A(\omega,\mathbf{k})$ is observed. In particular, the PDH structure was observed, for example, (i) from localized lattice polarons in insulating sodium tungsten bronze Na_xWO_3 [137]; (ii) from lattice polaron states near the dispersion kink in the electron-doped $Nd_{1.85}Ce_{0.15}CuO_4$ high temperature superconductor [138]; (iii) from lattice polarons at the Ti(100) surface [139]; (iv) from spin-orbital polaron [140] in misfit cobaltate $[Bi_2Ba_2O_4][CoO_2]_2$ [141]. Lattice polarons at bare $SrTiO_3$ surface are considered in Chap. 4, and those at the $LaAlO_3/SrTiO_3$ interface below in Sect. 6.3.2. Enhancement of EPI due to disorder was observed by Nie et al. [142].

Thorough theoretical studies of the ARPES results for high temperature superconductors reveal considerable EPI in these materials [143, 144]. For example, the extended t-J model [145–147], widely used for underdoped materials, well describes [148] the experimental ARPES dispersion [149] of the low energy peak in underdoped cuprates. However, while the unbiased Monte Carlo studies of the t-J model predict narrow width of this peak near the band bottom [150], the experiment finds the peak width larger than the whole dispersion bandwidth. Inclusion into these theoretical models of EPI extends correct predictions for the underdoped cuprates from the ARPES dispersion [151] to linewidth [87], as well as to temperature dependence of the ARPES spectra [152], kink phenomena [130] and optical conductivity [153].

6.3.2 Polaronic Nature of the LAO/STO Charge Carriers

6.3.2.1 Low Temperature Limit

We will analyze the low-temperature SX-ARPES data in Fig. 6.12 in terms of the one-electron spectral function $A(\omega,\mathbf{k})$, closely following Cancellieri et al. [86]. Figure 6.12 reproduces the above s-polarization data at the L_3- (a) and L_2- (d) resonances together with the corresponding $A(\omega,\mathbf{k})$ profiles for the d_{xy}-band along the waterfalls and for the d_{yz}-band (c and f, respectively). Remarkably, the experimental $A(\omega,\mathbf{k})$ reveals a pronounced peak-dip-hump (PDH) structure, where the peak reflects the QP and hump at lower energy its coupling to bosonic modes. Such modes are phonons coupling to electron excitations and forming a polaron (see Sect. 6.3) which is an electron (hole) dragging behind a local lattice distortion formed in response to strong EPI. The concomitantly increasing m^* of such composite charge carriers fundamentally limits their mobility $\mu \propto 1/m^*$ beyond the incoherent scattering processes. Therefore, the charge carriers at the LAO/STO interface have *polaronic* nature fundamentally limiting the 2DES mobility μ_{2DES}. The experimental d_{yz} dispersion shows the effective mass renormalization $m^*/m^0 \sim 2.5$ relative to the DFT mass m^0. Moreover, clear dispersion of the hump tracking the d_{yz}-band, seen in the second derivative plot of Fig. 6.12e, identifies a *large polaron* associated with long-range lattice distortions. The hump apex, located at ~118 meV below the QP peak, identifies the main coupling phonon frequency ω_0'. Based on DFT calculations of phonon modes [154] for bulk cubic STO shown in Fig. 6.13a, this phonon can be associated with the hard longitudinal optical mode LO3 at ~100 meV having the largest coupling constant λ

6 Electrons and Polarons at Oxide Interfaces ...

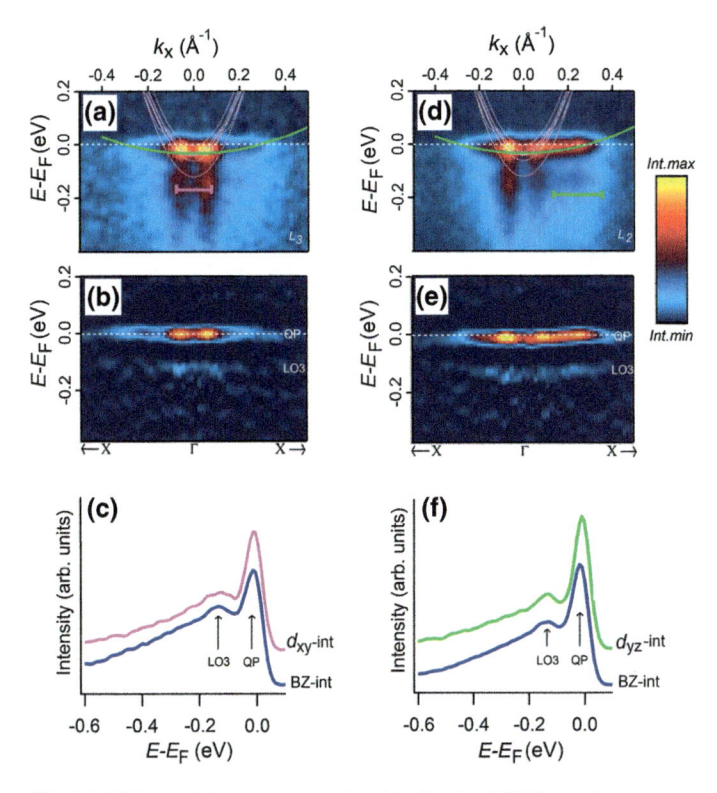

Fig. 6.12 SX-ARPES derived dispersions and $A(\omega,\mathbf{k})$ of the LAO/STO interface states: **a, d** ARPES images at the L_3 (**a**) and L_2 (**d**) resonances with theoretical d_{yz} (green) and d_{xy} (pink) bands. The panels **b, e** are their representation in (normalized) second derivative $-d^2I/dE^2 > 0$; **c, f** $A(\omega,\mathbf{k})$ reflecting the d_{xy} and d_{yz} bands integrated over the **k**-regions marked in (**a, c**). The characteristic PDH structure manifests polaronic nature of the interface charge carriers formed by the LO3 phonon, and the hump dispersion identifies a large polaron. Adapted from Cancelliri et al. [86]

among all LO modes [155, 156]. This mode is a breathing distortion of the octahedral cage around a Ti site (*b*).

Optical studies on bulk STO [157] have not only confirmed the involvement of the LO3 phonon in EPI but also found the corresponding effective mass renormalization $m^*/m^0 \sim 3.0$ close to our value. The same phonon was observed by Raman [158] and neutron spectroscopy [159] as well as by VUV-ARPES at the bare STO(001) surface [160–162] with the strength of the polaronic structure depending on n_s (see Chap. 4). The LO3 energy ~100 meV found in all these experiments perfectly matches the DFT calculations, Fig, 6.13a. At the LAO/STO interface however the LO3 frequency is somewhat different which can be attributed to coupling of the STO phonon modes to the LAO overlayer. Interestingly, a recent tunnelling study of oxygen isotopic effects at the LAO/STO interface [163] has again found the LO3 phonon (LO4 in the strict tetragonal nomenclature of this work) at ~100 meV. The difference may indicate the

Fig. 6.13 **a** DFT calculated phonon modes in doped bulk STO at various electron doping levels n_v (our LAO/STO case corresponds to $n_v \sim 0.12$); **b, c** Atomic displacements associated with the breathing LO3 mode and polar TO1 one, respectively [86]

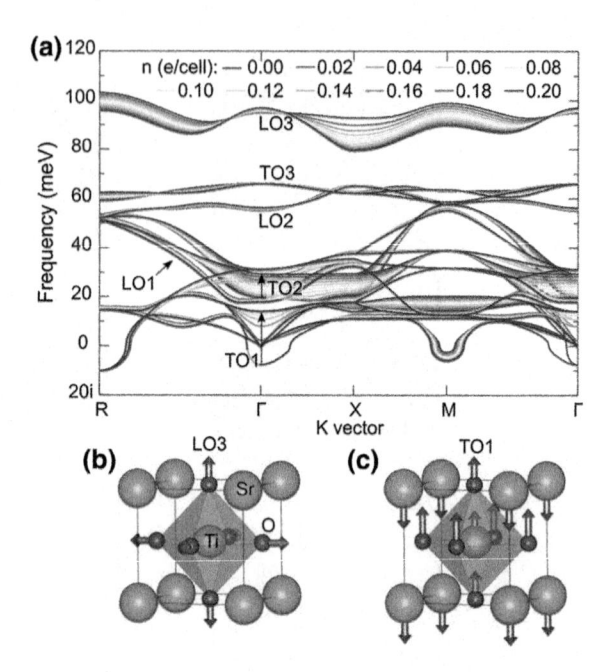

fact that whereas the SX-ARPES signal is formed only by coherent electrons coming from the interface region, the tunneling signal has a large contribution of incoherent electrons coming deep from the STO bulk. Nie et al. [142] have extended VUV-ARPES studies of polaronic effects to the quasi-2D material Sr_2LaTiO_4 from the STO-related Ruddlesden-Popper series $Sr_{n+1}Ti_nO_{3n+1}$. La-doping of this material to $Sr_{2-x}La_xTiO_4$ was found to set up an interesting interplay between EPI and disorder, reproduced by model theoretical analysis.

As we have seen above, the polaronic $A(\omega,\mathbf{k})$ can be expressed as $A(\omega,\mathbf{k}) = Z_0\delta[\omega - E(\mathbf{k})] + A_{SB}(\omega,\mathbf{k})$, where the first term represents the QP peak with dispersion $E(\mathbf{k})$ and the second one the polaronic hump. The experimental $A(\omega,\mathbf{k})$ in Fig. 6.12b, d shows integral weight of the QP peak in the total spectral intensity $Z_0 \sim 0.4$ for both d_{xy}- and d_{yz}-bands [although the d_{xy}-$A(\omega,\mathbf{k})$ shows slightly larger energy broadening resulting from the localization of these states near the defect-reach interface; this spectroscopic observation is consistent with larger defect scattering of the d_{xy} charge carriers found in Shubnikov–de Haas experiments [164]. Crucial for further analysis is the theoretical LO3 phonon dispersions in Fig. 6.13a. The LO3 frequency responds to the n_s variations mostly in the large-**q** region. This identifies a short-range character of the corresponding EPI described by the *Holstein-type polaron* (see Sect. 6.3.1.2). With moderate EPI strength, the Holstein-type EPI forms large polarons, similarly to the Fröhlich polarons driven by long-range EPI [165]. The short-range EPI constant $\lambda \sim 1 - Z_0$ is in our case ~0.6, which is too weak for the polaronic self-trapping [40] but can assist charge trapping on V_{OS} or shallow defects [122, 142] contributing to the polar field compensation in LAO [166].

Importantly, the QP mass renormalization for the Holstein polarons directly relates to the QP spectral weight as $m^*/m^0 = 1/Z_O$. In our case this relation returns $m^*/m^0 = 2.5$ exhausting the experimental band renormalization in Fig. 6.12a, c. This fact leaves not much room for the electron correlation effects in the interfacial 2DES, which is consistent with the tunneling data of Breitschaft et al. [76]. We note also that the significant occupied electronic bandwidth and LO3 dispersion result in significant deviations of the experimental $A_{SB}(\omega,\mathbf{k})$ from the ideal Poisson distribution (see Sect. 6.3.1.3).

6.3.2.2 Temperature Dependence

An intriguing peculiarity of the LAO/STO interface is a dramatic reduction of μ_{2DES} with temperature T [167] (see Chap. 3). We address this puzzle with a series of T-dependent L_3 angle-integrated spectra, Fig. 6.14a, because analysis of angle-integrated spectra is immune to relaxation of $\mathbf{k}$-selectivity with T [13]. The QP peak reduces with increase of T, and towards 200 K completely dissolves in the phonon hump of $A(\omega,\mathbf{k})$. The corresponding decrease of the QP weight Z_O, Fig. 6.14c, increases m^* and thus reduces the charge carriers mobility. This microscopic mechanism fully explains the puzzling 2DES mobility reduction observed in Hall effect, Fig. 6.14b. In line with our results, optical studies of the bulk STO [157] and LAO/STO interface [73] have also found reduction of the Drude weight with T.

The phonon mode behind the observed T-dependence can be identified by fitting of the experimental $Z_0(T)$ in Fig. 6.14c with the independent boson model [114] $Z_0(T) = e^{-2g(2N+1)} I_0 [2g(2N+1)]$, where $N = \left(e^{\omega_0/T} - 1\right)^{-1}$ is the Bose filling factor describing population of the phonon modes as a function of their frequency ω_0 and T. The fit identifies the frequency ω_0'' of the soft phonon dominating the EPI as ~18 meV in the low-T range and ~14 meV in the high-T range. This crossover can be linked to the second-order tetragonal to cubic phase transition in STO at 105 K [168]. The theoretical phonon modes in Fig. 6.13a suggest that the observed phonon is the polar TO1 one in bulk STO, Fig. 6.13c, sensitive to this phase transition and associated with long-range EPI. The corresponding coupling constant estimated as $\alpha \sim 2$ indicates moderate EPI consistent with the large polaron scenario. The TO1 phonon was also observed in kinks of ARPES dispersions at bare STO(100) [155] and its hardening due to presence of the 2DES by THz reflectivity spectroscopy [169]. This mode is inherent to the nearly ferroelectric nature of STO, being associated with its huge ϵ_{STO}.

The LAO/STO interfacial charge carriers have therefore polaronic nature involving at least two phonons with different energy and thermal activity. The LAO/STO superconductivity, if driven by a phonon mechanism [163], can be related to the discovered polaronic state. Whereas the hard LO3 phonon energy much exceeds the energy scale of the superconducting transition at 0.3 K, involved in the electron pairing may be the soft TO1 phonon. The polaronic activity is actually typical of TMO perovskites, reflecting their highly ionic character and easy structural transforma-

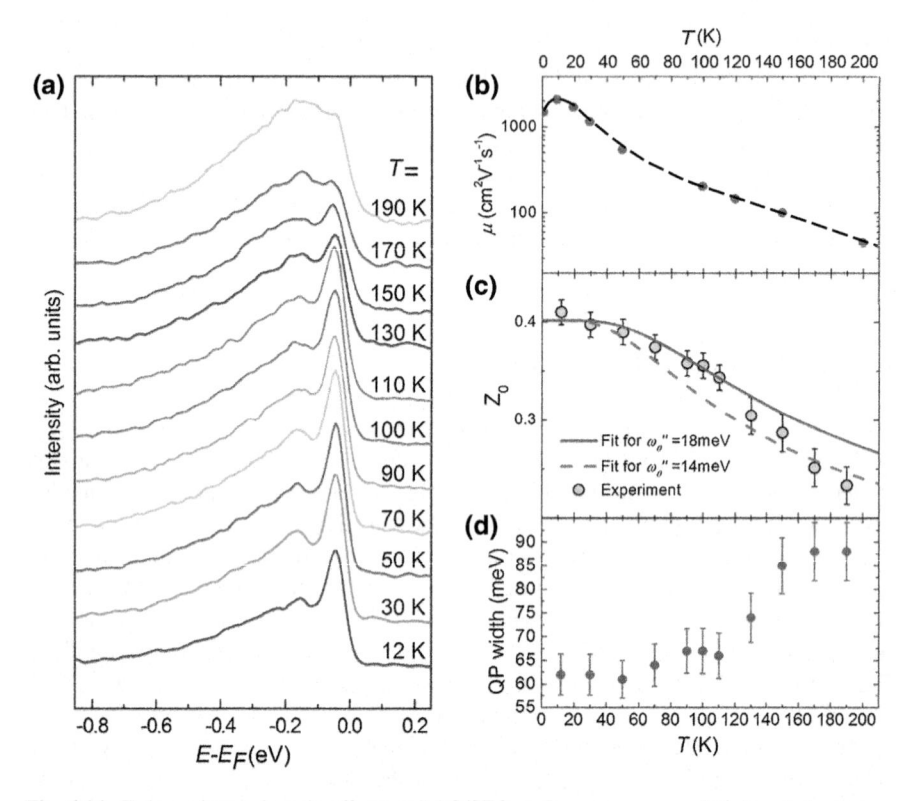

Fig. 6.14 T-dependent polaronic effects at LAO/STO. **a** L_3-resonance angle-integrated spectra. The QP peak dissolving into the hump towards ~190 K explains **b** the mobility drop with T; **c** QP spectral weight $Z_0(T)$ fitted by the independent boson model, identifies the soft TO1 phonon, Fig. 6.13c, sensitive to the phase transition in STO; **d** Corresponding QP energy width (Cancellieri et al. 2015)

tions [168]. For further details of polaronic physics at the LAO/STO interface the reader is referred to [86].

6.4 Oxygen Vacancies at LAO/STO

6.4.1 Signatures of Oxygen Vacancies in Photoemission

Oxygen (ox-) deficiency dramatically affects electronic and magnetic properties of the TMO systems. In particular, transport measurements evidence that the ox-deficiency dramatically increases both concentration and mobility of the interfacial 2DES [91, 170]. For all bulk, surface and interface STO-based systems, each oxygen vacancy (V_O) releases two electrons. In general, one of these electrons injected into the 2DES, whereas another stays near the Ti ion [166] to form there a localized state. Having a binding energy around 1.2 eV within the band gap of bulk STO, these states are commonly referred to as the in-gap states. Apart from the ARPES results discussed below, the coexistence of mobile and oxygen-derived localized electron states at the LAO/STO interface has been suggested by scanning tunneling spectroscopy [76, 171], resonant inelastic X-ray scattering [172], O K-edge XAS [173], etc.

In contrast to the stoichiometric STO where the Ti $3d$-e_g states are empty, the V_Os modify the local covalent bonding in its proximity and cause an orbital reconstruction where the e_g states shift down to become the in-gap states [174–176]. Tight electron localization in this state causes notable correlation effects [174, 177, 178] which necessitate the use of many-body theoretical methods like the DMFT for their analysis (see Chap. 8). Furthermore, electrons the in-gap states can hop to adjacent lattice sites under thermal or electric field activation [179–181] and may therefore be viewed as small polarons [166, 182] associated with a sizable distortion of the surrounding lattice. Finally, spins of the in-gap states can stabilize the ferromagnetic order (Sect. 6.4.3). Therefore, the VOs in the STO-based materials give rise to a duality of their electron system [178, 183] where delocalized 2DES quasiparticles (which are t_{2g} derived, weakly correlated, non-magnetic and forming large polarons, see Sect. 6.3.2) co-exist with localized in-gap states (e_g derived, more correlated, magnetic and forming small polarons).

Oxygen (ox-) deficient LAO/STO samples are usually grown at reduced O_2 pressure about 10^{-5} mBar without the standard post-annealing procedure. Typical ResPES response of these samples is shown in Fig. 6.15. Whereas the XAS spectrum (a) is hardly distinguishable from the annealed samples, Fig. 6.5, the ResPES intensity (b) immediately reveals the V_O derived in-gap states around $E_B = 1.2$ eV. The e_g character of these states manifests itself in advance of their resonant hv compared to the t_{2g}-derived 2DES. Interestingly, the in-gap peak slightly displaces in E_B as a function of excitation energy. This effect can trace back either to a multiplet structure of the in-gap states caused by electron correlations, or to different atomic configurations of the V_Os.

Prototypic of the TMO interfaces are their bare surfaces. VUV-ARPES experiments on TiO_2, STO and $BaTiO_3$ surfaces are reviewed in Chap. 4. Oxygen deficiency and thus conductivity of these surfaces can be achieved by their exposure to photons with energy larger than ~38 eV. In this case the formation of VOs involves excitation of Ti $3p$ core holes which are filled via an interatomic Auger process by electrons from the neighboring O $2p$ orbitals [184, 185]. An alternative method to create V_Os is deposition of a thin layer of Al [186] or other metals such as Ti, Nb, Pt, Eu and Sr [187]. The delocalized d_{yz} states at the ox-deficient STO interfaces and surfaces can extend to 200 Å and more into the STO bulk [74] that makes these states nearly three-dimensional [188]. Well controlled manipulation by V_Os to fine tune the electronic and magnetic properties of TMO heterostructures may assist engineering of future oxide electronic devices.

Fig. 6.15 ResPES of ox-deficient LAO/STO: **a** XAS spectrum; **b** Map of angle-integrated ResPES intensity. The in-gap states resonating around $E_B = 1.2$ eV are characteristic of the V_Os. A and B mark the V_Os and 2DES resonance regions, respectively, used in Fig. 6.17 for the phase separation analysis

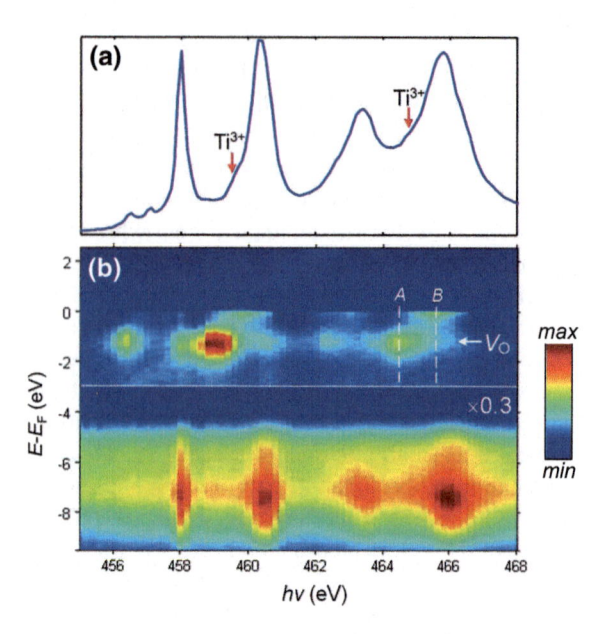

6.4.2 Tuning the Polaronic Effects

The polaronic reduction of μ_{2DES} fundamentally limits the application potential of the STO-based heterostructures. This limit can possibly be circumvented through manipulation of V_Os. As we have seen above, in general each V_O releases from the neighbour Ti ion two electrons. One of them stays in the impurity state, and another injects into the mobile 2DES to increases the electron screening and thus reduce the EPI strength [86].

The effect of the V_Os is illustrated by results of SX-ARPES measurements on oxygen deficient LAO/STO samples, Fig. 6.17, where increase of the V_O concentration gradually reduces the polaronic weight. This trend is consistent with VUV-ARPES studies at bare Ti(100) and STO(100) surfaces [139, 162] where n_s was varied in a wide range by exposure to VUV photons creating V_Os. The EPI changed in these experiments from very strong, resulting in overwhelming polaronic weight compared to the QP peak, to very weak, surviving in a kink structure of electron dispersions in vicinity of E_F. We note, on the other hand, that the V_Os can in principle assist the EPI due to charge trapping on the associated shallow defects [122, 142] as well as increase the defect scattering rate [189], both effects counteracting the above increase of μ_{2DES}.

The effect of the V_Os is actually beyond the simple doping picture restricted to changing of the band filling within the rigid band shift model. In particular, the V_Os increase the spatial extension of the $d_{xz/yz}$ bands into the STO bulk from ~50 Å for oxygen-annealed samples to ~150 Å and more [71, 72, 75] resulting in predominantly bulk conductivity and loss of the 2D nature of the LAO/STO interface system.

Furthermore, the manipulation of the V_Os is complicated by diffusion processes which are hard to precisely control. An interesting example of "defect engineering" is however the y-Al_2O_3/STO interface [190] where not only n_s increases but also low-temperature μ_{2DES} boosts by almost two orders of magnitute compared to LAO/STO. First-principles calculations suggest that in this case the V_Os diffuse out of the STO bulk and accumulate at the interfacial monolayer, reducing thereby their concentration and associated defect scattering in the deeper STO layers [191]. Further experiments on ox-deficient LAO/STO interfaces will bring better understanding of the role of V_Os and ways to optimize μ_{2DES}.

6.4.3 Interfacial Ferromagnetism

Ferromagnetism (FM) of TMO interfaces built up from non-magnetic constituents is one their most interesting functionalities (see, for example, [65, 192]. The origin of this phenomenon is still elusive even for the paradigm LAO/STO interface [193]. In one of the theories, so-called Hund's rule induced FM [64, 194], alignment of the motion of the d_{xy} electron spins is mediated by motion of the $d_{xz/yz}$ electrons. In this case the d_{xy} spins may rotate from site to site, giving thus rise to an exotic spin spiral state. An alternative scenario [79, 170, 175, 176] suggests that the interfacial FM is not an intrinsic property of 2DES and links it to the V_Os that spin split the 2DES energy levels. The decisive role of the V_Os is consistent with the fact that the interfacial FM quenches with annealing in oxygen.

Hints on the origin of FM at the LAO/STO were recently obtained from VUV spin-resolved ARPES (SARPES) experiments on bare STO surface Santander-Syro et al. 194]. They suggested that two d_{xy} bands possessed a Rashba-like spin splitting characterized by their opposite spin orientation, Fig. 6.16a, and helical winding around the corresponding FS sheets (b). This spin texture was interpreted as a giant Rashba splitting of ~100 meV enhanced by the antiferrodistortive corrugation of the bare STO(100) surface.as compared to the (field effect tunable) Rashba splitting of a few meV observed by magnetotransport for the LAO/STO interface [68, 195]. The most striking result of this study, a giant spin splitting of ~90 meV in the otherwise time-inversion protected Γ–point, was interpreted as an evidence of a FM order induced presumably by the V_Os. Subsequent theoretical analysis based on DFT slab calculations [183, 196] suggested a modified scenario: The Rashba-like spin splitting of the d_{xy} bands stays within a few meV consistent with the magnetotransport results, but magnetic moments of Ti ions spin split these bands with an energy difference of $\sim$ 100 meV at the Γ point consistent with the VUV-ARPES findings. While the magnetism tends to suppress the relativistic Rashba interaction effects, their signatures survive in complex spin textures of the 2DES. The giant spin splitting at the STO(100) surface was however not confirmed by another VUV-SARPES study by McKeown Walker et al. [197]. The conflicting results of the two studies can be reconciled based on a theoretical finding [177, 198] that while single–V_O configurations possess randomly oriented spins, multiple-V_O configurations can build up the

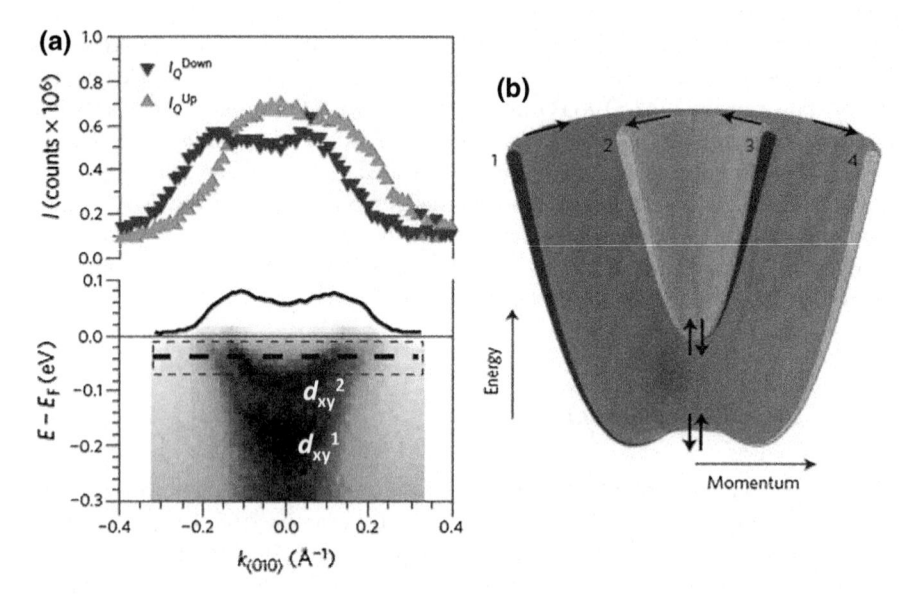

Fig. 6.16 VUV-SARPES of bare STO(100) surface: **a** Two d_{xy} bands (bottom) and their spin polarization (top); **b** Schematic spin texture. The Zeeman-like splitting of the d_{xy} bands in the Γ-point suggests a FM order connected with V_Os. Adapted from Santander-Syro et al. [194]

FM order. The PLD grown samples used by Santander-Syro et al. [194] could have larger concentration of V_Os (consistent with larger spectral broadening) compared to the cleaved samples used by McKeown Walker et al. [197] and thus could embed a larger fraction of the FM multi-V_O configurations. Spin-resolved ARPES will in near future be advanced to understand nature of the LAO/STO interfacial FM using ResPES measurements with multichannel spin detector iMott [199] as discussed in Sect. 6.5.

6.4.4 Phase Separation

An inherent feature of many TMOs and their interfaces is electronic phase separation (EPS) as spontaneous formation of micro- to nano-scale regions possessing different electronic and magnetic properties. This phenomenon plays crucial role, for example, in colossal magnetoresistance of manganites and stripe order in cuprates (for entries see Dagotto [200] and Shenoy et al. [201]). For the LAO/STO interface, evidence of this phenomenon has been observed by tunneling experiments [202, 203], scanning tunneling microscopy/spectroscopy [76, 171], atomic force microscopy [204], magnetoresistance and anomalous Hall effect [205], etc. The EPS has also been identified in a percolative character of the metal-to-superconductor transition with a significant fraction of 2DES resisting superconductivity down to the lowest temperature

[202, 206, 207]. On the magnetic side, FM puddles embedded in metallic phase have been observed by magnetotransport experiments [208] explaining the intriguing co-existence of the interfacial ferromagnetism and superconductivity. Finally, the EPS explains why n_s observed in transport at the LAO/STO interface always falls short of predictions of the mean-field theories. First ARPES study of the EPS has been performed by Dudy et al. [209] on bare STO(100) and (111) surfaces as prototypes of the LAO/STO interface accessible with VUV photons. They have found co-existence of metallic and insulating phases, and suggested that this inhomogeneity was driven by clustering of V_Os created under VUV irradiation.

Ox-deficient LAO/STO samples offer a convenient platform for studies of EPS in them with ARPES. For this purpose ox-deficient samples are cooled down to room temperature after the growth to stabilize to V_Os, and then ex situ post-annealed in oxygen at ~500 °C. The post-annealing quenches the V_Os, but they gradually recover under X-ray irradiation. This puzzling behavior of the post-annealed LAO/STO samples, contrasting them the standard in situ annealed samples immune to X-rays, is illustrated in Fig. 6.17 that shows angle-integrated ResPES intensity near the Ti L_3-edge at two $h\nu$ values enhancing the V_O (a) and 2DES (b) intensity (marked in Fig. 6.15) which depends on X-ray irradiation time (Strocov et al., unpublished). Scaling up of both V_O and 2DES peaks with irradiation reflects the formation of V_Os injecting mobile electrons into the 2DES The total number of mobile electrons in the system, reflected by the 2DES peak, progressively increases with irradiation. On the other hand, Fig. 6.17b shows the corresponding irradiation dependence of **k**-resolved spectral intensity at E_F whose maxima identify k_F. Surprisingly, for both d_{xy} and d_{yz} band the k_F values remain constant, identifying constant Luttinger count of the corresponding FS sheets and thus constant n_s in the system [this result does not change if k_F is determined by the gradient method [210]. The observed discrepancy between the increasing total number of mobile electrons and constant n_s identifies EPS at the interface, where the fraction of the conductive interface areas increases with irradiation while their local electronic structure stays constant (strictly speaking, electronic structure might in this experiment evolve but rapidly saturate already with small irradiation dose below the actual time sampling). Extrapolation of these spectroscopic results to the standard LAO/STO samples suggests significant EPS in them as well, in agreement with the above macroscopic evidences. Moreover, the conductive fraction in the standard samples is smaller compared to the ox-deficient ones: the corresponding ResPES data in Fig. 6.5a shows smaller 2DES to Ti $2p$ signal ratio compared to the ox-deficient samples, Fig. 6.15b.

Physics of EPS at the LAO/STO interface is a cooperative interplay of the electronic and V_O systems that is far from complete understanding. On the electronic side, recent theoretical analysis [211] suggested that the lateral confinement allows 2DES to avoid a thermodynamically unstable state with negative compressibility. This effect is particularly strong in the nearly ferroelectric STO due to its huge dielectric constant ϵ_{STO} screening the electron repulsion and thus allowing accumulation of large electron densities. Another electronic scenario invoked formation of a Jahn-Teller polaronic phase [212]. The electronic EPS can be tuned by external electric field [211]. On the V_O side of the EPS, coming into play in ox-deficient samples,

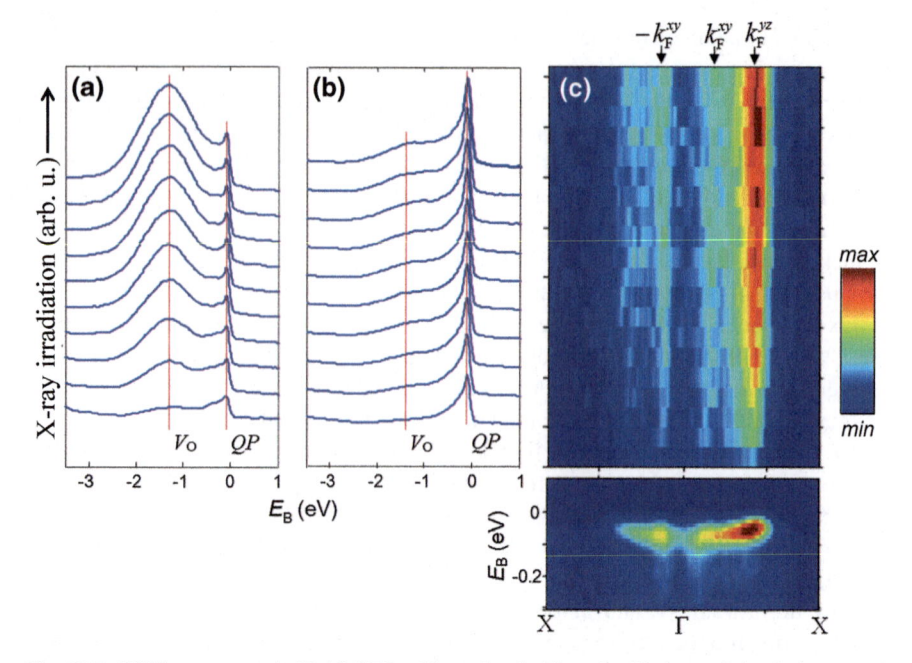

Fig. 6.17 EPS in post-annealed LAO/STO evidenced under X-ray irradiation: **a, b** Angle-integrated ResPES intensity for $h\nu = 464.6$ eV (**a**) and 466.2 eV (**b**) enhancing the V_O and 2DES intensity, respectively (*A* and *B* in Fig. 6.15); **c** Saturated $E(\mathbf{k})$ image (*bottom*) and **k**-resolved intensity at E_F (*top*). The increasing total number of mobile electrons expressed by the 2DES peak (**b**) juxtaposed with constant n_s expressed by the k_F values (**c**) identify the EPS (Strocov et al. unpublished)

theoretical analysis of interplay between the t_{2g} states of the 2DES with the orbitally reconstructed e_g states derived from the V_Os [176] suggested a complicated phase diagramm with regions of phase-separated magnetic states. Furthermore, the V_Os themselves have a tendency to form clusters, which will then accumulate the 2DES [175, 213]. Rationalizing of this intricate and diverse physics of the EPS requires further theoretical and experimental effort combining various area sensitive and local probe spectroscopies.

6.5 Prospects

SX-ARPES is a method of choice to explore a wealth of intriguing physics in the extremely reach functionality of TMO interfaces. Below we briefly sketch routes to apply SX-ARPES for investigation of the interfacial magnetism and field effect.

Interfacial spin structure. Ferromagnetism (FM) of TMO interfaces built up from non-magnetic constituents is one their most interesting functionalities. Understanding of this phenomenon will open new avenues for oxide spintronics such as ferromagnetic Josephson junctions (see, for example [214]) where both superconducting and FM constituents use the same TMO materials platform. Switching elements realized on such junctions promise power saving of up to 5 orders of magnitude compared to the nowadays semiconductor-based electronics.

Exploration of the TMO interfacial magnetism requires spin-resolved ARPES measurements (see Chap. 1) where the photoelectron spin is detected, in addition to energy and momentum. Small photoexcitation cross-section of valence states in the SX-ARPES energy range and attenuation of photoelectrons from the buried interfaces, on one side, combined with the immense intensity loss of at least 2 orders of magnitude associated with the spin resolution, on another side, make spin-resolved SX-ARPES experiments extremely difficult. We expect however that this challenge will in near future be resolved with advent of the angle and energy multichannel spin detectors such us the iMott [199] delivering an efficiency gain of a few orders of magnitude.

Field effect. This is a cornerstone property of the LAO/STO interface allowing, for example, fabrication of transistor heterostructures with enhanced functionalities [69, 70, 215]. In-operando SX-ARPES investigations of this effect should use the top-gate geometry, because bias applied from the back gate would hardly affects the QW region next to the interface hosting the d_{xy} states [211]. The latter is consistent with very weak back gating effect observed in soft- and hard-X-ray PE depth profiling [216]. The top-gate method suffers however from photoelectron attenuation in the top electrode. Ideal for this purpose would be graphene with its monolayer thickness. Recently, transfer of small (tens of μm) graphene flakes on LAO/STO structures has been demonstrated [217]. A method of TOABr-assisted electrochemical delamination [218] allows transfer of large (on the inch scale) graphene layers.

Complexity of the field effect at the LAO/STO interface goes much beyond the simple band filling picture typical of the conventional semiconductor heterostructures. First, while population of the deep d_{xy} bands will vary monotonously with bias, the $d_{xz/yz}$ bands placed in only ~40 meV from E_F can completely depopulate at certain negative bias (Lifshitz transition). Given different character of the planar and orthogonal d-orbitals, this transition will dramatically change the optical and transport properties of the LAO/STO interface [77]. Furthermore, variation of the LAO/STO interfacial n_s changes the EPI and thus effective mass and mobility of the charge carriers, resulting in non-linear dielectric response. Oxygen deficiency adds another dimension to this complexity. In this case the bias causes electromigration of the V_Os [179, 180] again affecting the interfacial n_s and EPI. Furthermore, the electromigration can affect multiple configurations of the V_Os with concomitant effect on their electron correlation and magnetic properties, see Sect. 6.4.3. We speculate that the latter can potentially open ways to multiferroic functionality of the LAO/STO interface, i.e. using electric field to manipulate its magnetic properties.

Further materials. The above application of high-resolution SX-ARPES to the LAO/STO interface taken place at the SLS in 2013 [77] has been the first time when **k**-resolved electron dispersions for a buried interface were determined. Having demonstrated its spectroscopic potential, SX-ARPES is now expanding to a wide range of interface and heterostructure systems actual for novel electronic and spintronic devices. We can mention, for example, QWs of STO embedded in LAO [219] where electron confinement between the two interfaces will form electron spectrum different from the sole LAO/STO interface and tunable by the STO thickness. Superstructures of such QWs (Li et al., unpublished) will allow creation of the artificial third dimension allowing fine tuning of physical properties of these systems. Another fascinating system is the $LaAlO_3/EuTiO_3/SrTiO_3$ heterostructure [220] where the ferromagnetic $EuTiO_3$ layer induces strong spin polarization of the embedded 2DES tunable by gate voltage in field effect transistor geometry. This system will be an obvious candidate for the spin-resolved SX-ARPES. Other interesting systems may be heterostructures of strongly correlated materials like $SrVO_3$ where tunable interplay between electron correlation and confinement effects can deliver conceptually new electronic devices such as Mott transistors [221]. SX-ARPES investigations of multiferroic interfaces are reviewed in Chap. 10. For further scientific and technological cases awaiting applications of (spin-resolved) SX-ARPES the reader is referred to a roadmap of novel oxide electronic materials compiled by [222].

6.6 Conclusions

This chapter has reviewed a number of diverse but interconnected scientific fields ranging from spectroscopic abilities of SX-ARPES to basics of polaron physics and to electronic structure of oxide interfaces. Crucial spectroscopic advantages of SX-ARPES for buried interfaces are its enhanced probing depth and chemical specificity achieved with resonant photoexcitation. We have demonstrated that its application to oxide interfaces delivers direct information on the most fundamental aspects of their electronic structure—momentum-resolved spectral function, band structure, Fermi surface. Focusing on the 2DES formed at the paradigm LAO/STO interface, we have demonstrated determination of its t_{2g} derived multi-orbital band structure and Fermi surface directly connected to the transport properties.

EPI is one of the key players in complex physics of oxides. At the LAO/STO interface, strong electron coupling to the hard LO3 breathing phonon mode forms Holstein-type large polarons. They manifesting themselves as pronounced peak-dip-hump structure of the experimental $A(\omega,\mathbf{k})$ where the quasiparticle peak is followed by broad phonon hump. The polaron formation fundamentally reduces mobility of the interface charge carriers. Furthermore, electron coupling to the soft TO1 polar mode results in dramatic reduction of mobility with temperature. Electron correlations in the interfacial 2DES are weaker compared to the EPI.

Oxygen deficiency adds another degree of freedom to the oxide systems. At the LAO/STO interface, the V_OS form a dual electron system where delocalized 2DES quasiparticles (t_{2g} derived, weakly correlated and non-magnetic) co-exist with localized in-gap states (e_g derived, strongly correlated and magnetic). Manipulation by V_OS allows therefore tuning of the polaronic coupling and thus mobility of the charge carriers through the 2DES density as well as tuning of the interfacial ferromagnetism critically depending on various atomic configurations of the V_OS.

Although SX-ARPES has already produced an impressive amount of fundamental knowledge about the oxide interfaces, this new spectroscopic technique is still in its adolescence period. Ahead are applications to a wide range of interfaces and complex heterostructures, spin-resolved SX-ARPES disclosing spin texture of interfaces, in-operando field effect experiments paving the way towards device applications utilizing the reach physics of oxide interfaces.

Acknowledgments We thank M.-A. Husanu, M. Kobayashi, L. L. Lev, V. A. Rogalev, U. Aschauer, A. Filippetti, J.–M. Triscone and others for their contribution to the main scientific cases discussed above, and P. R. Willmott, R. Claessen, M. Sing, M. Radović, F. Baumberger, O.S. Barišić and F. Lechermann for sharing fruitful discussions. Parts of this research were supported by the ImPACT Program of the Council for Science, Technology and Innovation (Cabinet Office, Government of Japan).

References

1. A. Sekiyama, S. Kasai, M. Tsunekawa, Y. Ishida, M. Sing, A. Irizawa, A. Yamasaki, S. Imada, T. Muro, Y. Saitoh, Y. Onuki, T. Kimura, Y. Tokura, S. Suga, Phys. Rev. B **70**, 060506(R) (2004)
2. S. Suga, A. Shigemoto, A. Sekiyama, S. Imada, A. Yamasaki, A. Irizawa, S. Kasai, Y. Saitoh, T. Muro, N. Tomita, K. Nasu, H. Eisaki, Y. Ueda, Phys. Rev. B **70**, 155106 (2004)
3. S. Suga, Ch. Tusche, J. Electron Spectr. Relat. Phenom. **200**, 119 (2015)
4. C.J. Powell, A. Jablonski, I.S. Tilinin, S. Tanuma, D.R. Penn, J. Electron Spectrosc. Relat. Phenom. **98/99**, 1 (1999). NIST Standard Reference Database 82, http://www.nist.gov/srd/nist82.cfm
5. V.N. Strocov, J. Electron Spectr. Relat. Phenom. **130**, 65 (2003)
6. V.N. Strocov, M. Shi, M. Kobayashi, C. Monney, X. Wang, J. Krempasky, T. Schmitt, L. Patthey, H. Berger, P. Blaha, Phys. Rev. Lett. **109**, 086401 (2012)
7. R. Eguchi, A. Chainani, M. Taguchi, M. Matsunami, Y. Ishida, K. Horiba, Y. Senba, H. Ohashi, S. Shin, Phys. Rev. B **79**, 115122 (2009)
8. L.L. Lev, J. Krempaský, U. Staub, V.A. Rogalev, T. Schmitt, M. Shi, P. Blaha, A.S. Mishchenko, A.A. Veligzhanin, Y.V. Zubavichus, M.B. Tsetlin, H. Volfová, J. Braun, J. Minár, V.N. Strocov, Phys. Rev. Lett. **114**, 237601 (2015)
9. S.L. Molodtsov, M. Richter, S. Danzenbächer, S. Wieling, L. Steinbeck, C. Laubschat, Phys. Rev. Lett. **78**, 142 (1997)
10. C.G. Olson, P.J. Benning, M. Schmidt, D.W. Lynch, P. Canfield, D.M. Wieliczka, Phys. Rev. Lett. **76**, 4265 (1996)
11. J.J. Yeh, I. Lindau, *Atomic Data and Nuclear Data Tables,* vol. 32, p. 1 (1985), http://ulisse.elettra.trieste.it/services/elements/WebElements.html
12. C. Solterbeck, W. Schattke, J.-W. Zahlmann-Nowitzki, K.-U. Gawlik, L. Kipp, M. Skibowski, C.S. Fadley, M.A. Van Hove, Phys. Rev. Lett. **79**, 4681 (1997)
13. J. Braun, J. Minár, S. Mankovsky, V.N. Strocov, N.B. Brookes, L. Plucinski, C.M. Schneider, C.S. Fadley, H. Ebert, Phys. Rev. B **88**, 205409 (2013)
14. F. Venturini, J. Minár, J. Braun, H. Ebert, N.B. Brookes, Phys. Rev. B **77**, 045126 (2008)
15. C. Papp, L. Plucinski, J. Minar, J. Braun, H. Ebert, C.M. Schneider, C.S. Fadley, Phys. Rev. B **84**, 045433 (2011)
16. Ph Hofmann, Ch. Søndergaard, S. Agergaard, S.V. Hoffmann, J.E. Gayone, G. Zampieri, S. Lizzit, A. Baraldi, Phys. Rev. B **66**, 245422 (2002)
17. S. Suga, A. Sekiyama, Eur. Phys. J. Special Topics **169**, 227 (2009)

18. C.S. Fadley, Synchr. Rad. News **25**, 26 (2012)
19. A.X. Gray, C. Papp, S. Ueda, B. Balke, Y. Yamashita, L. Plucinski, J. Minár, J. Braun, E.R. Ylvisaker, C.M. Schneider, W.E. Pickett, H. Ebert, K. Kobayashi, C.S. Fadley, Nat. Mater. **10**, 759 (2011)
20. V.N. Strocov, T. Schmitt, U. Flechsig, T. Schmidt, A. Imhof, Q. Chen, J. Raabe, R. Betemps, D. Zimoch, J. Krempasky, X. Wang, M. Grioni, A. Piazzalunga, L. Patthey, J. Synchr. Rad. **17**, 631 (2010)
21. C. S. Fadley, S.-H. Yang,, B.S. Mun, F.J. Garcia de Abajo (2003). *Solid-State Photoemission and Related Methods: Theory and Experiment*, ed. by W. Schattke, M.A. Van Hove (Wiley-VCH Verlag GmbH, Berlin) pp. 404–432
22. B.L. Henke, Phys. Rev. A **6**, 94 (1972)
23. V.N. Strocov, J. Synchrotron Rad. **20**, 517 (2013)
24. V.N. Strocov, X. Wang, M. Shi, M. Kobayashi, J. Krempasky, C. Hess, T. Schmitt, L. Patthey, J. Synchrotron Rad. **21**, 32 (2014)
25. V.N. Strocov, M. Kobayashi, X. Wang, L.L. Lev, J. Krempasky, V.A. Rogalev, T. Schmitt, C. Cancellieri, M. L. Reinle-Schmitt, Synchr. Rad. News **27** No. 2 (2014) 31
26. M. Kobayashi, I. Muneta, T. Schmitt, L. Patthey, S. Ohya, M. Tanaka, M. Oshima, V.N. Strocov, Appl. Phys. Lett. **101**, 242103 (2012)
27. L.L. Lev, D.V. Averyanov, A.M. Tokmachev, F. Bisti, V.A. Rogalev, V.N. Strocov, V.G. Storchak, J. Mater. Chem. C **5**, 192 (2017)
28. C. Caspers, A. Gloskovskii, M. Gorgoi, C. Besson, M. Luysberg, K.Z. Rushchanskii, M. Ležaić, C.S. Fadley, W. Drube, M. Müller, Sci. Rep. **6**, Art. No. 22912 (2016)
29. G.-X. Miao, J.S. Moodera, Phys. Chem. Chem. Phys. **17**, 751 (2015)
30. G. Berner, M. Sing, F. Pfaff, E. Benckiser, M. Wu, G. Christiani, G. Logvenov, H.-U. Habermeier, M. Kobayashi, V.N. Strocov, T. Schmitt, H. Fujiwara, S. Suga, A. Sekiyama, B. Keimer, R. Claessen, Phys. Rev. B **92**, 125130 (2015)
31. S. Nemšák, G. Conti, A.X. Gray, G.K. Palsson, C. Conlon, D. Eiteneer, A. Keqi, A. Rattanachata, A.Y. Saw, A. Bostwick, L. Moreschini, E. Rotenberg, V.N. Strocov, M. Kobayashi, T. Schmitt, W. Stolte, S. Ueda, K. Kobayashi, A. Gloskovskii, W. Drube, C.A. Jackson, P. Moetakef, A. Janotti, L. Bjaalie, B. Himmetoglu, C.G. Van de Walle, S. Borek, J. Minar, J. Braun, H. Ebert, L. Plucinski, J.B. Kortright, C.M. Schneider, L. Balents, F.M.F. de Groot, S. Stemmer, C.S. Fadley, Phys. Rev. B **93**, 245103 (2016)
32. A.M. Woolley, G. Wexler, J. Phys. C: Solid State Phys. **10**, 2601 (1977)
33. V.N. Strocov, H.I. Starnberg, P.O. Nilsson, H.E. Brauer, L.J. Holleboom, Phys. Rev. Lett. **79**, 467 (1997)
34. V.N. Strocov, E.E. Krasovskii, W. Schattke, N. Barrett, H. Berger, D. Schrupp, R. Claessen, Phys. Rev. B **74**, 195125 (2006)
35. K. Terashima, T. Sato, H. Komatsu, T. Takahashi, N. Maeda, K. Hayashi, Phys. Rev. B **68**, 155108 (2003)
36. J. Krempaský, V.N. Strocov, L. Patthey, P.R. Willmott, R. Herger, M. Falub, P. Blaha, M. Hoesch, V. Petrov, M.C. Richter, O. Heckmann, K. Hricovini, Phys. Rev. B **77**, 165120 (2008)
37. A. Damascelli, Z. Hussain, Z.X. Shen, Rev. Mod. Phys. **75**, 473 (2003)
38. C. Hartinger, F. Mayr, A. Loidl, T. Kopp, Phys. Rev. B **73**, 024408 (2006)
39. A.J. Millis, B.I. Shraiman, R. Mueller, Phys. Rev. Lett. **77**, 175 (1996)
40. A.J. Millis, Nature **392**, 147 (1998)
41. F. Bisti, V.A. Rogalev, M. Karolak, S. Paul, A. Gupta, T. Schmitt, G. Güntherodt, G. Sangiovanni, G. Profeta, V.N. Strocov, Phys. Rev. X **7**, 041067 (2017)
42. Y. Sassa, M. Månsson, M. Kobayashi, O. Götberg, V.N. Strocov, T. Schmitt, N.D. Zhigadlo, O. Tjernberg, B. Batlogg, Phys. Rev. B **91**, 045114 (2015)
43. V.A. Rogalev, O. Gröning, R. Widmer, J.H. Dil, F. Bisti, L.L. Lev, T. Schmitt, V.N. Strocov, Nat. Commun. **6**, 8607 (2015)
44. E. Razzoli, M. Kobayashi, V.N. Strocov, B. Delley, Z. Bukowski, J. Karpinski, N.C. Plumb, M. Radovic, J. Chang, T. Schmitt, L. Patthey, J. Mesot, M. Shi, Phys. Rev. Lett. **108**, 257005 (2012)

45. E. Razzoli, C.E. Matt, M. Kobayashi, X.-P. Wang, V.N. Strocov, A. van Roekeghem, S. Biermann, N.C. Plumb, M. Radovic, T. Schmitt, C. Capan, Z. Fisk, P. Richard, H. Ding, P. Aebi, J. Mesot, M. Shi, Phys. Rev. B **91**, 214502 (2015)
46. H.C. Xu, Y. Zhang, M. Xu, R. Peng, X.P. Shen, V.N. Strocov, M. Shi, M. Kobayashi, T. Schmitt, B.P. Xie, D.L. Feng, Phys. Rev. Lett. **112**, 087603 (2014)
47. M. Höppner, S. Seiro, A. Chikina, A. Fedorov, M. Güttler, S. Danzenbächer, A. Generalov, K. Kummer, S. Patil, S.L. Molodtsov, Y. Kucherenko, C. Geibel, V.N. Strocov, M. Shi, M. Radovic, T. Schmitt, C. Laubschat, D.V. Vyalikh, Nat. Commun. **4**, 1646 (2013)
48. G. Derondeau, F. Bisti, M. Kobayashi, J. Braun, H. Ebert, V.A. Rogalev, M. Shi, T. Schmitt, J. Ma, H. Ding, V.N. Strocov, J. Minár, Sci. Reports **7**, 8787 (2017)
49. B.Q. Lv, N. Xu, H.M. Weng, J.Z. Ma, P. Richard, X.C. Huang, L.X. Zhao, G.F. Chen, C.E. Matt, F. Bisti, V.N. Strocov, J. Mesot, Z. Fang, X. Dai, T. Qian, M. Shi, H. Ding, Nature Phys. **11**, 724 (2015)
50. N. Xu, H. Weng, B. Lv, C. Matt, J. Park, F. Bisti, V.N. Strocov, D.J. Gawryluk, E. Pomjakushina, K. Conder, N. Plumb, M. Radovic, G. Autès, O. Yazyev, Z. Fang, X. Dai, T. Qian, J. Mesot, H. Ding, M. Shi, Nat. Commun. **7**, 11006 (2016)
51. S.-Y. Xu, N. Alidoust, I. Belopolski, Z. Yuan, G. Bian, T.-R. Chang, H. Zheng, V.N. Strocov, D.S. Sanchez, G. Chang, C. Zhang, D. Mou, Y. Wu, L. Huang, C.-C. Lee, S.-M. Huang, B.-K. Wang, A. Bansil, H.-T. Jeng, T. Neupert, A. Kaminski, H. Lin, S. Jia, M.Z. Hasan, Nat. Phys. **11**, 748 (2015a)
52. S.-Y. Xu, I. Belopolski, D. S. Sanchez, C. Zhang, G. Chang, C. Guo, G. Bian, Z. Yuan, H. Lu, T.-R. Chang, P. P. Shibayev, M. L. Prokopovych, N. Alidoust, H. Zheng, C.-C. Lee, S.-M. Huang, R. Sankar, F. Chou, C.-H. Hsu, H.-T. Jeng, A. Bansil, T. Neupert, V. N. Strocov, H. Lin, S. Jia, M.Z. Hasan. Sci. Adv. **1**, e1501092 (2015b)
53. D. Di Sante, P.K. Das, C. Bigi, Z. Ergönenc, N. Gürtler, J.A. Krieger, T. Schmitt, M.N. Ali, G. Rossi, R. Thomale, C. Franchini, S. Picozzi, J. Fujii, V.N. Strocov, G. Sangiovanni, I. Vobornik, R.J. Cava, G. Panaccione, Phys. Rev. Lett. **119**, 026403 (2017)
54. B.Q. Lv, Z.-L. Feng, Q.-N. Xu, X. Gao, J.-Z. Ma, L.-Y. Kong, P. Richard, Y.-B. Huang, V.N. Strocov, C. Fang, H.-M. Weng, Y.-G. Shi, T. Qian, H. Ding, Nature **546**, 627 (2017)
55. G. Landolt, S.V. Eremeev, Y.M. Koroteev, B. Slomski, S. Muff, T. Neupert, M. Kobayashi, V.N. Strocov, T. Schmitt, Z.S. Aliev, M.B. Babanly, I.R. Amiraslanov, E.V. Chulkov, J. Osterwalder, J.H. Dil, Phys. Rev. Lett. **109**, 116403 (2012)
56. J. Krempaský, H. Volfová, S. Muff, N. Pilet, G. Landolt, M. Radović, M. Shi, D. Kriegner, V. Holý, J. Braun, H. Ebert, F. Bisti, V.A. Rogalev, V.N. Strocov, G. Springholz, J. Minár, J.H. Dil (2015). https://arxiv.org/abs/1503.05004
57. M. Kobayashi, I. Muneta, Y. Takeda, A. Fujimori, J. Krempasky, T. Schmitt, S. Ohya, M. Tanaka, M. Oshima, V.N. Strocov, Phys. Rev. B **89**, 205204 (2014)
58. C. Cancellieri, M.L. Reinle-Schmitt, M. Kobayashi, V.N. Strocov, T. Schmitt, P.R. Willmott, S. Gariglio, J.M. Triscone, Phys. Rev. Lett. **110**, 137601 (2013)
59. A.X. Gray, J. Minár, S. Ueda, P.R. Stone, Y. Yamashita, J. Fujii, J. Braun, L. Plucinski, C.M. Schneider, G. Panaccione, H. Ebert, O.D. Dubon, K. Kobayashi, C.S. Fadley, Nat. Mater. **11**, 957 (2012)
60. S. Ohya, K. Takata, M. Tanaka, Nat. Phys. **7**, 342 (2011)
61. A.D. Bouravleuv, L.L. Lev, C. Piamonteze, X. Wang, T. Schmitt, A.I. Khrebtov, Yu.B. Samsonenko, J. Kanski, G.E. Cirlin, V.N. Strocov, Nanotechnology **27**, 425706 (2016)
62. J. Krempaský, S. Muff, F. Bisti, M. Fanciulli, Volfová, A. Weber, N. Pilet, P. Warnicke, H. Ebert, J. Braun, F. Bertran, V.V. Volobuiev, J. Minár, G. Springholz, J.H. Dil, V.N. Strocov. Nat. Commun. **7**, 13071 (2016)
63. J.A. Krieger, C.-Z. Chang, M.-A. Husanu, D. Sostina, A. Ernst, M.M. Otrokov, T. Prokscha, T. Schmitt, A. Suter, M.G. Vergniory, E.V. Chulkov, J.S. Moodera, V.N. Strocov, Z. Salman, Phys. Rev. B **96**, 184402 (2017)
64. S. Banerjee, O. Erten, M. Randeria, Nature Phys. **9**, 626 (2013)

65. H.Y. Hwang, Y. Iwasa, M. Kawasaki, B. Keimer, N. Nagaosa, Y. Tokura, Nat. Mater. **11**, 103 (2012)
66. J. Mannhart, D.G. Schlom, Science **327**, 1607 (2010)
67. Y.-Y. Pai, A. Tylan-Tyler, P. Irvin, J Levy, Reports on Progress in Physics (2018). This is an accepted paper presently available at http://iopscience.iop.org/article/10.1088/1361-6633/aa892d. Also it was posted at https://arxiv.org/abs/1702.07690
68. A.D. Caviglia, M. Gabay, S. Gariglio, N. Reyren, C. Cancellieri, J.-M. Triscone, Phys. Rev. Lett. **104**, 126803 (2010)
69. M. Hosoda, Y. Hikita, H.Y. Hwang, C. Bell, Appl. Phys. Lett. **103**, 103507 (2013)
70. C. Woltmann, T. Harada, H. Boschker, V. Srot, P.A. van Aken, H. Klauk, J. Mannhart, Phys. Rev. Appl. **4**, 064003 (2015)
71. M. Basletic, J.-L. Maurice, C. Carrétéro, G. Herranz, O. Copie, M. Bibes, É. Jacquet, K. Bouzehouane, S. Fusil, A. Barthélémy, Nat. Mater. **7**, 621 (2008)
72. P. Delugas, A. Filippetti, V. Fiorentini, D.I. Bilc, D. Fontaine, P. Ghosez, Phys. Rev. Lett. **106**, 166807 (2011)
73. A. Dubroka, M. Rössle, K.W. Kim, V.K. Malik, L. Schultz, S. Thiel, C.W. Schneider, J. Mannhart, G. Herranz, O. Copie, M. Bibes, A. Barthélémy, C. Bernhard, Phys. Rev. Lett. **104**, 156807 (2008)
74. S. Gariglio, A. Fête, J.-M. Triscone, J. Phys.: Condens. Matter **27**, 283201 (2015)
75. W.-J. Son, E. Cho, B. Lee, J. Lee, S. Han, Phys. Rev. B **79**, 245411 (2009)
76. M. Breitschaft, V. Tinkl, N. Pavlenko, S. Paetel, C. Richter, J.R. Kirtley, Y.C. Liao, G. Hammerl, V. Eyert, T. Kopp, J. Mannhart, Phys. Rev. B **81**, 153414 (2010)
77. C. Cancellieri, M.L. Reinle-Schmitt, M. Kobayashi, V.N. Strocov, P.R. Willmott, D. Fontaine, Ph Ghosez, A. Filippetti, P. Delugas, V. Fiorentini, Phys. Rev. B **89**, R121412 (2014)
78. P.D.C. King, S. McKeown Walker, A. Tamai, A. de la Torre, T. Eknapakul, P. Buaphet, S.-K. Mo, W. Meevasana, M.S. Bahramy, F. Baumberger. Nat. Commun. **5**, 3414 (2014)
79. M. Salluzzo, S. Gariglio, D. Stornaiuolo, V. Sessi, S. Rusponi, C. Piamonteze, G.M. De Luca, M. Minola, D. Marré, A. Gadaleta, H. Brune, F. Nolting, N.B. Brookes, G. Ghiringhelli, Phys. Rev. Lett. **111**, 087204 (2013)
80. G. Berner, A. Müller, F. Pfaff, J. Walde, C. Richter, J. Mannhart, S. Thiess, A. Gloskovskii, W. Drube, M. Sing, R. Claessen, Phys. Rev. B **88**, 115111 (2013)
81. G. Berner, M. Sing, H. Fujiwara, A. Yasui, Y. Saitoh, A. Yamasaki, Y. Nishitani, A. Sekiyama, N. Pavlenko, T. Kopp, C. Richter, J. Mannhart, S. Suga, R. Claessen, Phys. Rev. Lett. **110**, 247601 (2013)
82. A. Koitzsch, J. Ocker, M. Knupfer, M.C. Dekker, K. Dörr, B. Büchner, P. Hoffmann, Phys. Rev. B **84**, 245121 (2011)
83. R.W. Paynter, J. Electron Spectrosc. Relat. Phenom. **169**, 1 (2009)
84. O. Copie, V. Garcia, C. Bödefeld, C. Carrétéro, M. Bibes, G. Herranz, E. Jacquet, J.-L. Maurice, B. Vinter, S. Fusil, K. Bouzehouane, H. Jaffrès, A. Barthélémy, Phys. Rev. Lett. **102**, 216804 (2009)
85. M. Sing, G. Berner, K. Goß, A. Müller, A. Ruff, A. Wetscherek, S. Thiel, J. Mannhart, S.A. Pauli, C.W. Schneider, P.R. Willmott, M. Gorgoi, F. Schäfers, R. Claessen, Phys. Rev. Lett. **102**, 176805 (2009)
86. C. Cancellieri, A.S. Mishchenko, U. Aschauer, A. Filippetti, C. Faber, O.S. Barisic, V.A. Rogalev, T. Schmitt, N. Nagaosa, V.N. Strocov, Nat. Commun. **7**, 10386 (2016)
87. F. Schoofs, M. Egilmez, T. Fix, J.L. MacManus-Driscoll, G. Mark, Blamire. Appl. Phys. Lett. **100**, 081601 (2012)
88. Z. Zhong, P.J. Kelly, Europhys. Lett. **84**, 27001 (2008)
89. N.C. Plumb, M. Kobayashi, M. Salluzzo, E. Razzoli, C.E. Matt, V.N. Strocov, K.J. Zhou, M. Shi, J. Mesot, T. Schmitt, L. Patthey, M. Radović, Appl. Surf. Sci. **412**, 271 (2017)
90. A. Fête, C. Cancellieri, D. Li, D. Stornaiuolo, A.D. Caviglia, S. Gariglio, J.-M. Triscone, Appl. Phys. Lett. **106**, 051604 (2015)

91. C. Cancellieri, N. Reyren, S. Gariglio, A.D. Caviglia, A. Fête, J.M. Triscone, Europhys. Lett. **91**, 17004 (2010)
92. J. Appel, in Solid State Physics, vol. 21, ed. by H. Ehrenreich, F. Seitz and D. Turnbull (Academic, New York, 1968)
93. H. Fröhlich, H. Pelzer, S. Zienau, Philos. Mag. **41**, 221 (1950)
94. T. Holstein, Ann. Phys. (N.Y.) **8**, 325 and 343 (1959)
95. S.A. Alexandrov (ed.), *Polarons in Advanced Materials* (Canopus/Springer, Bristol, 2007)
96. J.T. Devreese, A.S. Alexandrov, Rep. Prog. Phys. **72**, 066501 (2009)
97. J.P. Hague, P.E. Kornilovitch, A.S. Alexandrov, J.H. Samson, Phys. Rev. B **73**, 054303 (2006)
98. A.S. Mishchenko, N.V. Prokof'ev, A. Sakamoto, B.V. Svistunov, Phys. Rev. B **62**, 6317 (2000)
99. G. De Filippis, V. Cataudella, A.S. Mishchenko, C.A. Perroni, J.T. Devreese, Phys. Rev. Lett. **96**, 136405 (2006)
100. A.S. Mishchenko, N. Nagaosa, N.V. Prokof'ev, A. Sakamoto, B.V. Svistunov, Phys. Rev. Lett. **91**, 236401 (2003)
101. A.S. Mishchenko, N. Nagaosa, G. De Filippis, A. de Candia, V. Cataudella, Phys. Rev. Lett. **114**, 146401 (2015)
102. D.J.J. Marchand, G. De Filippis, V. Cataudella, M. Berciu, N. Nagaosa, N.V. Prokof'ev, A.S. Mishchenko, P.C.E. Stamp, Phys. Rev. Lett. **105**, 266605 (2010)
103. A. Heeger, S. Kivelson, J.R. Schrieffer, W.-P. Su, Rev. Mod. Phys. **60**, 781 (1988)
104. W.P. Su, J.R. Schrieffer, A.J. Heeger, Phys. Rev. Lett. **42**, 1698 (1979)
105. Z. Liu, E. Manousakis, Phys. Rev. B **45**, 2425 (1992)
106. F.M. Peeters, W. Xiaoguang, J.T. Devreese, Phys. Rev. B **33**, 3926 (1986)
107. P.B. Allen, Phys. Rev. B **16**, 5139 (1977)
108. H. Bilz, G. Guntherodt, W. Kleppmann, W. Kress, Phys. Rev. Lett. **43**, 1998 (1979)
109. K.A. Kikoin and A.S. Mishchenko, J. Phys,: Condens. Matt. **2**, 6491 (1990)
110. C. Slezak, A. Macridin, G.A. Sawatzky, M. Jarrell, T.A. Maier, Phys. Rev. B **73**, 205122 (2006)
111. G. Grimvall, *The Electron-Phonon Interaction in Metals* (North-Holland, Amsterdam, 1981)
112. C.J. Chandler, C. Prosko, F. Marsiglio, Sci. Reports **6**, 32591 (2016)
113. A.S. Mishchenko, N. Prokof'ev, N. Nagaosa, Phys. Rev. Lett. **113**, 166402 (2014)
114. G.D. Mahan, *Many Particle Physics* (Plenum, New York, 1981)
115. A. Troisi, G. Orlandi, Phys. Rev. Lett. **96**, 086601 (2006)
116. E.I. Rashba, in *Modern Problems in Condensed Matter Sciences*, vol. 2, ed. by V.M. Agranovich and A.A. Maradudin (North-Holland, Amsterdam, 1982), p. 543
117. M. Ueta, H. Kanzaki, K. Kobayashi, Y. Toyozawa, E. Hanamura, *Exciton Processes in Solids* (Springer, Berlin, 1986)
118. A.S. Mishchenko, N. Nagaosa, N.V. Prokof'ev, A. Sakamoto, B.V. Svistunov, Phys. Rev. B **66**, 020301(R) (2002)
119. M. Berciu, A.S. Mishchenko, N. Nagaosa, Europhys. Lett. **89**, 37007 (2010)
120. E. Burovski, H. Fehske, A.S. Mishchenko, Phys. Rev. Lett. **101**, 116403 (2008)
121. H. Ebrahimnejad, M. Berciu, Phys. Rev. B **85**, 165117 (2012)
122. A.S. Mishchenko, N. Nagaosa, A. Alvermann, H. Fehske, G. De Filippis, V. Cataudella, O.P. Sushkov, Phys. Rev. B. **79**, 180301(R) (2009)
123. Y. Shinozuka, Y. Toyozawa, J. Phys. Soc. Jpn. **46**, 505 (1979)
124. G.F. Koster, J.C. Slater, Phys. Rev. **96**, 1208 (1954)
125. A.S. Alexandrov, J. Ranninger, Phys. Rev. B **45**, 13109(R) (1992)
126. J. Ranninger, U. Thibblin, Phys. Rev. B **45**, 7730 (1992)
127. G.L. Goodvin, M. Berciu, Europhys. Lett. **92**, 37006 (2010)
128. O. Gunnarsson, O. Rosch, J. Phys.: Condens. Matter **20**, 043201 (2008)
129. M. Hohenadler, D. Neuber, W. von der Linden, G. Wellein, J. Loos, H. Fehske, Phys. Rev. B **71**, 245111 (2005)
130. A.S. Mishchenko, N. Nagaosa, K.M. Shen, Z.-X. Shen, X.J. Zhou, T.P. Devereaux, Europhys. Lett. **95**, 57007 (2011)

131. A. Lanzara, P.V. Bogdanov, X.J. Zhou, S.A. Kellar, D.L. Feng, E.D. Lu, T. Yoshida, H. Eisaki, A. Fujimori, K. Kishio, J.-I. Shimoyama, T. Noda, S. Uchida, Z. Hussain, Z.-X. Shen, Nature **412**, 510 (2001)
132. T. Cuk, F. Baumberger, D.H. Lu, N. Ingle, X.J. Zhou, H. Eisaki, N. Kaneko, Z. Hussain, T.P. Devereaux, N. Nagaosa, Z.-X. Shen, Phys. Rev. Lett. **93**, 117003 (2004)
133. T.P. Devereaux, T. Cuk, Z.-X. Shen, N. Nagaosa, Phys. Rev. Lett. **93**, 117004 (2004)
134. G.A. Sawatzky, Nature **342**, 480 (1989)
135. J.T. Devreese, L.F. Lemmens, J. van Royen, Phys. Rev. B **15**, 1212 (1977)
136. J.T. Devreese, S.N. Klimin, J.L.M. van Mechelen, D. van der Marrel, Phys. Rev. B **81**, 125119 (2010)
137. S. Raj, D. Hashimoto, H. Matsui, S. Souma, T. Sato, T. Takahashi, D.D. Sarma, P. Mahadevan, S. Oishi, Phys. Rev. Lett. **96**, 147603 (2006)
138. H. Liu, G. Liu, W. Zhang, L. Zhao, J. Meng, X. Jia, X. Dong, W. Lu, G. Wang, Y. Zhou, Y. Zhu, X. Wang, T. Wu, X. Chen, T. Sasagawa, Z. Xu, C. Chen, X.J. Zhou, EPJ Web of Conferences **23**, 00005 (2012)
139. S. Moser, L. Moreschini, J. Jaćimović, O.S. Barišić, H. Berger, A. Magrez, Y.J. Chang, K.S. Kim, A. Bostwick, E. Rotenberg, L. Forró, M. Grioni, Phys. Rev. Lett. **110**, 196403 (2013)
140. J. Chaloupka, G. Khaliullin, Phys. Rev. Lett. **99**, 256406 (2007)
141. A. Nicolaou, V. Brouet, M. Zacchigna, I. Vobornik, A. Tejeda, A. Taleb-Ibrahimi, P. Le Fevre, F. Bertran, S. Hebert, H. Muguerra, D. Grebille, Phys. Rev. Lett. **104**, 056403 (2010)
142. Y.F. Nie, D. Di Sante, S. Chatterjee, P.D.C. King, M. Uchida, S. Ciuchi, D.G. Schlom, K.M. Shen, Phys. Rev. Lett. **115**, 096405 (2015)
143. A.S. Mishchenko, Phys. Usp. **52**, 1193 (2009)
144. A.S. Mishchenko, Adv. Condens. Matter Phys. **2010**, 306106 (2009)
145. V.I. Belinicher, A.L. Chernyshev and V.A. Shubin, Phys. Rev. B **53**, 335 (1996)
146. V.I. Belinicher, A.L. Chernyshev and V.A. Shubin, Phys. Rev. B **54**, 14914 (1996)
147. F.C. Zhang, T.M. Rice, Phys. Rev. B **37**, 3759 (1988)
148. T. Xiang, J.M. Wheatley, Phys. Rev. B **54**, R12653 (1996)
149. B.O. Wells, Z.-X. Shen, A. Matsuura, D.M. King, M.A. Kastner, M. Greven, R.J. Birgeneau, Phys. Rev. Lett. **74**, 964 (1995)
150. A.S. Mishchenko, N.V. Prokof'ev, B.V. Svistunov, Phys. Rev. B **64**, 033101 (2001)
151. A.S. Mishchenko, N. Nagaosa, Phys. Rev. Lett. **93**, 036402 (2004)
152. V. Cataudella, G. De Filippis, A.S. Mishchenko, N. Nagaosa, Phys. Rev. Lett. **99**, 226402 (2007)
153. A.S. Mishchenko, N. Nagaosa, Z.-X. Shen, G. De Filippis, V. Cataudella, T.P. Devereaux, C. Bernhard, K.W. Kim, J. Zaanen, Phys. Rev. Lett. **100**, 166401 (2008)
154. U. Aschauer, N.A. Spaldin, J. Phys.: Condens. Matter **26**, 122203 (2014)
155. W. Meevasana, X.J. Zhou, B. Moritz, C.-C. Chen, R.H. He, S.-I. Fujimori, D.H. Lu, S.-K. Mo, R.G. Moore, F. Baumberger, T.P. Devereaux, D. van der Marel, N. Nagaosa, J. Zaanen, Z.-X. Shen, New J. Phys. **12**, 023004 (2010)
156. G. Verbista, F.M. Peetersa, J.T. Devreese, Ferroelectrics **130**, 27 (1992)
157. J.L.M. van Mechelen, D. van der Marel, C. Grimaldi, A.B. Kuzmenko, N.P. Armitage, N. Reyren, H. Hagemann, I.I. Mazin, Phys. Rev. Lett. **100**, 226403 (2010)
158. M. Cardona, Phys. Rev. A **140**, 651 (1965)
159. N. Choudhury, E.J. Walter, A.I. Kolesnikov, C.-K. Loong, Phys. Rev. B **77**, 134111 (2008)
160. Y.J. Chang, A. Bostwick, Y.S. Kim, K. Horn, E. Rotenberg, Phys. Rev. B **81**, 235109 (2010)
161. C. Chen, J. Avila, E. Frantzeskakis, A. Levy, M.C. Asensio, Nat. Commun. **6**, 8585 (2015)
162. Z. Wang, S. McKeown Walker, A. Tamai, Y. Wang, Z. Ristic, F.Y. Bruno, A. de la Torre, S. Riccò, N.C. Plumb, M. Shi, P. Hlawenka, J. Sánchez-Barriga, A. Varykhalov, T.K. Kim, M. Hoesch, P.D.C. King, W. Meevasana, U. Diebold, J. Mesot, B. Moritz, T.P. Devereaux, M. Radovic, F. Baumberger, Nat. Mater. **15**, 835 (2016)
163. H. Boschker, C. Richter, E. Fillis-Tsirakis, C.W. Schneider, J. Mannhart, Sci. Rep. **5**, Art. No. 12309 (2015)

164. A. Fête, S. Gariglio, C. Berthod, D. Li, D. Stornaiuolo, M. Gabay, J.-M. Triscone, New J. Phys. **16**, 112002 (2014)
165. V. Cataudella, G. De Filippis, C.A. Perroni, in *Polarons in Advanced Materials*, ed. by A.S. Alexandrov (Springer, Dordrecht, 2007)
166. X. Hao, Z. Wang, M. Schmid, U. Diebold, C. Franchini, Phys. Rev B **91**, 085204 (2015)
167. S. Gariglio, N. Reyren, A.D. Caviglia, J.-M. Triscone, J. Phys.: Condens. Matter **21**, 164213 (2009)
168. F.F.Y. Wang, K.P. Gupta, Metall. Trans. **4**, 2767 (1973)
169. A. Nucara, M. Ortolani, L. Baldassarre, W.S. Mohamed, U. Schade, P.P. Aurino, A. Kalaboukhov, D. Winkler, A. Khare, F. Miletto Granozio and P. Calvani. Phys. Rev. B **93**, 224103 (2016)
170. G. Herranz, M. Basletić, M. Bibes, C. Carrétéro, E. Tafra, E. Jacquet, K. Bouzehouane, C. Deranlot, A. Hamzić, J.-M. Broto, A. Barthélémy, A. Fert, Phys. Rev. Lett. **98**, 216803 (2007)
171. Z. Ristic, R. Di Capua, G.M. De Luca, F. Chiarella, G. Ghiringhelli, J.C. Cezar, N.B. Brookes, C. Richter, J. Mannhart, M. Salluzzo, Europhys. Lett. **93**, 17004 (2011)
172. K. Zhou, M. Radovic, J. Schlappa, V.N. Strocov, R. Frison, J. Mesot, L. Patthey, T. Schmitt, Phys. Rev. B **83**, 201402(R) (2011)
173. N. Palina, A. Annadi, T. Citra Asmara, C. Diao, X. Yu, M.B.H. Breese, T. Venkatesan, Ariando, A. Rusydi, Phys. Chem. Chem. Phys. **18**, 13844 (2016)
174. C. Lin, A.A. Demkov, Phys. Rev. Lett. **111**, 217601 (2013)
175. N. Pavlenko, T. Kopp, E.Y. Tsymbal, G.A. Sawatzky, J. Mannhart, Phys. Rev. B **85**, 020407(R) (2012)
176. N. Pavlenko, T. Kopp, J. Mannhart, Phys. Rev. B **88**, 201104 (2013)
177. F. Lechermann, L. Boehnke, D. Grieger, C. Piefke, Phys. Rev. B **90**, 085125 (2014)
178. F. Lechermann, H.O. Jeschke, A.J. Kim, S. Backes, R. Valentí, Phys. Rev. B **93**, 121103(R) (2016)
179. Y. Lei, Y. Li, Y.Z. Chen, Y.W. Xie, Y.S. Chen, S.H. Wang, J. Wang, B.G. Shen, N. Pryds, H.Y. Hwang, J.R. Sun, Nature Comm. **5**, 5554 (2014)
180. B.W. Veal, S.K. Kim, P. Zapol, H. Iddir, P.M. Baldo, J.A. Eastman, Nature Commun. **7**, 11892 (2016)
181. X. Wang, J.Q. Chen, A. Roy Barman, S. Dhar, Q.-H. Xu, T. Venkatesan and Ariando. Appl. Phys. Lett. **98**, 081916 (2011)
182. A. Janotti, J.B. Varley, M. Choi, C.G. Van de Walle, Phys. Rev. B **90**, 085202 (2014)
183. M. Altmeyer, H.O. Jeschke, O. Hijano-Cubelos, C. Martins, F. Lechermann, K. Koepernik, A.F. Santander-Syro, M.J. Rozenberg, R. Valentí, M. Gabay, Phys. Rev. Lett. **116**, 157203 (2016)
184. S. McKeown Walker, F.Y. Bruno, Z. Wang, A. de la Torre, S. Riccó, A. Tamai, T.K. Kim, M. Hoesch, M. Shi, M.S. Bahramy, P.D.C. King, F. Baumberger, Adv. Mater. **8**, 3894 (2015)
185. M.L. Knotek, P.J. Feibelman, Phys. Rev. Lett. **40**, 964 (1978)
186. T.C. Rödel, F. Fortuna, S. Sengupta, E. Frantzeskakis, P. Le Fèvre, F. Bertran, B. Mercey, S. Matzen, G. Agnus, T. Maroutian, P. Lecoeur, A.F. Santander-Syro, Adv. Mater. **28**, 1976 (2016)
187. A.B. Posadas, K.J. Kormondy, W. Guo, P. Ponath, J. Geler-Kremer, T. Hadamek, A.A. Demkov, J. Appl. Phys. **121**, 105302 (2017)
188. N.C. Plumb, M. Salluzzo, E. Razzoli, M. Månsson, M. Falub, J. Krempasky, C.E. Matt, J. Chang, M. Schulte, J. Braun, H. Ebert, J. Minár, B. Delley, K.-J. Zhou, T. Schmitt, M. Shi, J. Mesot, L. Patthey, M. Radović, Phys. Rev. Lett. **113**, 086801 (2014)
189. N.C. Bristowe, P.B. Littlewood, E. Artacho, Phys. Rev. B **83**, 205405 (2011)
190. Y.Z. Chen, N. Bovet, F. Trier, D.V. Christensen, F.M. Qu, N.H. Andersen, T. Kasama, W. Zhang, R. Giraud, J. Dufouleur, T.S. Jespersen, J.R. Sun, A. Smith, J. Nygård, L. Lu, B. Büchner, B.G. Shen, S. Linderoth, N. Pryds, Nat. Commun. **4**, 1371 (2013)
191. P. Schütz, D.V. Christensen, V. Borisov, F. Pfaff, P. Scheiderer, L. Dudy, M. Zapf, J. Gabel, Y.Z. Chen, N. Pryds, V.A. Rogalev, V.N. Strocov, C. Schlueter, T.-L. Lee, H.O. Jeschke, R. Valentí, M. Sing, R. Claessen, Phys. Rev. B **96**, 161409 (2017)

192. N. Reyren, S. Thiel, A.D. Caviglia, L. Fitting Kourkoutis, G. Hammerl, C. Richter, C.W. Schneider, T. Kopp, A.-S. Rüetschi, D. Jaccard, M. Gabay, D.A. Muller, J.-M. Triscone, J. Mannhart, Science **317**, 1196 (2008)
193. M. Gabay, J.-M. Triscone, Nat. Phys. **9**, 610 (2013)
194. A.F. Santander-Syro, F. Fortuna, C. Bareille, T.C. Rödel, G. Landolt, N.C. Plumb, J.H. Dil, M. Radović, Nat. Mater. **13**, 1085 (2015)
195. Z. Zhong, A. Tóth, K. Held, Phys. Rev. B **87**, 161102(R) (2013)
196. A.C. Garcia-Castro, M.G. Vergniory, E. Bousquet, A.H. Romero, Phys. Rev. B **93**, 045405 (2016)
197. S. McKeown Walker, S. Riccò, F.Y. Bruno, A. de la Torre, A. Tamai, E. Golias, A. Varykhalov, D. Marchenko, M. Hoesch, M.S. Bahramy, P.D.C. King, J. Sánchez-Barriga, F. Baumberger, Phys. Rev. B **93**, 245143 (2016)
198. M. Behrmann, F. Lechermann, Phys. Rev. B **92**, 125148 (2015)
199. V.N. Strocov, V.N. Petrov, J.H. Dil, J. Synchr. Rad. **22**, 708 (2015)
200. E. Dagotto, Science **309**, 257 (2005)
201. V.B. Shenoy, D.D. Sarma, C.N.R. Rao, Chem. Phys. Chem **7**, 2053 (2006)
202. D. Bucheli, S. Caprara, M. Grilli, Supercond. Sci. Technol. **28**, 045004 (2015)
203. C. Richter, H. Boschker, W. Dietsche, E. Fillis-Tsirakis, R. Jany, F. Loder, L.F. Kourkoutis, D.A. Muller, J.R. Kirtley, C.W. Schneider, J. Mannhart, Nature **502**, 528 (2013)
204. F. Bi, M. Huang, C.-W. Bark, S. Ryu, S. Lee, C.-B. Eom, P. Irvin, J. Levy, J. Appl. Phys. **119**, 025309 (2016)
205. A. Joshua, J. Ruhman, S. Pecker, E. Altman, S. Ilani, Proc. Natl. Acad. Sci. U.S.A. **110**, 9633 (2013)
206. S. Caprara, J. Biscaras, N. Bergeal, D. Bucheli, S. Hurand, C. Feuillet-Palma, A. Rastogi, R.C. Budhani, J. Lesueur, M. Grilli, Phys. Rev. B **88**, 020504(R) (2013)
207. S. Caprara, D. Bucheli, M. Grilli, J. Biscaras, N. Bergeal, S. Hurand, C. Feuillet-Palma, J. Lesueur, A. Rastogi, R.C. Budhani, SPIN **04**, 1440004 (2014)
208. Ariando, X. Wang, G. Baskaran, Z. Q. Liu, J. Huijben, J.B. Yi, A. Annadi, A. Roy Barman, A. Rusydi, S. Dhar, Y.P. Feng, J. Ding, H. Hilgenkamp, T. Venkatesan, Nat. Commun. **2**, 188 (2011)
209. L. Dudy, M. Sing, P. Scheiderer, J.D. Denlinger, P. Schütz, J. Gabel, M. Buchwald, C. Schlueter, T.-L. Lee, R. Claessen, Adv. Mater. **28**, 7443 (2016)
210. Th Straub, R. Claessen, P. Steiner, S. Hüfner, V. Eyert, K. Friemelt, E. Bucher, Phys. Rev. B **55**, 13473 (1997)
211. N. Scopigno, D. Bucheli, S. Caprara, J. Biscaras, N. Bergeal, J. Lesueur, M. Grilli, Phys. Rev. Lett. **116**, 026804 (2016)
212. B.R.K. Nanda, S. Satpathy, Phys. Rev. B **83**, 195114 (2011)
213. N. Mohanta, A. Taraphder, J. Phys.: Condens. Matter **26**, 215703 (2014)
214. P. Komissinskiy, G.A. Ovsyannikov, I.V. Borisenko, YuV Kislinskii, K.Y. Constantinian, A.V. Zaitsev, D. Winkler, Phys. Rev. Lett. **99**, 017004 (2007)
215. A.D. Caviglia, S. Gariglio, N. Reyren, D. Jaccard, T. Schneider, M. Gabay, S. Thiel, G. Hammerl, J. Mannhart, J.-M. Triscone, Nature **456**, 624 (2008)
216. M. Minohara, Y. Hikita, C. Bell, H. Inoue, M. Hosoda, H. K. Sato, H. Kumigashira, M. Oshima, E. Ikenaga and H. Y. Hwang (2014). https://arxiv.org/abs/1403.5594
217. I. Aliaj, I. Torre, V. Miseikis, E. di Gennaro, A. Sambri, A. Gamucci, C. Coletti, F. Beltram, F.M. Granozio, M. Polini, V. Pellegrini, S. Roddaro, APL Mater. **4**, 066101 (2016)
218. L. Koefoed, M. Kongsfelt, S. Ulstrup, A.G. Čabo, A. Cassidy, P.R. Whelan, M. Bianchi, M. Dendzik, F. Pizzocchero, B. Jørgensen, P. Bøggild, L. Hornekær, P. Hofmann, S.U. Pedersen, K. Daasbjerg, Phys. D: Appl. Phys. **48**, 115306 (2015)
219. D. Li, S. Gariglio, C. Cancellieri, A. Fête, D. Stornaiuolo, J.-M. Triscone, Appl. Phys. Lett. Mater. **2**, 012102 (2014)
220. D. Stornaiuolo, C. Cantoni, G. M. De Luca, R. Di Capua, E. Di. Gennaro, G. Ghiringhelli, B. Jouault, D. Marrè, D. Massarotti, F. Miletto Granozio, I. Pallecchi, C. Piamonteze, S. Rusponi, F. Tafuri and M. Salluzzo, Nat. Mater. **15**, 278 (2016)

221. Z. Zhong, M. Wallerberger, J.M. Tomczak, C. Taranto, N. Parragh, A. Toschi, G. Sangiovanni, K. Held, Phys. Rev. Lett. **114**, 246401 (2015)
222. M. Lorenz, M.S. Ramachandra Rao, T. Venkatesan, E. Fortunato, P. Barquinha, R. Branquinho, D. Salgueiro, R. Martins, E. Carlos, A. Liu, F. K. Shan, M. Grundmann, H. Boschker, J. Mukherjee, M. Priyadarshini, N. Das Gupta, D. J. Rogers, F.H. Teherani, E.V. Sandana, P. Bove, K. Rietwyk, A. Zaban, A. Veziridis, A. Weidenkaff, M. Muralidhar, M. Murakami, S. Abel, J. Fompeyrine, J. Zuniga-Perez, R. Ramesh, N.A. Spaldin, S. Ostanin, V. Borisov, I. Mertig, V. Lazenka, G. Srinivasan, W. Prellier, M. Uchida, M. Kawasaki, R. Pentcheva, P. Gegenwart, F. Miletto Granozio, J. Fontcuberta, N. Pryds, J. Phys. D: Appl. Phys. **49**, 433001 (2016)
223. K.M. Shen, F. Ronning, D.H. Lu, W.S. Lee, N.J.C. Ingle, W. Meevasana, F. Baumberger, A. Damascelli, N.P. Armitage, L.L. Miller, Y. Kohsaka, M. Azuma, M. Takano, H. Takagi, Z.-X. Shen, Phys. Rev. Lett. **93**, 267002 (2004)

Chapter 7
Standing-Wave and Resonant Soft- and Hard-X-ray Photoelectron Spectroscopy of Oxide Interfaces

Slavomír Nemšák, Alexander X. Gray and Charles S. Fadley

Abstract We discuss several new directions in photoemission that permit more quantitatively studying buried interfaces: going to higher energies in the multi-keV regime; using standing-wave excitation, created by reflection from either a multilayer heterostructure or atomic planes; tuning the photon energy to specific points near absorption resonances; and making use of near-total-reflection geometries. Applications to a variety of oxide and spintronic systems are discussed.

7.1 Basic Principles of Resonant Standing-Wave Soft- and Hard-X-ray Photoemission (SXPS, HXPS) and Angle-Resolved Photoemission (ARPES)

7.1.1 Standing-Wave Photoemission and Resonant Effects

In the Chaps. 5 and 6 the historical background, virtues and challenges of going upward from the typical limit in valence-band studies of ~150 eV into the soft X-ray (from a few hundred to ~2 keV) and hard (tender) X-ray regimes (~2–10 keV) in photoemission, including both core-level and valence-band angle-resolved

---

S. Nemšák (✉)
Peter-Grünberg-Institut PGI-6, Forschungszentrum Jülich GmbH, 52425 Jülich, Germany
e-mail: s.nemsak@fz-juelich.de

A. X. Gray
Department of Physics, Temple University, Philadelphia, PA 19122, USA
e-mail: axgray@temple.edu

C. S. Fadley
Department of Physics, University of California Davis, Davis, CA 95616, USA
e-mail: fadley@physics.ucdavis.edu

C. S. Fadley
Lawrence Berkeley National Laboratory, Materials Sciences Division, Berkeley, CA 94720, USA

photoemission (ARPES) measurements, have been discussed in detail, and we refer the reader to them, as well as to a prior overview article on ARPES by one of us [1], for background. There is also a recent book edited by Woicik that explores various aspect of hard X-ray photoemission [2]. We will use SXPS and HXPS (aka HAXPES) for these different energy regions.

In this chapter, we will add to these techniques the use of standing-wave (SW) excitation, resonant excitation to tune the SW properties, and the use of total reflection at buried interfaces to provide greatly enhanced resolution of buried interfaces, as well as bulk electronic structure. The SW method and its history is extensively discussed in various chapters in the book by Zegenhagen and Kazimirov [3].

SWs can be created by reflection from a synthetic multilayer, from atomic planes in single crystals or epitaxial overlayers, or in total reflection from buried interfaces. As photon energy is increased in these measurements, deeper interfaces or more truly bulk electronic structure can be studied. Prior reviews of these developments using multilayer reflection from our group provide additional background [4–7], including a detailed discussion of the X-ray optical theoretical modeling that is necessary to interpret experimental data [8].

Figure 7.1 illustrates the basic idea of SW photoemission. Strong Bragg reflection from either a multilayer heterostructure or atomic planes creates a standing wave inside of and above the sample. If the bilayer repeat spacing in a multilayer is d_{ML}, it is simple to show that, for first-order Bragg reflection, the standing wave period in $|E^2| \equiv \lambda_{SW} = d_{ML}$. The relevant Bragg equation is $n\lambda_x = 2d_{ML}sin\theta_{inc}$, where λ_x is the wavelength of the incident X-rays, n is the order of the reflection, and θ_{inc} is the angle of incidence relative to the multilayer. The same is true in Bragg reflection from atomic planes with spacing $d_{hk\ell}$, where $\lambda_{SW} = d_{hk\ell}$, with the corresponding Bragg equation $n\lambda_x = 2d_{hk\ell}sin\theta_{inc}$. The nodes and antinodes of the SW can be moved through the sample in two ways, as indicated: by changing the photon energy $h\nu$ through the Bragg reflection, or by scanning the incidence angle θ_{inc} over the Bragg reflection through a rocking curve (RC).

Another possibility is to tune into the Bragg reflection and scan the X-ray spot over a wedge profile sample grown on top of the mirror, which has been referred to as the SWEDGE method [9]. The first two methods permit scanning the standing wave scan over half of the SW period; the wedge scan can allow for multiple periods. From the basic SW formulas in Fig. 7.1, in which R is the reflectivity, φ is the phase shift between incident and reflected waves, z is the relative "coherent" position of a given atom under SW excitation and f the fraction of atoms that occupy a given coherent position, we can see that the modulation of the SW field strength will be proportional to $\pm 2\sqrt{R}$. Thus, for only a 1% reflectivity, one expects an overall 40% modulation in SW strength, making such measurements overall easier to perform than might at first sight be imagined.

An added interference effect often seen in such measurements is Kiessig (or Fresnel) fringes, which are associated with the X-ray reflection from the top layer(s) of a multilayer sample, and that from the bottom of it, where it meets the semi-infinite substrate on which the multilayer was grown. If the total thickness of the multilayer structure is D_{ML} then these interference peaks are described by $m\lambda_x = 2D_{ML}sin\theta_{inc}$,

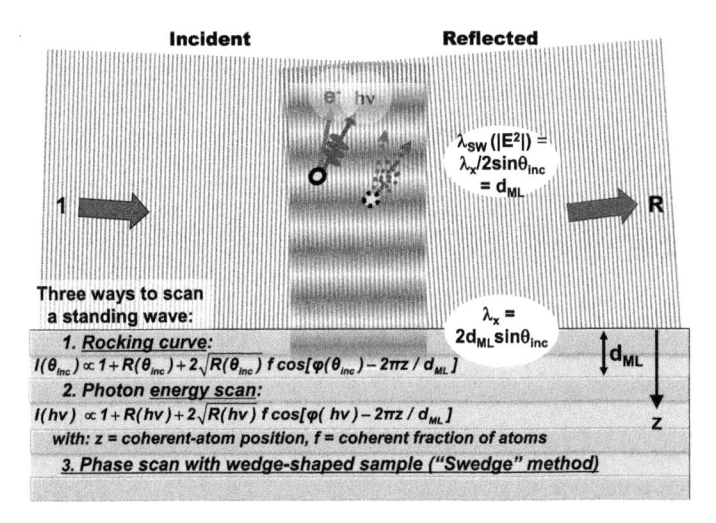

Fig. 7.1 Schematic diagram showing a standing-wave with the period λ_{SW} generated by a superlattice at the X-ray incidence angle corresponding to the first-order Bragg condition. Scanning the standing-wave along the direction perpendicular to the sample surface can be accomplished by either rocking the incidence angle (rocking curve), scanning the photon energy, or via a wedge scan. The equations describe the overall intensity for a given photon energy $h\nu$, reflectivity R, incidence angle θ_{inc}, phase shift on reflection of φ, position z of an atom of interest, and the fraction of atoms in coherent positions supporting the Bragg reflection

with much smaller angular separation than the Bragg peaks. These will be evident in some of the data discussed below, and are schematically illustrated in Fig. 7.2a.

Another possibility involving advanced SW manipulation uses optical resonances, which can enhance sample reflectivity significantly. In order to enhance reflectivity and thus SW effects, a photon energy close to the La M_5 resonance was used in the study of the $SrTiO_3/La_{0.7}Sr_{0.3}MnO_3$ interface in a multilayer sample by Gray et al. [10]. The photon energy in this case was selected to be just below the La M_5 resonance, with a resultant three-fold increase in reflectivity from off-resonance, or an $\sim\sqrt{3}$ increase in the SW modulation, as illustrated in Fig. 7.2. Figure 7.2a shows the sample configuration, with various quantities defined. In Fig. 7.2b is shown the La M_5 absorption spectrum, and in Fig. 7.2c the reflectivity in first-order Bragg reflection for various photon energies over the resonance. The presence of Kiessig fringes is also indicated in Fig. 7.2c.

A second possibility for tuning the SW is shown in Fig. 7.3. Figure 7.3a shows a plot of the real and imaginary parts of the index of refraction (δ and $\beta \propto$ absorption, respectively) of Gd over its analogous M_5 resonance, and Fig. 7.3b and c the depth distributions of the simulated standing-wave electric-field intensities $|E^2|$ as functions of X-ray incidence angle for a similar $SrTiO_3/GdTiO_3$ multilayer [11]. By tuning the photon energy from 1181 eV, which is just below the Gd M_5 resonance peak, to just above the peak at 1187 eV, the positions of the standing-wave antinodes ($|E^2|$ maxima) at the Bragg condition move from a location near the top of the $SrTiO_3$

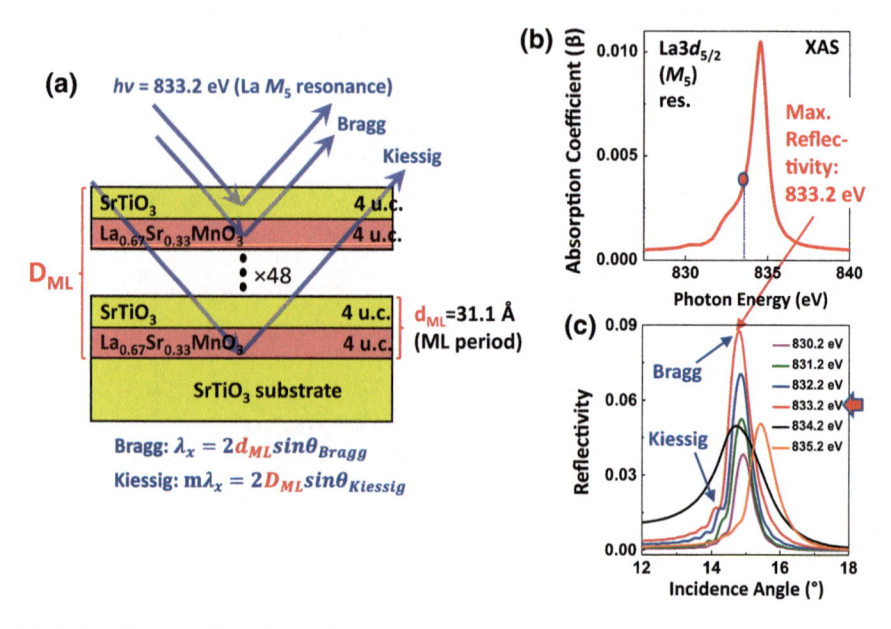

Fig. 7.2 a The sample and experimental configuration for a standing-wave XPS and ARPES study of a $La_{0.67}Sr_{0.33}MnO_3/SrTiO_3$ (LSMO/STO) multilayer sample, with dimensions indicated, together with expressions describing both first-order Bragg reflection and Kiessig fringes. b The absorption coefficient, (expressed as β, the imaging part of the index of refraction), of this multilayer for photon energies scanning through the La $3d_{5/2}$ = La M_5 absorption resonance. c The reflectivity on scanning angle through the Bragg reflection as a function of photon energies going over the resonance. From these data, 833.2 eV was chosen to maximize reflectivity and thus SW modulation. Experimental data from the Advanced Light Source (ALS). (From [10])

layer (Fig. 7.3b) to the buried interface between $SrTiO_3$ and $GdTiO_3$ (Fig. 7.3c). This effect was first pointed out in Bragg reflection from crystal planes [12], and has been used by Nemšák et al. in a recent SW study of a two-dimensional electron gas (2DEG) near the interface in this system [11], as discussed below.

Thus, reflectivity and the SW modulation amplitude, as well as the SW vertical position, can be tuned by selecting photon energies at appropriate positions below or above a strong absorption maximum, providing capabilities that we will illustrate in example applications below.

7.1.2 Near-Total Reflection Measurements

Total reflection in SXPS at ~1 keV was first studied in detail by Henke [13], who pointed out the decreased depth of X-ray penetration for low incidence angles, and the existence of an enhanced peak in intensity just before total reflection. This was pointed out as a technique for surface analysis by Mehta and Fadley [14] and later

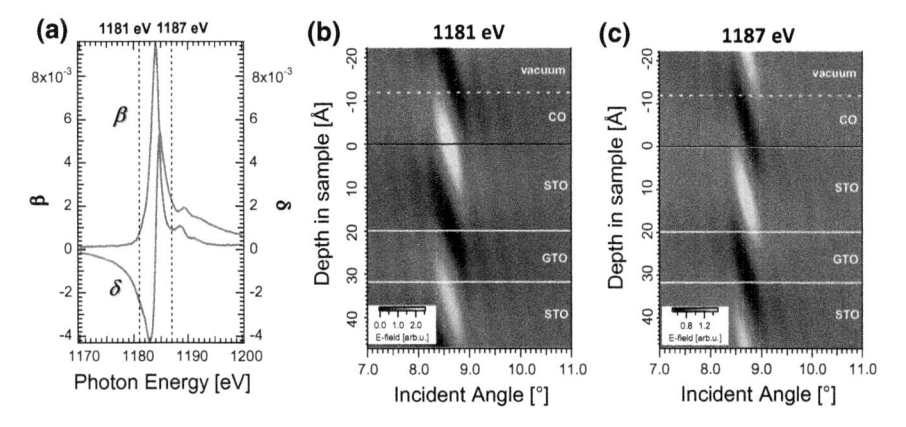

Fig. 7.3 X-ray optical simulations (see [8]) demonstrating a technique for shifting the phase of the standing-wave inside a 20-period SrTiO₃/GdTiO₃ superlattice by tuning the incident X-ray photon energy above and below and above the Gd M₅ absorption resonance, as shown in (**a**). **a** X-ray optical constants (δ and β) obtained experimentally via XAS (β) and Kramers-Kronig formalism (δ). **b, c** Electric field intensity of the standing wave inside the sample as a function of depth and incidence angle for the two photon energies, below (1181 eV) and above (1187 eV) the Gd M₅ resonance. Note the vertical shift in the antinode position of the SW, from sensitivity to the STO surface at 1181 eV to sensitivity to the GTO/STO interface at 1187 eV. Kiessig fringes are also evident in these calculations as the nearly vertical ripples in the field intensity

amplified for this purpose by Jach and co-workers [15] and by Kawai et al. [16], including a recent comprehensive review [17]. Interference fringes in near-total-reflection (NTR) from overlayers on thick substrates have also been pointed out [17]; these in effect consist of standing waves of varying periods as θ_{inc} is scanned into total reflection. The use of such NTR effects to characterize the different depths in multilayer structures has also been discussed, e.g. in [8]. We return below to making use of NTR measurements to study charge accumulation at a buried oxide interface [18].

7.1.3 ARPES in the Soft- and Hard-X-ray Regimes

As discussed in the chapter by Strocov, Cancellieri, and Mishchenko, extending such measurements into the soft and hard X-ray regimes requires considering both the effects of phonon excitation during photoemission, which can smear the momentum resolution in ARPES measurements, and lattice recoil, which can shift and smear the energy resolution in any sort of spectrum.

The effects of phonon excitation in reducing the fraction of simply-interpretable direct transitions (DTs) can be estimated from a photoemission Debye-Waller factor,

$$W(T) \approx exp\left[-g_{hk\ell}^2 \langle u^2(T) \rangle\right], \qquad (1)$$

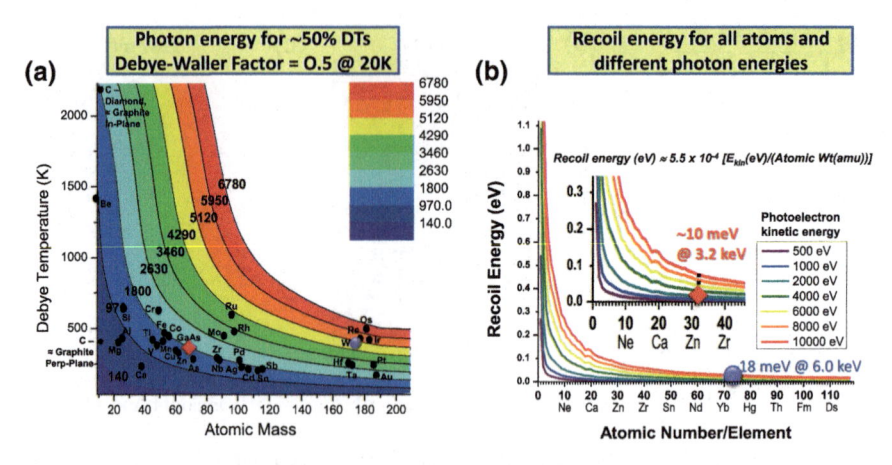

Fig. 7.4 Calculated Debye-Waller factors for different atomic mass and temperatures. **a** Contours for a photoemission Debye-Waller factor W(T) of 0.5 at a typical LHe cryocooling temperature of 20 K, and **b** the recoil energy for all atomic masses as a function of photon energy. Values for W and GaAs studied with hard X-rays [22] are highlighted. (From [19])

with g_{hkl} the magnitude of the reciprocal lattice vector involved in a given direct transition (DT) and $u^2(T)$ the one-dimensional mean-squared vibrational displacement at temperature T.

As an illustration of how important these effects might be, Fig. 7.4a shows a plot of the photon energy for which 50% of the transitions remain direct at 20 K (a reasonable criterion for being able to carry out ARPES), as a function of the two relevant parameters of Debye temperature and atomic mass, with points for various elements indicated [19]. These are calculated using the Debye model. This plot can be used to estimate the feasibility of a given ARPES experiment as energy is increased, although prior experiments make it clear that such simple estimates can be on the conservative side. One likely reason for this is that the Debye model does not include correlation of vibrations for near-neighbor atoms, that is, that nearest and next-nearest neighbor atoms will vibrate less than atoms further away, being more rigid in their motion with respect to a given atom [20]. A more accurate method of modeling such phonon effects has been developed by Braun et al. [21]; this makes use of the coherent potential approximation (CPA) to model the effects of atomic displacements, and is the first quantitative modeling of the temperature dependence of soft- and hard-X-ray ARPES.

Recoil can also be estimated by assuming a single-atom of mass M recoils in the lattice, again a conservative estimate, which yields the energy shift as

$$E_{recoil} \approx \frac{h\nu}{2M} \approx 5.5 \times 10^{-4} \left[\frac{E_{kin}(eV)}{M(amu)} \right]. \qquad (2)$$

Figure 7.4b shows a family of curves of recoil energy, as a function of photon energy and mass, again permitting an estimate of the shifting and smearing expected for a given system. Of course, such recoil effects are most significant for lighter elements, but they also depend crucially on the Debye temperature, or more precisely the rigidity of the vibrational potentials associated with a given atom. The Debye-Waller factor in (1) can in fact be used to estimate the fraction of transitions that occur with no recoil, in the same spirit as in the analysis of Mössbauer spectra. But different local correlated vibrational environments for different atoms will no doubt lead to deviations from the simple estimates of Fig. 7.4. For example, Fig. 7.4b notes that the Debye temperature for in-plane vibrations in graphite, or equivalently in graphene, is much higher than that perpendicular-to-plane. Thus, ARPES for graphite/graphene connected with in-plane dispersions is expected to be possible at much higher photon energies (ca. 2000 eV from the figure) and/or temperatures. The two plots in Fig. 7.4 are highlighted for W and GaAs, the two materials for which it was first demonstrated by Gray et al. [22] that ARPES could be performed at up to 6 keV and 3 keV, respectively.

Finally, we note that, in the limit of high photon energies and/or high temperatures and/or a high angular integration in the electron spectrometer, VB spectra converge to what can to a first approximation be considered as a matrix-element-weighted density of states (MEWDOS) limit, or more simply, what has often been referred to as the "XPS limit".

7.2 Applications to Various Oxide Systems and Spintronics Materials

7.2.1 Overview of Past Studies—Standing Waves from Multilayer Reflection

Standing-wave spectroscopy from multilayer samples began with work by Bedzyk et al. [23] using X-ray fluorescence detection, by Kortright et al. [24, 25] using X-ray magnetic circular dichroism, and by our collaborators using photoemission [26]. These studies have by now involved sample which have been grown *as* the multilayer or on top of a suitable multilayer, with applications to a wide range of materials systems, some of which we list below, before focusing on some much more recent studies that demonstrate the full potential of the technique:

- **The Fe/Cr giant magnetoresistive interface**: The interface depth profiles of concentration and magnetic order were determined, the latter being via photoemission magnetic circular dichroism (PMCD) measurements [9]. This study made use of the standing-wave wedge (SWEDGE) method illustrated in Fig. 7.1.
- **A CoFeB/CoFe/Al_2O_3 magnetic tunnel junction**: The depth-dependent variation of the buried-layer density of states was determined, and related to tunneling properties [27].

- **Co microdots on a silicon substrate**: The use of scanned-photon-energy SW excitation to add depth resolution in photoelectron microscopy was demonstrated [28].
- **The Fe/MgO magnetic tunnel junction system**: The variations of atomic concentrations, densities of state, and Fe magnetization through the interface were determined, again using PMCD and the SWEDGE method [29].
- **The Ta/CoFeB/MgO magnetic tunnel junction system**: The diffusion of B with annealing of a Ta/CoFeB/MgO magnetic tunnel junction was determined with standing-wave hard X-ray photoemission [30].

7.2.2 Overview of Past Studies—Standing Waves from Atomic-Plane Reflection

Standing-wave spectroscopy with Bragg reflection from atomic planes has a longer history, with first measurements again involving X-ray fluorescence detection [31, 32], and photoemission detection coming some time later [33].

A significant additional development involving atomic-plane Bragg reflection, of which we will make use below, is the combination of core-level rocking curve or energy-scan measurements with analogous valence-band (VB) data to decompose the VB data into element-specific components, as pioneered by Woicik et al. [34, 35]. This will be discussed in more detail below.

7.2.3 Multilayer Standing-Wave Soft- and Hard-X-ray Photoemission and ARPES from the Interface Between a Half-Metallic Ferromagnet and a Band Insulator: $La_{0.67}Sr_{0.33}MnO_3/SrTiO_3$

In a combined soft X-ray/hard X-ray study using standing-wave excitation from a multilayer of $SrTiO_3$ (STO) and $La_{0.7}Sr_{0.3}MnO_3$ (LSMO), Gray et al. [10, 36] demonstrated the power of the SW approach for determining the depth profiles of concentration, chemical state and valence-band electronic structure through buried interfaces. Some of these results are summarized in Figs. 7.5, 7.6, 7.7 and 7.8. Figure 7.2a shows the sample configuration, and Fig. 7.2b, c the method of tuning the photon energy to maximize the SW strength. The photon energy 833.2 eV just below the La M_5 edge was chosen to maximize reflectivity and thus the SW modulation. Figure 7.5a illustrates the excellent theoretical fit to core-level rocking-curve results at 833 eV and 5.96 keV for all of the elements in the sample, Fig. 7.5b the resulting depth distribution of bilayer thickness, and Fig. 7.5c the depth distribution of concentration and soft X-ray optical constants. Of particular note is that the bilayer thickness drifted by ca. 6% over the 48-bilayer thickness, a finding confirmed later by TEM/EELS measurements. TEM/EELS also confirmed the findings in Fig. 7.5c, although it is not sensitive to a change in the soft X-ray optical constants of LSMO near the interface that are revealed by the SW photoemission analysis. Note also in Fig. 7.5b that the rocking curves show features due to both the 1st-order Bragg reflection from the multilayer and Kiessig interference oscillations due to X-ray reflection from the top and bottom of the multilayer, with these being much stronger compared to the Bragg peak with hard X-ray excitation at ~6 keV. A likely explanation for this is that the higher-energy photons can more readily penetrate the full multilayer and reflect back from the substrate interface. Beyond this, Bragg reflection has been enhanced in the soft X-ray data by tuning the photon energy, as shown in Fig. 7.2b, and this may be another cause of the higher relative intensity of the Kiessig fringes at higher energy, where no resonance is involved.

Figure 7.6 further shows the detailed interface-sensitive SW core-level spectroscopy that is possible, with Mn $3p$, but not Mn $3s$, showing a small shift near the interface. Figure 7.6a shows the variation in binding energy of Mn $3p$, Ti $3p$, and Mn $3s$ as angle is scanned over the Bragg condition. Only Mn $3p$ shows a small, but reproducible shift, as further seen in Fig. 7.6b, c. Figure 7.6d shows calculations of the wave field as a function of angle, verifying that Mn $3p$ shifts to higher binding energy by about 0.3 eV when the SW selectively probes the interface. These results have been explained in terms of an Anderson Impurity Model as being due to a Jahn-Teller distortion of the MnO_6 octahedra near the interface that does not affect Mn $3s$ or its well-known multiplet splitting (also shown in Fig. 7.6c).

Figure 7.7 illustrates another groundbreaking aspect of these studies by Gray et al. [36] in which SW ARPES (SWARPES) measurements were made on the same system at two different angles of incidence to selectively look into the "bulk" of LSMO and at the LSMO/STO interface. Figure 7.7a shows the angle-integrated ARPES spectra, which should represent a matrix-element weighted density of states, with different regions labelled as to their origins: 1 strongly Mn e_g, 2 strongly Mn t_{2g}, 3 and 4 strongly STO states (due to its being the top layer of the sample), and 5 probably the lowest bands in LSMO. Figure 7.7b–d are $k_x - k_y$ SWARPES maps for enhanced "bulk" LSMO sensitivity, enhanced interface sensitivity, and the difference between the two, respectively. With the difference, one is able to selectively look at how the k-resolved interface electronic structure differs from that further into the LSMO. Particularly for region 5, but also for the LSMO Mn $3d$ regions 1 and 2, there are significant differences that indicate these data represent the first time that k-resolved changes in electronic structure near an interface have been measured. Theoretical calculations shown elsewhere, including using the most accurate one-step time-reversed LEED approach, are at least qualitatively in agreement with these data [36].

The work discussed in this section thus indicates the high sensitivity of SW soft- and hard-X-ray photoemission to the bonding configuration of atoms at interfaces, to the detailed character of the multilayer, including the depth distributions of all species and the index of refraction, and via SWARPES to the momentum-resolved electronic structure at the interfaces.

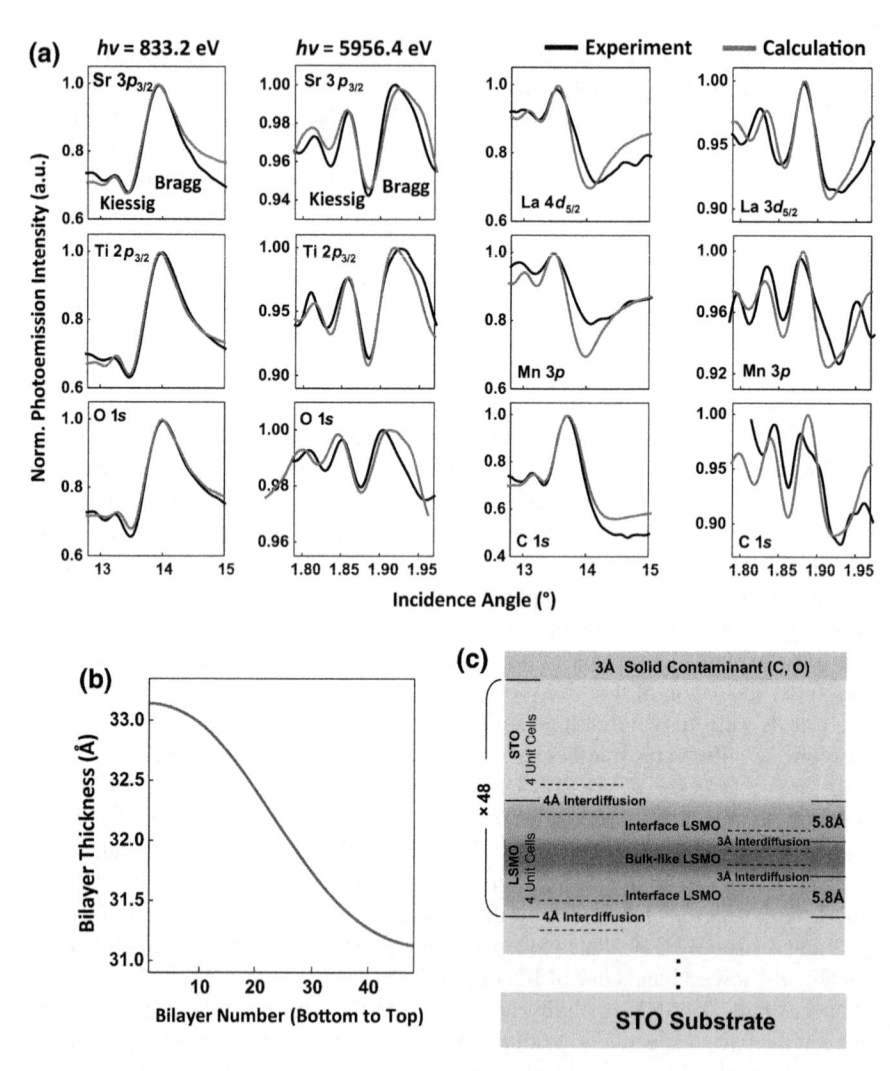

Fig. 7.5 a X-ray optical analysis and best fits of the standing-wave rocking curves of the core-level peak intensities for every element in the LSMO/STO superlattice, measured at soft and hard X-ray energies. b LSMO/STO bilayer thickness profile diminishes by approximately 6% with depth, from bottom to top. Detailed chemical/interdiffusion profile of the interface is shown in (c). The result shown in (b) and (c) have been quantitatively confirmed by standing transmission electron microscopy with electron energy loss spectroscopy. Experimental data from the ALS and SPring-8. (From [10, 36])

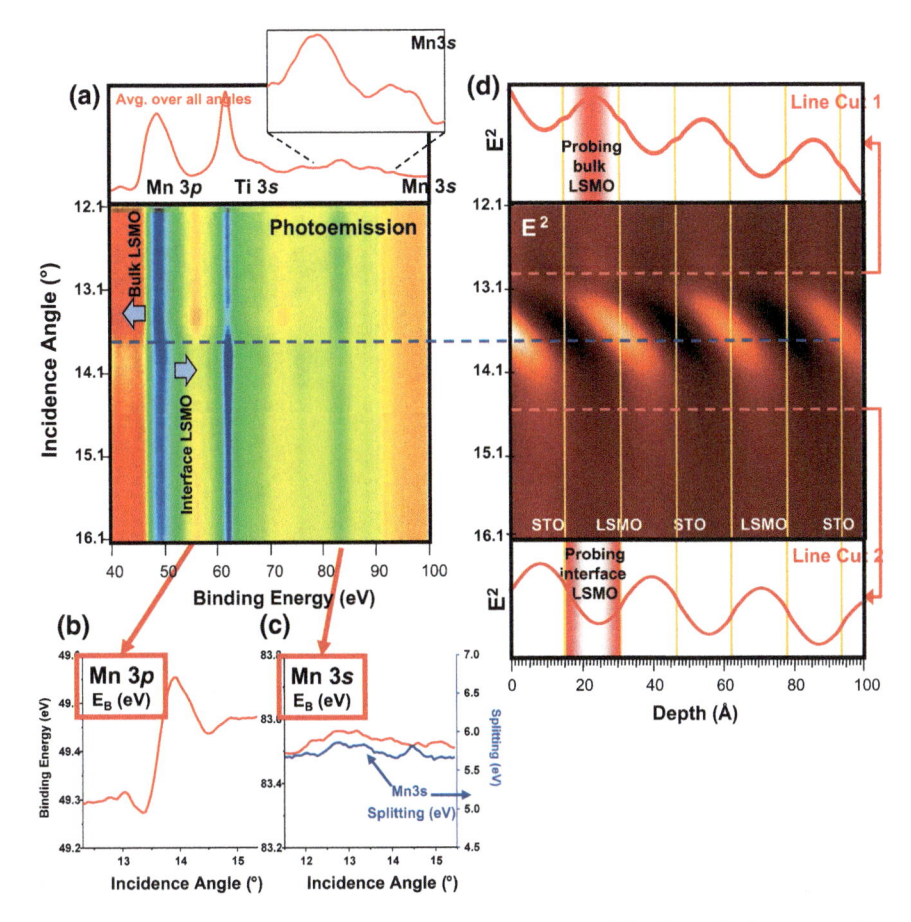

Fig. 7.6 Standing-wave rocking-curve spectroscopy with an 833 eV excitation energy just below the La M_5 resonance to enhance reflectivity from the 48-bilayer multilayer of $SrTiO_3/La_{0.7}Sr_{0.3}MnO_3$ shown in Fig. 7.2a. **a** The evolution of Mn 3s and 3p and Ti 3s core-level binding energies and intensities through a rocking curve. **b, c** The change in binding energy and multiplet splitting for Mn 3p and Mn 3s. **d** The variation of the standing-wave field strength through the same angle range as (**a**). It is concluded that Mn 3p only shows a shift near the LSMO/STO interface. Experimental data from the ALS. (From [10])

7.2.4 Multilayer Standing-Wave Soft X-ray Photoemission and ARPES Study of the Two-Dimensional Electron Gas at the Interface Between a Mott Insulator and a Band Insulator: *GdTiO₃/SrTiO₃*

As another example illustrating the power of the standing-wave technique, we show the results of a soft X-ray SW-XPS and SW-ARPES study of the $SrTiO_3/GdTiO_3$ (STO/GTO) superlattice in Fig. 7.8. This complex oxide system has received much

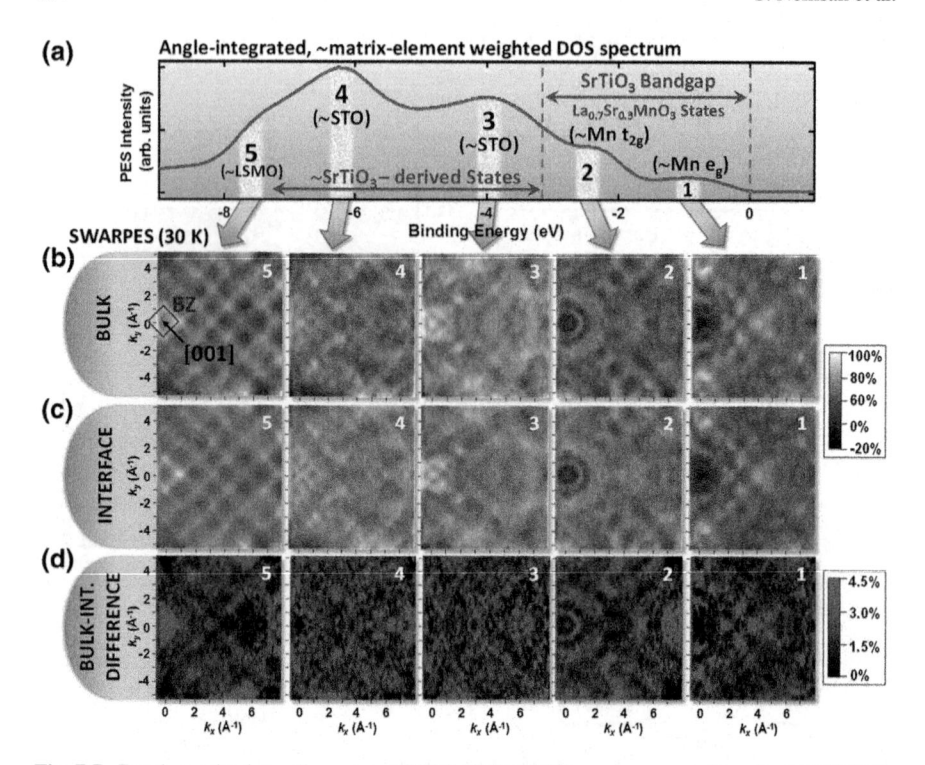

Fig. 7.7 Depth-resolved standing-wave ARPES (SWARPES) measurements from the LSMO/STO superlattice shown in Fig. 7.2a, again at 833 eV photon energy. **a** An angle-integrated valence band spectrum; the distinctive spectral features are labeled 1–5, with their origins and approximate atomic-orbital characters indicated. **b** SWARPES scans divided into these five energy windows in a SW geometry enhancing bulk-LSMO signal. Shown are XPD-normalized angle-resolved (k_x, k_y) photoemission intensity maps of the Mn 3d e_g (1), Mn 3d t_{2g} (2) derived states, the largely STO-derived states (3 and 4), and the valence-band bottom LSMO states (5). **c** As (**b**), but for the SW geometry enhancing signal from the LSMO/STO-interface. **d** Difference (k_x, k_y) maps calculated by subtracting maps in (**c**) from maps shown in (**b**). The most significant differences are from the Mn 3d e_g and t_{2g} bands near Fermi level and from the bottom of the valence band (feature 5). The color bars on the right indicate the relative amplitudes of the effects. (From [36])

attention because of the recent observation of a two-dimensional electron gas (2DEG) at the STO/GTO interface by Stemmer et al. [37]. In this particular study, Nemšák et al. addressed the question of whether the 2DEG can be detected using soft X-ray ARPES, and whether its depth distribution can be directly measured using SW excitation. The sketch of the multilayer sample used for the study is shown in Fig. 7.8a. As depicted, the incident angle of the light was varied, which leads to a change in the SW position. In order to enhance the signal of the Ti 3d components in the valence spectra, the ARPES data shown in Fig. 7.8b were measured at a Ti 2p-3d resonant energy of ~465 eV. Two distinctive features near the Fermi level are identified—a

7 Standing-Wave and Resonant Soft- and Hard-X-ray Photoelectron … 165

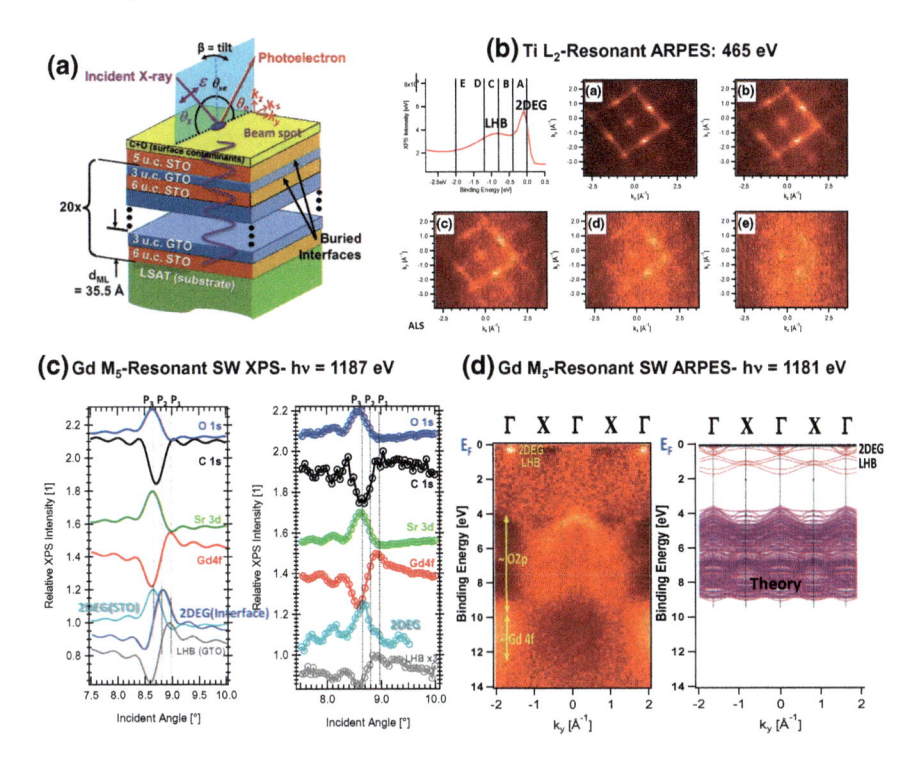

Fig. 7.8 Experimental and theoretical results from a combined standing-wave and resonant photoemission study of a SrTiO$_3$/GdTiO$_3$ multilayer. **a** The sample configuration. **b** Normal-emission Ti L$_2$-resonant ARPES at 465 eV over the 2DEG and LHB features near the Fermi level, as binned in five separate regions A to E. **d** Experimental (right panel) and simulated (left panel) Gd M$_5$ resonant standing-wave photoemission at 1187 eV, emphasizing the interface (cf. Fig. 7.3c), including rocking curves for O 1s, C 1s (surface impurity), Sr 3d, Gd 4f, and valence-level intensity for the 2DEG and the LHB, as derived by peak fitting spectra such as those in Fig. 7.8b. Simulations are shown for the 2DEG for its being distributed throughout the STO layer (turquoise), and for it occupying only one unit cell near the interface (blue). **e** Binding energy-k_x SW-ARPES image at an interface sensitive Gd M$_5$-resonant energy of 1187 eV, compared to theoretical calculations for the multilayer using hybrid functionals. Experimental data from ALS and the Swiss Light Source (SLS). (From [11])

sharp peak in the immediate proximity of the E$_F$ and another broader one at ca. 0.7 eV binding energy. The sharp feature near the Fermi level exhibits a dispersion in $k_x - k_y$ that is characteristic of a 2DEG. The broader feature can tentatively be identified as the lower Hubbard band (LHB) expected in GTO. The panels A through E here represent different energy intervals over which the ARPES images were taken. Both features have a strong Ti 3d character, as revealed by the resonant photoemission. Also, since all of the panels A–E show very similar dispersion patterns, there is a strong mixing in the character of these two states, which is confirmed by LDA calculations [11].

The question remains, however, as to whether the 2DEG is actually localized at the buried interface, or whether it originates at the surface of the sample, specifically in the top STO layer, as has been observed in previous studies via low-energy VUV-ARPES [38, 39]. This question can be easily answered by recording two sets of rocking curves, with incident X-ray energies of 1181 and 1187 eV just below and just above the Gd M_5 absorption threshold. X-ray optical simulations of the standing-wave electric-field intensities within the sample shown in Fig. 7.3 suggest that at the incident photon energy of 1187 eV the standing-wave antinode is positioned at the buried interface between STO and GTO. Figure 7.8c shows the experimental and best-fit calculated rocking curves recorded at this energy. Comparison between the experimental and simulated rocking curves reveals that the 2DEG states are localized near the buried STO/GTO interface, and that the 2DEG extends through the entire STO layer. For example, the phase (angular position) of the rocking curve associated with the 2DEG is completely different from those of C 1s (originating from the surface contamination on top of the STO layer) and that of Gd 4f from the bulk of the GTO layer. It is, however, essentially identical to the Sr 3d RC. We can therefore conclude that the 2DEG is localized in STO near the buried interface with GTO, and that it is not a surface-specific 2DEG, as previously observed with ARPES [38, 39]. Further comparing the experimental RC in Fig. 7.8c-right panel with theoretical RCs for the 2DEG assuming it exists throughout the full STO layer and one assuming it exists within only the first unit cell above the interface (Fig. 7.8c-left panel) confirms that the 2DEG essentially extends through the entire STO layer, consistent with prior resonant tunneling measurements [37]. The RC of the peak that we tentatively assigned to the GTO lower Hubbard band is very similar to that Gd 4f RC, thus confirming this assignment. Additional experimental and theoretical RC results of this type at 1181 eV confirm these conclusions [11].

Finally, Fig. 7.8d compares a binding energy-k_x resonant ARPES plot with theoretical calculations based on hybrid functionals [11], and there is good agreement as to the general features. Theory also confirms our assignment of the LHB and the 2DEG.

In overview of this study, a combination of resonant effects was employed to obtain detailed information on the 2DEG in the GTO/STO multilayer system. Enhancement of the signal from the valence electron contribution of a given atom was achieved via standard resonant photoemission. In another part of the experiment, tuning to resonant photon energies was used to enhance the X-ray reflectivity, which also enhances the SW effects. Last, but not least, the photon energy was tuned below and above the absorption edge to move the position of SW more dramatically. It should be possible to apply an analogous set of experiments can be to a multitude of other oxide multi-layer systems to obtain otherwise inaccessible information on depth distributions of species.

7.2.5 Multilayer Standing-Wave Photoemission Determination of the Depth Distributions at a Liquid/Solid Interface: Aqueous NaOH and CsOH on Fe_2O_3

Very recently, it has been demonstrated that standing-wave excited photoemission can be used for direct Ångstrom-level depth-resolved studies of the buried interface between solids and gases and/or solids and liquids [40]. Such interfaces are of critical importance in numerous areas of energy-related and environmental research, since they host a wide variety of surface and catalytic reactions, as well as electrochemical processes. An archetypal example of such a system is the electrochemical double-layer, which, despite of over 100 years of studies, is still not completely understood [41, 42].

The first such standing-wave ambient pressure photoemission (SWAPPS) study has provided depth-dependent insight on the reaction between water, NaOH and CsOH at the surface of a thin Fe_2O_3 film. The film was grown on a Si/Mo multilayer mirror, acting as a standing-wave generator, and the measurements were carried out at a pressure and temperature which facilitate a "wet" surface with a thin "liquid-like" film on top [40]. Figure 7.9a shows the sample configuration, with excitation at 910 eV, and a solution of NaOH and CsOH that was prepared by drop casting onto Fe_2O_3 in air and then inserting into a "humid" vacuum environment. Analysis of the relative core-level peak intensities using the SESSA XPS simulation program [43] reveals that for the temperature of 2.5 °C and the in situ pressure of 400 mTorr we have used, the thickness of the adsorbed surface water layer is approximately 10 Å. A typical spectrum for the O 1s core level is shown in Fig. 7.9b. Four distinct chemical components, including the one originating from water in the gas phase above the surface, can be clearly resolved. What is even more remarkable, is that all four features exhibit unique RC shapes, which are clearly modulated differently as the standing wave scans through the sample and the surface. Figure 7.9d furthermore compares the RCs for Cs 4d and Na 2p core-level intensities, which exhibit a clear shift of 0.04° evident in the steeply sloping parts of the two spectra near the Bragg condition, and also a different shape both below and above the Bragg peak, with higher intensities for the Cs 4d rocking curve. These differences immediately indicate that the two ions have a different depth distribution. Fully analyzing the rocking curves from all of the resolvable chemical species using a specially-written program [8] leads to the excellent fits of theory to experiment shown in Fig. 7.9e and the depth distributions shown in Fig. 7.9f.

A subsequent SWAPPS study has made use of hard X-ray excitation to look through a thin film of electrolyte from an operating electrochemical cell in which Ni was being oxidized in a KOH solution [44], using what has been termed variously the "meniscus" or "dip-and-pull" or "dipstick" method [45]. In this approach, an active electrode is pulled from a working cell, leaving a thin layer of electrolyte on the surface that is in equilibrium with that in the cell, thus permitting photoemission and SWAPPS measurement in *operando*.

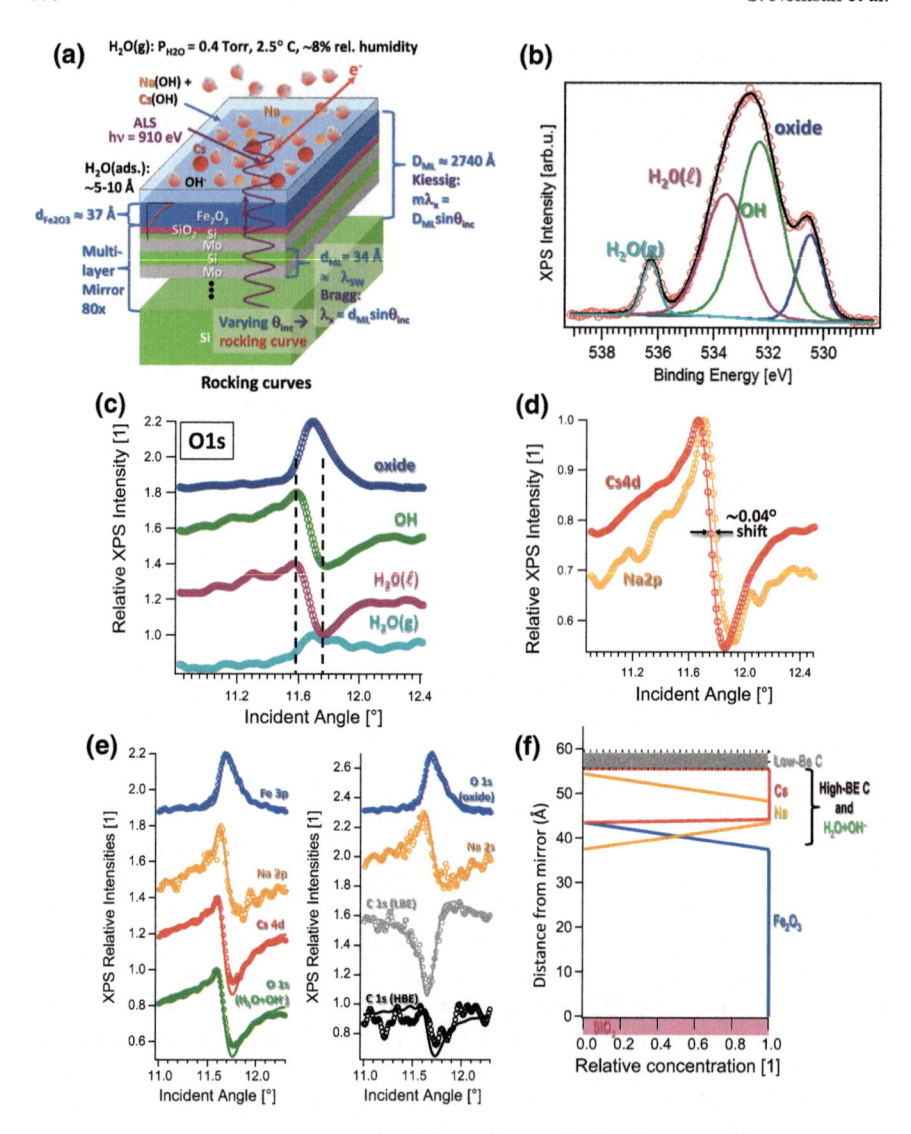

Fig. 7.9 Standing-wave ambient pressure photoemission (SWAPPS) measurements of a Fe_2O_3 film grown on a Si/Mo multilayer mirror. Water solution of CsOH, NaOH was drop-casted in air and later rehydrated in the UHV chamber by a water vapor at 400 mTorr and 2.5 °C. **a** The sample configuration including the SW profile, which is moved through the surface by scanning the incidence angle (cf. Fig. 7.1). **b** Four distinct oxygen species identified in the O 1s spectrum. **c** The variation of the O 1s intensities for the four components resolved in the panel (**b**) as a function of incident angle. Distinctive differences in rocking curve shapes are visible for a different species. **d** A comparison of the rocking curves of Cs and Na, indicating a shift in their phase of ~0.04° between them. **d** Comparison of experimental and calculated rocking curves for all elements and chemically-resolved peaks, for the final sample structure, as shown in (**f**). Experimental data are from ALS. (From [40])

The two studies discussed in this section show the potential of SWAPPS measurements, particularly with hard X-ray excitation, which is then able to probe thicker electrolyte layers and/or to penetrate thin-film windows enclosing the cell. The higher kinetic energies of the measured photoelectrons lead to an increased information depth of the measurements, so the standing wave modulation is then needed to maintain the interface sensitivity. Both soft and hard X-ray SWAPPS are new powerful tools for liquid/solid interface studies of elemental/chemical and electrical field gradients, which can be examined with an unprecedented detail. A broad range of SWAPPS applications lie in the energy, catalysis and environmental research fields.

7.2.6 Near-Total Reflection Measurement of the Charge Accumulation at the Interface Between a Ferroelectric and a Doped Mott Insulator: $BiFeO_3/(Ca_{1-X}Ce_X)MnO_3$

We have already mentioned that carrying out measurements in the total reflection regime can be used very effectively for enhancing surface sensitivity, as well as reducing inelastic backgrounds in photoemission, as shown in several prior studies [13–17]. In this case, additional X-ray optical effects, such as interferences associated with reflection from a buried interface, can also provide useful structural information about the sample in the total reflection regime. Utilization of this method is particularly advantageous because it does not require growth of multilayer samples, and can, in principle, be applied to any heterostructure or trilayer.

We illustrate the NTR approach in Fig. 7.10 with some HXPS experimental data and X-ray optical simulations for a bilayer sample of ferroelectric $BiFeO_3$ (BFO) on top of a Ce-doped Mott insulator $(Ca_{0.96}Ce_{0.04})MnO_3$ (CCMO). The system has drawn attention due to a Mott metal-insulator transition in the 2DEG between BFO and CCMO, which can be ferroelectrically controlled [46]. The excitation photon energy was in the tender X-ray regime (2.8 keV), thus permitting the study of a buried interface in a sample of 10 nm BFO on 210 nm of CCMO, grown on a $YAlO_3$ substrate, as shown in Fig. 7.10c. Core-level spectra were collected as a function of incident angle for Ca $2p$ (Fig. 7.10a), C $1s$ (in a surface contaminant layer) and Bi $4f$ (originating from BFO). Ca $2p$ is interesting in showing two components, which have a different intensity ratio depending on the angle of incidence. Using the X-ray optical simulation of the electric field strength (Fig. 7.10b), we see that the two angles in Fig. 7.10a and highlighted in Fig. 7.10b represent the signal being enhanced either at the BFO/CCMO interface (~0.7°) or in the bulk of CCMO (~1.5°). Further analysis of the angular dependent intensity profiles for all four above mentioned core-levels and their comparison to theory are shown in Fig. 7.10e. All four curves show a gradual decay to zero for the shallow incident angle, as total reflection conditions are reached. On the high incident angle side, intensity curves exhibit a series of oscillations, due to SWs created by reflection from the buried interface. The comparison of the experimental data and the theoretical calculations by fitting the NTR experimental results for the two Ca components to theory yields excellent agreement for an optimized sample geometry. The simulations reveal the presence of an ~1 nm charge accumulation region in the CCMO at the interface derived. These results were further confirmed by a separate analysis using the STEM-EELS Mn L_3 near-edge features.

The oscillations in Fig. 7.10d, e arise due to the interference and multiple reflections at the surface and the interfaces, and the increase in intensities at approximately 0.9° due to the spreading of the X-ray beam along the spectrometer entrance slit direction, as well as the concentration of E-field near the surface, as observed and described by Henke [13]. Shifts in the oscillations phases between Bi and Ca (see Fig. 7.10d) are instrumental in optimizing the fits between the experiment and theory, and thus deriving the depth-dependent chemical profile.

As general background, it is also worth noting that SW effects in hard X-ray reflection and emission of the type shown in Fig. 7.8 have been used previously, e.g. to study the distribution of ionic species in solution above oxide surfaces [12, 47], but with our photoemission approach having the advantage of chemical- and spin-state specificity through core level shifts and fine structure, as well as the ability to look directly at VB DOS changes at interfaces.

7.2.7 Atomic-Plane Bragg Reflection Standing-Wave Hard X-ray ARPES: Element- and Momentum-Resolved Electronic Structure of GaAs and the Dilute Magnetic Semiconductor $Ga_{1-x}Mn_xAs$

Mn-doped GaAs (doping level ≈ 0.03–0.06) is one of the prototypical dilute magnetic semiconductors (DMS) and until recently there was some controversy connected to the origin of its ferromagnetism. Two scenarios were discussed—one being the so-called double-exchange mechanism, in which Mn-induced states would form an impurity band clearly separated from the p-bands of Ga and As and a second being the *p-d* exchange mechanism, in which these states are merged with the GaAs impurity bands. The first hard X-ray angle resolved study of a sample with composition $Ga_{0.97}Mn_{0.03}As$ was performed by Gray et al. [48]. The samples were cleaned only by the HCl etching in air to remove surface oxide, illustrating a key advantage of more bulk sensitive hard X-ray excited photoelectrons. The normalized angle-resolved experimental data together with one-step theory calculations are shown in Fig. 7.11a–d. Figure 7.11e, f shows deeper and relatively flat As $4s$ bands that exhibits X-ray photoelectron diffraction (XPD). The angular intensity distributions of the $Ga_{0.97}Mn_{0.03}As$ are smeared out in both experiment and theory relative to those of the undoped GaAs. This can be explained by the presence of the Mn atoms, which due to their random spatial distribution disturb the long-range periodicity. One-step photoemission theory shows a remarkable agreement with experiment, with further theoretical results and analysis presented elsewhere [48]. Further analysis of the

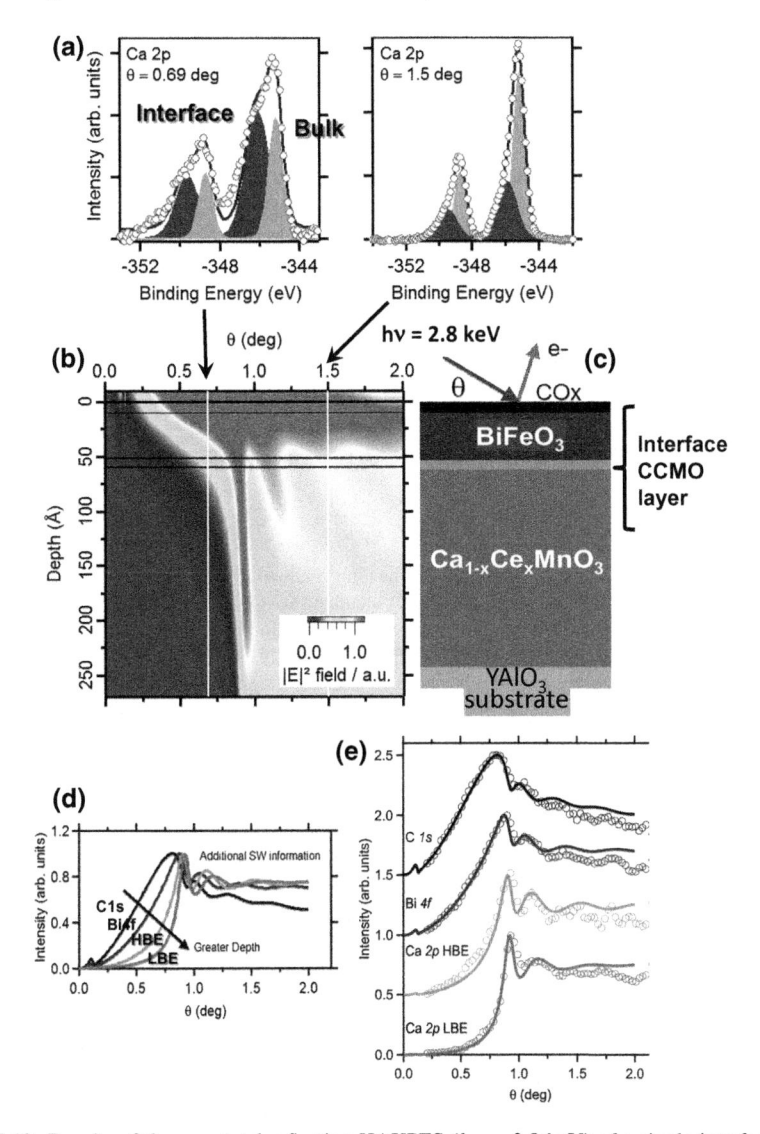

7 Standing-Wave and Resonant Soft- and Hard-X-ray Photoelectron ... 171

Fig. 7.10 Results of the near-total reflection HAXPES ($h\nu = 2.8$ keV) of a single interface in a bilayer BiFeO$_3$/Ce-doped CaMnO$_3$/YAlO$_3$ substrate complex oxide heterostructure. **a** Ca 2p core-level spectra recorded in two experimental geometries, emphasizing the interface (high-E$_b$) and bulk (low-E$_b$) regions in the buried Ce-doped CaMnO$_3$ layer. **b** X-ray optical calculation of the standing-wave electric field intensity inside the sample (shown in **c**) as a function of X-ray incidence angle, obtained by fitting incidence angle-resolved Bi 2p, Ca 2p and C 1s core-level intensities. **d** The normalized experimental intensities of Bi 4f, C 1s and both components of the Ca 2p core levels as functions of the incidence angle over the near-total reflection region, showing the systematic trend in the approach to total reflection, with peaks deeper within the sample turning off first. **e** Optimized best fits of the X-ray optical simulations to the experimental data collected at the ALS. (From [18])

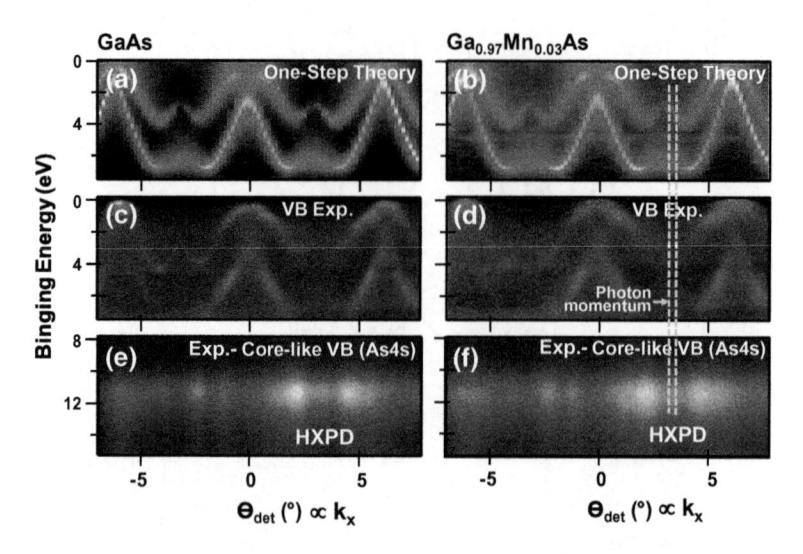

Fig. 7.11 HXPS and HARPES from GaAs and the dilute magnetic semiconductor Mn-doped GaAs. A photon energy of 3.2 keV was used. **a, b** show calculated HARPES patterns for the two materials in the same experimental geometries as the experimental results of (**c**) and (**d**). **e, f** Show the same kind of experimental data for the highly localized and nearly purely As 4s band, which shows no band dispersions in energy, and exhibits only hard X-ray photoelectron diffraction (HXPD) variations in angle. The dashed lines indicate the slight shift between the center of symmetry of the HXPD, which is linked to the surface normal of the sample, and the center of symmetry in k_x, which is shifted due to the photon momentum. Experimental data from SPring-8. (From [48])

k-resolved and angle-integrated results, with special attention to the region very near E_F, permitted concluding that both p-d exchange and double-exchange interactions must be considered to explain ferromagnetism in this material. This conclusion has also been essentially confirmed by soft X-ray resonant ARPES [49].

It is also interesting to note in these data the shift in the symmetry center of the ARPES results compared to the symmetry center of the HXPD pattern of the As 4s band; this is due to the photon momentum, a non-dipole effect that must be included in the conservation of k, as discussed above and by Gray et al. [22].

This was the first application of HARPES to a material system whose electronic and magnetic properties were under debate. It furthermore suggested a wide area of application in the future for studying the bulk electronic structure of complex materials. The future perspectives and limitations of HARPES have been discussed in more detail elsewhere [1, 22].

Beyond this, we now discuss standing-wave HARPES (SW-HARPES) making use of Bragg reflection from atomic planes, so as to assess the possibility of using this technique to derive element- and momentum-resolved electronic structure, possibly also as a function of spatial position within the unit cell. Going to energies above about 2 keV for which the X-ray wavelength is less than 6 Å permits creating standing-waves due to reflection from crystal planes separated by greater than 3 Å, with the

SW then scanning by half a cycle through the unit cell and along the [$hk\ell$] direction as photon energy is moved over the appropriate Bragg condition.

We have noted that several prior studies have used Bragg reflection from atomic planes to create standing waves that are then scanned through the unit cell via photon energy scans, with combined core- and valence-data then being used to deconvolute densities of states into their element-specific contributions [34, 35, 50]. These prior measurements have all been carried out in the Brillouin-zone-averaged MEWDOS or XPS limit. As shown before, with cryogenic cooling, such experiments are also feasible in an angle-resolved mode, in which case the HARPES results could represent element- and momentum-resolved sampling of different points in the Brillouin zone. Although soft X-ray resonant photoemission can provide somewhat similar information on element- and k-resolved electronic structure [49], the spatial phase information provided by the SW, coupled with core level intensities, should permit a much more quantitative decomposition of the VB intensities into their atomic components and spatial distribution that is furthermore more truly bulk sensitive.

We now present the first proof-of-principle experiments of this kind for the same two systems of GaAs (001) and the DMS Ga(Mn)As (001), which were measured at the soft- and hard-XPS facility at Diamond Light Source Beamline I09 [51]. The results are displayed in Fig. 7.12. In Fig. 7.12a, we illustrate one of two experimental geometries used: (111) reflection, with the SW then scanned along the [111] direction, and four possible positions of its nodes/antinodes indicated. In Fig. 7.12b, the energy scans of As $3d$ and Ga $3d$ are shown, and they are clearly very different, due to their different positions in the unit cell along the [111] direction; these results also agree well with prior core-level experiments for GaAs (111) by Woicik et al. [35]. In Fig. 7.12c, analogous scans for a sample consisting of ~100 nm of $Ga_{0.95}$ $Mn_{0.05}As$ deposited on a GaAs (001) substrate, are shown. Two different Bragg reflections are seen here, a narrower structure from the substrate and a wider structure for the finite-thickness $Ga_{0.95}$ $Mn_{0.05}As$, with differences between the Ga and As curves associated with both of them. Figure 7.12d shows similar data for the same sample, but with (311) reflection, and with the addition of an energy scan for Mn $2p$; although noisy due to the low concentration of Mn, it is clear that the Mn curve agrees with that for Ga, indicating a substitutional position of the Mn atoms.

In Fig. 7.13, we consider adding elemental sensitivity to HARPES via such SW measurements. In Fig. 7.13a we show HARPES results for GaAs and $Ga_{0.95}$ $Mn_{0.05}As$, over a wider angular range than the data in Fig. 7.11, for a photon energy of 2719 eV in the middle of the energy scan. Over plotted on these data are light blue curves calculated for direct transitions from the ground-state electronic structure of both materials, with CPA again being used for the doped material, to a strictly free-electron final state (FEFS), a very useful method of initial analysis of HARPES data [22, 48] that permits determining the exact orientation of the sample. This method is accurate to less than 0.5°, including a small tilt of the sample for the present case and a slight difference in geometry for the two samples. In Fig. 7.13b, we show the experimental data again, but after a procedure making use of the core-level energy scans for Ga $3d$ and As $3d$ that permits identifying the element-resolved contributions to each pixel. This method proceeds in the following

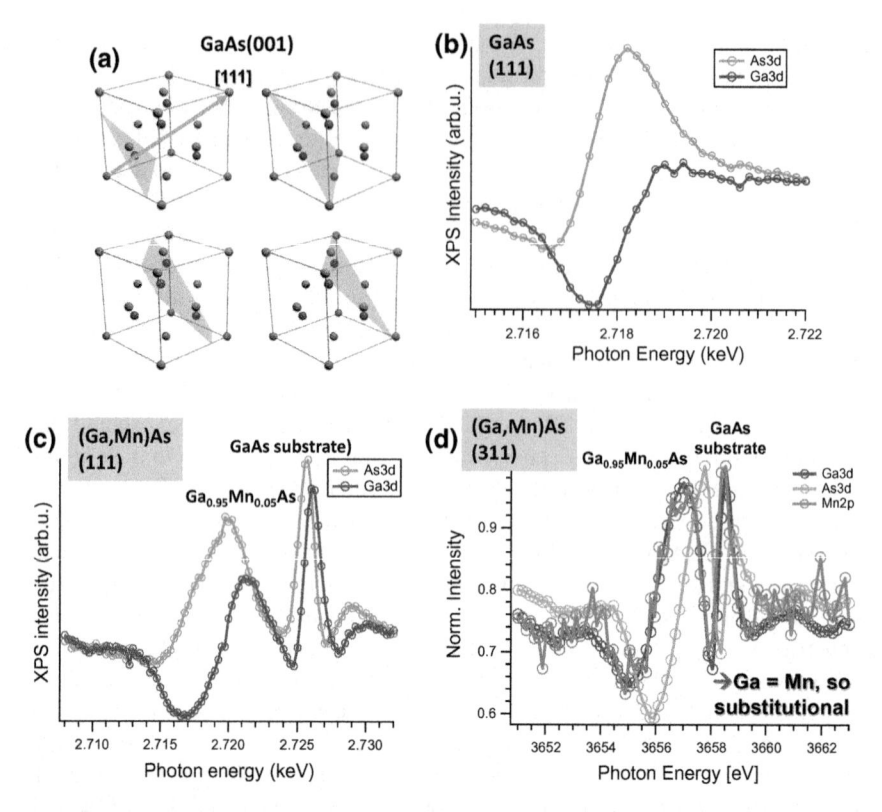

Fig. 7.12 Standing-wave HXPS and HARPES from GaAs(001) and the dilute magnetic semiconductor Mn-doped GaAs (Ga, Mn)As also in (001) surface orientation. Photon energies were 2.7 and 3.6 keV. **a** The schematic scanning of the SW along the [111] direction as the photon energy is scanned over the (111) Bragg diffraction condition. **b** The variation of As 3d and Ga 3d intensities through such an energy scan, with obvious strong differences in behavior. **c** A similar energy scan for a sample with ~100 nm of Ga$_{0.95}$Mn$_{0.05}$As grown on GaAs(001). Two Bragg peaks are observed, due to the slight difference in the lattice constant of the (Ga, Mn)As, with similar differences between As 3d and Ga 3d in going over both. **d** A similar energy scan for (Ga, Mn)As but with (311) reflection, and Mn 2p included. The near identity of the curves for Ga and Mn indicates a high degree of substitutional sites for Mn. Experimental data from Diamond. (From [51])

way. We first assume that the intensity in each HARPES pixel can be described as a superposition of a contribution from As and from Ga (or Mn) as:

$$I_{HARPES}(E_B, \vec{k}, h\nu) \approx I_{As}(E_B, \vec{k}, h\nu) + I_{Ga(Mn)}(E_B, \vec{k}, h\nu). \quad (3)$$

Then, the energy-dependence of intensity for each pixel is projected into fractional As and Ga(Mn) components by using a least-squares comparison to the superposition of core-level intensities as:

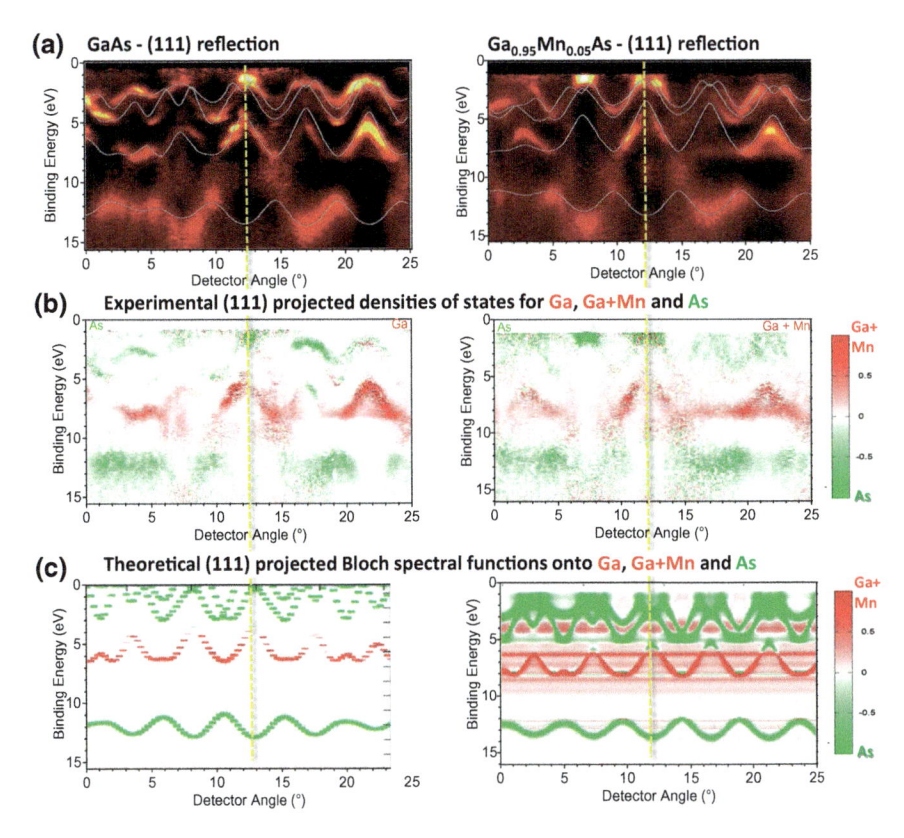

Fig. 7.13 Projection of SW-HARPES data into element-resolved components. **a** Experimental data for GaAs and (Ga, Mn)As in a (111) reflection geometry. The light blue curves are free-electron final-state calculations used to determine the exact orientation in $\vec{k}$-space, which was slightly different for the two samples. **b** Experimental decomposition into Ga + Mn and As components using core-level intensities and Equations (3) and (4). **c** Local-density calculations of element-resolved Bloch spectral functions using the coherent potential approximation (CPA) for (Ga, Mn)As, with the same color scale of maximum (red) Ga, Mn = 1.0 and maximum (green) As = −1.0 as in (**b**). Experimental data from Diamond. (From [51])

$$I_{Proj}(E_B, \vec{k}, h\nu) = f_{As}(E_B, \vec{k})I_{As3d}(h\nu) + f_{Ga(Mn)}(E_B, \vec{k})I_{Ga3d(Mn2p)}(h\nu), \quad (4)$$

where $f_{Ga(Mn)} = 1 - f_{As}$. It is important to point out that it has, in fact, been long realized through various theoretical studies that higher excitation energies, the cross sections and matrix elements for valence-band photoemission are increasingly controlled by the inner spatial regions of the atoms involved [52–54]. Therefore, using core-level photoemission intensities from the same atoms at nearly the same kinetic energies is a good approximation for such standing-wave projection procedures. This sort of projection has been used for SW MEWDOS spectra [34, 35, 50], but not with k resolution. Once the two f quantities have been determined, they are applied to each pixel, and a color scale going from red +1.0 = maximum Ga(Mn)/minimum As to green -1 = maximum As/minimum Ga(Mn) is applied to the data. The result is the plot shown in Fig. 7.13b, which reveals a drastic difference between the top and bottom sets of bands, which are strongly As and the middle bands, which are strongly Ga(Mn). For the doped sample, one can even see evidence of what is probably weak Mn intensity in the upper bands over 0–6 eV, consistent with prior conclusions that Mn affects the entire band structure [48, 49]. So we now have an element- and k-resolved band structure, directly derived from experiment.

Theoretical calculations also support this method, as shown in Fig. 7.13c. Here, we show element-resolved Bloch spectral functions, computed for the same trajectory in k-space as derived from the FEFS calculations in Fig. 7.13a, with CPA used for the doped material, and in the same color scale. There is excellent qualitative agreement as to the major elemental ingredients in each band. We do not expect fully quantitative agreement, as matrix elements are not included in the theoretical calculations, but future progress with one-step theory should permit going to this limit as well.

We thus view these results as most positive for the future use of SW-HARPES to study the element- and momentum-resolved bulk electronic structure of many multicomponent materials. Making use of simultaneous analysis of multiple Bragg reflections, as in prior MEWDOS-level studies [50], will improve the accuracy of localizing the electronic structure, both in momentum and within the unit cell.

7.3 Conclusions and Future Outlook

In conclusion, the use of higher-energy excitation in photoemission, either in the soft X-ray (from a few hundred to ca. 2 keV) or hard X-ray (from 2 keV and up to ca. 10 keV) permits looking at buried interfaces and/or bulk electronic structure that are not accessible with the lower-energies of traditional ARPES. Adding standing-wave excitation through either X-ray reflection from a multilayer heterostructure or atomic planes, enhances the depth resolution of the measurement significantly. Tuning the photon energy to various positions near strong absorption resonances can significantly increase the standing-wave modulation and also permits tuning its phase and thus anti-node position.

Through the example studies presented here, we believe it is clear that such techniques should permit studying a very wide variety of oxide and other interfaces in the future, including even the liquid-solid interface. As more intense and more highly focused radiation sources become available, the precision of such measurements will be enhanced, including the ability to add lateral resolution through simultaneous photoelectron microscopy in one of its several modalities. Adding time resolution through free-electron laser or high-harmonic generation sources will also be an exciting new direction. Varying polarization will also permit studying magnetic systems with higher precision, as for example, the depth variation of in-plane and perpendicular-to-plane magnetization at interfaces. Simultaneous spin detection will permit doing what one might call the complete photoemission experiment, providing resolution in energy, momentum, and spin, in three spatial dimensions through the interface, or through the unit cell, and in time.

Acknowledgements The specific sources of funding for the various studies presented here are listed in the publications cited. Beyond this, C.S.F. has also been supported during the writing of this chapter for salary by the Director, Office of Science, Office of Basic Energy Sciences (BSE), Materials Sciences and Engineering (MSE) Division, of the U.S. Department of Energy under Contract No. DE-AC02-05CH11231, through the Laboratory Directed Research and Development Program of Lawrence Berkeley National Laboratory, and through a DOE BES MSE grant at the University of California Davis from the X-ray Scattering Program under Contract DE-SC0014697. A.X.G acknowledges support during the writing of this chapter from the U.S. Army Research Office, under Grant No. W911NF-15-1-0181.

References

1. C.S. Fadley, Synchrotron Radiat. News **25**, 26 (2012)
2. *Hard X-ray Photoelectron Spectroscopy (HAXPES)*, ed. by J.C. Woicik. Springer Springer Series in Surface Sciences, vol. 59 (2016)
3. *The X-ray Standing Wave Technique-Principles and Applications*, ed. by J. Zegenhagen, A. Kazimirov (World Scientific Publishing, 2013)
4. C.S. Fadley, J. Electron Spectrosc. **178–179**, 2–32 (2010)
5. C.S. Fadley, J. Electron Spectrosc. **190**, 165–179 (2013), Special issue dedicated to hard X-ray photoemission, ed. by W. Drube
6. A.X. Gray, J. Electron Spectrosc. **195**, 399 (2014)
7. C.S. Fadley, S. Nemšák, in special issue of the Journal of Electron Spectroscopy dedicated to Structure Determination and Wave-Function Analysis, ed. by H. Daimon, A. Hishikawa, C. Miron, J. Electron Spectrosc. **195**, 409–422 (2014)
8. S.-H. Yang, A.X. Gray, A.M. Kaiser, B.S. Mun, J.B. Kortright, C.S. Fadley, J. Appl. Phys. **113**, 073513 (2013)
9. S.-H. Yang, B.S. Mun, N. Mannella, S.-K. Kim, J.B. Kortright, J. Underwood, F. Salmassi, E. Arenholz, A. Young, Z. Hussain, M.A. Van Hove, C.S. Fadley. J. Phys. Cond. Matt. **14**, L406 (2002)
10. A.X. Gray, C. Papp, B. Balke, S.-H. Yang, M. Huijben, E. Rotenberg, A. Bostwick, S. Ueda, Y. Yamashita, K. Kobayashi, E.M. Gullikson, J.B. Kortright, F.M.F. de Groot, G. Rijnders, D.H.A. Blank, R. Ramesh, C.S. Fadley, Phys. Rev. B **82**, 205116 (2010)
11. S. Nemšák, G. Pálsson, A.X. Gray, D. Eiteneer, A.M. Kaiser, G. Conti, A.Y. Saw, A. Perona, A. Rattanachata, A. Conlon, A. Bostwick, V. Strocov, M. Kobayashi, W. Stolte, A. Gloskovskii, W. Drube, M.-C. Asencio, J. Avila, J. Son, P. Moetakef, C. Jackson, A. Janotti, C.G. Van de Walle, J. Minar, J. Braun, H. Ebert, J.B. Kortright, S. Stemmer, C.S. Fadley, Phys. Rev. B **93**, 245103 (2016)
12. M.J. Bedzyk, G. Materlik, M.V. Kovalchukas, Phys. Rev. B **30**, 2453 (1984)
13. B.L. Henke, Phys. Rev. **6**, 94 (1972)
14. M. Mehta, C.S. Fadley, Chem. Phys. Lett. **46**, 225 (1977)
15. M.J. Chester, T. Jach, S. Thurgate, J. Vac. Sci. Technol. B **11**, 1609 (1993); M.J. Chester, T. Jach, Phys. Rev. B **48**, 17262 (1993)
16. J. Kawai, S. Hayakawa, Y. Kitajima, Y. Gohshi, Adv. X-ray Chem. Anal. Jpn. 26s 97 (1995)
17. J. Kawai, Rev. J. Electron Spectrosc. 178–179, 268–272 (2010)
18. M. Marinova, J.E. Rault, A. Gloter, S. Nemšák, G.K. Pálsson, J.-P. Rueff, C.S. Fadley, C. Carrétéro, H. Yamadag, K. March, V. Garcia, S. Fusil, A. Barthélémy, O. Stéphan, C. Colliex, M. Bibes, Nano Lett. **15**, 2533–2541 (2015)
19. C. Papp, L. Plucinski, J. Minar, J. Braun, H. Ebert, C.S. Fadley, Phys. Rev. B **84**, 045433 (2011)
20. M. Sagurton, E.L. Bullock, C.S. Fadley, Surf. Sci. **182**, 287 (1987)
21. J. Braun, J. Minar, S. Mankovsky, L. Plucinski, V.N. Strocov, N.B. Brookes, C.M. Schneider, C.S. Fadley, H. Ebert, Phys. Rev. B **88**, 205409 (2013)

22. A.X. Gray, C. Papp, S. Ueda, B. Balke, Y. Yamashita, L. Plucinski, J. Minár, J. Braun, E.R. Ylvisaker, C.M. Schneider, W.E. Pickett, H. Ebert, K. Kobayashi, C.S. Fadley, Nat. Mat. **10**, 759 (2011)
23. M.J. Bedzyk, D. Bilderback, J. White, H.D. Abrufia, M.G. Bommarito, J. Phys. Chem. **90**, 4926 (1986)
24. J.B. Kortright, A. Fischer-Colbrie, J. Appl. Phys. **61**, 1130 (1987)
25. S.-K. Kim, J.B. Kortright, Phys. Rev. Lett. **86**, 1347 (2001)
26. S.-H. Yang, B.S. Mun, A.W. Kay, S.-K. Kim, J.B. Kortright, J.H. Underwood, Z. Hussain, C.S. Fadley, Surf. Sci. Lett. **461**, L557 (2000)
27. S.-H. Yang, B.S. Mun, N. Mannella, A. Nambu, B.C. Sell, S.B. Ritchey, F. Salmassi, S.S.P. Parkin, C.S. Fadley, J. Phys. C Solid State **18**, L259–L267 (2006)
28. A.X. Gray, F. Kronast, C. Papp, S.-H. Yang, S. Cramm, I.P. Krug, F. Salmassi, E.M. Gullikson, D.L. Hilken, E.H. Anderson, P.J. Fischer, H.A. Dürr, C.M. Schneider, C.S. Fadley, Appl. Phys. Lett. **97**, 062503 (2010)
29. S.-H. Yang, B. Balke, C. Papp, S. Döring, U. Berges, L. Plucinski, C. Westphal, C.M. Schneider, S.S.P. Parkin, C.S. Fadley, Phys. Rev. B **84**, 184410 (2011)
30. A.A. Greer, A.X. Gray, S. Kanai, A.M. Kaiser, S. Ueda, Y. Yamashita, C. Bordel, G. Palsson, N. Maejima, S.-H. Yang, G. Conti, K. Kobayashi, S. Ikeda, F. Matsukura, H. Ohno, C.M. Schneider, J.B. Kortright, F. Hellman, C.S. Fadley, Appl. Phys. Lett. **101**, 202402 (2012)
31. B.W. Batterman, Phys. Rev. **133**, A759 (1964)
32. P.I. Cowan, J.A. Golovchenko, M.F. Robbins, Phys. Rev. Lett. **44**, 1680 (1980)
33. W. Drube, A. Lessmann, G. Materlik, Rev. Sci. Instrum. **63**, 1138 (1992)
34. J.C. Woicik, E.J. Nelson, P. Pianetta, Phys. Rev. Lett. **84**, 773 (2000)
35. J.C. Woicik, E.J. Nelson, D. Heskett, J. Warner, L.E. Berman, B.A. Karlin, I.A. Vartanyants, M.Z. Hasan, T. Kendelewicz, Z.X. Shen, P. Pianetta, Phys. Rev. B **64**, 125115 (2001)
36. A.X. Gray, J. Minár, L. Plucinski, M. Huijben, A. Bostwick, E. Rotenberg, S.-H. Yang, J. Braun, A. Winkelmann, D. Eiteneer, A. Rattanachata, A. Greer, G. Rijnders, D.H.A. Blank, D. Doennig, R. Pentcheva, J.B. Kortright, C.M. Schneider, H. Ebert, C.S. Fadley, Europhys. Lett. **104**, 17004 (2013)
37. P. Moetakef, J.Y. Zhang, A. Kozhanov, B. Jalan, R. Seshadri, S.J. Allen, S. Stemmer, Appl. Phys. Lett. **98**, 112110 (2011); P. Moetakef, T.A. Cain, D.G. Ouellette, J.Y. Zhang, D. O. Klenov, A. Janotti, C.G. Van de Walle, S. Rajan, S.J. Allen, S. Stemmer, Appl. Phys. Lett. **99**, 232116 (2011); C. Jackson, P. Moetakef, S.J. Allen, S. Stemmer, Appl. Phys. Lett. **100**, 232106 (2012); P. Moetakef, J.R. Williams, D. G. Ouellette, A. P. Kajdos, D. Goldhaber-Gordon, S.J. Allen, S. Stemmer, Phys. Rev. X **2**, 021014 (2012)
38. W. Meevasana, P.D.C. King, R.H. He, S.-K. Mo, M. Hashimoto, A. Tamai, P. Songsiriritthigul, F. Baumberger, Z.-X. Shen, Nat. Mat. **10**, 114 (2011)
39. A.F. Santander-Syro, O. Copie, T. Kondo, F. Fortuna, S. Pailhès, R. Weht, X.G. Qiu, F. Bertran, A. Nicolaou, A. Taleb-Ibrahimi, P. Le Fèvre, G. Herranz, M. Bibes, N. Reyren, Y. Apertet, P. Lecoeur, A. Barthélémy, M.J. Rozenberg, Nature **469**(7329), 189 (2011)
40. S. Nemšák, A. Shavorskiy, O. Karslioglu, I. Zegkinoglou, A. Rattanachata, C.S. Conlon, A. Keqi, P.K. Greene, K. Liu, F. Salmassi, E.M. Gullikson, H. Bluhm, C.S. Fadley, Nat. Comm. **5**, 5441 (2014)
41. H. Ohno, *Electrochemical Aspects of Ionic Liquids* (Wiley, 2011)
42. G.E. Brown, G. Calas, Mineral-aqueous solution interfaces and their impact on the environment. Geochem. Persp. **1**, 483–742, with special discussion of the mineral-electrolyte double layer over pp. 552–557
43. W. Smekal, W.S.M. Werner, C.J. Powell, Surf. Interface Anal. **3**, 1059 (2005); W.S.M. Werner, W. Smekal, T. Hisch, J. Himmelsbach, C.J. Powell, J. Electron Spectrosc. **190**, 137 (2013)
44. O. Karslıoğlu, S. Nemšák, I. Zegkinoglou, A. Shavorskiy, M. Hartl, F. Salmassi, E.M. Gullikson, M.L. Ng, Ch. Rameshan, B. Rude, D. Bianculli, A.A. Cordones-Hahn, S. Axnanda, E.J. Crumlin, P.N. Ross, C.M. Schneider, Z. Hussain, Z. Liu, C.S. Fadley, H. Bluhm, Faraday Soc. Disc. **180**, 35 (2015)

45. S. Axnanda, E.J. Crumlin, B. Mao, S. Rani, R. Chang, P.G. Karlsson, M.O.M. Edwards, M. Lundqvist, R. Moberg, P.N. Ross, Z. Hussain, Z. Liu, Sci. Rep. **5**, 09788 (2015)
46. H. Yamada, M. Marinova, P. Altuntas, A. Crassous, L. Bégon-Lours, S. Fusil, E. Jacquet, V. Garcia, K. Bouzehouane, A. Gloter, J.E. Villegas, A. Barthélémy, M. Bibes, Sci. Rep. **3**, 2834 (2013)
47. M.J. Bedzyk, in Surface X-ray and neutron scattering, in *Springer Proceedings in Physics*, ed. by H. Zabel, I.K. Robinson I, vol. 61, 51, pp. 113–117 (Springer, 1992) (and references therein)
48. A.X. Gray, J. Minar, S. Ueda, P.R. Stone, Y. Yamashita, J. Fujii, J. Braun, L. Plucinski, C.M. Schneider, G. Panaccione, H. Ebert, O.D. Dubon, K. Kobayashi, C.S. Fadley, Nat. Mat. **11**, 957 (2012)
49. M. Kobayashi, I. Muneta, Y. Takeda, Y. Harada, A. Fujimori, J. Krempasky, T. Schmitt, S. Ohya, M. Tanaka, M. Oshima, V.N. Strocov, Phys. Rev. B **89**, 205204 (2014)
50. S. Thiess, T.-L. Lee, F. Bottin, J. Zegenhagen, Solid State Commun. **150**, 553 (2010)
51. S. Nemšák, M. Gehlmann, C.T. Kuo, S.C. Lin, C. Schlueter, E. Mlynczak, T.L. Lee, L. Plucinski, H. Ebert, I. Di Marco, J. Minár, C.M. Schneider, C.S. Fadley, Unpublished results
52. U. Gelius, K. Siegbahn, Discuss. Faraday Soc. **54**, 257 (1972)
53. C.S. Fadley, in *Electron Spectroscopy: Theory, Techniques, and Applications*, ed. by C.R. Brundle, A.D. Baker, vol. II, Chap. 1 (Academic Press, London, 1978), pp. 56–57
54. C. Solterbeck, W. Schattke, J.-W. Zahlmann-Nowitzki, K.-U. Gawlik, L. Kipp, M. Skibowski, C.S. Fadley, M.A. Van Hove, Phys. Rev. Lett. **79**, 4681 (1997)

Chapter 8
Ab-Initio Calculations of TMO Band Structure

A. Filippetti

Abstract We review the fundamental aspects related to ab-initio band structure calculations for the SrTiO$_3$/LaAlO$_3$ interface, analyzing capabilities and limits of the most advanced approaches, using available experiments as a reference. In particular, we discuss accuracy and failures for what concern the description of electronic, transport, and thermoelectric properties of oxide heterostructures. Despite evident shortcomings, our overview assesses the usefulness and the satisfying quality of ab-initio methods as an efficient approach for oxide heterostructure design and analysis.

8.1 Fundamentals of 2DEG Formation in SrTiO$_3$/LaAlO$_3$ According to Ab-Initio Calculations

This section is dedicated to describe from an ab-initio viewpoint the fundamental aspects concerning the electron gas formation in STO/LAO. In particular, in Sect. 8.1.1 we will see how, assuming a perfectly stoichiometric, ideal heterostructure, ab-initio calculations precisely describe the polar discontinuity at the interface, the polar catastrophe, the electronic charge transfer and confinement in the STO side of the interface, in conformity with the scenario drawn by simple electrostatic modeling, but with some dissonant aspects with respect to the observations. In Sect. 8.1.2 the most important non-stoichiometric mechanisms of polarity compensation will be considered, analyzing virtues and drawbacks of the ab-initio approach for problems whose complexity surmounts the limits of the present computing power capability.

A. Filippetti (✉)
CNR-IOM Cagliari and Physics Department, University of Cagliari,
09042 Monserrato, CA, Italy
e-mail: alessio.filippetti@dsf.unica.it

8.1.1 Ab-Initio Description of Polar Catastrophe: Role of Polar Discontinuity and Band Alignment

Soon after the first observations of a 2D electron gas (2DEG) at the interface between $SrTiO_3$ (STO) and $LaAlO_3$ (LAO) [1–3], a massive activity based on state-of-art ab-initio calculations started, in the attempt to unveil the secret of the 2DEG formation.

Theoretical scientists knew very well, however, that several methodological difficulties could hamper, or make it difficult, an accurate description of the experimental findings. In particular, it was worthy of consideration the well known inaccuracy of standard energy functionals based on local density approximation (LDA) and generalized-gradient approximation (GGA) in treating correlated oxides, especially for what concern band gap and band alignment in heterostructures, which are crucial aspects in the phenomenology of charge confinement. Although the impact of electronic correlation in the STO/LAO 2DEG was a point of debate, the potential failure of standard methods motivated scientists to apply a broad range of beyond-standard energy functionals, which were popularly used in the study of strongly-correlated materials, such as the LDA+U or GGA+U [4], the pseudo-self-interaction corrected energy functional (PSIC) [5, 6], and several hybrid functionals (B1-WC [7], HSE [8]) and tight-binding potentials. A very incomplete list of methods based on the Density Functional Theory and close variants, applied to STO/LAO, includes LDA [9–12], GGA [13–16], LDA+U [17–21], GGA+U [22], tight-binding [23], PSIC [24], hybrid functionals B1-WC [24–26], and HSE [27, 28], and Hubbard-type models [29, 30].

However, the most severe limitation for ab-initio calculations concerns not so much the theoretical fundamentals of the method, but rather the system size, which must be typically limited to a few (one or two at most) hundred atoms per elemental cell, periodically repeated in the infinite space when working in periodic boundary conditions. To stay in these strict limits, the actual structure must be as much as possible 'idealized', thus cutting-off several aspects which in fact may have a role in the observations. As an example, the near totality of ab-initio simulations for the STO/LAO [001] interface assumes 1×1 periodicity in the (001) plane, which implies discarding possible octahedral distortions and tiltings in the interface plane, and largely prevents the occurrence of possible orbital, charge, and magnetic ordering, not to mention the formation of polaronic states or localized states. And even in the interface-orthogonal direction, the supercell simulation could be hardly longer than a few nm, thus not including more than a dozen STO unit cells. While this is sufficient to recover bulk-like behavior in the insulating state, the 2DEG extension may easily outstretch this size, resulting in significant uncertainties for what concern structural rumpling, polarization, and charge redistribution along the substrate. In the analysis of the ab-initio description, the consequence of these simplifying assumptions concerning the adopted simulation supercell must always be taken in mind.

Albeit in these rather narrow computational limits, it turned out that ab-initio calculations depict a landscape quite coherent with the simple electrostatic picture based on the argument of polarization catastrophe and Zener breakdown (ZB) mechanism. The ZB concept is schematically illustrated in Fig. 8.1a, b: due to the polar

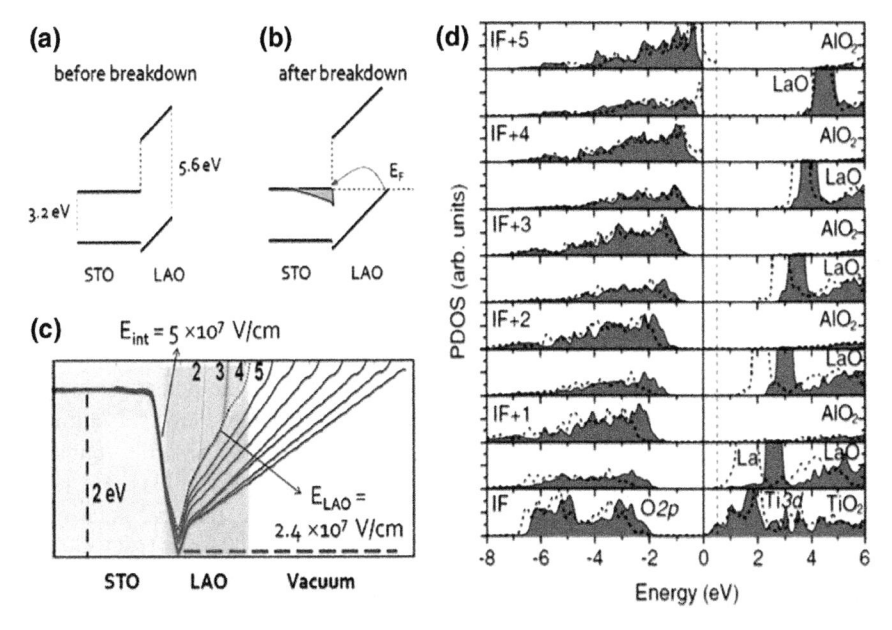

Fig. 8.1 a Schematic band alignment before breakdown; b band alignment after breakdown; c planar-averaged electrostatic potential across the STO/LAO slab calculated ab-initio with the B1-WC hybrid energy functional; each line is for an increasing amount of LAO layers: red lines are for 5 or more LAO layers; the red-dashed line indicate the infinite-thickness limit, when polarity is fully compensated and the field completely erased. d Layer-resolved DOS calculated for STO/LAO with 5 LAO layers. Dashed and shaded lines are for unrelaxed and relaxed structures (see text). Panel (d) is a reprinted figure with permission from [14]. Copyright 2009 by the American Physical Society

discontinuity along the $z = (001)$ axis, the electrostatic potential rises in LAO with the number of LAO units from the interface; accordingly, the band energies in the LAO side progressively rise until, for a sufficiently thick film, the LAO valence band top (VBT) overcomes the conduction band bottom (CBB) of the STO side, causing electron charge transfer from the surface to the interface (n-type doping). In Fig. 8.1c the calculated planar-averaged electrostatic potential seen by the electrons is shown as a function of the number of LAO layers. The potential in LAO is roughly linear, and in turn, the electric field E_{LAO} nearly constant. For an increasing number of LAO layers up to 4, the situation remains unchanged and the field E_{LAO} is pinched to its insulating value $\sim 2.4 \times 10^7$ V/cm. After breakdown, charge starts flowing from the LAO surface to the interface, and E_{LAO} progressively falls. Notice that only in the limit of infinite thickness the field is completely erased (dashed red line); this situation corresponds to a compensating charge transfer of 1/2 electron per unit interface area. Furthermore, there is a strong built-in electrostatic field at the interface (E_{int}), which act to confine the Ti 3d charge towards the interface. This aspect is often overlooked in the analysis, but in fact it is instrumental to understand why the injected electrons stay confined at the interface instead of spreading deeper into the substrate. A detailed electrostatic modeling of 2DEG formation in STO/LAO can be found in [31], while an alternative viewpoint is proposed in [32] by Stengel and Vanderbilt, who described the polarization at the interface of two polar insulators based on the concept of formal polarization, as defined by the Berry phase approach [33].

The most explicative representation of the electronic properties of an heterostructure based on ab-initio calculations is furnished by the density of states (DOS) resolved layer-by-layer along the direction orthogonal to the interface. In Fig. 8.1d we report one of the best examples of this kind, obtained within GGA by Pentcheva et al. [14]. Assuming ideal, unrelaxed atomic positions (dashed lines), the potential in LAO rises quickly, and at the 3rd LAO layer, the ZB occurs. On the other hand, allowing atomic relaxations along the z axis (shaded lines), the potential slope is reduced and the ZB only occurs at a critical thickness of 4 LAO layers. Central to these large structural relaxations is the so called rumpling (i.e. the O disalignment with respect to the cation positions) which develops along the z axis [34]. The rumpling generates an induced polarization which counteracts the nominal polarity of the LAO layer, thus reducing the energy cost associated to the field in LAO.

The description of the observed critical thickness for the occurrence of the 2DEG was considered a striking success of the ab-initio approach, and has represented a key validating aspect of the so-called 'intrinsic' hypothesis. Nevertheless, looking more in detail to the specific features of the calculations, several important quantitative discrepancies with respect to the observations can be revealed. First, the critical thickness of 4–5 LAO layers overestimates the 3-layer thickness observed in the experiments. This is not a major failure, since a number of factors not included in the calculations, from structural distortions to defects, could modify the potential slope and in turn, the threshold for ZB. Also, methodological aspects should be considered: in Fig. 8.1d we notice that the GGA band gaps for STO ($\sim$2 eV, see the lowest panel) and LAO ($\sim$3.5 eV, as seen in the second-topmost panel) largely underestimate the experimental counterparts (3.2 eV, and 5.6 eV, respectively). This aspect can be corrected using one of the various beyond-LDA approaches mentioned above. More critical, on the other hand, is the fact that ZB implies the presence of an equal amount of electron charge at the interface and hole charge at the surface. The latter, however, has never been observed, either in Hall resistivity, magnetoresistivity, or optical measurements. This in itself does not necessarily rule out the polarization catastrophe, but suggests that the breakdown should be complemented by further assumptions. For example, the hole charge at the surface could be stuck in localized defects or compensated by atoms absorbed at the surface. Hole localization at the LAO surface will be discussed in Sect. 8.1.2.

But the most important discordance between observations and calculations probably concerns the amount of electron charge confined at the interface, and the related charge-transfer mechanism. As seen from the B1-WC ab-initio calculations (see Fig. 8.2b, from [25]), for an increasing number of LAO layers the charge is progressively accumulated at the interface, exponentially approaching the 2D density limit $n_{2D} = 3.4 \times 10^{14}$ cm^{-2} corresponding to the 1/2 electron per unit area required to reach full compensation according to ZB. Correspondingly, electric (Fig. 8.2a) and

Fig. 8.2 Electric field (**a**), charge density (**b**), and strain field (**c**), calculated for STO/LAO interfaces as a function of LAO unit cell number n. For charge and strain, experimental values (black squares) derived by transport and electrostriction measurements are also reported. The scales of calculated (top horizontal axis) and experimental (bottom axis) number of LAO layers are shifted to eliminate the offset due to the different critical thickness. Reprinted figure with permission from [25]. Copyright 2011 by the American Physical Society

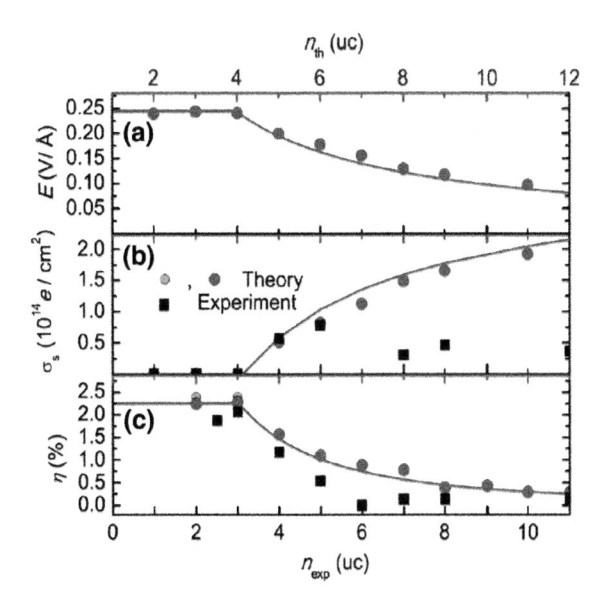

strain (Fig. 8.2c) fields in LAO due to the polar discontinuity are progressively relieved. Experimentally, the picture appears different: a charge $n_{2D} = 3-6 \times 10^{13}$ cm^{-2} (thus about an order of magnitude lower than the theoretical 1/2 electron limit) is transferred all at once at the critical thickness, and remains roughly of the same amount for thicker LAO layers; the measured strain is also relieved more abruptly than according to the calculations.

For what concern the question of the 'missing' charge, a long-standing and still unresolved debate has been developed over the years. Recent evidences of polaronic behavior [35] and charge localization [36] suggest the presence of additional electronic charge with respect to what usually seen in transport or photoemission measurements. Coherently, optical experiments [37, 38] estimate a larger charge at the interface ($n_{2D} \sim 10^{14}$ cm^{-2}), albeit still much lower than the 1/2 electron expected according to the ZB. We remark that these discrepancies may not necessarily derive by a failure of the theory in itself, but could be consequence of our limited computing power: in fact, when the 1×1 planar periodicity is assumed, we constrain charge transfer from surface to interface to occur in charge fractions, thus both interface and surface must be metallic. However, in a hypothetical $N \times N$ planar geometry with huge N, nothing would prevent, in principle, the electronic ground-state to be a mixture of localized (flat band) and delocalized (dispersed band) states, or to have an integer number of electrons, corresponding to a fractions of 0.5 electrons per unit area, transferred all at once, thus abruptly compensating the polar discontinuity and yet retaining an insulating character at the surface. Alas, these are likely destined to remain suggestive but unproved hypotheses until the advent of a quantum leap forward in our computing capability. In fact, methods like B1-WC or PSIC have demonstrated to be able to describe charge localization in a number of

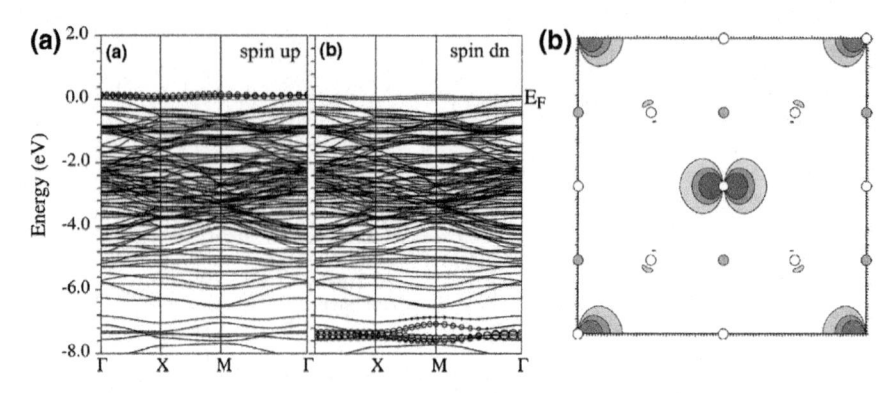

Fig. 8.3 LDA+U calculations for the p-type non-stoichiometric STO/LAO interface with 2 holes in excess per interface, and antiferromagnetic alignment of O(p) spin moments in the plane. **a** Band structure calculation for majority and minority spin populations. **b** Corresponding Fermi surfaces. Reprinted figure with permission from [20]. Copyright 2006 by the American Physical Society

low-dimensional bulk systems, such as superconducting cuprates [39, 40] and multiferroic manganites [41, 42], but in these systems localization is strong and may occur even in the limit of high carrier concentration, where relatively small supercells can be used.

In conclusion, the experimental evidence shows quite clearly that the purely electronic ZB alone is hardly capable to give a complete and precise explanation of the observed 2DEG characteristics. On the other hand, the polar discontinuity can be compensated by a number of alternative 'extrinsic' mechanisms, such as oxygen vacancies, surface adsorptions, cation intermixing, eventually coexisting with the electronic ZB effects. These effects will be discussed in the next section.

8.1.2 Ab-Initio Description of Non-stoichiometric Mechanisms: Oxygen Doping and Cation Mixing

Going beyond simple stoichiometric conditions in order to include additional extrinsic effects such as charge localization, oxygen doping and cation mixing, is quite a challenging enterprise for ab-initio simulations. Here we describe some of the few brave attempts of this kind presented in literature.

The first attempt to describe charge localization in STO/LAO was, at our knowledge, proposed by Pentcheva and Pickett [20]. In order to investigate hole localization and polaron formation, they considered a p-type STO/LAO interface (with SrO/AlO$_2$ termination) simulated using a 2 × 2 non-stoichiometric supercell with 2 holes in excess at the interface, i.e. 0.5 hole charge per unit area. For the calculations the LDA+U functional was employed, with $U_p = 7$ eV to ensure robust correlation on the O(p) states. They found that in case of anti-paired (i.e. antiferromagnetic) spin alignment, the hole states form very flat spin-split bands (Fig. 8.3a) located 50 meV above the O(p) valence band top. On the other hand, for the spin-paired ferromagnetic

8 Ab-Initio Calculations of TMO Band Structure

solution (not shown), the hole bands are more dispersed and the system is metallic, just like in the non-correlated case described by LDA. Here the adopted supercell is arguably too small to describe a realistically doped configuration, and the absence of structural relaxations implies a limited predictive capability. Nevertheless, it is conceptually instructive to see that electron correlation and antiferromagnetic ordering (which may be intended as a crude approximation of a more realistic paramagnetic state) can easily lead to hole localization and insulating behavior at the LAO surface. The capability of beyond-LDA approaches to describe these aspects including static correlations draws a larger horizon than what is usually assumed for ab-initio simulations, which are sometime improperly misinterpreted as uncorrelated single-particle theories.

For what concern the effect of O-vacancies (V_O), which has been considered the most credited alternative hypothesis to ZB for the 2DEG formation [44, 45], it is worthy to mention the work by Zhang et al. [43] where a remarkably large 4×4 cell is employed to simulate, by GGA calculations, n-type and p-type STO/LAO interfaces (consistent results are also obtained by Li et al. [46] using GGA). Removing one O from this structure adds $\sim 8 \times 10^{13}$ electrons per cm^2, which is not much larger than the density actually observed. For both structures, authors calculated the formation energy of the vacancy as a function of the layer where the vacancy is located (Fig. 8.4). Interestingly, for the p-type interface the most favorable V_O location is right at the

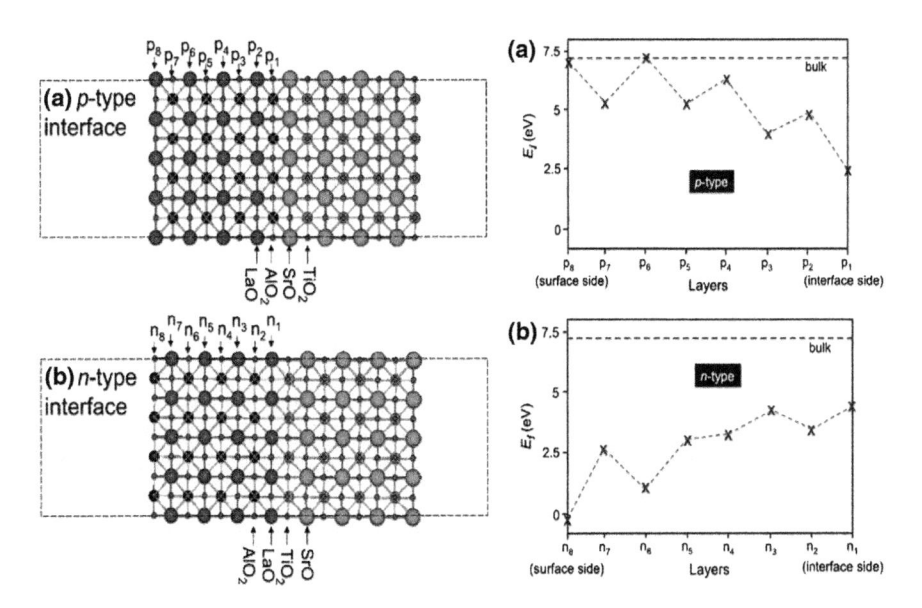

Fig. 8.4 Ab-initio calculation of the formation energy for oxygen vacancies in each single monolayer of a 4 u.c. LAO slab, for both p-type and n-type STO/LAO interfaces. Left panels: p-type and n-type structures considered for the calculations. Right panels: corresponding formation energies for various vacancy positions (indicated in the structural plot). Reprinted figures with permission from [43]. Copyright 2010 by the American Physical Society

interface layer, while for the n-type structure, the vacancy is preferentially located at the surface. It follows that for the former an excess hole charge (also located at the SrO/AlO$_2$ interface) eventually due to the ZB mechanism can recombine with the electron charge induced by V_O, thus explaining the observed insulating behavior. On the other hand, for the n-type structure V_O stays at the surface, and delivers electrons at the interface. To do that, a further condition to be matched is that the V_O donor level must be higher in energy than the CBB of the STO side.

The 2DEG scenario based on surface oxygen vacancy was thoroughly investigated by Bristowe et al. [31, 47]. Using an electrostatic modeling based on ab-initio parameters, they showed that the V_O formation energy decreases linearly with the number of LAO unit cells, with a slope depending on the vacancy concentration, and that there is a critical thickness coherent with the observations for the presence of the charge at the interface. Later on, Yu and Zunger [28] calculated by HSE the formation energy of a variety of possible defects, and showed that the key condition of having a V_O donor level higher than the STO CBB is indeed satisfied, see Fig. 8.5b: single and double-ionized V_O^+/V_O^{++} defect levels are higher in energy than the STO CBB, and donate part of the electrons to STO, and part to localized holes due to Ti → Al substitutions. Also, they showed that the V_O defect formation energy is negative only for a number of LAO layers larger than 4, thus justifying the presence of a critical thickness.

The presence of oxygen vacancies at the surface not only appears as a viable alternative to the pure electronic ZB mechanism, but also justifies the absence of mobile hole charge at the surface. Of course, this is only one of the various non-intrinsic scenarios which can be envisaged to compensate the LAO polarity. For example, the adsorption of H atoms at the surface (for this case ab-initio calculations are reported in [48]) could equally well compensate the surface polarity by creating $(AlO_2)^{1-}H_1^+$ complexes at the surface and donate electrons at the interface, provided that the localized holes introduced by the additional H could be higher in energy than the STO CBB at the interface.

Another important ingredient of the 2DEG phenomenology is represented by the metal contact, as discussed by a joint experimental and theoretical study in [13]: they showed that a Co layer deposited on top of LAO can produce in itself a charge transfer to the CBB interface, effectively reducing the critical thickness of the insulating-metal transition. On the other hand, in [49] it was shown by ab-initio calculations that a Pt overlayer on LAO counteracts the ZB mechanism and, in absence of bias, completely hinders the transfer of charge to the interface. These aspects are obviously crucial for the interpretation of electric transport and field-effect measurements, and suggest that the device characteristics may disguise or alter the intrinsic properties of the pristine heterostructure.

Finally, we devote just some hints to another major interpretation of the 2DEG formation, alternative or complementary to the pure ZB, that is cation mixing: it is clearly revealed in XAS experiments [34, 50, 51] that a few layers across the interface are in fact characterized by Ti/Al and La/Sr mixing, which works in favor of an effective reduction of the polar discontinuity, counteracting the polarization catastrophe, and in turn increasing the LAO critical thickness relative to ZB, as shown in

8 Ab-Initio Calculations of TMO Band Structure

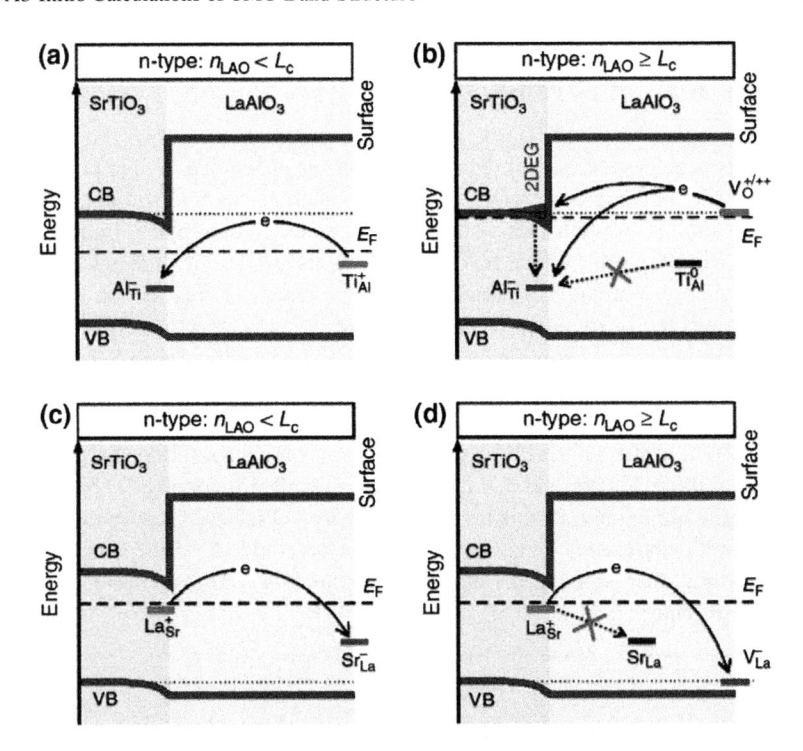

Fig. 8.5 Schematic band diagrams relative to various defects and vacancy calculations for STO/LAO, reported by Yu and Zunger [28]. See text for the description of each case. Reprinted by permission from Macmillan Publishers Ltd: Nature Communications [28], copyright 2014

[26] where a solid solution $Sr_{1-x}La_xTi_{1-y}Al_yO_3$ is grown and characterized. Furthermore, a non-stoichiometric excess of La atoms over Sr or Ti over Al could explain in itself the presence of some additional electron charge at the interface, although in this case it is not obvious why this cation-mixing charge should only appear above a universal critical thickness. In [28] the case of a stoichiometric Al/Ti mixing across the LAO slab is studied as well, deducing that below this critical thickness (Fig. 8.5a) the formation energy do favors acceptor-like Ti → Al substitutions at the interface, and donor Al → Ti substitution at the surface; alas, the defect levels are too low to donate charge to the STO CBB. The same occurs for Sr → La substitutions at the interface coupled with Al → Sr at the surface (Fig. 8.5c) or coupled with La vacancies at the surface (Fig. 8.5d): even if energetically stable, none of these configurations appear capable to donate electrons to the STO CBB. Of course, even these results, albeit quite conceptually stimulating, cannot be taken from granted, since transition-state theory may be affected by a number of uncertainties, first of all the assumed chemical potentials, and the fact that actual structural morphology is not or very partially included in the calculations, due to the supercell limitations described above.

8.2 Band Structure and Related Properties: Spectroscopy, Thermal, and Transport Properties

The ab-initio calculation of the electronic properties of the 2DEG is in principle an extremely valuable tool to complement the experimental characterization. In fact, an impressive amount of published work concerning electronic, optical, transport, and thermoelectric properties of the STO/LAO presents a synergic analysis of experiments and theory. Without any pretence of completeness, in this section we furnish an overview of the band structure theoretical description, and the level of qualitative and quantitative agreement with the experiments. Specifically, in Sect. 8.2.1 the basic features of the electronic properties are described, and in Sect. 8.2.2 we furnish a compared analysis of band structure calculations and ARPES experiments. In Sect. 8.2.3 a useful approach to the calculation of transport and thermoelectric properties of the gas is presented, with validating results for doped STO bulk, and in Sect. 8.2.4 the application of this method to STO/LAO is described, in comparison with transport experiments. Finally, Sect. 8.2.5 is devoted to describe the calculation of phonon-drag, that is one of the most spectacular observations reported so far for this heterostructure.

8.2.1 Basic Aspects of Band-Structure Calculation at the STO/LAO Interface

To start with the analysis of the STO/LAO electronic properties, a paramount aspect should be taken in mind: rigid band approximation is overly inadequate for this system, since both structural and electronic properties present a remarkable evolution with the amount of electron charge confined at the interface. So, we are in the need to specify how much charge should be included in the calculation, a problem which is deeply related to the choice of our simulation supercell.

Basically, two main choices have been practiced over the years in the ab-initio simulations of this system: (i) a stoichiometric $(STO)_n/(LAO)_m$ structure with some vacuum layer on top of LAO, eventually doubled (i.e. symmetrized) along the [001] axis to avoid spurious fields across STO and the vacuum region; (ii) a non-stoichiometric $(STO)_n/(LAO)_{m+1/2}$ superlattice, with one additional monolayer of LaO which dope each of the two equivalent TiO_2/LaO interfaces in the cell by 1/2 electron charge. The first configuration is closer to a realistic situation, and can describe the metal-insulating transition at the critical thickness, but also displays some unwanted features, such as the hole charge at the surface, and a not well defined amount of electron charge which progressively grows with the number of LAO layers, as seen in Fig. 8.2. The second choice is in practice a fully-compensated delta-doping situation, with no built-in field in the LAO side and no LAO surface. Alas, the 1/2 electron charge is an ideal limit never actually observed in the experiments. It follows that neither of

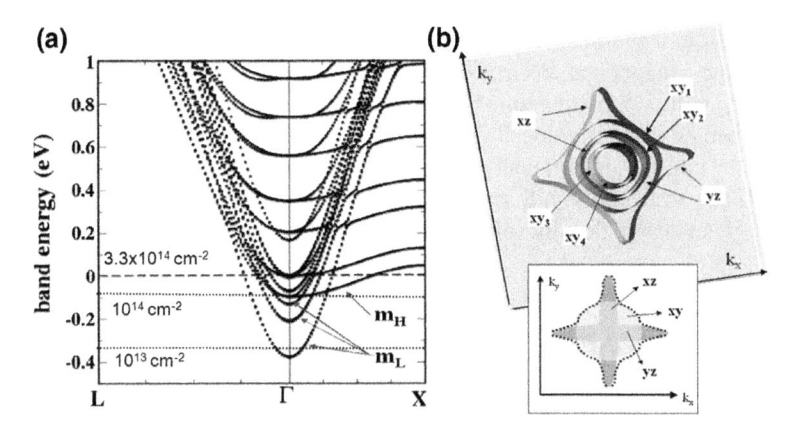

Fig. 8.6 Band structure (**a**) and Fermi surfaces (**b**) for the 2DEG in STO/LAO calculated by the PSIC approach for the non-stoichiometric, fully-compensated structure with one additional LaO layer at the surface and 1/2 electron included. Adapted from [24]

the two gives a representation perfectly matching the observed characteristics of the gas.

In Fig. 8.6 we display bands and Fermi surfaces calculated by PSIC functional in [24] for the fully-compensated cell, with 1/2 electron charge confined at the interface. The main characteristics described below are common to almost all the ab-initio descriptions, and not much dependent on the specific functional used for the calculation (which instead is crucial to describe the band gap). We can see that at the conduction bottom there are several partially occupied Ti 3d t_{2g} bands of d_{xy} character and mostly lying within the (x, y) plane, the lowest of which located about 0.37 eV below the Fermi energy. Higher in energy are the t_{2g} bands with d_{xz} and d_{yz} character, thus with one light and one heavy mass in the plane of the interface. The heavy mass direction (Γ-X for d_{yz} and Γ-Y for d_{xz}) is easily recognizable in Fig. 8.6a. The calculated light and heavy effective masses in the (x, y) plane are $\sim 0.7\,m_e$, and $\sim 8\,m_e$, respectively, according to ab-initio calculations. In Fig. 8.6b the corresponding Fermi surfaces are shown: the d_{xy} Fermi surfaces are circles centered at Γ, while d_{xz} and d_{yz} are cigar-shaped, with the long arm lying along the corresponding heavy-mass direction. Roughly speaking, about two thirds of the half-electron charge is hosted by the planar d_{xy} states, and the remaining one third by the d_{xz} and d_{yz} states.

The general features described above agree qualitatively, but not quantitatively, with the experiments. In fact, there is overwhelming evidence from ARPES [52, 53] that the bottom of conduction band of d_{xy} character is located just about 100 meV below E_F, while the lowest d_{xz}/d_{yz} bands appears a few tens of meV below E_F. Consistently, XAS experiments [54, 55] indicate an energy separation ~ 50 meV between the lowest d_{xy} and the d_{xz}/d_{yz} states, whereas, according to the calculations in Fig. 8.6, the t_{2g} splitting is as large as 0.3 eV. Furthermore, magnetotransport indicate one-band or two-band behavior [56], which is not coherent with the several partially filled bands appearing in Fig. 8.6a. Since the band shape is substantially well described in the calculation, it follows that the discrepancy with the experiments cannot but derive from the exaggerated electron charge in the bands. In Fig. 8.6a the dotted lines indicate E_F values corresponding to some lower charge densities. However, a mere E_F rigid-band downshift relocating the charge in the range of Hall-measured values (2–6×10^{13} cm^{-2}) is insufficient to recover a band structure in satisfying agreement with ARPES; instead, the full structural and electronic calculation must be redone for each given charge density value. This argument is developed in the next section.

8.2.2 Calculated Band Structures and Comparison with Spectroscopy

In [53] a compared analysis of band structure calculations and ARPES measurements is reported, with the aim of highlighting the evolution of the electronic properties with the charge density confined at the interface. In the article it is put in evidence for the first time that, whatever the source, if an amount of charge similar to what is measured in Hall experiments is introduced at the STO/LAO interface, the calculated band energies are substantially coherent with what is observed in photoemission experiments. In other words, once the charge is confined at the interface, the 2DEG characteristics can be monitored irrespectively on the charge derivation, whether due to ZB, O-vacancy, or induced by gate field. This is an important aspect since it allows to leave aside the questions related to the 'intrinsic' versus 'extrinsic' debate, and concentrate to the detailed description of the 2DEG properties.

Experimentally it is well known that the charge density can be tuned, to a certain extent, by the choice of growth temperature; in [53] ARPES spectra of STO/LAO samples grown at different temperatures were compared with band energy calculations at variable charge density, performed as following: starting from the stoichiometric STO/LAO interface in the insulating state (i.e. with a number of LAO layers lower than the ZB threshold), specific fractions of electron charge were included in the supercell, then leaving the system to relax to the structural and electronic ground state relative to that charge density value. The calculated band structures for an increasing amount of charge density are reported in Fig. 8.7.

We see that the band structure evolves in a very non-rigid band fashion: for zero charge, only one band of d_{xy} character (red-dotted line in Fig. 8.7a) is split by 30 meV from the bottom of the t_{2g} band manifold (see Fig. 8.7b), as a consequence of the symmetry breaking induced by the interface geometry. Then, the t_{2g} splitting progressively increases with the charge density. For densities up to $n_{2D} \sim 1.5 \times 10^{13}$ cm^{-2} only the lowest d_{xy} state is occupied, while above this value a second d_{xy} state starts to host some charge as well; then above $n_{2D} \sim 3.5 \times 10^{13}$ cm^{-2} the heavier t_{2g} bands finally set in. For $n_{2D} \sim 5 \times 10^{13}$ cm^{-2} the lowest d_{xy} band is descended to ~ 90 meV below E_F, and the d_{xz}/d_{yz} band bottom is ~ 10 meV below E_F. If this theoretical picture is correct, we may expect to see a regime transition between single-band light-mass to multi-band heavy-mass behavior at some critical threshold

8 Ab-Initio Calculations of TMO Band Structure

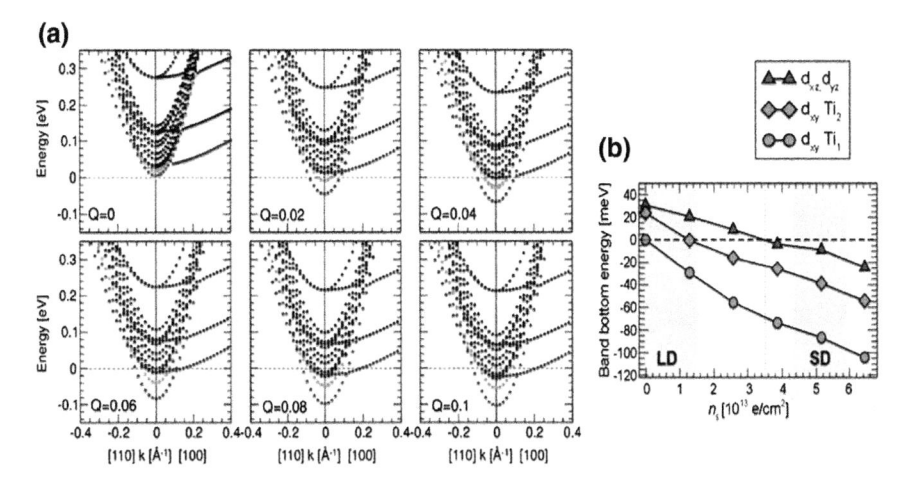

Fig. 8.7 Band structure calculated by PSIC approach as a function the charge density confined in the Ti 3d states at the STO/LAO interface. **a** Conduction band energies near E_F, for an increasing Q (charge per unit area); the dotted line indicates E_F. **b** Band bottoms for the two lowest d_{xy} bands, and the lowest d_{xz}, d_{yz} doublet. LD and SD stand for 'low-density' and 'standard-density' regions. Reprinted figure with permission from [53]. Copyright 2014 by the American Physical Society

of the order of some 10^{13} cm^{-2}. In fact, the occurrence of a regime transition around this density is confirmed quite clearly by transport experiments, where the charge density is varied either by changing the growth temperature [57] or more directly by field-effect tuning [58], although not all the features extracted from transport can be unambiguously reconnected with the calculations (these aspects will be analyzed in Sect. 8.2.4).

In Fig. 8.8 the detailed comparison of calculated band structure and ARPES obtained in [53] is displayed. Two STO/LAO samples are considered, with Hall-measured charge density of 0.1 and 0.04 electrons per unit cell; the spectra are juxtaposed to the Fermi surfaces calculated for equivalent charge densities. The ARPES spectra are s-polarized, thus only the d_{xy} and d_{yz} orbitals appear, having odd-parity with respect to the direction [010] orthogonal to the mirror plane. For standard density, both d_{xy} and d_{yz} are well visible, spanning energy and k-space regions in substantial agreement with the calculated band structure. At low density, on the other hand, only the d_{xy} is clearly visible, in agreement with the band structure evolution depicted by the calculations.

This remarkable evolution of the band structure with the charge density indicates a certain degree of electron correlation of the t_{2g} bands. In fact, orbital polarization (here we leave aside orbital magnetization) is governed by the Coulomb repulsion which causes the splitting of occupied and unoccupied orbitals. At the lowest order of approximation, this splitting can be parametrized as $\Delta t_{2g} = \Delta t_{2g}^0 + U_{at}\Delta p$, where Δp is the orbital occupancy difference, and Δt_{2g}^0 the splitting at zero occupancy (∼30 meV according to the PSIC results). In Fig. 8.9 we compare this expression

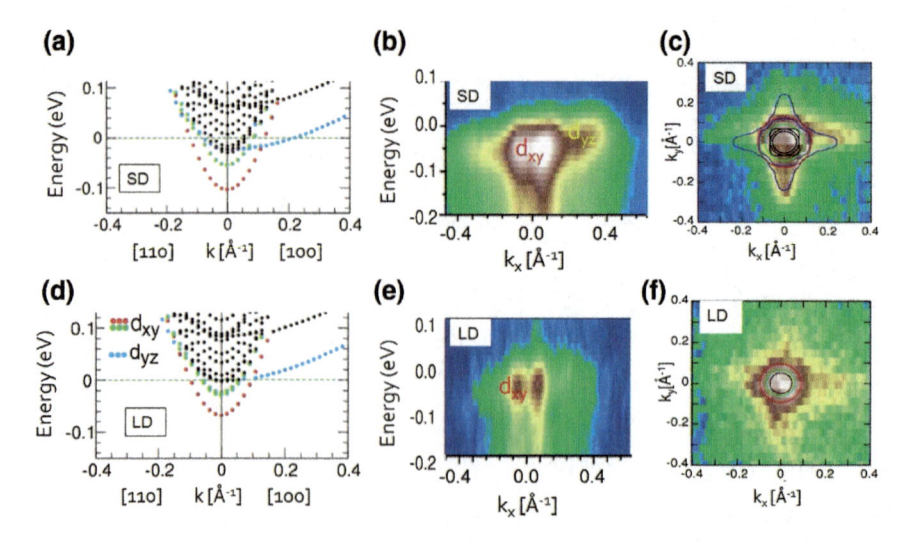

Fig. 8.8 PSIC-calculated conduction band energies near E_F for two different charge densities (standard density (SD) and low density (LD)) are compared with ARPES measurements for two samples of corresponding density. **a, d** Band energies calculated at SD and LD; **b, e** corresponding ARPES spectra; **c, f** calculated and measured Fermi surfaces juxtaposed for SD and LD samples. Reprinted figure with permission from [53]. Copyright 20164 by the American Physical Society

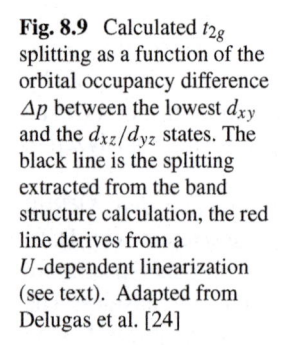

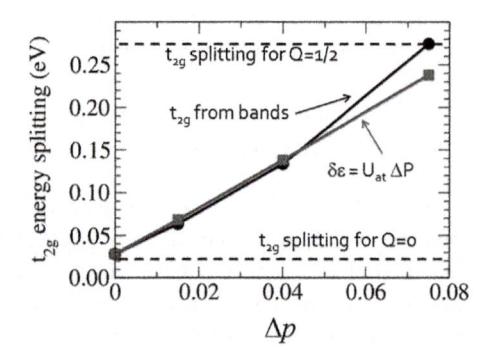

Fig. 8.9 Calculated t_{2g} splitting as a function of the orbital occupancy difference Δp between the lowest d_{xy} and the d_{xz}/d_{yz} states. The black line is the splitting extracted from the band structure calculation, the red line derives from a U-dependent linearization (see text). Adapted from Delugas et al. [24]

with the splitting obtained from the PSIC band structure in Fig. 8.7b; a agreement between bands and linear model is obtained across the whole occupancy range for $U_{at} = 2.8$ eV, which indicates a moderate but sensible amount of electronic correlation.

The band structure illustrated so far did not include any spin-orbit (SO) contribution. Typically, SO effects for 3d atoms like titanates are not particularly relevant and can be safely discarded with respect to, e.g. the inter-band transition energies which govern the optical properties. Nevertheless, if we are interested in fine-tuning the transport properties for certain specific charge density values, even small SO splitting can be important to give a correct interpretation to the observed phenomenology.

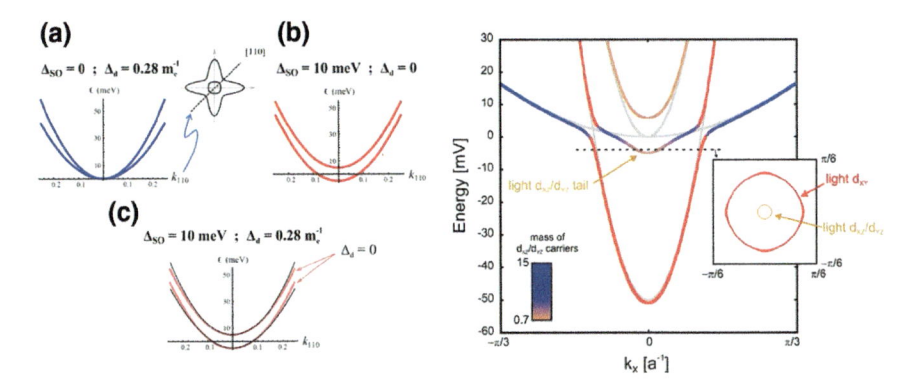

Fig. 8.10 Band calculations with SO coupling included, by Joshua et al. **a, b, c** Model band energy calculations for d_{xz} and d_{yz} orbitals, corresponding to 3 different types of splitting energy along the [110] direction (see text description). **d** Ab-initio band energies with (colored) and without (gray) SO splitting. Reprinted by permission from Macmillan Publishers Ltd: Nature Communications [58], copyright 2012

Among the vast literature reporting the SO description, here we focus on [58] where the direct connection with transport properties is carried out with clear evidence.

In Fig. 8.10 the band structure obtained with and without spin-orbit coupling are displayed. To illustrate more in detail the SO coupling, they also report a d_{xz}/d_{yz} model splitting along the [110] direction obtained by different splitting contributions: in Fig. 8.10a the splitting is only due to a quadratic diagonal hopping $\sim \Delta_d k_x k_y$, which does not split the band bottom; in Fig. 8.10b the splitting only includes an atomic SO contribution $\Delta_{SO} = 10$ meV, and in Fig. 8.10c both splittings are included, resulting in a almost k-independent shift of the two heavy-mass bands. A further SO contribution, due to the asymmetric $k \rightarrow -k$ reflection (Rashba effect) is too small to be visible on the scale. However, Rashba effects are important for the interpretation of magnetoresistance and superconductivity at low-T under an applied gate field, as shown in [59, 60].

The ab-initio calculations with and without SO included is shown in Fig. 8.10d. As expected, the most relevant SO effects occurs in the band crossing regions. In particular, d_{xz} and d_{yz} not only split with respect to the SO-free representation (gray lines in the figure) but also mix their light and heavy characters along the k_x axis. Thus, while the lowest light-mass band of purely d_{xy} character is basically unmodified with respect to the non-SO case, the second lowest band now presents a rather light character at the very bottom, with $m = 2/(m_h^{-1} + m_l^{-1}) \sim 1.5\, m_e$, and then it recovers the usual heavy character of d_{xz} and d_{yz} while moving above in energy. In other words, this second-lowest band displays an effective mass which changes with the Fermi energy (represented by the orange-to-blue color scale in Fig. 8.10d) in the critical charge density region which characterize the transition from single-band to double-band behavior. This light-mass, low-density band value is coherent with the scenario depicted by Shubnikov-de Haas experiments [56, 58] which give a 1/B frequency period of the resistivity oscillations corresponding to a small density band $n_{2D} \sim 1\text{--}5 \times 10^{12}$ cm^{-2}, and mass $\sim 2\,m_e$. Alternatively, this low-density band could be also attributed to another d_{xy} band, which appears in the SO-free treatment described in Fig. 8.7b, eventually 'fattened' by Anderson localization or polaronic behavior.

8.2.3 Thermoelectric and Transport Properties: The Bloch-Boltzmann Approach

Thermoelectric, transport and magnetotransport characterizations are perhaps the most practiced experimental techniques for the investigation of 2DEG systems in oxide heterostructures, and a large amount of these data have been delivered for the STO/LAO interface since its discovery in 2005. Having solid theoretical approaches to the calculation of these quantities is thus an invaluable help to the interpretation of the fundamental phenomenology on the one hand, and to the design of new materials with enhanced capabilities on the other. Here in particular we review one approach which in the last few years was applied to a series of oxide heterostructures, including STO/LAO, with apparently satisfying results. This approach is based on the Bloch-Boltzmann Theory (BBT) [61], as implemented in the freeware BoltZTraP code [62], and specifically suited to describe diffusive transport in relaxation-time approximation. This approach to electronic transport is based on the observation (due to Bloch in the 30's) that, as long as the diffusion length of the electron is much longer than the characteristic electronic wavelength, the static electronic band structure is still meaningful even under non-equilibrium conditions, and the scattering of the electrons with phonons or impurity sources results in jumps across two Bloch states with different band energies and/or crystalline momentum. Assuming specific conditions, an average 'relaxation' time τ between two consecutive scatterings can be defined, and the BBT expressions for electrical conductivity σ_{ij}, thermopower (i.e. Seebeck coefficient) S_{ij}, and Hall conductivity σ_{ijk}^H are derived as following:

$$\sigma_{ij} = \left(\frac{e^2}{V}\right) \sum_{n\mathbf{k}} \left(-\frac{\partial f}{\partial \epsilon_{n\mathbf{k}}}\right) v_{n\mathbf{k},i} v_{n\mathbf{k},j} \qquad (8.1)$$

$$S_{ij} = -\left(\frac{e}{VT}\right) \frac{1}{\sigma_{il}} \sum_{n\mathbf{k}} \tau_{n\mathbf{k}} \left(-\frac{\partial f}{\partial \epsilon_{n\mathbf{k}}}\right) (\epsilon_{n\mathbf{k}} - \mu) v_{n\mathbf{k},l} v_{n\mathbf{k},j} \qquad (8.2)$$

$$\sigma_{ijk}^H = \left(\frac{e^3}{V}\right) \sum_{n\mathbf{k}} \tau_{n\mathbf{k}}^2 \left(-\frac{\partial f}{\partial \epsilon_{n\mathbf{k}}}\right) e_{klm} v_{n\mathbf{k},i} M_{n\mathbf{k},jl}^{-1} v_{n\mathbf{k},m} \qquad (8.3)$$

Here i, j, k, l, m are Cartesian indices, e_{ijk} the Levi-Civita tensor, $v_{n\mathbf{k},i}$ and $M_{n\mathbf{k},jl}$ band velocity and effective mass tensor, and f the Fermi-Dirac occupancy.

From (8.1) and (8.3), the Hall coefficient commonly extracted from the experiments is obtained as:

$$R_{ijk}^H = \sigma_{il}^{-1}\sigma_{lmk}^H\sigma_{mj}^{-1} \qquad (8.4)$$

Expressions (8.1)–(8.4) require two sets of input data: (i) ab-initio band energies $\epsilon_{n\mathbf{k}}$, band velocities $v_{n\mathbf{k},m}$, and effective mass tensor $M_{n\mathbf{k},jl}$, carefully interpolated over a very dense $\mathbf{k}$-space grid in order to reach a sufficient level of accuracy and convergence; (ii) the relaxation time $\tau_{n\mathbf{k}}$, which includes all the information concerning the carrier scattering. For the latter, the simplest and most common choice is taking $\tau_{n\mathbf{k}}$ as a constant; in this case, it follows immediately from (8.2) and (8.4) that both Seebeck and Hall coefficient becomes τ-independent, which results in a huge simplification of the calculations. However, this is a rather crude assumption, since the temperature dependence is largely suppressed (an example will be shown later on). A simple but effective alternative is assuming $\tau_{n\mathbf{k}} = \tau(\epsilon_{n\mathbf{k}})$, and modeling the energy-dependence in terms of a power λ whose value depends on the dominant scattering mechanism:

$$\tau(T,\epsilon) = F(T)\left(\frac{\epsilon - \epsilon_0}{K_B T}\right)^\lambda \qquad (8.5)$$

For $F(T)$ simple analytical forms can be employed, with parameters optimized for the specific material [63, 64]. Alternatively, more accurate numerical models can be used for $\tau(\epsilon)$ written on the basis of fundamental parameters which can be estimated from ab-initio calculations. As an example, the relaxation time expression for the electron-acoustic phonon scattering, which dominates the conductivity at room-T in case of conventional, non-polar semiconductors, is:

$$\tau_{AP}^{-1}(T,\epsilon) = \frac{(2\tilde{m})^{3/2} K_B T D^2 \epsilon^{1/2}}{2\pi \hbar^4 \rho v_s^2} \qquad (8.6)$$

Here key parameters to be determined by separate calculations are the sound velocity v_s, the geometrically averaged effective mass $\tilde{m}$, and the deformation potential D, i.e. the linear change of electron energy at the band edge for an applied homogeneous strain, while ρ is the mass density of the system. Expressions for this and other scattering types can be found in [65–71].

As a validation of the BBT calculations for oxides, an obvious test case is the STO bulk, for which a massive amount of transport experiments exist in literature. In Fig. 8.11 the comparison between calculated and measured Seebeck coefficients is reported for electron-doped STO bulk. The experiments from [72] concern a series of $Sr_{1-x}La_x TiO_3$ samples, corresponding to Hall-measured charge densities $n_{3D} = 8.8 \times 10^{19}$ cm^{-3} (x = 0), 2.3×10^{20} cm^{-3} (x = 0.015), 1.2×10^{21} cm^{-3} (x = 0.05), and 1.9×10^{21} cm^{-3} (x = 0.1). For the calculations, on the other hand, doping is treated at the level of rigid band approximation, which is sufficient in 3D isotropic bulk systems for these low doping concentrations. To appreciate the efficiency of the scattering model, BBT results obtained within constant-τ approximation are

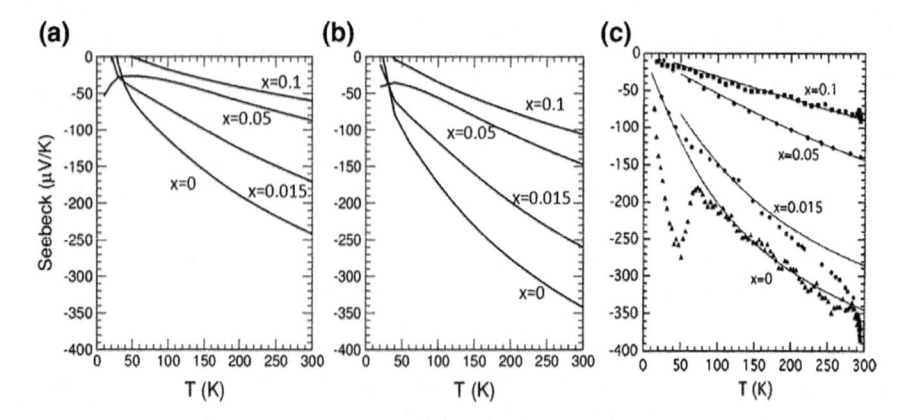

Fig. 8.11 Calculated versus measured Seebeck for doped STO bulk. **a** BBT calculation assuming constant τ; **b** BBT calculations with phenomenological $\tau(\epsilon)$ modeling described in 8.5 with $\lambda = 3/2$; **c** Seebeck measurements for La-doped STO. Panel (**c**) is a reprinted figure with permission from [72]. Copyright 2001 by the American Physical Society

reported as well (Fig. 8.11a). The latter delivers flatter curves, which substantially underestimate the measured values at room-T. On the other hand, using the simple $\tau(\epsilon)$ model reported in (8.5) with $\lambda = 3/2$, a nice agreement with the experiment is restored for all the considered doping concentrations, except for the peak at low-temperature measured for the low-doped (x = 0) sample, which is due to phonon-drag and requires a separate treatment (described later on). In Fig. 8.11c the measured values are also well interpolated by an effective mass model (solid lines) which includes the same $\lambda = 3/2$ scattering dependence of the BBT calculations: thus, we can deduce that in order to obtain a good Seebeck description, the scattering model in this case is more crucial than the actual ab-initio description of the bands. Clearly, this is only valid in the specific conditions of low doping regime and 3D isotropic band topology, which makes the single-band effective mass model adequate to the aim.

The inclusion of an energy-dependent scattering rate is also instrumental to describe the temperature dependence of conductivity: in Fig. 8.12 the conductivity of an electron-doped STO bulk is calculated by the BBT approach, and compared with that measurements by Ohta et al. [73] for several La-doped and Nb-doped STO samples at high temperature. Again, we report both constant-τ (Fig. 8.12a) and model-τ (Fig. 8.12b) calculations. The former decently reproduces the conductivity magnitude at room-T, but delivers a very flat temperature-dependent behavior. Using the scattering rate model, on the other hand, a satisfying agreement with the measurements is obtained, for the various samples.

Clearly, in order to have a meaningful evaluation of transport and thermoelectric properties, a reliable estimate of the charge concentration present in the sample is essential. Experimentally, the most common practice is the evaluation of the Hall resistivity coefficient as $R_{ijk}^H = E_j/J_i B_k$, i.e. the ratio between the steady-state

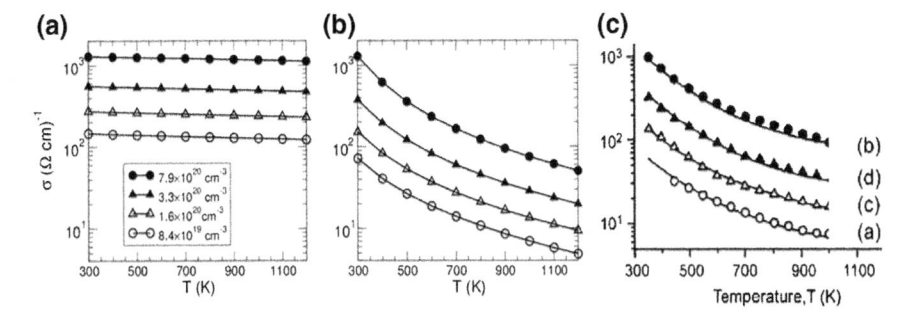

Fig. 8.12 Calculated versus measured conductivity for electron-doped STO bulk. **a** BBT calculation assuming constant $\tau = 7$ fs; **b** BBT calculations with phenomenological $\tau(\epsilon)$ modeling described in (8.5) with $\lambda = 3/2$; **c** Conductivity measurements from Ohta et al., adapted from [73]. In the experiment, circlets and triangles are for La-doped and Nb-doped samples, respectively

transverse electric field E due to the Lorentz force, and the current in the drift direction J times the applied magnetic field B. It can be shown that in the Drude model (we drop Cartesian indices for simplicity) it is simply $R^H = \pm 1/(en)$, where the sign $\pm$ holds for holes and electrons, respectively, thus the measurement of R^H is a useful way to estimate both the concentration and the type of mobile carriers in the sample. The deviation from the Drude model can be included in the more general expression $n = \pm r^H/(eR^H)$, where r^H is called Hall factor. From the Hall coefficient, the Hall mobility is determined as $\mu_{ij}^H = R_{ilm}^H \sigma_{mj}$, where $\mu_{ij}^H = \pm r^H \mu_{ij}$. In practice, since the experimental evaluation of r^H is difficult and in most situations it is not much different from unity, the equivalence between Hall and electron mobility is postulated. In fact, r^H depends on the scattering regime and may fluctuate by 20–30%, depending on the specific case. From the theoretical viewpoint, on the other hand, using (8.1)–(8.4) both Hall and electron mobilities (and in turn the Hall factor) can be independently calculated. An example of this capability of the BBT approach is furnished in [63], whose main results are reported in Fig. 8.13.

Here two Nb-doped STO samples (sample I and II) are considered; the charge measured from inverse Hall resistivity is reported in the left panels (blue squared symbols). Let's focus on sample I, for reference: as a function of temperature, $(eR^H)^{-1}$ displays a sinusoidal behavior, spanning a density range between 1.8 and 2.3×10^{19} cm^{-3}; in order to reproduce this behavior, authors first calculated $(eR^H)^{-1}$ and r^H from (8.1) to (8.4) for a range of fixed n_{3D} values spanning the above mentioned experimental limits (red lines in the theoretical panels). It results that r^H can sensibly differ from unity in the range between 0 and room T, while its charge dependence is weak and can be discarded, at least in the considered range of charge densities. By rescaling the experimental $(eR^H)^{-1}$ with the calculated charge-averaged r^H, the actual $n_{3D} = r^H/(eR^H)$ versus temperature can be obtained (solid blue line in the experimental panel). Finally, $(eR^H)^{-1}$ is recalculated again using as input this $n_{3D}(T)$ (open blue circles). We can see that the latter is in striking agreement with

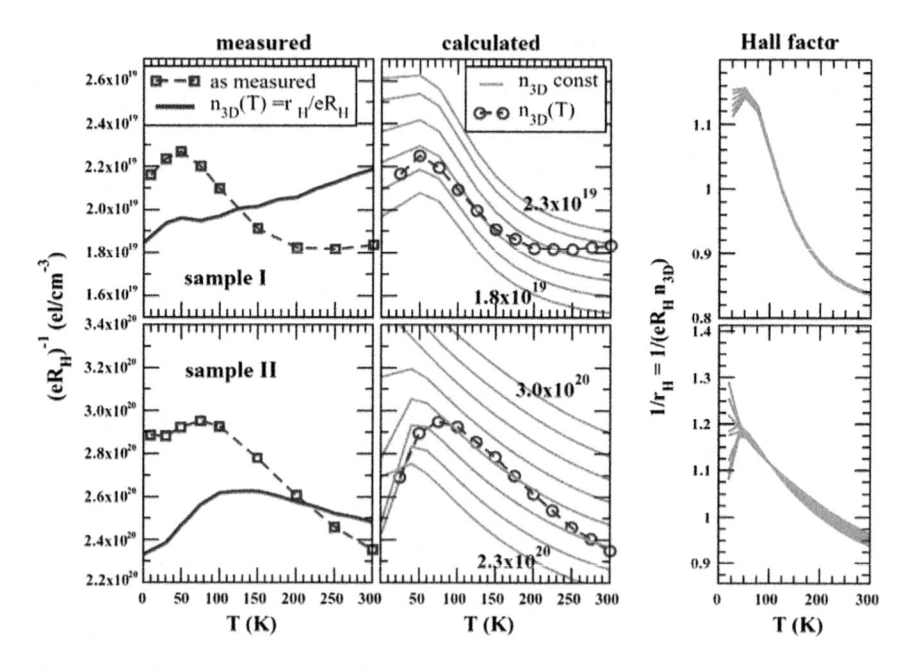

Fig. 8.13 Left panels: measured and calculated inverse Hall resistivity for two Nb-doped STO samples. Right panels: calculated Hall factors for the two samples. Reprinted figure with permission from [63]. Copyright 2013 by the American Physical Society

the 'as measured' Hall resistivity. This accord is not fortuitous, since the same level of agreement is also recovered for sample II in Fig. 8.13.

In summary, these tests validate the application of the BBT plus model scattering approach for the study of doped oxide systems, giving evidence that this method represents a useful complement to the analysis based on Hall, transport, and thermoelectric measurements.

8.2.4 Transport and Thermoelectric Properties of STO/LAO

The transport properties of STO/LAO are subject of an impressive amount of literature, too long, in fact, to be mentioned with any pretence of completeness. Here we limit our interest to assess the qualitative accuracy of the BBT results, and their usefulness in terms of fundamental interpretation of the STO/LAO phenomenology. In particular, we are interested to discuss the transport properties in terms of their charge density dependence. To the aim, we can use as experimental reference the results reported in [57], where transport experiments are presented for a series of STO/LAO samples fabricated using different growth temperature, in order to obtain

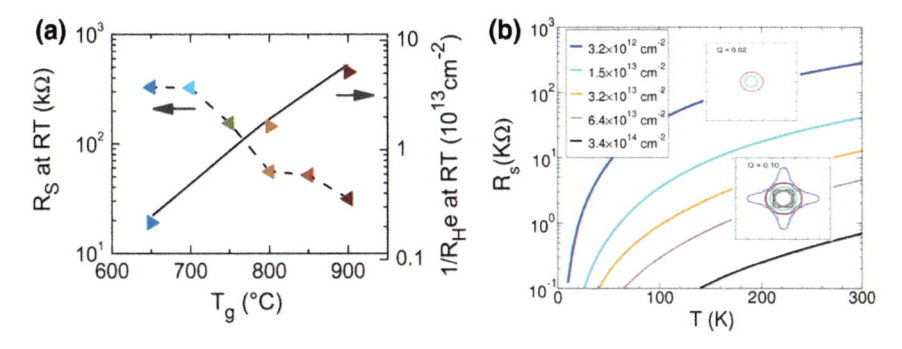

Fig. 8.14 a 2D sheet resistivity (R_s, left vertical axis) and Hall-measured charge density (right vertical axis) measured at room temperature for several STO/LAO samples, ordered along the horizontal axis in terms of their growth temperature T_g. From [57]. **b** BBT-calculations for STO/LAO with a variable amount of 2D charge density, reported in the legend. In the regions of low and high density, the Fermi surfaces calculated at the corresponding densities are juxtaposed. Panel (**a**) is reprinted from [57], with the permission of AIP Publishing

2DEG's with a variety of charge densities. The experimental results are shown in Fig. 8.14a.

At typical ("standard") growth temperature T_g of about 900°, STO/LAO displays its usual value $n_{2D} \sim 5 \times 10^{13}$ cm^{-2}; for a lower $T_g = 650°$ the charge density is reduced to 2×10^{12} cm^{-2}, i.e. more than an order of magnitude smaller than the standard one. Correspondingly, the sheet resistivity rises by an order of magnitude, from 30 kΩ of the standard sample up to 300 kΩ of the low-density sample, as expected from the inversely linear density dependence of resistivity. The BBT-calculated sheet resistivity reported in Fig. 8.14b for a series of electron density levels, agrees well, at least in terms of order of magnitude, with the measurements: at room temperature, for $n_{2D} = 3.2 \times 10^{12}$ cm^{-2} is $R_s = 284$ kΩ, and for $n_{2D} = 3.2 \times 10^{13}$ cm^{-2} $R_s = 13$ kΩ.

Most of all, the theoretical analysis offers a simple interpretation of the charge-density dependence of the samples in terms of band filling: according to [57], the low-density sample has a much higher low-T mobility than the standard sample. This is coherent with the description given in Sect. 8.2.2 of the progressive band filling with the increasing charge density in the 2DEG, according to which below a n_{2D} threshold value $\sim 1.5 \times 10^{13}$ cm^{-2}, only the energy-lowest, light-mass d_{xy} band is filled, while above 3.5×10^{13} cm^{-2} the heavier d_{xz} and d_{yz} bands also come into play, thus effectively reducing the overall mobility of the gas. However, this aspect remains controversial since Shubnikov-de Haas measurements on STO/LAO [56] suggest that even for the low-density samples, the contribution of d_{xz} and d_{yz} to the 2DEG may be present. Indeed, fitting the oscillations of conductance versus $1/B$, they extract an effective mass of $2.2\,m_e$ which could be more compatible with the light-mass band-bottom of d_{xz}/d_{yz} mixed orbitals resulting from the SO inclusion, described in Sect. 8.2.2. For larger 2DEG density, on the other hand, the multi-band behavior (at least for a certain range of temperatures) is evidenced by

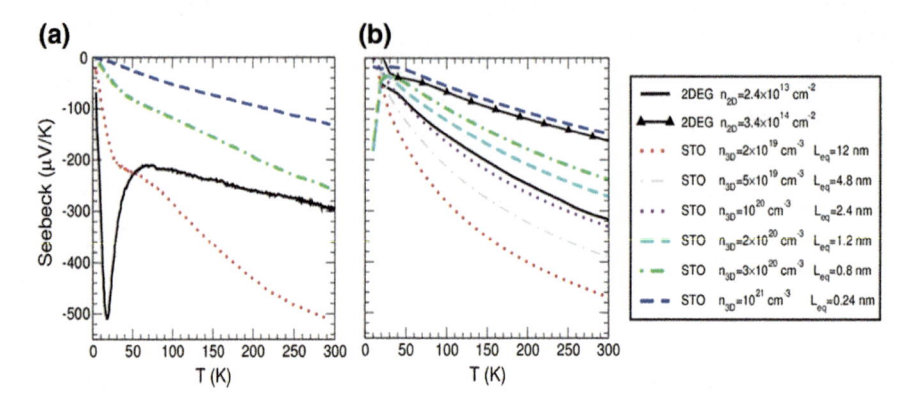

Fig. 8.15 Measured (**a**) versus calculated (**b**) thermopower for the 2DEG and several electron-doped STO bulk samples. In the theoretical panel, two curves (in black) are present for STO/LAO: the black-arrowed is for the half-electron charged 2DEG; the black solid is for the interface with charge density equal to that effectively measured by Hall experiments in the STO/LAO sample (the density is reported in the legend). Figure adapted from [64]

magnetotransport experiments showing non-linear Hall resistance versus magnetic field [58, 74–77]. In [58], for example, it is shown that the transition from single-band (linear Hall resistivity at $B \sim 0$) to multi-band behavior can be controlled by switching the gate field across a threshold value.

The thermoelectric properties of 2DEG systems have become a fascinating subject since when a relevant enhancement of thermoelectric efficiency was revealed for 2D nanostructured tellurides, with respect to their 3D counterparts [78, 79]. While this effect should be primarily attributed to the suppression of thermal conductivity [80], works based on effective-mass model calculations speculated that the reduced dimensionality could also produce a relevant increment of Seebeck coefficient, as due to quantum confinement effects, thus stimulating thermopower measurements in various oxide heterostructures [73, 81–83]. In particular, measurements on Nb-doped STO superlattices [81] reported some large room-T Seebeck (order of 10^3 μV/K), thus inspiring analogous analysis for the STO/LAO interface [75, 76, 84, 85].

In [64] a compared BBT and experimental analysis of transport and thermoelectric properties for the STO/LAO interface is delivered. The most important results are summarized in Fig. 8.15. For reference, STO/LAO (black lines) is compared with several electron-doped STO bulk samples (colored lines). Again, we can appreciate the good qualitative agreement between measurements and calculations for the STO samples of various charge densities. For the 2DEG, two BBT calculations are reported: one (black-arrowed line) concerns the polar-compensated system with 1/2 electron charge per unit interface; another curve (black solid line) is for STO/LAO with a charge density $n_{2D} = 2.4 \times 10^{13}$ cm^{-2} matching that measured for the sample shown in Fig. 8.15a. We see that the former largely underestimates the Seebeck amplitude at any temperature, while the latter is close to that measured in the high-T regime. Nevertheless, at low-T the measured value shows a huge deep associated with the phonon-drag effect, which is not included in the calculation. This result is in line with what previously seen for the electronic properties, i.e. the 2DEG phenomenology is reproduced by the calculations only if the correct amount of charge density is included in the system, while the fully compensated interface largely deviates from the observations.

A second important aspect concerns the effect of reduced dimensionality: apparently, the Seebeck amplitude for the 2DEG is not particularly incremented with respect to the bulk values. Notice, however, that an unbiased comparison between 2D and 3D is problematic since thermopower is strictly dependent on the Fermi level and hence on the charge density. The conversion between n_{2D} and n_{3D} requires the knowledge of the gas thickness L ($n_{2D} = n_{3D}L$), whose precise determination is quite cumbersome, as evidenced by the range of widely scattered values reported in literature ($L \sim 1-10$ nm) [86, 87]. Since n_{2D} for STO/LAO and n_{3D} for bulk samples are known, it is possible to define an 'equivalent' length value $L_{eq} = n_{2D}/n_{3D}$ (reported in the legend) which uniquely links STO/LAO with its doping-equivalent STO value. Assuming this criterion of doping equivalence, from Fig. 8.15 we see that only for extremely confined gases ($L_{eq} < 2$ nm) the 2DEG Seebeck overcomes its 3D analog, and even in this case the value is substantially of the same order of magnitude. Thus, the claim of possible thermopower enhancement due to quantum confinement appears unjustified, at least for what concern room-T.

At low temperature, however, the situation is quite different: for the 2DEG it is observed a narrow deep peak centered around T = 50 K, which is only barely visible in the STO bulk samples. This peak, called phonon-drag, has a different origin with respect to the diffusive thermopower which dominates at room-T: the latter is characterized by a roughly linear temperature dependence, and is due to the carrier diffusion in direct response to the temperature gradient; on the other hand, phonon-drag is an indirect, additional effect, due to the diffusion of phonons, which, in situation of strong electron-phonon coupling, drag the electrons, thus causing an additional thermoelectric response.

8.2.5 Analysis of Phonon-Drag in 2D Systems

The exceptional enhancement of the phonon-drag peak with respect to the 3D analog at equivalent doping appears as the most evident signature of 2D quantum confinement. To get more insights on this crucial aspect, in [88] a compared experimental and theoretical analysis was performed for several STO/LAO samples and some Nb-doped STO bulk samples as a counterpart. From the theoretical side, the formulation of phonon-drag was originally presented by Bailyn for 3D [89], and then adapted to 2D systems by Cantrell and Butcher [90, 91], based on the Boltzmann transport equations for coupled electrons and phonons. The phonon-drag thermopower in the direction j for a 2D system can be written:

$$S_j^{pd} = -\frac{e\,v_s^2}{(2\pi)^3\,\sigma_j K_B T^2} \sum_n \sqrt{\frac{m_{nx}m_{ny}}{m_{nj}^2}} \int_{\epsilon_n^0}^{\epsilon_n^0+W_n} d\epsilon\, f(\epsilon)\,\tau_n(\epsilon)$$

$$\times \int_0^{q_0} dq_p\, q_p^3 \int_{-q_0}^{q_0} dq_z \frac{N_q(1-f(\epsilon+\hbar\omega_q))}{\sqrt{C_n^2 - X_n^2}} \frac{A(q)F_n(q_z)}{\tau_{ph}^{-1}(q)} \qquad (8.7)$$

Now σ_j is the 2D conductivity (thus in Ω^{-1}); ϵ_n^0 and W_n are and electron band bottoms and bandwidths; N_q, ω_q and $\tau_{ph}(q)$ are phonon occupancy, frequency, and relaxation time, respectively; q_p and q_z are in-plane and orthogonal phonon wavevector components; $A(q)F_n(q_z)$ is the 2D electron-acoustic phonon coupling amplitude in the deformation potential approach; C_n and X_n are coefficients depending on electron and phonon energies. Here the crucial parameter is the electron-phonon coupling amplitude over phonon scattering ratio, appearing at the end of the formula: the electron-phonon scattering must be large with respect to phonon-phonon scattering, in order to have large phonon-drag. This occurs only in a narrow temperature range, since if T is too low the number of activated phonons and in turn the electron-phonon scattering is also low, while if T is too large the phonon-phonon scattering becomes dominant. In 2D, the phonon-drag enhancement is governed by the form factor:

$$F_n(q_z) = \left| \int_t dz\, \psi_n^2(z)\, e^{iq_z z} \right|^2 \qquad (8.8)$$

that is the Fourier transform of the squared electronic wavefunction of the 2D electrons confined within a slice of thickness t. In the $t \to \infty$ limit, $F_n \to \delta_{q_z,0}$, i.e. the 3D case is restored, with q_z fully determined by the crystal momentum conservation, while in the $t \to 0$ limit all q_z's up to the Debye wavelength contribute to the coupling, and $F_n = 1$.

The most important results of [88] are summarized in Fig. 8.16. Here panels (a) and (b) are Seebeck measurements for three STO/LAO interfaces and three STO samples with different doping charges, respectively; the interfaces are characterized by slightly different charge densities and mobilities, while the STO samples show large variation in doping concentration. It is very clear that the phonon-drag peak is well pronounced only at the interfaces, while in the STO bulk it is barely visible or completely absent. Also, depending on the specific interface, peak values from 500 μV/K up to well beyond 1000 μV/K can be reached, while the peak position in temperature is almost independent on the sample, and always located around 20 K.

The most important features of the 2DEG which determine the phonon-drag characteristics are well explained by the simulations reported in Fig. 8.16c, d, obtained according to the expressions in (8.7) and (8.8). We can see that the major qualitative aspects are reproduced. Most importantly, the simulations highlight two fundamental aspects: phonon-drag is amplified by both the increase of confinement (the 2DEG

8 Ab-Initio Calculations of TMO Band Structure

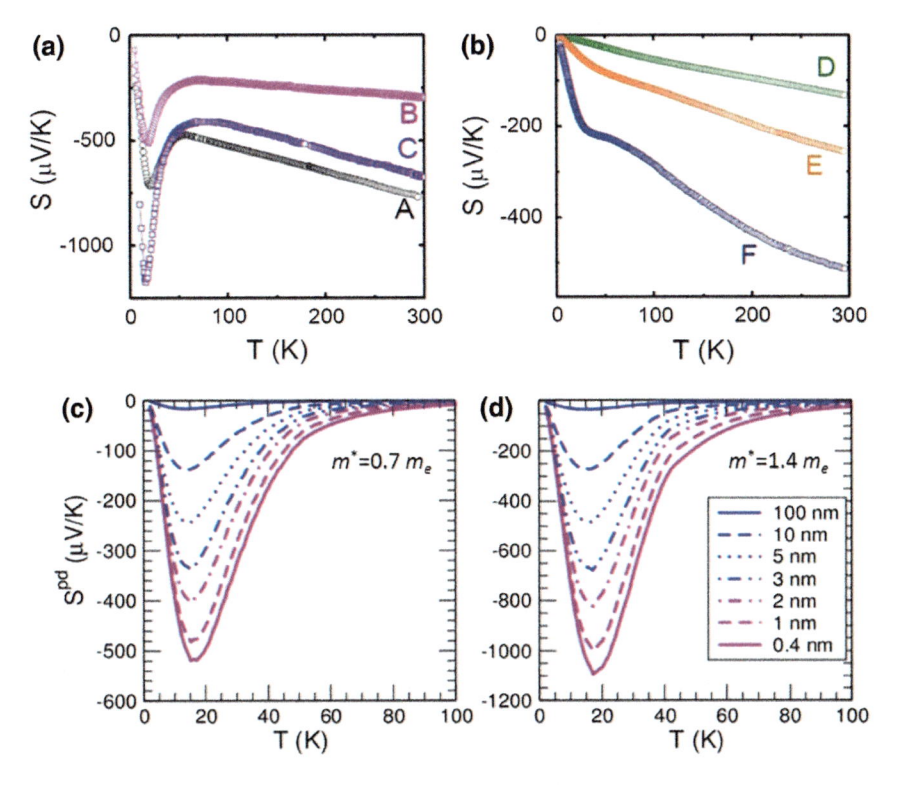

Fig. 8.16 a Seebeck measurement for several STO/LAO interfaces and **b** STO bulk samples with different doping concentrations. **c, d** Phonon-drag calculations for the 2DEG with different thickness (indicated in the legend) and effective masses. Panels (**a**) and (**b**) are reprinted figures with permission from [88]. Copyright 2016 by the American Physical Society

thickness reported in the legend of panel d), or the increase of the effective masses, which governs the in-plane localization. The phonon-drag peak is very sensitive to both t and m, while its temperature position is substantially unchanged. According to this analysis, phonon-drag emerges as a useful parameter to investigate the degree of charge localization in 2DEG systems.

The possible presence of localized charges in STO/LAO is one of the major subjects of discussion, and still an unsolved matter of debate for what concerns the 2DEG phenomenology. In the last few years, several experimental evidences were furnished on the presence of heavy-mass polaronic states lying below the 2DEG mobility edge [35, 92]. A direct evidence of these states was revealed by a very peculiar thermopower behavior obtained at low temperature under large negative gate field V_G, i.e. in the heavily charge-depleted regime [36]: Seebeck was observed to oscillate with regular frequency along with V_G, and simultaneously diverge to order of 10–100 mV/K for increasing negative field, until currents too small to be measured were reached. In Fig. 8.17a the measured Seebeck under a negative V_G

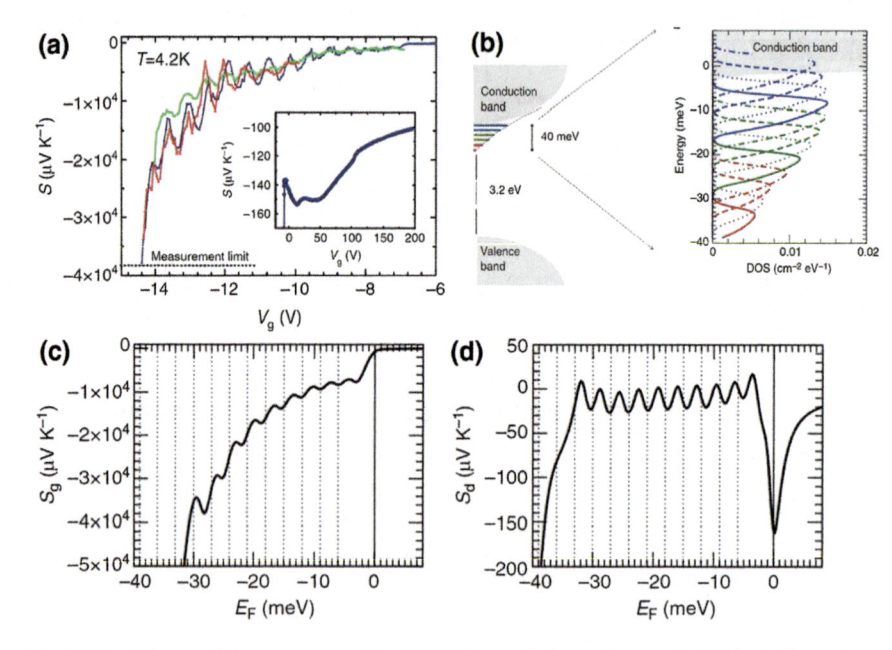

Fig. 8.17 a Measured thermopower at T = 4.2 K for applied negative gate field (depletion regime). Different colors refer to different samples. Inset: thermopower measured for positive gate field (accumulation regime). **b** Model density of states which simulate an array of polaronic bands located in a 40 meV region below the conduction band edge. **c** Calculated phonon-drag versus Fermi energy for the model system illustrated in (**b**); **d** calculated diffusive Seebeck versus E_F for the model system in (**b**). Figures from [36] distributed under a Creative Commons CC-BY license

applied in bottom-gated mode (thus exploring the depleted regime) is reported for three STO/LAO samples corresponding to different colors; in the inset, Seebeck under positive V_G (accumulation regime) is also reported. Only in the depletion conditions the oscillatory regime is found, while in accumulation Seebeck displays an ordinary behavior. The rationale of this finding was attributed to a gigantic phonon-drag effect, induced by the crossing of the Fermi level through a series of polaronic states (Fig. 8.17b) regularly distributed within a 40 meV energy range, with very large effective masses, bandwidths of a few meV, and charge densities smaller than 10^{11} cm^{-2}. Notice that in case of wider bands, phonon-drag would be order-of-magnitude smaller than the observed. This huge phonon-drag is an intriguing manifestation of the effect discussed above, and resides in the enhanced ratio of electron-acoustic phonon scattering over phonon-phonon scattering.

The phonon-drag calculation versus Fermi energy (Fig. 8.17c) obtained for this model band structure reproduces all the main qualitative features of the measurement, at variance with the diffusive component displayed in Fig. 8.17d for the same model. In fact, by definition, the diffusive contribution must always oscillate around zero when EF crosses isolated bands. In other words, diffusive Seebeck can oscillate or even diverge, but never oscillate and diverge simultaneously, as found in the experiments. It should be emphasized that this experiment is an excellent example of the fact that thermoelectric measurements under gate field can be a very accurate and powerful form of spectroscopy to probe fine details of the electronic structure at low-temperature for oxide heterostructures and 2DEG systems in general.

8.3 Conclusions

The STO/LAO interface is certainly one of the most studied materials of the last 30 years in physics; this enormous interest is justified not only by the huge potential for future nanoelectronic applications, but also by the extraordinary range of unexpected fundamental properties revealed for this material, which only in minimal portion have been recollected in this review. This richness of phenomenology has represented over the years a formidable opportunity but also a challenge for the most advanced ab-initio theoretical approaches, which struggled to correctly reproduce and interpret the variety of observations reported for STO/LAO. After more than ten years since its discovery, we can safely attest that these theories are capable to describe with satisfying accuracy a range of fundamental aspects concerning structural and electronic properties, but only if the pristine, stoichiometric interface is assumed. Alas, many of the reported characterizations are probably related to non-stoichiometric, extrinsic features, which only on a very limited extent can be included in the ab-initio treatment, for the reasons briefly mentioned at the beginning of this review. It follows that a number of fundamental aspects are almost completely uncovered, or covered on a very limited extent, by realistic simulations: of those, the most crucial probably concern oxygen vacancies and cation mixing at realistic concentrations, surface reconstruction, and structural disorder in general. Some of the most striking phenomena observed in STO/LAO, such as charge localization (e.g. polarons, defects), magnetism and superconductivity, are only barely approached by the calculations, and certainly with simplifications which make predictions very questionable. For some specific properties, like electrical transport, thermopower, and phonon-drag, we gave evidence here that a satisfying compromise between accuracy and feasibility can be achieved if the ab-initio band-energy results are complemented by the semi-classical Bloch-Boltzmann transport theory, and by analytical modeling for the formulation of electron and phonon scattering rates. We believe that this multi-methodological approach mixing ab-initio and model theories can be, at least in the short run, the most efficient way to attack some of the properties of oxide heterostructures, at least until a huge quantum leap in computational capability will be achieved, thus allowing full ab-initio simulations on a very large atomic scale.[93–95]

Acknowledgements A.F. warmly thanks Sbastien Lemal and Philippe Ghosez for their careful critical reading of the manuscript, and acknowledges financial support under Project PON-NETERGIT, and computing support from CRS4 Computing Centre (Loc. Piscina Manna, Pula, Italy).

References

1. M. Huijben, G. Rijnders, D.H. Blank, S. Bals, S. Van Aert, J. Verbeeck, G. Van Tendeloo, A. Brinkman, H. Hilgenkamp, Electronically coupled complementary interfaces between perovskite band insulators. Nat. Mater. **5**, 556–560 (2006), http://www.nature.com/doifinder/10.1038/nmat1675
2. N. Nakagawa, H.Y. Hwang, D.A. Muller, Why some interfaces cannot be sharp. Nat. Mater. **5**, 204–209 (2006), http://www.nature.com/doifinder/10.1038/nmat1569
3. A. Ohtomo, H.Y. Hwang, A high-mobility electron gas at the LaAlO$_3$/SrTiO$_3$ heterointerface. Nature **427**, 423–427 (2004), http://www.nature.com/doifinder/10.1038/nature04773
4. V.I. Anisimov, J. Zaanen, O.K. Andersen, Band theory and Mott insulators: Hubbard U instead of Stoner I. Phys. Rev. B **44**, 943 (1991), https://doi.org/10.1103/PhysRevB.44.943
5. A. Filippetti, V. Fiorentini, A practical first-principles band-theory approach to the study of correlated materials. Eur. Phys. J. B **71**, 139–183 (2009), http://www.springerlink.com/index/10.1140/epjb/e2009-00313-2
6. A. Filippetti, C.D. Pemmaraju, S. Sanvito, P. Delugas, D. Puggioni, V. Fiorentini, Variational pseudo-self-interaction-corrected density functional approach to the ab initio description of correlated solids and molecules. Phys. Rev. B **84**, 195127 (2011), http://link.aps.org/doi/10.1103/PhysRevB.84.195127
7. D.I. Bilc, R. Orlando, R. Shaltaf, G.M. Rignanese, J. Íñiguez, Ph. Ghosez, Hybrid exchange-correlation functional for accurate prediction of the electronic and structural properties of ferroelectric oxides. Phys. Rev. B **77**, 165107 (2008), http://link.aps.org/doi/10.1103/PhysRevB.77.165107
8. J. Heyd, G.E. Scuseria, M. Ernzerhof, Hybrid functionals based on a screened Coulomb potential. J. Chem. Phys. **118**, 8207–8215 (2003), http://scitation.aip.org/content/aip/journal/jcp/118/18/10.1063/1.1564060
9. J.-M. Albina, M. Mrovec, M.B. Meyer, C. Elssser, Structure, stability, and electronic properties of SrTiO$_3$/LaAlO$_3$ and SrTiO$_3$/SrRuO$_3$ interfaces. Phys. Rev. B **76**, 165103 (2007), http://link.aps.org/doi/10.1103/PhysRevB.76.165103
10. N.C. Bristowe, E. Artacho, P.B. Littlewood, Oxide superlattices with alternating p and n interfaces. Phys. Rev. B **80**, 045425 (2009), http://link.aps.org/doi/10.1103/PhysRevB.80.045425
11. M.S. Park, S.H. Rhim, A.J. Freeman, Charge compensation and mixed valency in LaAlO$_3$/SrTiO$_3$ heterointerfaces studied by the FLAPW method. Phys. Rev. B **74**, 205416 (2006), http://link.aps.org/doi/10.1103/PhysRevB.74.205416
12. W.-J. Son, E. Cho, B. Lee, J. Lee, S. Han, Density and spatial distribution of charge carriers in the intrinsic n-type LaAlO$_3$/SrTiO$_3$ interface. Phys. Rev. B **79**, 245411 (2009), http://link.aps.org/doi/10.1103/PhysRevB.79.245411
13. E. Lesne, N. Reyren, D. Doennig, R. Mattana, H. Jaffrs, V. Cros, F. Petroff, F. Choueikani, P. Ohresser, R. Pentcheva, A. Barthlmy, M. Bibes, Suppression of the critical thickness threshold for conductivity at the LaAlO$_3$/SrTiO$_3$ interface. Nat. Commun. **5**, 4291 (2014), http://www.nature.com/doifinder/10.1038/ncomms5291
14. R. Pentcheva, W.E. Pickett, Avoiding the polarization catastrophe in LaAlO$_3$ Overlayers on SrTiO$_3$ (001) through polar distortion. Phys. Rev. Lett. **102**, 107602 (2009), http://link.aps.org/doi/10.1103/PhysRevLett.102.107602
15. Z.S. Popovic, S. Satpathy, R.M. Martin, Origin of the two-dimensional electron gas carrier density at the LaAlO$_3$ on SrTiO$_3$ interface. Phys. Rev. Lett. **101**, 256801 (2008), http://link.aps.org/doi/10.1103/PhysRevLett.101.256801
16. U. Schwingenschlgl, C. Schuster, Interface relaxation and electrostatic charge depletion in the oxide heterostructure LaAlO$_3$/SrTiO$_3$. Europhysics Lett. **6**, 27005 (2009), http://stacks.iop.org/0295-5075/86/i=2/a=27005?key=crossref.2c314eb51de14f37348a226e7c16b631
17. K. Janicka, J.P. Velev, E.Y. Tsymbal, Quantum nature of two-dimensional electron gas confinement at LaAlO$_3$/SrTiO$_3$ interfaces. Phys. Rev. Lett. **102**, 106803 (2009), http://link.aps.org/doi/10.1103/PhysRevLett.102.106803

18. J. Lee, A. Demkov, Charge origin and localization at the n-type SrTiO$_3$/LaAlO$_3$ interface. Phys. Rev. B **78**, 193104 (2008), http://link.aps.org/doi/10.1103/PhysRevB.78.193104
19. N. Pavlenko, T. Kopp, Structural relaxation and metal-insulator transition at the interface between SrTiO$_3$ and LaAlO$_3$. Surf. Sci. **605**, 11141121 (2011), http://linkinghub.elsevier.com/retrieve/pii/S003960281100118X
20. R. Pentcheva, W.E. Pickett, Charge localization or itinercy at LaAlO$_3$/SrTiO$_3$ interfaces: hole polarons, oxygen vacancies, and mobile electrons. Phys. Rev. B **74**, 035112 (2006), http://link.aps.org/doi/10.1103/PhysRevB.74.035112
21. Z. Zhong, P.J. Kelly, Electronic-structure induced reconstruction and magnetic ordering at the LaAlO$_3$|SrTiO$_3$ interface. Europhys. Lett. **84**, 27001 (2008), http://stacks.iop.org/0295-5075/84/i=2/a=27001?key=crossref.aa208f51927f0f59a7826061a6531da2
22. R. Pentcheva, W.E. Pickett, Ionic relaxation contribution to the electronic reconstruction at the n-type LaAlO$_3$/SrTiO$_3$ interface. Phys. Rev. B **78**, 205106 (2008), http://link.aps.org/doi/10.1103/PhysRevB.78.205106
23. M. Stengel, First-principles modeling of electrostatically doped perovskite systems. Phys. Rev. Lett. **106**, 136803 (2011), http://link.aps.org/doi/10.1103/PhysRevLett.106.136803
24. P. Delugas, A. Filippetti, V. Fiorentini, D.I. Bilc, D. Fontaine, Ph. Ghosez, Spontaneous 2-dimensional carrier confinement at the n-type SrTiO$_3$/LaAlO$_3$ interface. Phys. Rev. Lett. **106**, 166807 (2011), http://link.aps.org/doi/10.1103/PhysRevLett.106.166807
25. C. Cancellieri, D. Fontaine, S. Gariglio, N. Reyren, A.D. Caviglia, A. Fête, S.J. Leake, S.A. Pauli, P.R. Willmott, M. Stengel, Ph. Ghosez, J.-M. Triscone, Electrostriction at the LaAlO$_3$/SrTiO$_3$ interface. Phys. Rev. Lett. **107**, 056102 (2011), http://link.aps.org/doi/10.1103/PhysRevLett.107.056102
26. M.L. Reinle-Schmitt, C. Cancellieri, D. Li, D. Fontaine, M. Medarde, E. Pomjakushina, C.W. Schneider, S. Gariglio, Ph. Ghosez, J.-M. Triscone, P.R. Willmott, Tunable conductivity threshold at polar oxide interfaces. Nat. Commun. **3**, 932 (2012), http://www.nature.com/doifinder/10.1038/ncomms1936
27. F. Cossu, U. Schwingenschlogl, V. Eyert, Metal-insulator transition at the LaAlO$_3$/SrTiO$_3$ interface revisited: a hybrid functional study. Phys. Rev. B **88**, 045119 (2013), http://link.aps.org/doi/10.1103/PhysRevB.88.045119
28. L. Yu, A. Zunger, A polarity-induced defect mechanism for conductivity and magnetism at polar nonpolar oxide interfaces. Nat. Commun. **5**, 5118 (2014), http://www.nature.com/doifinder/10.1038/ncomms6118
29. W.-C. Lee, A.H. MacDonald, Electronic interface reconstruction at polar-nonpolar Mott-insulator heterojunctions. Phys. Rev. B **76**, 075339 (2007), http://link.aps.org/doi/10.1103/PhysRevB.76.075339
30. K. Yada, S. Onari, Y. Tanaka, J.-I. Inoue, Electrically controlled super-conducting states at the heterointerface SrTiO$_3$/LaAlO$_3$. Phys. Rev. B **80**, 140509 (2009), http://link.aps.org/doi/10.1103/PhysRevB.80.140509
31. N.C. Bristowe, Ph. Ghosez, P.B. Littlewood, E. Artacho, The origin of two-dimensional electron gases at oxide interfaces: insights from theory. J. Phys. Condens. Matter **26**, 143201 (2014), http://stacks.iop.org/0953-8984/26/i=14/a=143201?key=crossref.f7d9ccb17419daab8029060acecb0b0e
32. M. Stengel, D. Vanderbilt, Berry-phase theory of polar discontinuities at oxide-oxide interfaces. Phys. Rev. B **80**, 241103 (2009), http://link.aps.org/doi/10.1103/PhysRevB.80.241103
33. R.D. King-Smith, D. Vanderbilt, Theory of polarization of crystalline solids. Phys. Rev. B **47**, 1651–1654 (1993), http://link.aps.org/doi/10.1103/PhysRevB.47.1651
34. S.A. Pauli, S.J. Leake, B. Delley, M. Bjorck, C.W. Schneider, C.M. Schleputz, D. Martoccia, S. Paetel, J. Mannhart, P.R. Willmott, Evolution of the interfacial structure of LaAlO$_3$ on SrTiO$_3$. Phys. Rev. Lett. **106**, 036101 (2011), http://link.aps.org/doi/10.1103/PhysRevLett.106.036101
35. C. Cancellieri, A.S. Mishchenko, U. Aschauer, A. Filippetti, C. Faber, O.S. Barii, V.A. Rogalev, T. Schmitt, N. Nagaosa, V.N. Strocov, Polaronic metal state at the LaAlO$_3$/SrTiO$_3$ interface. Nat. Commun. **7**, 10386 (2016), http://www.nature.com/doifinder/10.1038/ncomms10386

36. I. Pallecchi, F. Telesio, D. Li, A. Fête, S. Gariglio, J.-M. Triscone, A. Filippetti, P. Delugas, V. Fiorentini, D. Marré, Giant oscillating thermopower at oxide interfaces. Nat. Commun. **6**, 6678 (2015), http://www.nature.com/doifinder/10.1038/ncomms7678
37. G. Berner, S. Glawion, J. Walde, F. Pfaff, H. Hollmark, L.-C. Duda, S. Paetel, C. Richter, J. Mannhart, M. Sing, R. Claessen, LaAlO/SrTiO oxide heterostructures studied by resonant inelastic X-ray scattering. Phys. Rev. B **82**, 241405 (2010), http://link.aps.org/doi/10.1103/PhysRevB.82.241405
38. A. Dubroka, M. Rössle, K.W. Kim, V.K. Malik, L. Schultz, S. Thiel, C.W. Schneider, J. Mannhart, G. Herranz, O. Copie, M. Bibes, A. Barthlmy, C. Bernhard, Dynamical response and confinement of the electrons at the $LaAlO_3/SrTiO_3$ interface. Phys. Rev. Lett. **104**, 156807 (2010), http://link.aps.org/doi/10.1103/PhysRevLett.104.156807
39. A. Filippetti, V. Fiorentini, Double-exchange driven ferromagnetic metal-paramagnetic insulator transition in Mn-doped CuO. Phys. Rev. B **74**, 220401 (2006), http://link.aps.org/doi/10.1103/PhysRevB.74.220401
40. A. Filippetti, V. Fiorentini, Magnetic ordering under strain and spin-Peierls dimerization in $GeCuO_3$. Phys. Rev. Lett. **98**, 196403 (2007), http://link.aps.org/doi/10.1103/PhysRevLett.98.196403
41. G. Colizzi, A. Filippetti, V. Fiorentini, Multiferroicity and orbital ordering in $Pr0.5Ca0.5MnO3$ from first principles. Phys. Rev. B **82**, 140101 (2010), http://link.aps.org/doi/10.1103/PhysRevB.82.140101
42. D. Puggioni, A. Filippetti, V. Fiorentini, Ordering and multiple phase transitions in ultrathin nickelate superlattices. Phys. Rev. B **86**, 195132 (2012), http://dx.doi.org/10.1103/PhysRevB.86.195132
43. L. Zhang, X.-F. Zhou, H.-T. Wang, J.-J. Xu, J. Li, E.G. Wang, S.-H. Wei, Origin of insulating behavior of the p-type $LaAlO_3/SrTiO_3$ interface: polarization-induced asymmetric distribution of oxygen vacancies. Phys. Rev. B **82**, 125412 (2010), http://link.aps.org/doi/10.1103/PhysRevB.82.125412
44. A. Kalabukhov, R. Gunnarsson, J. Borjesson, E. Olsson, T. Claeson, D. Winkler, Effect of oxygen vacancies in the $SrTiO_3$ substrate on the electrical properties of the $LaAlO_3/SrTiO_3$ interface. Phys. Rev. B **75**, 121404 (2007), http://link.aps.org/doi/10.1103/PhysRevB.75.121404
45. W. Siemons, G. Koster, H. Yamamoto, W. Harrison, G. Lucovsky, T.H. Geballe, D.H. Blank, M.R. Beasley, Origin of charge density at $LaAlO_3$ on $SrTiO_3$ heterointerfaces: possibility of intrinsic doping. Phys. Rev. Lett. **98**, 196802 (2007), http://link.aps.org/doi/10.1103/PhysRevLett.98.196802
46. Y. Li, S.N. Phattalung, S. Limpijumnong, J. Kim, J. Yu, Formation of oxygen vacancies and charge carriers induced in the n-type interface of a $LaAlO3$ overlayer on $SrTiO3$ (001). Phys. Rev. B **84**, 245307 (2011), http://link.aps.org/doi/10.1103/PhysRevB.84.245307
47. N.C. Bristowe, P.B. Littlewood, E. Artacho, Surface defects and conduction in polar oxide heterostructures. Phys. Rev. B **83**, 205405 (2011), http://link.aps.org/doi/10.1103/PhysRevB.83.205405
48. W.-J. Son, E. Cho, J. Lee, S. Han, Hydrogen adsorption and carrier generation in $LaAlO_3$-$SrTiO_3$ heterointerfaces: a first-principles study. J. Phys. Condens. Matter **22**, 315501 (2010), http://stacks.iop.org/0953-8984/22/i=31/a=315501?key=crossref.b074a735f655a7211eedde5a0f98bc00
49. C. Cazorla, M. Stengel, First-principles modeling of $Pt/LaAlO_3/SrTiO_3$ capacitors under an external bias potential. Phys. Rev. B **85**, 075426 (2012), http://link.aps.org/doi/10.1103/PhysRevB.85.075426
50. V. Vonk, J. Huijben, D. Kukuruznyak, A. Stierle, H. Hilgenkamp, A. Brinkman, S. Harkema, Polar-discontinuity-retaining A-site intermixing and vacancies at $SrTiO_3/LaAlO_3$ interfaces. Phys. Rev. B **85**, 045401 (2012), http://link.aps.org/doi/10.1103/PhysRevB.85.045401
51. P.R. Willmott, S.A. Pauli, R. Herger, C.M. Schleputz, D. Martoccia, B.D. Patterson, B. Delley, R. Clarke, D. Kumah, C. Cionca, Y. Yacoby, Structural basis for the conducting interface between $LaAlO_3$ and $SrTiO_3$. Phys. Rev. Lett. **99**, 155502 (2007), http://link.aps.org/doi/10.1103/PhysRevLett.99.155502

52. G. Berner, A. Mller, F. Pfaff, J. Walde, C. Richter, J. Mannhart, S. Thiess, A. Gloskovskii, W. Drube, M. Sing, R. Claessen, Band alignment in LaAlO/SrTiO oxide heterostructures inferred from hard X-ray photoelectron spectroscopy. Phys. Rev. B **88**, 115111 (2013), http://link.aps.org/doi/10.1103/PhysRevB.88.115111
53. C. Cancellieri, M.L. Reinle-Schmitt, M. Kobayashi, V.N. Strocov, P.R. Willmott, D. Fontaine, Ph. Ghosez, A. Filippetti, P. Delugas, V. Fiorentini, Doping-dependent band structure of LaAlO$_3$/SrTiO$_3$ interfaces by soft X-ray polarization-controlled resonant angle-resolved photoemission. Phys. Rev. B **89**, 121412 (2014), http://link.aps.org/doi/10.1103/PhysRevB.89.121412
54. M. Salluzzo, J.C. Cezar, N.B. Brookes, V. Bisogni, G.M. De Luca, C. Richter, S. Thiel, J. Mannhart, M. Huijben, A. Brinkman, G. Rijnders, G. Ghiringhelli, Orbital reconstruction and the two-dimensional electron gas at the LaAlO$_3$/SrTiO$_3$ interface. Phys. Rev. Lett. **102**, 166804 (2009), http://link.aps.org/doi/10.1103/PhysRevLett.102.166804
55. M. Salluzzo, S. Gariglio, X. Torrelles, Z. Ristic, R. Di Capua, J. Drnec, M. Moretti Sala, G. Ghiringhelli, R. Felici, N.B. Brookes, Structural and electronic reconstructions at the LaAlO$_3$/SrTiO$_3$ interface. Adv. Mater. **25**, 2333–2338 (2013), http://doi.wiley.com/10.1002/adma.201204555
56. A. Fête, S. Gariglio, C. Berthod, D. Li, D. Stornaiuolo, M. Gabay, J.M. Triscone, Large modulation of the Shubnikov-de Haas oscillations by the Rashba interaction at the LaAlO$_3$/SrTiO$_3$ interface. New J. Phys. **16**, 112002 (2014), http://stacks.iop.org/1367-2630/16/i=11/a=112002?key=crossref.36462db5f9e2db63f1445d97256dabe0
57. A. Fête, C. Cancellieri, D. Li, D. Stornaiuolo, A.D. Caviglia, S. Gariglio, J.-M. Triscone, Growth-induced electron mobility enhancement at the LaAlO$_3$/SrTiO$_3$ interface. Appl. Phys. Lett. **106**, 051604 (2015), http://scitation.aip.org/content/aip/journal/apl/106/5/10.1063/1.4907676
58. A. Joshua, S. Pecker, J. Ruhman, E. Altman, S. Ilani, A universal critical density underlying the physics of electrons at the LaAlO$_3$/SrTiO$_3$ interface. Nat. Commun. **3**, 1129 (2012), http://www.nature.com/doifinder/10.1038/ncomms2116
59. A.D. Caviglia, M. Gabay, S. Gariglio, N. Reyren, C. Cancellieri, J.-M. Triscone, Tunable Rashba spin-orbit interaction at oxide interfaces. Phys. Rev. Lett. **104**, 126803 (2010), http://link.aps.org/doi/10.1103/PhysRevLett.104.126803
60. A. Fête, S. Gariglio, A.D. Caviglia, J.-M. Triscone, M. Gabay, Rashba induced magnetoconductance oscillations in the LaAlO$_3$-SrTiO$_3$ heterostructure. Phys. Rev. B **86**, 201105 (2012), http://link.aps.org/doi/10.1103/PhysRevB.86.201105
61. P.B. Allen, Boltzmann theory and resistivity of metals, in *Quantum Theory of Real Materials*, ed. J.R. Chelikowsky, S.G. Louie (Kluwer, Boston, 1996), pp. 219–250
62. G. Madsen, D. Singh, BoltzTraP, a code for calculating band-structure dependent quantities. Comput. Phys. Commun. **175**, 67 (2006), http://linkinghub.elsevier.com/retrieve/pii/S0010465506001305
63. P. Delugas, A. Filippetti, M.J. Verstraete, I. Pallecchi, D. Marré, V. Fiorentini, Doping-induced dimensional crossover and thermopower burst in Nb-doped SrTiO$_3$ superlattices. Phys. Rev. B **88**, 045310 (2013), http://link.aps.org/doi/10.1103/PhysRevB.88.045310
64. A. Filippetti, P. Delugas, M.J. Verstraete, I. Pallecchi, A. Gadaleta, D. Marré, D.F. Li, S. Gariglio, V. Fiorentini, Thermopower in oxide heterostructures: the importance of being multiple-band conductors. Phys. Rev. B **86**, 195301 (2012), http://link.aps.org/doi/10.1103/PhysRevB.86.195301
65. S. Altunz, H. Celik, M. Cankurtaran, Temperature and electric field dependences of the mobility of electrons in vertical transport in GaAs/Ga1-yAlyAs barrier structures containing quantum wells. Cent. Eur. J. Phys. **6**, 479–490 (2008), http://www.springerlink.com/content/q60113w8rl723807/
66. D.R. Anderson, N.A. Zakhleniuk, M. Babiker, B.K. Ridley, C.R. Bennett, Polar-optical phonon-limited transport in degenerate GaN-based quantum wells. Phys. Rev. B **63**, 245313 (2001), http://link.aps.org/abstract/PRB/v63/e245313

67. P. Delugas, V. Fiorentini, A. Mattoni, A. Filippetti, Intrinsic origin of two-dimensional electron gas at the (001) surface of SrTiO$_3$. Phys. Rev. B **91**, 115315 (2015), http://link.aps.org/doi/10.1103/PhysRevB.91.115315
68. K. Kaasbjerg, K.S. Thygesen, A.-P. Jauho, Acoustic phonon limited mobility in two-dimensional semiconductors: deformation potential and piezoelectric scattering in monolayer MoS2 from first principles. Phys. Rev. B **87**, 235312 (2013), http://link.aps.org/doi/10.1103/PhysRevB.87.235312
69. S. Su, J. Ho You, C. Lee, Electron transport at interface of LaAlO$_3$ and SrTiO$_3$ band insulators. J. Appl. Phys. **113**, 093709 (2013), http://scitation.aip.org/content/aip/journal/jap/113/9/10.1063/1.4794057
70. M.P. Vaughan, B.K. Ridley, Solution of the Boltzmann equation for calculating the Hall mobility in bulk GaNxAs1-x. Phys. Rev. B **72**, 075211 (2005), http://link.aps.org/doi/10.1103/PhysRevB.72.075211
71. J. Zhou, X. Li, G. Chen, R. Yang, Semiclassical model for thermoelectric transport in nanocomposites. Phys. Rev. B **82**, 115308 (2010), http://link.aps.org/doi/10.1103/PhysRevB.82.115308
72. T. Okuda, K. Nakanishi, S. Miyasaka, Y. Tokura, Thermoelectric response of metallic perovskites. Phys. Rev. B **63**, 113104 (2001), http://link.aps.org/doi/10.1103/PhysRevB.63.113104
73. S. Ohta, T. Nomura, H. Ohta, K. Koumoto, High-temperature carrier transport and thermoelectric properties of heavily La- or Nb-doped SrTiO$_3$ single crystals. J. of Appl. Phys. **97**, 034106 (2005), http://scitation.aip.org/content/aip/journal/jap/97/3/10.1063/1.1847723
74. M. Ben Shalom, A. Ron, A. Palevski, Y. Dagan, Shubnikov-de Haas oscillations in SrTiO$_3$/LaAlO$_3$ interface. Phys. Rev. Lett. **105**, 206401 (2010), http://link.aps.org/doi/10.1103/PhysRevLett.105.206401
75. A. Jost, V.K. Guduru, S. Wiedmann, J.C. Maan, U. Zeitler, S. Wenderich, A. Brinkman, H. Hilgenkamp, Transport and thermoelectric properties of the LaAlO$_3$/SrTiO$_3$ interface. Phys. Rev. B **91**, 045304 (2015), http://link.aps.org/doi/10.1103/PhysRevB.91.045304
76. S. Lerer, M. Ben Shalom, G. Deutscher, Y. Dagan, Low-temperature dependence of the thermomagnetic transport properties of the SrTiO$_3$/LaAlO$_3$ interface. Phys. Rev. B **84**, 075423 (2011), http://link.aps.org/doi/10.1103/PhysRevB.84.075423
77. H.J. Harsan Ma, Z. Huang, W.M. Lu, A. Annadi, S.W. Zeng, L.M. Wong, S.J. Wang, T. Venkatesan, Tunable bilayer two-dimensional electron gas in LaAlO$_3$/SrTiO$_3$ superlattices. Appl. Phys. Lett. **105**, 011603 (2014), http://scitation.aip.org/content/aip/journal/apl/105/1/10.1063/1.4887235
78. T.C. Harman, P.J. Taylor, M.P. Walsh, B.E. LaForge, Quantum dot superlattice thermoelectric materials and devices. Science **297**, 2229 (2002), http://www.sciencemag.org/cgi/doi/10.1126/science.1072886
79. R. Venkatasubramanian, E. Siivola, T. Colpitts, B. O'Quinn, Thin-film thermoelectric devices with high room-temperature figures of merit. Nature **413**, 597 (2001), http://www.nature.com/doifinder/10.1038/35098012
80. K. Biswas, J. He, I.D. Blum, C.-I. Wu, T.P. Hogan, D.N. Seidman, V.P. Dravid, M.G. Kanatzidis, High-performance bulk thermoelectrics with all-scale hierarchical architectures. Nature **489**, 414 (2012), http://www.nature.com/doifinder/10.1038/nature11439
81. H. Ohta, S. Kim, Y. Mune, T. Mizoguchi, K. Nomura, S. Ohta, T. Nomura, Y. Nakanishi, Y. Ikuhara, M. Hirano, H. Hosono, K. Koumoto, Giant thermoelectric Seebeck coefficient of a two-dimensional electron gas in SrTiO$_3$. Nat. Mater. **6**, 129–134 (2007), http://www.nature.com/doifinder/10.1038/nmat1821
82. H. Ohta, Y. Masuoka, R. Asahi, T. Kato, Y. Ikuhara, K. Nomura, H. Hosono, Field-modulated thermopower in SrTiO$_3$-based field-effect transistors with amorphous 12CaO 7Al$_2$O$_3$ glass gate insulator. Appl. Phys. Lett. **95**, 113505 (2009), http://scitation.aip.org/content/aip/journal/apl/95/11/10.1063/1.3231873
83. S. Shimizu, S. Ono, T. Hatano, Y. Iwasa, Y. Tokura, Enhanced cryogenic thermopower in SrTiO$_3$ by ionic gating. Phys. Rev. B **92**, 165304 (2015), http://link.aps.org/doi/10.1103/PhysRevB.92.165304

84. I. Pallecchi, M. Codda, E. Galleani d'Agliano, D. Marré, A.D. Caviglia, N. Reyren, S. Gariglio, J.-M. Triscone, Seebeck effect in the conducting LaAlO$_3$/SrTiO$_3$ interface. Phys. Rev. B **81**, 085414 (2010), http://link.aps.org/doi/10.1103/PhysRevB.81.085414
85. A. Rastogi, S. Tiwari, J.J. Pulikkotil, Z. Hossain, D. Kumar, R.C. Budhani, Doped LaAlO$_3$-SrTiO$_3$ interface: electrical transport and characterization of the interface potential. Europhys. Lett. **106**, 57002 (2014), http://stacks.iop.org/0295-5075/106/i=5/a=57002?key=crossref.25dd2df3a410e21729789524c7dc9874
86. M. Basletic, J.-L. Maurice, C. Carrtro, G. Herranz, O. Copie, M. Bibes, E. Jacquet, K. Bouzehouane, S. Fusil, A. Barthlmy, Mapping the spatial distribution of charge carriers in LaAlO$_3$/SrTiO$_3$ heterostructures. Nat. Mater. **7**, 621 (2008), http://www.nature.com/doifinder/10.1038/nmat2223
87. O. Copie, V. Garcia, C. Bdefeld, C. Carrétéro, M. Bibes, G. Herranz, E. Jacquet, J.-L. Maurice, B. Vinter, S. Fusil, K. Bouzehouane, H. Jaffrs, Barthlmy, Towards two-dimensional metallic behavior at LaAlO$_3$/SrTiO$_3$ interfaces. Phys. Rev. Lett. **102**, 216804 (2009), http://link.aps.org/doi/10.1103/PhysRevLett.102.216804
88. I. Pallecchi, F. Telesio, D. Marré, D. Li, S. Gariglio, J.-M. Triscone, A. Filippetti, Large phonon-drag enhancement induced by narrow quantum confinement at the LaAlO$_3$/SrTiO$_3$ interface. Phys. Rev. B **93**, 195309 (2016), http://link.aps.org/doi/10.1103/PhysRevB.93.195309
89. M. Bailyn, Phonon-drag part of the thermoelectric power in metals. Phys. Rev. **157**, 480–485 (1967), https://doi.org/10.1103/PhysRev.157.480
90. D.G. Cantrell, P.N. Butcher, A calculation of the phonon-drag contribution to the thermopower of quasi-2D electrons coupled to 3D phonons. I. General theory. J. Phys. C: Solid State Phys. **28**, 1087 (1985), http://stacks.iop.org/0022-3719/20/i=13/a=014?key=crossref.8bab983124e4b525af037ec6f81b5eb2
91. D.G. Cantrell, P.N. Butcher, A calculation of the phonon-drag contribution to the thermopower of quasi-2D electrons coupled to 3D phonons: II. Applications. J. Phys. C: Solid State Phys. **28**, 1993–2003 (1987), http://stacks.iop.org/0022-3719/20/i=13/a=014?key=crossref.8bab983124e4b525af037
92. Y. Yamada, H.K. Sato, Y. Hikita, H.Y. Hwang, Y. Kanemitsu, Measurement of the femtosecond optical absorption of LaAlO$_3$/SrTiO$_3$ heterostructures: evidence for an extremely slow electron relaxation at the interface. Phys. Rev. Lett. **111**, 047403 (2013), http://link.aps.org/doi/10.1103/PhysRevLett.111.047403
93. W. Liu, S. Gariglio, A. Fte, D. Li, M. Boselli, D. Stornaiuolo, J.-M. Triscone, Magneto-transport study of top- and back-gated LaAlO$_3$/SrTiO$_3$ heterostructures. APL Mater. **3**, 062805 (2015), http://scitation.aip.org/content/aip/journal/aplmater/3/6/10.1063/1.4921068
94. Y. Mune, H. Ohta, K. Koumoto, T. Mizoguchi, Y. Ikuhara, Enhanced Seebeck coefficient of quantum-confined electrons in SrTiO$_3$/SrTi0.8Nb0.2O3 superlattices. Appl. Phys. Lett. **91**, 192105 (2007), http://scitation.aip.org/content/aip/journal/apl/91/19/10.1063/1.2809364
95. B.K. Ridley, Polar-optical-phonon and electron-electron scattering in large-bandgap semiconductors. J. Phys.: Condens. Matter **10**, 6717–6726 (1998), http://stacks.iop.org/0953-8984/10/i=30/a=011?key=crossref.62f4e46fbffe8e1ec0e9dbdece463b64

Chapter 9
Dynamical Mean Field Theory for Oxide Heterostructures

O. Janson, Z. Zhong, G. Sangiovanni and K. Held

Abstract Transition metal oxide heterostructures often, but by far not always, exhibit strong electronic correlations. State-of-the-art calculations account for these by dynamical mean field theory (DMFT). We discuss the physical situations in which DMFT is needed, not needed, and where it is actually not sufficient. By means of an example, $SrVO_3/SrTiO_3$, we discuss step-by-step and figure-by-figure a density functional theory (DFT) + DMFT calculation. The second part reviews DFT + DMFT calculations for oxide heterostructure focusing on titanates, nickelates, vanadates, and ruthenates.

9.1 Introduction

The extraordinary progress to grow heterostructures of transition metal oxides, atomic-layer-by-atomic-layer, cherished hopes of discovering novel physical phenomena that are non-existing in conventional semiconductor heterostructures. Arguably the most remarkable difference and breeding ground for these hopes is the fact that transition metal oxides are often strongly correlated materials. Thence unconventional and unexpected states or colossal responses are imaginable. Hand-in-hand with the experimental analysis of prospective correlation phenomena we need a theoretical tool to address electronic correlations in oxide heterostructures.

O. Janson (✉) · K. Held
Institute for Solid State Physics, 1040 TU Wien, Vienna, Austria
e-mail: olegjanson@gmail.com

K. Held
e-mail: held@ifp.tuwien.ac.at

Z. Zhong
Max Planck Institute for Solid State Physics, 70569 Stuttgart, Germany
e-mail: zhong@fkf.mpg.de

G. Sangiovanni
Institut Für Theoretische Physik Und Astrophysik, Universität Würzburg,
97074, Würzburg, Am Hubland, Germany
e-mail: sangiovanni@physik.uni-wuerzburg.de

© Springer International Publishing AG, part of Springer Nature 2018
C. Cancellieri and V. Strocov (eds.), *Spectroscopy of Complex Oxide Interfaces*, Springer Series in Materials Science 266,
https://doi.org/10.1007/978-3-319-74989-1_9

Obviously this requires going beyond density functional theory (DFT), see Chap. 8. It is instead the realm of dynamical mean field theory (DMFT) [1–3] and its merger with DFT for real materials calculations [4, 5].

Before turning to this method, let us here in the Introduction address the questions: For which oxide heterostructures are electronic correlations important so that DMFT is needed? Actually, for many oxide heterostructures studied experimentally electronic correlations are not that strong. In particular the LaAlO$_3$/SrTiO$_3$ prototype [6], which has been at the focus of the experimental efforts at the dawn of the research field, is not strongly correlated. The reason for this is that SrTiO$_3$ is a band insulator with empty Ti-d orbitals. Through the polar catastrophe or oxygen defects, these Ti-bands are doped, but only slightly. With only a few charge carriers in the Ti-d orbitals we are far away from an integer filling of the d orbitals, and the d electrons can quite freely move around the interface without often seeing a Ti site already occupied with a d-electron and hence prone to a strong Coulomb interaction. Therefore not surprisingly DFT or even a simple tight binding modeling [7] are sufficient to reproduce or predict the angular resolved photoemission spectroscopy (PES, ARPES) spectra [8–10]. Let us note that there might be localization of (some of) the d-electrons at oxygen defects [11–14], cf. Chap. 5. For this localization electronic correlations play a role.

In general, electronic correlations are strong whenever we are at or close to an integer filling of the d-orbitals in some of the layers or at some of the sites. In this situation the Coulomb repulsion U is strong and hence important. It suppresses the mobility of the charge carriers, leads to a strongly correlated metal with strong quasiparticle renormalization or even to a Mott insulating state [15]. Correlations may be stronger [16, 17] or weaker [18] than in the corresponding bulk state. This kind of physics is included in DMFT and described by its local but dynamic correlations. Another situation where electronic correlations are important is a Hund's metal [19, 20] where several d electrons form a local magnetic moment; the Hund's exchange J, and not the Coulomb repulsion U plays the decisive role. Here, the effect of electronic correlations is less pronounced in the one-particle spectrum, but is strongly reflected in two-particle correlation functions such as e.g. the magnetic susceptibility.

Electronic correlations can give rise to magnetic and/or orbital ordering. Indeed magnetism and possible spintronic applications is one of the prospective advantages of oxide heterostructures. Such an ordering can already be described by the simpler DFT + U [21] which is a Hartree-Fock treatment and hence does not include genuine correlations. Indeed, a fully polarized ferromagnet is a single Slater determinant and such a ground state is perfectly accounted for in DFT + U. What DMFT additionally covers is the correlated paramagnetic state which competes with the magnetic state. This competition is most relevant for the question whether there is ordering or not. Hence DMFT is superior to DFT + U in its predictive power regarding ordering and also allows us to calculate the critical temperature without further approximations or without an adjustment of U.

For the PES also the excited states are important even if the ground state is a single Slater determinant. In this respect, DMFT e.g. describes the extra spin-polaron peaks [22] in the spectrum of an antiferromagnet which is beyond the Hartree-Fock physics of DFT + U. A symmetry broken DFT + U calculation is often employed for mimicking a localized state also in the paramagnetic phase. An example are the localized states at oxygen vacancies mentioned above which can be described in both, DMFT [11] and DFT + U [14, 23, 24]. For the PES, DFT + U is at least a valid first approximation. Some aspects such as the spin-polaron peaks are missing, but a gap in the spectrum and a localized state can be described this way.

An important aspect of magnetism is the screening of the magnetic moment. Even though we might have a magnetic moment on the femtosecond time scale, this moment and its direction might fluctuate in time [25]. It is screened. Obviously this is an effect of dynamic correlations included in DMFT, but neither in DFT nor DFT + U. This suppression of the long-time magnetic moment is important for magnetism [20, 26]. Indeed it is one physical reason why there is no magnetism even if it is predicted by DFT which usually underestimates correlations and tendencies towards magnetic ordering in transition metal oxides. Note that DFT + U, on the other hand, grossly overestimates tendencies of magnetic ordering because its only way to avoid the interaction U is ordering. The screening is a further reason why DMFT is more reliable regarding predictions of magnetic ordering. Let us add that experimentally the short-time local moment is discernible in fast, e.g., X-ray absorption, experiments, whereas no moment will be seen in experiments on a longer time scale, e.g., when measuring the magnetic susceptibility.

DMFT includes local dynamic (quantum) correlations, but neglects spatial correlations. Such non-local correlations can be more important at lower temperatures and for lower dimensional systems. Thence they may be also important for oxide heterostructures which are intrinsically two-dimensional. Non-local correlations give rise to additional physical phenomena beyond the realm of DMFT; and one should always be aware of the limitations of the method employed. Physical phenomena that might be relevant for oxide heterostructures and rely on such non-local correlations are: excitons and further vertex corrections to the conductivity such as weak localization, spin fluctuations that might suppress the DMFT-calculated critical temperature for magnetic ordering, and unconventional superconductivity. Often such phenomena can be understood in terms of simple ladder diagrams in orders of U, a treatment which is however not sufficient if correlations are truly strong. Cluster [27] and diagrammatic extensions [28–32] of DMFT which include all the local DMFT correlations but also non-local correlations beyond are a promising way to include such effects. The latter diagrammatic extensions describe a similar physics understood by the aforementioned ladder diagrams at weak coupling, but now for a strongly correlated system including all the DMFT correlations, e.g., the quasiparticle renormalization. These non-local approaches might, in the future, help us understand non-local correlations, but for the time being DMFT is state-of-the-art for correlations in oxide heterostructures and it will remain the method of choice whenever local correlations play the dominant role.

In this chapter, we review the DFT + DMFT approach and its application to oxide heterostructures. Section 9.2 is devoted to methodological aspects. After explaining the advantages of a DFT + DMFT treatment (Sect. 9.2.1), we guide the reader through the main steps of a DFT + DMFT calculation for a typical correlated heterostructure—a bilayer of $SrVO_3$ on a $SrTiO_3$ substrate (Sect. 9.2.2). We also briefly discuss how DFT + DMFT results can be compared with the experimental spectra (Sect. 9.2.3).

Section 9.3 reviews the results obtained so far by DFT + DMFT for oxide heterostructures, focusing on titanates, nickelates, vanadates and ruthenates. We start in Sect. 9.3.1 with titanates, for which electronic reconstruction can lead to a metallic interface but also oxygen vacancies are relevant as a competing mechanism and can give rise to localized states. In Sect. 9.3.2 we turn to nickelates which were the first oxide heterostructure studied in DFT + DMFT. Here, heterostructuring might give rise to a cuprate-like Fermi surface or topological states depending on the direction of stacking. Results for vanadates which hint at possible applications of oxide heterostructures as solar cells or as Mott transistors follow in Sect. 9.3.3. Section 9.3.4 is devoted to ruthenates which are arguably most promising for ferromagnetism and spintronic applications. Finally Sect. 9.4 summarizes the chapter and provides a brief outlook.

9.2 Steps of a DFT + DMFT Calculation Illustrated by SVO/STO Heterostructures

To illustrate the DFT + DMFT method, we select a numerically tractable, yet instructive correlated compound—the cubic perovskite $SrVO_3$ (Fig. 9.1). Being a rare example of a metallic V^{4+} compound, it shows distinct fingerprints of a correlated metal: the pronounced lower Hubbard band observed in PES [33], the quasiparticle peak seen in ARPES [34], and the upper Hubbard band in X-ray absorption [35]. From the computational viewpoint, the high crystal symmetry and the d^1 electronic configuration render $SrVO_3$ a convenient material for testing new numerical techniques.

The strength of electronic correlations can be drastically affected by a dimensional reduction, as it is indeed the case for ultrathin $SrVO_3$ layers grown on $SrTiO_3$. Here, the reduction of the $SrVO_3$ film thickness down to two monolayers leads to a metal-insulator transition [36] and the formation of quantum well states with an anomalous effective mass [10]. In this chapter, we will use such a 2 SVO / 4 STO heterostructure, consisting of a $SrVO_3$ bilayer on four $SrTiO_3$ substrate layers, as a model system to guide the reader through the main steps of a DFT + DMFT calculation.

9.2.1 Motivation for DFT + DMFT: The Electronic Structure of Bulk $SrVO_3$

Before introducing the method, we briefly explain the advantages of DFT + DMFT over conventional DFT techniques and DFT + U. To this end, we consider bulk $SrVO_3$. V atoms in its crystal structure are located in $1b$ Wyckoff positions with the point group symmetry $m\bar{3}m$. For d electrons ($l = 2$), the irreducible representations are E_g (twofold degenerate) and T_{2g} (threefold degenerate). The electrostatic repulsion

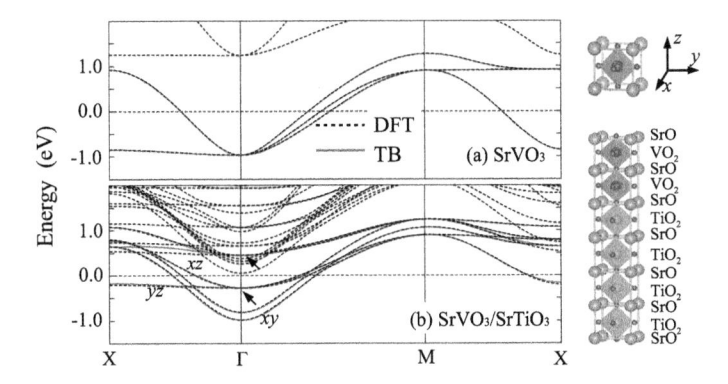

Fig. 9.1 Left: DFT band structure (black dashed lines) and the V t_{2g} bands (red solid lines) calculated for bulk SrVO$_3$. Right top: SrVO$_3$ perovskite cell with V (blue), Sr (green) and O (red) atoms are shown on the right. Right bottom: SrVO$_3$/SrTiO$_3$ heterostructure containing two SrVO$_3$ layers on a substrate of four SrTiO$_3$ and 10 Å of vacuum. The spatial confinement along z leads to quantized energy levels indicated by arrows for the vanadium yz/xz orbitals at Γ in the left panel

between the V d and O p electrons (the crystal field) pushes the e_g orbitals higher in energy, and the single d electron of V is distributed over the three degenerate t_{2g} orbitals.

The DFT band structure of bulk SrVO$_3$ is shown in Fig. 9.1a. At the Γ point, the three t_{2g} bands are degenerate in accord with the space group representation. For an arbitrary k-point, e.g. on the Γ-X (π/a, 0, 0) path, this degeneracy is partially lifted, and one of the bands (corresponding to the yz orbital along Γ-X) shows a minute dispersion ($\sim$0.12 eV), whereas the two degenerate bands (xy and xz orbitals) have a sizable dispersion of $\sim$1.9 eV since the orbital lobes are extended in the x direction. If we integrate the DFT bands over the Brillouin zone to obtain the density of states (DOS), it becomes clear that DFT fails to describe the experimental spectral features: the DFT DOS in Fig. 9.2a lacks the Hubbard bands, whereas the width of the quasiparticle peak is considerably overestimated (Fig. 9.2d).

The simplest DFT-based scheme which accounts for on-site interactions in a static mean-field way is DFT + U. By construction, this method favors integer orbital occupations [37] and therefore is particularly suitable for orbitally-ordered insulators, but its applicability to correlated metals is at best limited [38]. Indeed, for SrVO$_3$ DFT + U not only fails to describe the band renormalization, but even yields a spurious magnetic ground state (Fig. 9.2b).

The idea of DFT + U is to add correlation effects to the DFT on the Hartree-Fock level. In this way, the DFT Hamiltonian is supplemented with a purely real and constant self-energy Σ for V d states. DMFT may be seen as a dynamical version of DFT + U, i.e. it accounts also for local scattering processes, and the self-energy Σ in DMFT becomes complex and frequency-dependent. This describes (temperature-dependent) scattering processes, which lead to a finite life time. Thus, instead of the DOS, the spectrum of an interacting system is described by the spectral function $A(k, \omega)$. The room-temperature DFT + DMFT spectral function for bulk SrVO$_3$ is shown Fig. 9.2c. By comparing it with the experimental spectra in Fig. 9.2d,

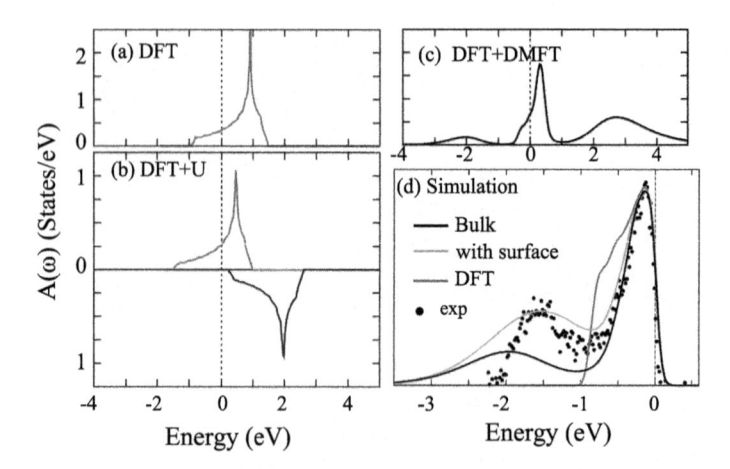

Fig. 9.2 V t_{2g} DOS in **a** DFT yielding a nonmagnetic ground state, **b** DFT + U ($U_d = 2$ eV) yielding a half-metallic magnetic state with half-filled majority states (red), and **c** DFT + DMFT yielding a nonmagnetic solution with lower Hubbard band, the upper Hubbard band, and the quasiparticle peak. The DFT + DMFT results show good agreement with the experiment [33] in (**d**), which can be further improved by including SrVO$_3$ surface spectral contributions (grey)

we find a considerable improvement over DFT: all spectral features are reproduced, and the width of the quasiparticle peak is in good agreement with the ARPES data.

9.2.2 Workflow of a DFT + DMFT Calculation

The main steps of a DFT + DMFT calculation are[1]: (i) construction of the unit cell, (ii) a DFT calculation, (iii) construction of the low-energy Hamiltonian, (iv) mapping the lattice Hamiltonian onto a set of single-site impurity problems (DMFT), (v) a numerical solution of the resulting single-site models, (vi) adding the double-counting correction, and (vii) a necessary postprocessing to compute the observables. The way these basic blocks are combined into a computational scheme depends on the implementation and the convergence criterion. In the standard DFT + DMFT scheme, the step (ii) is repeated to converge the electronic density on the DFT level and the steps (v) and (vi) are iterated until the self-energy is converged [4]. In charge-self-consistent calculations, the sequence of steps (ii)–(vi) is repeated until charge redistributions become small [39–41]. Most computationally demanding schemes involve structural relaxations: the charge redistributions yielded by DMFT are used to calculate forces and provide a new structural input, going back to the step (i) [42]. Such a calculation terminates once the changes in the crystal structure become small enough.

---

[1] For further details and the theoretical background we refer the reader to [4].

Next, we illustrate all steps in more detail using as an example two SrVO$_3$ layers on top of a SrTiO$_3$ substrate (modeled by four layers), following [17]. The starting point of a DFT + DMFT calculation is the construction of a unit cell. For (001) heterostructures, it is convenient to think of the SrVO$_3$ perovskite structure as a periodic alternation of SrO and VO$_2$ monolayers. By stacking such monolayers on top of each other, we construct a unit cell with two VO$_2$ layers, four TiO$_2$ layers, and seven SrO layers. Both terminal layers are SrO monolayers, which ensures that transition metal atoms (V and Ti) have an octahedral coordination. Finally, we make the system effectively two-dimensional by adding a sufficiently thick (10 Å) vacuum layer along the z direction, leading to an elongated unit cell shown in the right panel of Fig. 9.1b. The lateral lattice constant a of this tetragonal cell is fixed to that of the SrTiO$_3$ substrate ($a = 3.92$ Å), while all the internal coordinates are optimized in DFT. This latter step is crucial, because surface and strain effects can have a big impact on the crystal structure. For instance, a DFT relaxation of 2 SVO/4 STO yields an inward drift of the surface Sr atom by $\sim$0.25Å.

The next step is a DFT calculation. Here, we consider the simplest DFT + DMFT scheme: DFT convergence is reached before going to the next step. At present, a plethora of very accurate DFT codes exists [43]; here, we choose the all-electron full potential augmented plane-wave method implemented in Wien2k [44]. It is important to keep in mind that DFT + DMFT results can in some cases sensibly depend on the chosen DFT functional. Most DFT + DMFT studies employ either the local density approximation (LDA) [45, 46] or the generalized gradient approximation (GGA) [47]; here we will use the latter. Finally, DFT calculations are performed on a k-mesh, and heterostructures have the advantage of the reduced dimensionality. DFT calculations presented here are performed on a $10 \times 10 \times 1$ k-mesh.

The GGA band structure of the 2 SVO/4 STO heterostructure is shown in Fig. 9.1b. The thicket of bands above the Fermi energy are mostly unoccupied Ti t_{2g} states that will be neglected later in the low-energy model [48]. Moreover, the unit cell has now two inequivalent VO$_2$ monolayers, which doubles the number of V bands. But the main difference is that the electronic structure becomes two-dimensional: The insulating SrTiO$_3$ substrate on one side and the vacuum on the other side confine the V d electrons to move in the plane of the bilayer. The partially broken translational symmetry reduces the V point group to $4mm$, and the t_{2g} bands at Γ split into a_{1g} (xy) and e'_g (yz and xz) manifolds. Moreover, the dissimilar potential felt by V atoms in the surface and subsurface VO$_2$ layers lifts the degeneracy between the V sites.

DFT results can be used to construct a low-energy model and parameterize its tight-binding part. To this end, the DFT Hamiltonian H^{DFT} is projected onto a set of localized orbitals. The choice of this correlated subspace depends on the nature of the compound as well as on the problem at hand. For instance, to describe the spectrum of a charge-transfer insulator, one needs to include ligand p states. But in the case of SrVO$_3$, a natural choice for the minimal model is to select the subspace of V t_{2g} orbitals. Since there are two inequivalent V atoms in the 2 SVO/4 STO heterostructure, the number of orbitals, and hence the dimension of the Hamiltonian matrix are doubled. Using this basis, we search for a tight-binding Hamiltonian H^{TB}, which reproduces the DFT band structure in the corresponding energy range.

Table 9.1 Leading transfer integrals $t_{ij}(\mathbf{R}_{ij})$ (in meV) for bulk SVO and a 2 SVO/4 STO heterostructure: onsite terms (first row) and nearest neighbor hopping (along x; second row). Due to the mutual orthogonality of t_{2g} orbitals, only neighboring terms with the same orbital character are nonzero

Transfer integral	bulk SrVO$_3$			2 SVO/4 STO heterostructure					
				Surface V			Subsurface V		
	xy	yz	xz	xy	yz	xz	xy	yz	xz
Onsite energies ($|\mathbf{R}_{ij}|=0$)	579	579	579	399	576	576	540	574	574
Nearest-neighbors along x ($\mathbf{R}_{ij}=\mathbf{a}$)	−259	−26	−259	−243	−37	−180	−241	−33	−262

We therefore construct V-centered maximally localized Wannier orbitals [49–51] using the Wannier90 code [52] and the wien2wannier [53] interface which has been integrated into Wien2k recently.

From the maximally localized Wannier functions, we obtain matrix elements of the 6×6 $H^{TB}(\mathbf{k})$ matrix that have the form $\sum_{ij} t_{ij} \exp(i\mathbf{R}_{ij} \cdot \mathbf{k})$, where t_{ij} is the transfer integral (hopping) between the orbitals i, and $\mathbf{R}_{ij}$ connects their centers. The leading t_{ij} terms are listed in Table 9.1. The main effect of the confinement is the drastic reduction of the onsite energy for the V xy orbital of the surface layer, ensuing from the surface reconstruction (the inward drift of surface Sr atoms).

$H^{TB}(\mathbf{k})$ describes transfer processes of the low-energy model. In the spirit of the Hubbard model, we supplement this Hamiltonian with an onsite interaction term. In the general case, the onsite interaction is described by the interaction vertex U_{lmno}, where l, m, n, and o and orbital indices. For practical purposes, simplified versions of this form are typically used [19], e.g. the reduction to the density-density interaction [$l = m$ (same spin), and $n = o$ (same spin)]. Here, we use the rotationally invariant form given by Kanamori ($l = m$ and $n = o$), which in addition to density-density interactions, accounts accounts for spin exchange and pair hopping processes. The Kanamori Hamiltonian has two parameters: the intra-orbital Coulomb repulsion U and the Hund's exchange J, while the inter-orbital Coulomb repulsion $U' = U - 2J$ follows from the symmetry. For V d orbitals, we adopt $U' = 3.55$ eV from constrained LDA calculations [33, 54] and $J = 0.75$ eV, which is a standard value for early 3d metal oxides.

The resulting Hamiltonian represents a multi-orbital Hubbard model on a two-dimensional lattice. DMFT provides an approximate solution of it, by performing a mapping onto a set of single-site Anderson impurity problems. The unit cells of heterostructures often consist of many (two in the case of 2 SVO/4 STO) correlated atoms, and therefore this mapping involves several steps. First, we denote with n and m the number of atoms in the cell and the number of orbitals per atom, respectively (for the SVO bilayer and V t_{2g} orbitals we have $n = 2$ and $m = 3$). Depending on the number and on the type of Wannier functions chosen, the m orbitals associated to each atom can be either considered all as correlated (e.g. the three t_{2g}-orbitals here) or can be further split in a subset of correlated ones and a subset of "ligands" (e.g. p-orbitals). In the latter case the ligands are formally associated to one of the n atoms

9 Dynamical Mean Field Theory for Oxide Heterostructures

in the cell, but this is only an arbitrary assignment. In the standard treatment, the interaction term affects only the d correlated orbitals and it is applied locally to each of the n atoms.

We first define the local Green's function G_{loc}:

$$G_{\text{loc}}(\omega) = \sum_{\mathbf{k}} [(\omega + \mu)\mathbf{I} - H(\mathbf{k}) - \Sigma(\omega) - H_{\text{DC}}]^{-1}, \quad (9.1)$$

where ω is the complex frequency, μ is the chemical potential, $\mathbf{I}$ is the unitary matrix and $\Sigma(\omega)$ the self energy matrix of dimension $mn \times mn$. The latter is usually set to zero in the first DMFT cycle, and H_{DC} is the double-counting correction, which will be discussed later.

We now consider the i-th atom and start to construct the corresponding impurity problem. The i-th $m \times m$ block along the diagonal of G_{loc} is denoted as $G_{\text{loc},i}$ ($0 < i \leq n$). The i-th impurity Weiss' field $\mathscr{G}_i$ is constructed by inverting the corresponding $G_{\text{loc},i}$:

$$\mathscr{G}_i(\omega) = \left[(G_{\text{loc},i})^{-1} + \Sigma_i(\omega)\right]^{-1}, \quad (9.2)$$

where $\Sigma_i(\omega)$ is now a $m \times m$ diagonal matrix containing the self-energy of that block. It is important to note that $\mathscr{G}_i(\omega)$ contains nonlocal contributions stemming from the off-block-diagonal elements of H^{TB}, that enter G_{loc} (and hence $G_{\text{loc},i}$) by matrix inversion in (9.1).

The impurity Green's function $\mathscr{G}_i(\omega)$ allows us to formulate the corresponding Anderson impurity problem, amenable to a numerical solution. Various techniques can be used [2, 4], yet presently the method of choice is the continuous time quantum Monte Carlo in the hybridization expansion (CT-HYB, see [55] for a review; an implementation for the Kanamori Hamiltonian is discussed in [56]). The main parameters of a CT-HYB calculation are the inverse temperature β ($\beta = 39.6\,\text{eV}^{-1}$ corresponds to room temperature), the interaction parameters and the number of Monte Carlo sweeps. After the CT-HYB calculations (n of them or less, in case some of the atoms are locally equivalent) we access the set of self-energies $\Sigma_i(\omega)$ ($0 < i \leq n$). This in turn allows for the construction of a new local Green's function G_{loc} for the whole heterostructure, as in (9.1). Let us stress again that in DMFT the self-energy is frequency-dependent, but does not depend on $\mathbf{k}$, because the impurity model is a local problem. With the new Green's function, we can start another DMFT cycle. This procedure is repeated until convergence.

The only term that we have not discussed so far is the double-counting correction H_{DC} (9.1). The underlying idea is to subtract those contributions to Σ_i that are already accounted for by the GGA. Unfortunately, GGA lacks a diagrammatic description, and such contributions can not be determined rigorously. Several forms of H_{DC} have been proposed, but no general solution to this conundrum exists, a problem also relevant for DFT + U. However, in our calculation with t_{2g}-orbitals only, the symmetry is close to cubic, and H_{DC} is equivalent to a trivial shift of the chemical potential μ [4].

Several remarks concerning the performance should be made. For typical (not too large) supercells, the computationally most time-consuming step is solving the DMFT impurity problems. Here, the crucial parameter is the number of orbitals (m), because for CT-HYB the computational time scales exponentially with m. In contrast, the number of sites (n) defines the number of impurity problems to solve, and the computational time scales only linearly with n. For larger supercells the matrix inversion in (9.1) which scales like $(mn)^3$ (or best $\sim (mn)^{2.373}$) becomes the bottleneck.

9.2.3 Comparison with the Experiment

The final step of a DFT + DMFT calculation is postprocessing. CT-QMC calculates Green's functions $G(i\omega)$ on the imaginary frequency (Matsubara) axis and an analytical continuation is required in order to compare with the experimental $A(\omega)$ spectra. Although there exists a one-to-one correspondence between $G(\omega)$ and $G(i\omega)$, a straightforward solution is practically impossible. The root cause of the problem are statistical errors in $G(i\omega)$ that give rise to huge differences in the analytically-continued $G(\omega)$. A standard solution to this problem is the maximum entropy method [57].

After an analytical continuation of the self-energy, the spectral function $A(\mathbf{k}, \omega)$ can be evaluated as

$$A(k, \omega) = -\frac{1}{\pi} \operatorname{Im} G(k, \omega) = -\frac{1}{\pi} \operatorname{Im}[(\omega + \mu) \mathbf{I}_m \otimes \mathbf{I}_n - H(k) - \Sigma(\omega) - H_{\mathrm{DC}}]^{-1}. \tag{9.3}$$

Orbital occupations can be obtained by integrating the respective diagonal elements of $A(\mathbf{k}, \omega)$ over the Brillouin zone in the negative frequency range up to the Fermi energy.

We are now in the position to compare the DFT + DMFT spectral functions with the DOS and in this way evaluate the effect of electronic correlations. The left panels of Fig. 9.3 show the layer- and orbital-resolved DOS for the 2 SVO/4 STO heterostructure, with the nonzero DOS at the Fermi level indicating that both layers are metallic. The xy DOS is similar to bulk SrVO$_3$, but for the yz/xz orbitals the confinement in the z direction leads to two peaks in the Fig. 9.3a, b and a narrowing of the DOS. The yz/xz DOS center of mass is shifted upwards to higher energies compare the onsite terms in Table 9.1. Figure 9.3 (right) shows the room-temperature DFT + DMFT spectral functions for 2 SVO/4 STO. The effect of correlations is dramatic [17]: In the surface layer, they trigger the orbital polarization which renders the xy orbital half-filled. At half-filling, the intraorbital Coulomb repulsion is particularly efficient, and it splits the spectrum into lower and upper Hubbard bands, stabilizing the Mott insulating state. The subsurface layer also becomes insulating, albeit with a weaker orbital polarization and a much narrower gap.

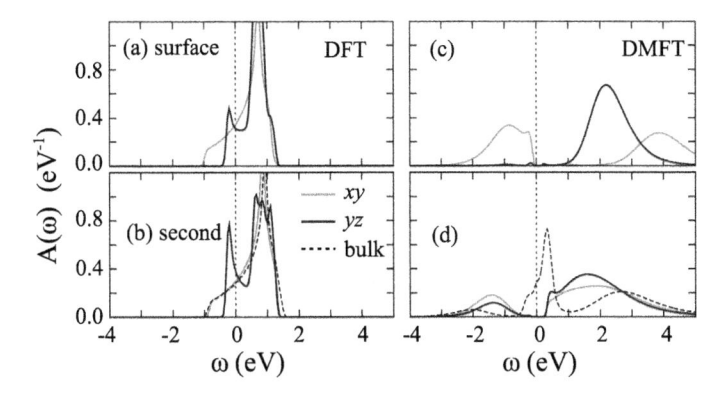

9 Dynamical Mean Field Theory for Oxide Heterostructures

Fig. 9.3 Layer- and orbital-resolved DOS (left) and the respective DFT + DMFT spectral functions (right) of the 2 SVO/4 STO heterostructure, compared to bulk SrVO$_3$ (dashed line) (adapted from [17])

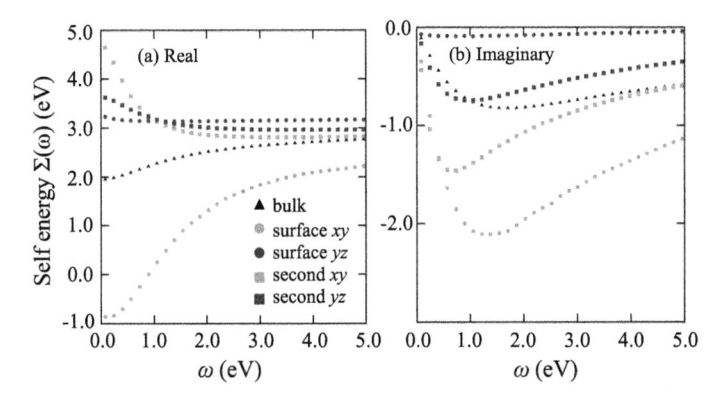

Fig. 9.4 a Real and b imaginary part of the layer- and orbital-resolved DMFT self-energy for 2 SVO/4 STO as a function of real frequency ω. Results for bulk SrVO$_3$ are shown to illustrate the effect of dimensional reduction

Figure 9.4 shows the self-energy $\Sigma(\omega)$ of the 2 SVO/4 STO heterostructure. Correlations in the xy orbital are particularly strong, as reflected in the sizable frequency dependence of Re Σ. At higher frequencies, the DMFT self-energy recovers the static Hartree shift, but it is the low energy part of Re Σ which deviates strongly for the different orbitals, enhances the DFT crystal field splitting and leads to insulating SVO layers, similar as in [58].

From the metallic-like behavior of Im Σ, we can conclude that the 2 SVO/4 STO heterostructure is on the verge of a Mott transition. Switching between metallic and insulating regimes can be achieved by strain, doping, applying external electric field, or capping the surface layer. DFT + DMFT is an excellent computational tool to explore these possibilities and evaluate the optimal way to tune the physics of this interesting system [17].

9.3 Applications

After the discovery of the emergent metallicity at the $LaTiO_3/SrTiO_3$ interfaces, one of the major challenges for theory was to explain the mechanism underlying the formation of the metallic state at the interface between a band insulator ($SrTiO_3$) and a Mott insulator ($LaTiO_3$). This dichotomy motivated Okamoto and Millis to build up the concept of "electronic reconstruction", extending earlier ideas for fullerenes [59] to oxides. The first model calculations [60, 61] along these lines included correlation effects on the Hartree-Fock level. A DMFT model calculation for the $LaTiO_3/SrTiO_3$ heterostructure followed [62], showing a "leakage" of electrons from the Mott- to the band-insulating side. This gives rise to a partially filled d-shell at the interface layer which facilitates conductivity. Even though the low-energy model comprised a single orbital only, these results strongly contributed to the understanding of the electronic reconstruction mechanism. Later realistic DFT + DMFT calculations for oxide heterostructures followed, starting with the work of Hansmann et al. for nickelates [63]. In the following four Sections we review DFT + DMFT calculations for oxide heterostructures, which focused hitherto on titanates, nickelates, vanadates, and ruthenates.

9.3.1 Titanates

Titanates are by far the most studied class of oxide heterostructures. At the same time, they are arguably also the best understood class, and the main ingredients of their emergent behaviors are known from detailed DFT + DMFT studies, pioneered by Lechermann et al. [11, 64–66]. In particular, this is the case for $LaTiO_3/SrTiO_3$ (LTO/STO) where DFT + DMFT is needed to properly describe the Mott insulator LTO. In the $LaAlO_3/SrTiO_3$ heterostructure on the other hand, which is experimentally most frequently studied, $LaAlO_3$ (LAO) is instead a simple band insulator. The metallic state induced by a slight doping due to oxygen vacancies or electronic reconstruction, is hence only weakly correlated. This situation in LAO/STO can be well described by DFT, except for states localized at the oxygen vacancies.

State-of-the-art DFT + DMFT(CT-HYB) calculations for the LTO/STO heterostructure started with Lechermann et al. [64] who reported the formation of a quasiparticle peak primarily of d_{xy}-orbital character, due to the lifted t_{2g}-degeneracy and in line with the experimental observations. A subsequent study focused on the magnetic properties and revealed that ferromagnetism is stabilized by the joint effect of electronic correlations and oxygen vacancies [65]. The role of the latter is a question of fundamental importance, and DFT + U (see e.g. [24]) and DFT + DMFT [11, 66, 67] studies are at the forefront of the research. DFT + DMFT indicates the formation of in-gap states in LAO/STO [65–67], in good agreement with PES measurements, cf. Chaps. 5 and 6. This is definitely one of the most promising directions of investigation, even though the presence of oxygen vacancies increases the computational effort tremendously. The strongly correlated nature of such defect states gives rise to a strong sensitivity of the electronic properties to small changes in the interaction parameters or local distortions ensuing from the structural relaxation.

Another important aspect are local distortions and rotations of the oxygen octahedra. Their role in stabilizing an insulating state in bulk d^1 titanates has been advocated by Pavarini et al. [68, 69]. Heterostructuring strongly affects these distortions and rotations because of strain and breaking translational invariance. Dymkowski and Ederer found in DFT + DFMT that compressive strain reduces the t_{2g}-splitting which is intimately related to the collective tilts and rotations, and eventually turns LaTiO$_3$ into a metal [70].

More recently a promising route towards new electronic behaviors, the so-called δ-doping, has been proposed, i.e., adding a single impurity layer into a layered superlattice. To this end, one of the LaO layers in a large LaTiO$_3$ supercell has been replaced by an SrO layer and studied using DFT + DMFT [71]. It was found that depending on the distance from the interface, titanate layers exhibit three distinct electronic states [71].

9.3.2 Nickelates

Bulk rare-earth nickelates RNiO$_3$ with trivalent Ni^{3+} exhibit interesting electronic, magnetic and transport properties [72–74]. One of the most remarkable phenomena is the metal-insulator transition observed in all RNiO$_3$ materials except for LaNiO$_3$ (LNO), which remains metallic down to the lowest temperatures [73]. The nature of this metal-insulator transition has been addressed by experiment [75–78] and theory [76, 79–81], but the discussion is not yet settled.

In the simplest ionic model, Ni^{3+} has a d^7 configuration in LNO. The octahedral crystal field splits the d levels into t_{2g} and e_g manifolds, with the low-lying t_{2g} states completely filled. Hence all electronic, orbital and spin degrees of freedom pertain to the e_g states. Close similarities between nickelates and cuprates were noticed in the early days of high-temperature superconductors. Indeed, the e_g shell of d^7 Ni with its single electron is seemingly a sibling of d^9 Cu having a single hole. Thus, a hypothetical nickelate with half-filled $x^2 - y^2$ and empty $3z^2 - r^2$ orbitals would have a cuprate-like Fermi surface and eventually become superconducting upon doping. But bulk Ni^{3+} systems do not exhibit such an orbital polarization [82].

Chaloupka and Khaliulin [83] suggested to resort instead to nickelate heterostructures, with the idea in mind that the z axis confinement leads to a reduced bandwidth of the $3z^2 - r^2$ orbital which hence gets Mott insulating, leaving behind a $x^2 - y^2$ Fermi surface. In realistic DFT + DMFT calculations, Hansmann et al. [63] found that it is instead the correlation enhanced crystal field splitting between $x^2 - y^2$ and $3z^2 - r^2$ orbitals which pushes the $3z^2 - r^2$ orbital higher in energy and gives rise to single-sheet Fermi surface of $x^2 - y^2$ character. However, strain or a PrScO$_3$ substrate is needed to achieve this, or unrealistically large U values [63]. This can also trigger a metal-insulator transition which is very different for the $x^2 - y^2$ and

$3z^2 - r^2$ orbital [84]: While Im $\Sigma_{x^2-y^2}(\omega)$ diverges for $\omega \to 0$ as in the single-band Hubbard model, Im $\Sigma_{3z^2-r^2}(\omega)$ retains a seemingly metallic-like behavior. The insulating state is instead induced by the change in Re $\Sigma_{3z^2-r^2}(\omega)$ relative to Re $\Sigma_{x^2-y^2}(\omega)$ at $\omega = 0$, i.e., by the aforementioned enhancement of the crystal field splitting.

Later, Han et al. [85] challenged this picture and vouched for an explicit inclusion of the oxygen ligand orbitals into the calculation. Their study found electronic correlations to decrease the orbital polarization disregarding the initial crystal-field splitting. The main difference between the two DMFT studies is the choice of the basis: Hansmann et al. employed a d-only calculation including the Ni e_g states which requires a minimal number of interaction parameters. In contrast, Han et al. [85] employed a d-p basis, where the O $2p$ orbitals are included explicitly. The choice of the d-p model is motivated by the high oxidation state of Ni^{3+}, which can be unstable towards the formation of a ligand hole (L) and the transfer of an electron to the Ni d shell. In the d-p calculation, which allows for such charge-transfer processes, the strong hybridization between Ni d and O p states and the Hund's exchange in the d-shell reduce the orbital polarization [85]. In principle, including additionally the O p states should be better as it allows for charge transfer processes. However, such calculations are very sensitive to the double counting or d-p interaction and a commonly accepted scheme still needs to be established.

An extensive comparative DMFT study of d and d-p models allowed Parragh et al. [86] to pinpoint the difference between the two models [86], see Fig. 9.5. Parragh et al. found that the deviations are ruled by the filling of the correlated d-shell. In the d-only model the filling is d^7 or one electron per site in the two e_g orbitals so that shifting the second $3z^2 - r^2$ orbital upwards and reducing its occupation is energetically favorable (Fig. 9.5a). For the d-p model instead an essentially d^8 (ligand hole) configuration brings the Hund's exchange into the game. This favors the two e_g electrons to form a spin $S = 1$, equally occupying the two different $3z^2 - r^2$ and $x^2 - y^2$ orbitals (Fig. 9.5b). This actually leads to a downward instead of an upward shift of the $3z^2 - r^2$ orbital [86]. This has dramatic consequences for the topology of the Fermi surfaces: while the d-only model features a single-sheet cuprate-like Fermi surface (Fig. 9.5c), a second sheet always appears in d-p model calculations (Fig. 9.5d). Later Peil et al. applied DFT + DMFT to study strain effects in nickelate superlattices [87]. As in [63] they found that strain can induce a sizable orbital polarization. However they point out that the Hund's exchange can reduce the orbital polarization for both the $d^8 L$ and d^7 configuration. The latter d^7 result revises common knowledge and is an indirect consequence of the intersite hopping. Noteworthy, their DMFT orbital polarization is smaller than in DFT.

Which of the two calculations, d-only or d-p, provides a more appropriate framework for the fermiology and spin-state of nickelate heterostructures? From the physical perspective, the d-p model is clearly superior to the d-only model: Owing to the larger energy window used for Wannier projections, the resulting Wannier functions are more localized and atomic-like. In addition, the occupation of the d-shell in d-p models is flexible, and hence such models can capture charge transfer. However, the price to pay is the drastic increase in the number of free parameters, including the on-site interactions on the O $2p$ orbitals as well as d-p interaction whose experimen-

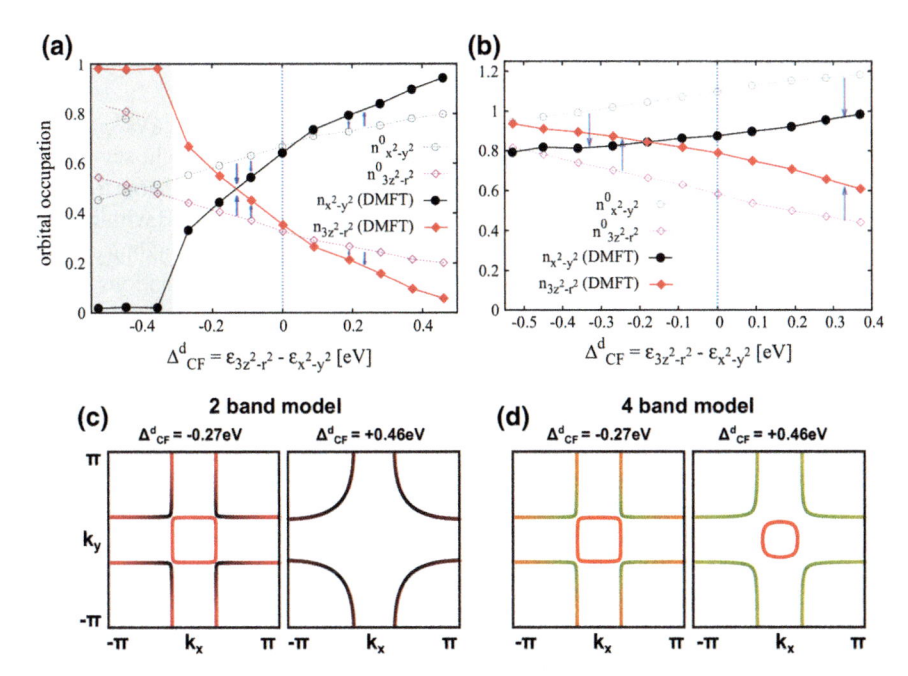

Fig. 9.5 a Orbital occupations versus crystal field splitting for a d-only (2 band) model of a LNO monolayer without interaction (open symbols) and within DMFT (filled symbols; $U = 5.5$ eV, $J = 0.75$ eV, $T = 1165$ K, $n_e = 1$ electron/site). The shaded region denotes the Mott-Hubbard insulating phase. **b** Same for a d-p (4 band) model ($U = 10$ eV, $n_e = 5$). **c-d** Shape and orbital character of the Fermi surface as calculated by DMFT for (**c**) the d-only and (**d**) the d-p model. The color coding is as follows: black for $x^2 - y^2$, red for $3z^2 - r^2$, yellow for O p. For the d-only model electronic correaltions enhance the orbital disproportionation (**a**) while for the d-p model it is reduced, preventing a single-sheet Fermi surface (**d**) as in (**c**) (adapted from [86])

tal estimation is very challenging. Or, alternatively, a still tentative double counting or an ad hoc adjustment of the O p level to experiment is necessary.

On the experimental forefront, the situation remains controversial. In LAO/LNO heterostructures, orbital polarizations (but only moderate ones) have been found. For 4 LNO/4 LAO heterostructures, soft X-ray reflectivity measurements reveal a small orbital polarization of 7(4)% for the outer (inner) LNO monolayers [88]. On the other hand, by choosing a different non-correlated spacer, such as e.g. GdScO$_3$, an orbital polarization as high as 25% can be achieved [89]. Recently grown three-component LTO/LNO/LAO heterostructures show even larger polarizations [90]. The origin of sizable orbital polarization in these heterostructures has been addressed in a very recent DFT + DMFT study [91]. In the layered bulk nickelate Eu$_{2x}$Sr$_x$NiO$_4$, Uchida et al. [92] found a single cuprate-like Fermi surface in ARPES which even evidences a pseudogap. Experimentally the smoking gun would be to determine the local magnetic moment which should be ~ 2 and ~ 1 μ_B for the $d^8 L$ and d^7 configuration, respectively.

9.3.3 Vanadates

In this section we review DFT + U and DFT + DMFT calculations of LaVO$_3$ (LVO) heterostructures. Please note that results for SrVO$_3$/SrTiO$_3$ which might serve as a Mott transistor [17] have already been discussed in Sect. 9.2.2; also the charge transfer in SrVO$_3$/SrMnO$_3$ has been analyzed [93]. Bulk LVO has been analyzed within DFT + U by Fang and Nagaosa [94] and within DFT + DMFT by De Raychaudhury, et al. [95]. The latter calculation revealed the existence of strong orbital fluctuations above the Néel temperature T_N, which coincides with the structural transition temperature. These quantum effects reduce quite rapidly the orbital polarization, in contrast to YVO$_3$, for instance, where a pronounced orbital polarization persists up to temperatures of the order of 1000 K, i.e. well above T_N. The magnetism in LVO is of C-type, i.e. ferromagnetically stacked planes with antiferromagnetic order within each plane ($q = \pi \pi 0$). This ordering is accompanied by an orbital pattern of an intermediate kind between G- and C-type, due to the competition between the Jahn-Teller of the oxygen octahedra and the GdFeO$_3$-type distortion [95]. Note that in SVO/LVO heterostructures also a ferromagnetic state is possible. The structural transition at 140 K is from a high-temperature orthorhombic to a low-temperature monoclinic phase.

Hotta et al. [96] were able to grow LVO epitaxially on STO and found a metallic n-type interface for thick enough LVO films, which they interpreted to originate from the polar discontinuity. Motivated by these experiment and also intrigued by the advantageous size of the gap of LVO ($\sim$1 eV) for photovoltaic applications, Assmann et al. studied LVO thin films with DFT + U [97] and DFT + DMFT [98].

Figure 9.6 (left panel) shows the scheme proposed in [97] for a solar cell with transition-metal oxide perovskites as absorbing materials. The main idea of the LaVO$_3$-based solar cells is that electrons and holes produced by the incoming photon can be efficiently separated and harvested due to the polarity of the heterostructure with its intrinsic potential gradient (indicated as an electric field E in the left panel of Fig. 9.6). The DFT + U results for 4 LVO layers on top of a STO substrate are shown in Fig. 9.6 (right panel). The position of the lower and upper LVO Hubbard band shifts from layer to layer, indicating the polar electric field which can separate photoelectrically induced electrons and holes. Above a critical thickness of four layers, the valence and conduction d-bands of vanadium touch the Fermi level, which leads to an electronic reconstruction and a metallic interface and surface layer. The latter might need protection from a capping layer to suppress disorder effects and achieve a sufficient electron mobility. In a solar cell, these metallic layers may be exploited to transport the electron and holes to the power consumer.

The optical absorption which indicates the efficiency of the photoelectric field in the LVO|STO heterostructure was calculated in [97]. The results compare quite favorably with the most modern thin-film solar cell materials. Particularly appealing is that oxide heterostructures can overcome the infamous Schockley-Queisser limit of 38% efficiencies in two ways: (i) As indicated in Fig. 9.6 (left panel) different transition metals M and M' may lead to different gaps (LaFeO$_3$ and LaVO$_3$ for instance were considered in [97]). Such a "band-grading" is very flexible in oxide

9 Dynamical Mean Field Theory for Oxide Heterostructures 231

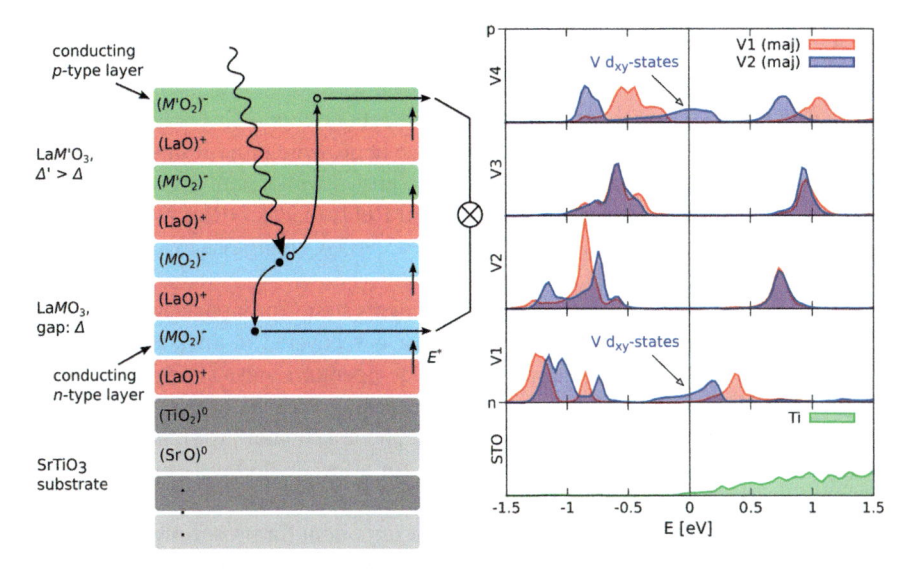

Fig. 9.6 Left: Proposal of solar cell made out of an oxide heterostructure with a polar discontinuity. The emerging electric field E separates electrons and holes that are generated through the photoelectric effect in the LaMO$_3$. Using different layers with e.g. M = V, M' = Fe allows for a gap grading. Right: DFT + U calculation for four layer of LVO on top of a STO substrate (only the topmost STO layer is shown). The layer-resolved DOS shows the polar field as a shift of the upper and lower Hubbard bands and an electronic reconstruction giving rise to metallic interface and surface layers (adapted from [97])

heterostructures and can overcome loosing the excess photon energy beyond the band gap to photons (heating up the solar cell instead of gaining electric energy). (ii) A genuine correlation effect, impact ionization, creates additional electron-hole pairs on the 10 fs time scale [99] if the photon energy is larger than twice the band gap.

Recently, Wang, et al. [100] reported the realization of an actual solar cell, where LaVO$_3$ serves as the light absorber. Further work on the quality of the heterostructure samples is needed to enhance the efficiency of the solar cell which still suffers from a poor mean free path and is only comparable to the dawn of Si solar cells. Also for the other proposed material, LaFeO$_3$ on SrTiO$_3$, Nakamura et al. [101] reported its photovoltaic applicability.

While it is unclear at present whether oxide heterostructures may become a viable alternative to Si solar cells, the proposal and experimental confirmation already served another purpose: It confirms the existence of a polar field. The latter has been controversially debated recently, in particular in view of the fact that oxygen vacancies counteract the polar field and also may induce a metallic interface layer, e.g. in LAO/STO. Demonstrating an actually working solar cell proves that a polar field survives in various oxide heterostructures.

9.3.4 Ruthenates

Bulk SrRuO$_3$ (SRO) is a rare example of a ferromagnetic conducting oxide [102], and there is a long standing experimental experience in growing films of this perovskite: thin single-domain films of high quality can be grown with a pure SrO termination, which is assisted by the volatility of RuO$_3$ and RuO$_4$ oxides [103]. A typical substrate is STO with its cubic lattice constant $a = 3.905$ Å giving rise to to a moderate compressive strain ($\sim$0.45 %) in SRO [102].

Resistivity measurements on ultrathin SRO layers show a pronounced dependence on the layer thickness: while the bulk behavior is recovered in thick slabs of $\gtrsim$15 SRO layers, fewer layers show a reduced Curie temperature and a strongly enhanced resistivity [104]. Drastic changes occur upon reaching the critical thickness of four monolayers, at which SRO turns into a non-magnetic insulator [105]. Experimental attempts to stabilize ferromagnetism in single (or a few) SRO layers by capping or applying compressive/tensile strain remain unsuccessful so far.

Since such ultrathin ferromagnetic films are important for prospective (e.g. spintronics) applications, understanding the nature of the metal-insulator transition and the absence of ferromagnetism is mandatory. Quite remarkably, neither the electronic ground state, nor an antiferromagnetism or otherwise non-ferromagnetic phase are captured by DFT: it yields a ferromagnetic state for all slabs thicker than one SRO monolayer and fails to reproduce the insulating state [106]. Similarly, DFT + U yields a metal even for sizable U values [106]. Later studies showed that an antiferromagnetic insulator can be stabilized in DFT + U for the spurious RuO$_2$ termination [107], which is however in contrast to the experiment [108]. Moreover, DFT + U yields antiferromagnetism for layers of up to eight SRO monolayers, i.e. substantially overestimates the experimental critical thickness.

These apparent shortcomings of DFT and DFT + U call for a more realistic treatment of electronic correlations within DFT + DMFT. The latter accounts for the behavior of bulk SrRuO$_3$ [109, 110], and although DMFT is a cruder approximation in 2D than in 3D, we can still expect it to capture the physics of thin SRO layers including its magnetism. The influence of the SRO layer thickness onto the electronic and magnetic ground states was studied very recently in [111]. All DFT calculations were done using a $\sqrt{2} \times \sqrt{2} \times 6$ supercell, which allows us to study SRO/STO compositions ranging from a SRO monolayer (1:5) to a four-layer slab (4:2) and is compatible with both ferromagnetic and antiferromagnetic in-plane order. The out-of-plane unit cell parameter as well as the internal atomic coordinates were optimized using DFT, while keeping the in-plane lattice constant equal to that of the STO substrate. The wien2wannier [53]-derived Hamiltonian was supplemented with a Coulomb repulsion $U = 3.5$ eV and Hund's exchange $J = 0.3$ eV as derived by the constrained random phase approximation (cRPA) approach. The resulting interacting Hamiltonians were treated in DMFT and solved using a CT-HYB solver.

For SRO monolayers and bilayers, DFT + DMFT readily yields an insulating antiferromagnetic state at room temperature (Fig. 9.7d). The microscopic origin of this state lies in the sizable orbital polarization: the xy orbital is occupied by two

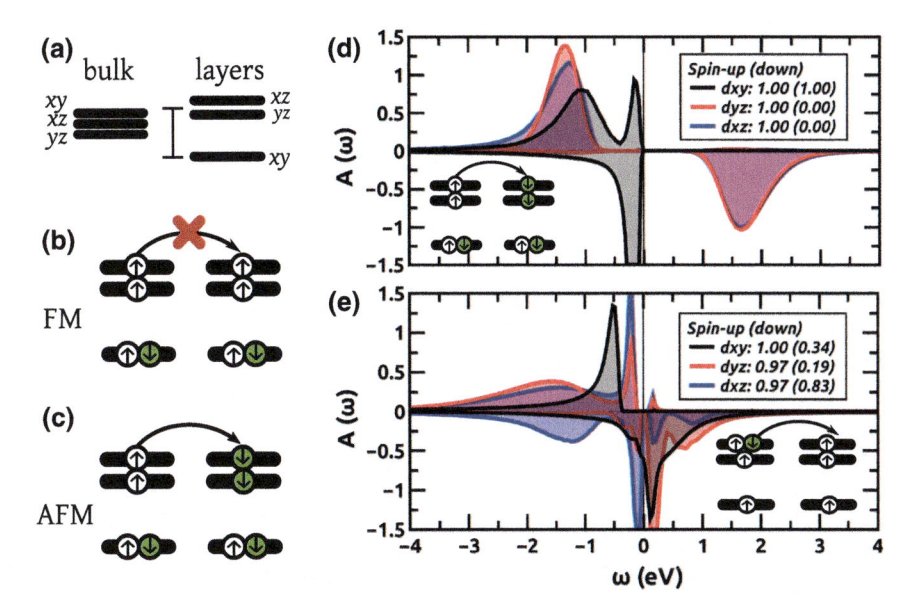

Fig. 9.7 Ramifications of the dimensional reduction in SrRuO$_3$ thin films. **a** The degeneracy of the three t_{2g} in the bulk is lifted for ultrathin layers. **b** For a ferromagnetic configuration there is hence no hopping, while for **c** an antiferromagnetic configuration virtual (superexchange) hoppings within the t_{2g} manifold are allowed. This explains why antiferromagnetism is energetically favorable in SRO thin films. **d–e** Orbital-resolved DFT + DMFT spectral function A as a function of real frequency ω for (**d**) one SRO monolayer on a STO substrate revealing an antiferromagnetic insulating state, and (**e**) an electron-doping with 4.3 electrons/Ru turning the SRO film into a half-metallic ferromagnet (adapted from [111])

electrons, while the xz and yz orbitals are half-filled, see Fig. 9.7a–c. In accord with the Hund's rule, the spins of the latter two orbitals align parallel, giving rise to a localized magnetic moment of $\sim 2\,\mu_B$. While in a ferromagnetic configuration the xz and yz electrons are essentially immobile (Fig. 9.7b), in the antiferromagnetic phase they can hop between the neighboring Ru sites gaining kinetic energy through superexchange processes (Fig. 9.7c). Hence DMFT does not only agree with experiment, but also explains the metal-insulator and ferromagnetic-antiferromagnetic transition upon reducing the thickness of SRO films.

Unlike DFT and DFT + U, DMFT allows us to study the temperature evolution of electronic and magnetic states. In this way, it was shown that SRO monolayers retain a sizable orbital polarization even in the paramagnetic state (~ 1000 K). Indeed, a strong tendency towards orbital polarization is visible already at the DFT level, marked by the difference between the on-site energies of the xy orbital and the degenerate xz and yz orbitals. This situation is typical for thin films confined along the z direction but apparently different from bulk SRO, where the three t_{2g} orbitals are degenerate (Fig. 9.7a). This effect is somewhat countered by the elongation of the topmost RuO$_6$ octahedra, which in the crystal-field picture lowers the energy of

xz and yz orbitals. However, this effect is weak and the xy orbital always lies lower in energy [111].

Can DFT + DMFT calculations give a clue to the quest of growing ultrathin *ferromagnetic* SRO films? A standard way to manipulate physical properties of thin films is the strain induced by the lattice mismatch between the film and the substrate. Unfortunately, DMFT calculations reveal that for SRO neither compressive nor tensile strain stabilizes an FM state, at least for realistic values of the on-site Coulomb repulsion [111]. Another possibility to bring on-site energies closer together is capping SRO with STO layers, which should to a certain extent restore the bulk behavior. Although the level splitting is indeed reduced in DFT, DMFT still yields an antiferromagnetic state [111], albeit with now somewhat more balanced orbital occupations. Thus, the antiferromagnetism is a robust feature of SRO layers: neither strain nor capping can stabilize the desired ferromagnetic state. This explains why experimental attempts to fabricate ferromagnetic SRO layers were hitherto unsuccessful.

There is however an alternative route to restore ferromagnetism: doping. To verify this scenario, additional DFT calculations were performed within the virtual crystal approximation, mimicking the Ru d-level occupation of 3.7 electrons. For this filling, DMFT indeed yields a half-metallic ferromagnet (Fig. 9.7e). Quite remarkably, for both, electron and hole doping, the magnetic moments and the magnetic ordering temperatures is higher than in bulk SRO [111].

The good agreement between DFT + DMFT and experiment for the undoped phase leaves little doubt that DMFT provides a realistic description of thin SRO layers. Still, there are many open issues that deserve further studies with computational methods, including DMFT. The first and arguably most relevant is the effect of oxygen vacancies. Real materials are never perfect, and particularly thin films are often subject to surface contaminations, growth defects and vacancies. Oxygen vacancies represent a severe experimental challenge: their concentration is difficult to estimate, while the presence of a single vacancy can dramatically alter the properties of the surrounding atoms. Structural relaxations invoked by vacancies can be studied by DFT, yet the alteration of the Ru $4d$-level filling and its ramifications pertain to correlated electrons and require an appropriate many-body treatment. DFT + DMFT is a state-of-the-art tool for such problems.

9.4 Conclusion and Outlook

DMFT is state-of-the-art for dealing with electronic correlations which are of prime importance for many oxide heterostructures; and its merger with DFT [4, 5] allows us to do realistic materials calculations. This way quantum correlations in time are included, corresponding to a local self energy which might differ from site to site. Physically, DFT + DMFT reliably describes among others, quasiparticle-renormalizations, metal-insulator transitions, correlation enhanced crystal field splittings, charge transfers, and magnetism. On the other hand, non-local correlations are neglected. These are important in the vicinity of phase transitions or to describe the physics of excitons and weak localization corrections to the conductivity. In the future, cluster [27] or diagrammatic extensions [30, 32] of DMFT, which are computationally more demanding, and hence hitherto focused on model Hamiltonians, will also be applied to oxide heterostructures.

In the first part, Sect. 9.2, we illustrated all steps of a DFT + DMFT calculation for a showcase heterostructure: two layers of SVO on an STO substrate. This starts with a DFT calculation, here using Wien2k [44]. Subsequently a projection onto Wannier orbitals and a corresponding low energy Hamiltonian is performed with the help of wien2wannier [53] (integrated in the most recent Wien2k version) and Wannier90 [52]. This may result in a d-only Hamiltonian but also additional e.g. oxygen p orbitals can be included. The Wannier projection also defines what is (site-)local in DMFT. For the subsequent DMFT calculation we used w2dynamics [56], which solves the DMFT impurity problem by continuous-time quantum Monte Carlo simulations [55] in the hybridization expansion. This yields the DFT + DMFT self energy and spectrum. For calculating (optical) conductivities and thermal responses, postprocessing with woptics [112] is needed; a DFT + DMFT charge self consistency is possible [113].

In the second part, Sect. 9.3, we reviewed previous DFT + DMFT calculations for oxide heterostructures. Not only experimentally but also regarding DMFT calculations arguably most well studied are titanates, in particular, LAO/STO and LTO/STO. Here the former is only weakly correlated whereas electronic correlations are strong for the latter as LTO is a Mott insulator. DMFT describes the formation of a two dimensional electron gas at the interface due to an electronic reconstruction, but also the localization of electrons at oxygen vacancies if these are included in the supercell.

Nickelate-based heterostructures may be used to turn the nickel Fermi surface into a cuprate-like one [84]. This might however be prevented by charge transfer processes: i.e. oxygen ligand holes may lead to a d^8L instead of a d^7 configuration [85]. The physics is completely different [86], spin-1 instead of spin-1/2, and the theoretical uncertainty regarding the oxygen-p position is too large to reliably predict which scenario prevails. The smoking gun experiment in this respect is measuring the short-time magnetic moment (spin) e.g. by neutron scattering or X-ray absorption spectroscopy.

Turning to vanadates, STO/LVO is a promising candidate for efficient solar cells because of the size of the LVO Mott-Hubbard gap and the polar electric field of the heterostructure which has been suggested [97] and experimentally verified [100] to separate photovoltaically generated electron-hole pairs. SVO on the other hand is a strongly correlated metal in the bulk. With a dimensional reduction it turns insulating below a critical thickness of $\sim$3 layers [10, 36]. DFT + DMFT [17] could trace back the microscopic origin of this metal-insulator transition to the correlation enhanced crystal field splitting and showed that this transition can also be triggered by a small gate voltage. This makes SVO/STO an ideal candidate for a Mott transistor.

Ruthenates such as SRO/STO are promising for heterostructures that are both metallic and ferromagnetic. DFT + DMFT [111] could explain why the experimental efforts to get ultrathin ferromagnetic SRO/STO failed so-far, despite predictions of the contrary: While bulk SRO is a metallic ferromagnet in DMFT, the correlation enhanced crystal field splitting makes a competing antiferromagnetic phase favorable. This antiferromagnetic phase is quite stable against strain, capping layers etc. The most promising route to metallic ferromagnetism is electron doping.

We have only seen the beginning of DFT + DMFT calculations for oxide heterostructures. In the future, these calculation will help us to better understand experiment and to predict novel physics, which already led to some successes in the past. While the numerical effort is larger than DFT, it is still much more efficient and cheaper than experiment to scan the myriad of possible combinations of heterostructure slabs by DFT + DMFT. In the past the theoretical and experimental focus has been on heterostructures with layers perpendicular to the (001) direction. Now both experiment and theory turn to other geometries: confinement in the (110) direction shows a much more complicated quantization and flat bands [9]; bilayers in a (111) stacking on the other hand are interesting regarding prospective topological states [114]. With the close cooperation between experiment on the one side and DFT as well as DMFT theory on the other side, a bright future in the research area of oxide heterostructures lies ahead.

We thank our colleagues and coauthors E. Assmann, S. Bhandary, P. Blaha, P. Hansmann, R. Laskowski, G. Li, S. Okamoto, N. Parragh, L. Si, J. M. Tomczak, A. Toschi, and M. Wallerberger for useful discussions and joint efforts. This work was supported financially by European Research Council under the European Union's Seventh Framework Program (FP/2007-2013)/ERC grant agreement n. 306447, the Austrian Science Fund (FWF) through SFB ViCoM F41, project I-610, and Lise Meitner programme, project M2050, as well as research group FOR1346 of the Deutsche Forschungsgemeinschaft (DFG). Calculations reported were performed on the Vienna Scientific Cluster (VSC).

References

1. A. Georges, G. Kotliar, Hubbard model in infinite dimensions. Phys. Rev. B **45**, 6479–6483 (1992). https://doi.org/10.1103/PhysRevB.45.6479
2. A. Georges, G. Kotliar, W. Krauth, M.J. Rozenberg, Dynamical mean-field theory of strongly correlated fermion systems and the limit of infinite dimensions. Rev. Mod. Phys. **68**, 13–125 (1996). https://doi.org/10.1103/RevModPhys.68.13
3. W. Metzner, D. Vollhardt, Correlated lattice fermions in d = ∞ dimensions. Phys. Rev. Lett. **62**, 324–327 (1989). https://doi.org/10.1103/PhysRevLett.62.324
4. K. Held, Electronic structure calculations using dynamical mean field theory. Adv. Phys. **56**, 829–926 (2007). https://doi.org/10.1080/00018730701619647
5. G. Kotliar, S.Y. Savrasov, K. Haule, V.S. Oudovenko, O. Parcollet, C.A. Marianetti, Electronic structure calculations with dynamical mean-field theory. Rev. Mod. Phys. **78**, 865–951 (2006). https://doi.org/10.1103/RevModPhys.78.865
6. A. Ohtomo, H.Y. Hwang, A high-mobility electron gas at the $LaAlO_3/SrTiO_3$ heterointerface. Nature (London) **427**, 423–426 (2004). https://doi.org/10.1038/nature02308
7. Z. Zhong, Q. Zhang, K. Held, Quantum confinement in perovskite oxide heterostructures: tight binding instead of a nearly free electron picture. Phys. Rev. B **88**, 125,401 (2013). https://doi.org/10.1103/PhysRevB.88.125401

8. A.F. Santander-Syro, O. Copie, T. Kondo, F. Fortuna, S. Pailhés, R. Weht, X.G. Qiu, F. Bertran, A. Nicolaou, A. Taleb-Ibrahimi, P.L. Févre, G. Herranz, M. Bibes, N. Reyren, Y. Apertet, P. Lecoeur, A. Barthélémy, M.J. Rozenberg, Two-dimensional electron gas with universal subbands at the surface of $SrTiO_3$. Nature (London) **469**, 189–193 (2011). https://doi.org/10.1038/nature09720
9. Z. Wang, Z. Zhong, X. Hao, S. Gerhold, B. Stoger, M. Schmid, J. Sanchez-Barriga, A. Varykhalov, C. Franchini, K. Held, U. Diebold, Anisotropic two-dimensional electron gas at $SrTiO_3(110)$ protected by its native overlayer. Proc. Nat. Acad. Sci. **333**, 3933 (2014). https://doi.org/10.1073/pnas.1318304111
10. K. Yoshimatsu, K. Horiba, H. Kumigashir, T. Yoshida, A. Fujimori, M. Oshima, Metallic quantum well states in artificial structures of strongly correlated oxide. Science **333**, 319–322 (2011). https://doi.org/10.1126/science.1205771
11. M. Behrmann, F. Lechermann, Interface exchange processes in $LaAlO_3$ / $SrTiO_3$ induced by oxygen vacancies. Phys. Rev. B **92**, 125,148 (2015). https://doi.org/10.1103/PhysRevB.92.125148
12. G. Berner, A. Müller, F. Pfaff, J. Walde, C. Richter, J. Mannhart, S. Thiess, A. Gloskovskii, W. Drube, M. Sing, R. Claessen, Band alignment in $LaAlO_3/SrTiO_3$ oxide heterostructures inferred from hard x-ray photoelectron spectroscopy. Phys. Rev. B **88**, 115,111 (2013). https://doi.org/10.1103/PhysRevB.88.115111
13. G. Berner, M. Sing, H. Fujiwara, A. Yasui, Y. Saitoh, A. Yamasaki, Y. Nishitani, A. Sekiyama, N. Pavlenko, T. Kopp, C. Richter, J. Mannhart, S. Suga, R. Claessen, Direct k-space mapping of the electronic structure in an oxide-oxide interface. Phys. Rev. Lett. **110**, 247,601 (2013). https://doi.org/10.1103/PhysRevLett.110.247601
14. H.O. Jeschke, J. Shen, R. Valentí, Localized versus itinerant states created by multiple oxygen vacancies in $SrTiO_3$. New J. Phys. **17**, 023,034 (2015). https://doi.org/10.1088/1367-2630/17/2/023034
15. M. Imada, A. Fujimori, Y. Tokura, Metal-insulator transitions. Rev. Mod. Phys. **70**, 1039–1263 (1998). https://doi.org/10.1103/RevModPhys.70.1039
16. A. Liebsch, Surface versus bulk Coulomb correlations in photoemission spectra of $SrVO_3$ and $CaVO_3$. Phys. Rev. Lett. **90**, 096,401 (2003). https://doi.org/10.1103/PhysRevLett.90.096401
17. Z. Zhong, M. Wallerberger, J.M. Tomczak, C. Taranto, N. Parragh, A. Toschi, G. Sangiovanni, K. Held, Electronics with correlated oxides: $SrVO_3/SrTiO_3$ as a Mott transistor. Phys. Rev. Lett. **114**, 246,401 (2015). https://doi.org/10.1103/PhysRevLett.114.246401
18. G. Lantz, M. Hajlaoui, E. Papalazarou, V.L.R. Jacques, A. Mazzotti, M. Marsi, S. Lupi, M. Amati, L. Gregoratti, L. Si, Z. Zhong, K. Held, Surface effects on the Mott-Hubbard transition in archetypal V_2O_3. Phys. Rev. Lett. **115**, 236,802 (2015). https://doi.org/10.1103/PhysRevLett.115.236802
19. A. Georges, L. de' Medici, J. Mravlje, Strong electronic correlations from Hund's coupling. Ann. Rev. Condens. Matter Phys. **4**, 137 (2013). https://doi.org/10.1146/annurev-conmatphys-020911-125045
20. Z.P. Yin, K. Haule, G. Kotliar, Kinetic frustration and the nature of the magnetic and paramagnetic states in iron pnictides and iron chalcogenides. Nat. Phys. **10**, 932 (2011). https://doi.org/10.1038/nmat3120
21. V.I. Anisimov, J. Zaanen, O.K. Andersen, Band theory and Mott insulators: Hubbard U instead of Stoner I. Phys. Rev. B **44**, 943–954 (1991). https://doi.org/10.1103/PhysRevB.44.943
22. G. Sangiovanni, A. Toschi, E. Koch, K. Held, M. Capone, C. Castellani, O. Gunnarsson, S.K. Mo, J.W. Allen, H.D. Kim, A. Sekiyama, A. Yamasaki, S. Suga, P. Metcalf, Static versus dynamical mean-field theory of Mott antiferromagnets. Phys. Rev. B **73**, 205,121 (2006). https://doi.org/10.1103/PhysRevB.73.205121
23. D.D. Cuong, B. Lee, K.M. Choi, H.S. Ahn, S. Han, J. Lee, Oxygen vacancy clustering and electron localization in oxygen-deficient $SrTiO_3$: LDA+U study. Phys. Rev. Lett. **98**, 115,503 (2007). https://doi.org/10.1103/PhysRevLett.98.115503

24. Z. Zhong, P.X. Xu, P.J. Kelly, Polarity-induced oxygen vacancies at LaAlO$_3$SrTiO$_3$ interfaces. Phys. Rev. B **82**, 165,127 (2010). https://doi.org/10.1103/PhysRevB.82.165127
25. P. Hansmann, R. Arita, A. Toschi, S. Sakai, G. Sangiovanni, K. Held, Dichotomy between large local and small ordered magnetic moments in iron-based superconductors. Phys. Rev. Lett. **104**, 197,002 (2010). https://doi.org/10.1103/PhysRevLett.104.197002
26. A. Galler, C. Taranto, M. Wallerberger, M. Kaltak, G. Kresse, G. Sangiovanni, A. Toschi, K. Held, Screened moments and absence of ferromagnetism in FeAl. Phys. Rev. B **92**, 205,132 (2015). https://doi.org/10.1103/PhysRevB.92.205132
27. T. Maier, M. Jarrell, T. Pruschke, M.H. Hettler, Quantum cluster theories. Rev. Mod. Phys. **77**, 1027–1080 (2005). https://doi.org/10.1103/RevModPhys.77.1027
28. A.A. Katanin, A. Toschi, K. Held, Comparing pertinent effects of antiferromagnetic fluctuations in the two- and three-dimensional Hubbard model. Phys. Rev. B **80**, 075,104 (2009). https://doi.org/10.1103/PhysRevB.80.075104
29. H. Kusunose, Influence of spatial correlations in strongly correlated electron systems: extension to dynamical mean field approximation. J. Phys. Soc. Jpn. **75**, 054,713 (2006). https://doi.org/10.1143/JPSJ.75.054713
30. A.N. Rubtsov, M.I. Katsnelson, A.I. Lichtenstein, Dual fermion approach to nonlocal correlations in the Hubbard model. Phys. Rev. B **77**, 033,101 (2008). https://doi.org/10.1103/PhysRevB.77.033101
31. C. Slezak, M. Jarrell, T. Maier, J. Deisz, Multi-scale extensions to quantum cluster methods for strongly correlated electron systems. J. Phys.: Condens. Matter **21**, 435,604 (2009). https://doi.org/10.1088/0953-8984/21/43/435604
32. A. Toschi, A.A. Katanin, K. Held, Dynamical vertex approximation: a step beyond dynamical mean-field theory. Phys. Rev. B **75**, 045,118 (2007). https://doi.org/10.1103/PhysRevB.75.045118
33. A. Sekiyama, H. Fujiwara, S. Imada, S. Suga, H. Eisaki, S.I. Uchida, K. Takegahara, H. Harima, Y. Saitoh, I.A. Nekrasov, G. Keller, D.E. Kondakov, A.V. Kozhevnikov, T. Pruschke, K. Held, D. Vollhardt, V.I. Anisimov, Mutual experimental and theoretical validation of bulk photoemission spectra of Sr$_{1-x}$Ca$_x$VO$_3$. Phys. Rev. Lett. **93**, 156,402 (2004). https://doi.org/10.1103/PhysRevLett.93.156402
34. T. Yoshida, K. Tanaka, H. Yagi, A. Ino, H. Eisaki, A. Fujimori, Z.X. Shen, Direct observation of the mass renormalization in SrVO$_3$ by angle resolved photoemission spectroscopy. Phys. Rev. Lett. **95**, 146,404 (2005). https://doi.org/10.1103/PhysRevLett.95.146404
35. I.H. Inoue, I. Hase, Y. Aiura, A. Fujimori, K. Morikawa, T. Mizokawa, Y. Haruyama, T. Maruyama, Y. Nishihara, Systematic change of spectral function observed by controlling electron correlation in Ca$_{1x}$Sr$_x$VO$_3$ with fixed $3d^1$ configuration. Physica C **235–240**, 1007–1008 (1994). https://doi.org/10.1016/0921-4534(94)91728-0
36. K. Yoshimatsu, T. Okabe, H. Kumigashira, S. Okamoto, S. Aizaki, A. Fujimori, M. Oshima, Dimensional-crossover-driven metal-insulator transition in SrVO$_3$ ultrathin films. Phys. Rev. Lett. **104**, 147,601 (2010). https://doi.org/10.1103/PhysRevLett.104.147601
37. E.R. Ylvisaker, W.E. Pickett, K. Koepernik, Anisotropy and magnetism in the LSDA + U method. Phys. Rev. B **79**, 035,103 (2009). https://doi.org/10.1103/PhysRevB.79.035103
38. A.G. Petukhov, I.I. Mazin, L. Chioncel, A.I. Lichtenstein, Correlated metals and the LDA + U method. Phys. Rev. B **67**, 153,106 (2003). https://doi.org/10.1103/PhysRevB.67.153106
39. O. Grånäs, I.D. Marco, P. Thunström, L. Nordström, O. Eriksson, T. Björkman, J.M. Wills, Charge self-consistent dynamical mean-field theory based on the full-potential linear muffin-tin orbital method: Methodology and applications. Comput. Mater. Sci. **55**, 295–302 (2012). https://doi.org/10.1016/j.commatsci.2011.11.032
40. F. Lechermann, A. Georges, A. Poteryaev, S. Biermann, M. Posternak, A. Yamasaki, O.K. Andersen, Dynamical mean-field theory using Wannier functions: a flexible route to electronic structure calculations of strongly correlated materials. Phys. Rev. B **74**, 125,120 (2006). https://doi.org/10.1103/PhysRevB.74.125120
41. H. Park, A.J. Millis, C.A. Marianetti, Computing total energies in complex materials using charge self-consistent DFT+DMFT. Phys. Rev. B **90**, 235,103 (2014). https://doi.org/10.1103/PhysRevB.90.235103

42. I. Leonov, V.I. Anisimov, D. Vollhardt, First-principles calculation of atomic forces and structural distortions in strongly correlated materials. Phys. Rev. Lett. **112**, 146,401 (2014). https://doi.org/10.1103/PhysRevLett.112.146401
43. K. Lejaeghere, G. Bihlmayer, T. Björkman, P. Blaha, S. Blügel, V. Blum, D. Caliste, I.E. Castelli, S.J. Clark, A. Dal Corso, S. de Gironcoli, T. Deutsch, J.K. Dewhurst, I. Di Marco, C. Draxl, M. Dułak, O. Eriksson, J.A. Flores-Livas, K.F. Garrity, L. Genovese, P. Giannozzi, M. Giantomassi, S. Goedecker, X. Gonze, O. Grånäs, E.K.U. Gross, A. Gulans, F. Gygi, D.R. Hamann, P.J. Hasnip, N.A.W. Holzwarth, D. Iuşan, D.B. Jochym, F. Jollet, D. Jones, G. Kresse, K. Koepernik, E. Küçükbenli, Y.O. Kvashnin, I.L.M. Locht, S. Lubeck, M. Marsman, N. Marzari, U. Nitzsche, L. Nordström, T. Ozaki, L. Paulatto, C.J. Pickard, W. Poelmans, M.I.J. Probert, K. Refson, M. Richter, G.M. Rignanese, S. Saha, M. Scheffler, M. Schlipf, K. Schwarz, S. Sharma, F. Tavazza, P. Thunström, A. Tkatchenko, M. Torrent, D. Vanderbilt, M.J. van Setten, V. Van Speybroeck, J.M. Wills, J.R. Yates, G.X. Zhang, S. Cottenier, Reproducibility in density functional theory calculations of solids. Science **351**(6280), 1415 (2016). https://doi.org/10.1126/science.aad3000
44. K. Schwarz, P. Blaha, G. Madsen, Electronic structure calculations of solids using the WIEN2k package for material sciences. Comp. Phys. Commun. **147**, 71–76 (2002). https://doi.org/10.1016/S0010-4655(02)00206-0
45. W. Kohn, L.J. Sham, Self-consistent equations including exchange and correlation effects. Phys. Rev. **140**, A1133–A1138 (1965). https://doi.org/10.1103/PhysRev.140.A1133
46. J.P. Perdew, Y. Wang, Accurate and simple analytic representation of the electron-gas correlation energy. Phys. Rev. B **45**, 13,244–13,249 (1992). https://doi.org/10.1103/PhysRevB.45.13244
47. J.P. Perdew, K. Burke, M. Ernzerhof, Generalized gradient approximation made simple. Phys. Rev. Lett. **77**, 3865–3868 (1996). https://doi.org/10.1103/PhysRevLett.77.3865
48. Z. Zhong, P. Wissgott, K. Held, G. Sangiovanni, Microscopic understanding of the orbital splitting and its tuning at oxide interfaces. EPL **99**, 37,011 (2012). https://doi.org/10.1209/0295-5075/99/37011
49. N. Marzari, A.A. Mostofi, J.R. Yates, I. Souza, D. Vanderbilt, Maximally localized Wannier functions: theory and applications. Rev. Mod. Phys. **84**, 1419–1475 (2012). https://doi.org/10.1103/RevModPhys.84.1419
50. N. Marzari, I. Souza, D. Vanderbilt, An introduction to maximally-localized Wannier functions. In: Ψ_k Newsletter (Highlight 57) (2003)
51. N. Marzari, D. Vanderbilt, Maximally localized generalized Wannier functions for composite energy bands. Phys. Rev. B **56**, 12,847–12,865 (1997). https://doi.org/10.1103/PhysRevB.56.12847
52. A.A. Mostofi, J.R. Yates, Y.S. Lee, I. Souza, D. Vanderbilt, N. Marzari, wannier90: A tool for obtaining maximally-localised Wannier functions. Comp. Phys. Commun. **178**, 685–699 (2008). https://doi.org/10.1016/j.cpc.2007.11.016
53. J. Kuneš, R. Arita, P. Wissgott, A. Toschi, H. Ikeda, K. Held, Wien2wannier: from linearized augmented plane waves to maximally localized Wannier functions. Comput. Phys. Commun. **181**, 1888 (2010). https://doi.org/10.1016/j.cpc.2010.08.005
54. I.A. Nekrasov, K. Held, G. Keller, D.E. Kondakov, T. Pruschke, M. Kollar, O.K. Andersen, V.I. Anisimov, D. Vollhardt, Momentum-resolved spectral functions of $SrVO_3$ calculated by LDA + DMFT. Phys. Rev. B **73**, 155,112 (2006). https://doi.org/10.1103/PhysRevB.73.155112
55. E. Gull, A.J. Millis, A.I. Lichtenstein, A.N. Rubtsov, M. Troyer, P. Werner, Continuous-time Monte Carlo methods for quantum impurity models. Rev. Mod. Phys. **83**, 349–404 (2011). https://doi.org/10.1103/RevModPhys.83.349
56. M. Wallerberger, A. Hausoel, P. Gunacker, A. Kowalski, N. Parragh, F. Goth, K. Held, G. Sangiovanni, W2dynamics: local one- and two-particle quantities from dynamical mean field theory, (2018). arXiv:1801.10209
57. M. Jarrell, J. Gubernatis, Bayesian inference and the analytic continuation of imaginary-time quantum Monte Carlo data. Phys. Rep. **269**, 133–195 (1996). https://doi.org/10.1016/0370-1573(95)00074-7

58. G. Keller, K. Held, V. Eyert, D. Vollhardt, V.I. Anisimov, Electronic structure of paramagnetic V_2O_3: Strongly correlated metallic and mott insulating phase. Phys. Rev. B **70**(20), 205,116 (2004). https://doi.org/10.1103/PhysRevB.70.205116
59. R. Hesper, L.H. Tjeng, A. Heeres, G.A. Sawatzky, Photoemission evidence of electronic stabilization of polar surfaces in K_3C_{60}. Phys. Rev. B **62**, 16,046–16,055 (2000). https://doi.org/10.1103/PhysRevB.62.16046
60. S. Okamoto, A.J. Millis, Electronic reconstruction at an interface between a Mott insulator and a band insulator. Nature (London) **428**, 630–633 (2004). https://doi.org/10.1038/nature02450
61. S. Okamoto, A.J. Millis, N.A. Spaldin, Lattice relaxation in oxide heterostructures: $LaTiO_3/SrTiO_3$ superlattices. Phys. Rev. Lett. **97**, 056,802 (2006). https://doi.org/10.1103/PhysRevLett.97.056802
62. S. Okamoto, A.J. Millis, Spatial inhomogeneity and strong correlation physics: a dynamical mean-field study of a model Mott-insulator–band-insulator heterostructure. Phys. Rev. B **70**, 241,104 (2004). https://doi.org/10.1103/PhysRevB.70.241104
63. P. Hansmann, X. Yang, A. Toschi, G. Khaliullin, O.K. Andersen, K. Held, Turning a nickelate Fermi surface into a cupratelike one through heterostructuring. Phys. Rev. Lett. **103**, 016,401 (2009). https://doi.org/10.1103/PhysRevLett.103.016401
64. F. Lechermann, L. Boehnke, D. Grieger, Formation of orbital-selective electron states in $LaTiO_3/SrTiO_3$ superlattices. Phys. Rev. B **87**, 241,101 (2013). https://doi.org/10.1103/PhysRevB.87.241101
65. F. Lechermann, L. Boehnke, D. Grieger, C. Piefke, Electron correlation and magnetism at the $LaAlO_3/SrTiO_3$ interface: A DFT + DMFT investigation. Phys. Rev. B **90**, 085,125 (2014). https://doi.org/10.1103/PhysRevB.90.085125
66. F. Lechermann, H.O. Jeschke, A.J. Kim, S. Backes, R. Valentí, Electron dichotomy on the $SrTiO_3$ defect surface augmented by many-body effects. Phys. Rev. B **93**, 121,103 (2016). https://doi.org/10.1103/PhysRevB.93.121103
67. M. Altmeyer, H.O. Jeschke, O. Hijano-Cubelos, C. Martins, F. Lechermann, K. Koepernik, A.F. Santander-Syro, M.J. Rozenberg, R. Valentí, M. Gabay, Magnetism, spin texture, and in-gap states: atomic specialization at the surface of oxygen-deficient $SrTiO_3$. Phys. Rev. Lett. **116**, 157–203 (2016). https://doi.org/10.1103/PhysRevLett.116.157203
68. E. Pavarini, S. Biermann, A. Poteryaev, A.I. Lichtenstein, A. Georges, O.K. Andersen, Mott transition and suppression of orbital fluctuations in orthorhombic $3d^1$ perovskites. Phys. Rev. Lett. **92**, 176,403 (2004). https://doi.org/10.1103/PhysRevLett.92.176403
69. E. Pavarini, A. Yamasaki, J. Nuss, O.K. Andersen, How chemistry controls electron localization in $3d^1$ perovskites: a Wannier-function study. New J. Phys. **7**, 188 (2005). https://doi.org/10.1088/1367-2630/7/1/188
70. K. Dymkowski, C. Ederer, Strain-induced insulator-to-metal transition in $LaTiO_3$ within DFT+DMFT. Phys. Rev. B **89**, 161,109 (2014). https://doi.org/10.1103/PhysRevB.89.161109
71. F. Lechermann, M. Obermeyer, Towards Mott design by δ-doping of strongly correlated titanates. New J. Phys. **17**, 043,026 (2015). https://doi.org/10.1088/1367-2630/17/4/043026
72. G. Catalan, Progress in perovskite nickelate research. Phase Transitions **81**, 729–749 (2008). https://doi.org/10.1080/01411590801992463
73. J.L. García-Muñoz, J. Rodríguez-Carvajal, P. Lacorre, J.B. Torrance, Neutron-diffraction study of $RNiO_3$ (R =La, Pr, Nd, Sm): Electronically induced structural changes across the metal-insulator transition. Phys. Rev. B **46**, 4414–4425 (1992). https://doi.org/10.1103/PhysRevB.46.4414
74. J.B. Torrance, P. Lacorre, A.I. Nazzal, E.J. Ansaldo, C. Niedermayer, Systematic study of insulator-metal transitions in perovskites $RNiO_3$ (R =Pr, Nd, Sm, Eu) due to closing of charge-transfer gap. Phys. Rev. B **45**, 8209–8212 (1992). https://doi.org/10.1103/PhysRevB.45.8209
75. J.A. Alonso, J.L. García-Muñoz, M.T. Fernández-Díaz, M.A.G. Aranda, M.J. Martínez-Lope, M.T. Casais, Charge disproportionation in $RNiO_3$ perovskites: Simultaneous metal-insulator and structural transition in $YNiO_3$. Phys. Rev. Lett. **82**, 3871–3874 (1999). https://doi.org/10.1103/PhysRevLett.82.3871

76. E.A. Nowadnick, J.P. Ruf, H. Park, P.D.C. King, D.G. Schlom, K.M. Shen, A.J. Millis, Quantifying electronic correlation strength in a complex oxide: a combined DMFT and ARPES study of LaNiO$_3$. Phys. Rev. B **92**, 245,109 (2015). https://doi.org/10.1103/PhysRevB.92.245109
77. J. Rodríguez-Carvajal, S. Rosenkranz, M. Medarde, P. Lacorre, M.T. Fernandez-Díaz, F. Fauth, V. Trounov, Neutron-diffraction study of the magnetic and orbital ordering in ^{154}SmNiO$_3$ and ^{153}EuNiO$_3$. Phys. Rev. B **57**, 456–464 (1998). https://doi.org/10.1103/PhysRevB.57.456
78. U. Staub, G.I. Meijer, F. Fauth, R. Allenspach, J.G. Bednorz, J. Karpinski, S.M. Kazakov, L. Paolasini, F. d'Acapito, Direct observation of charge order in an epitaxial NdNiO$_3$ film. Phys. Rev. Lett. **88**, 126,402 (2002). https://doi.org/10.1103/PhysRevLett.88.126402
79. B. Lau, A.J. Millis, Theory of the magnetic and metal-insulator transitions in RNiO$_3$ bulk and layered structures. Phys. Rev. Lett. **110**, 126,404 (2013). https://doi.org/10.1103/PhysRevLett.110.126404
80. S. Lee, R. Chen, L. Balents, Landau theory of charge and spin ordering in the nickelates. Phys. Rev. Lett. **106**, 016,405 (2011). https://doi.org/10.1103/PhysRevLett.106.016405
81. S. Lee, R. Chen, L. Balents, Metal-insulator transition in a two-band model for the perovskite nickelates. Phys. Rev. B **84**, 165,119 (2011). https://doi.org/10.1103/PhysRevB.84.165119
82. V.I. Anisimov, D. Bukhvalov, T.M. Rice, Electronic structure of possible nickelate analogs to the cuprates. Phys. Rev. B **59**, 7901–7906 (1999). https://doi.org/10.1103/PhysRevB.59.7901
83. J. Chaloupka, G. Khaliullin, Orbital order and possible superconductivity in LaNiO$_3$/LaMO$_3$ superlattices. Phys. Rev. Lett. **100**, 016,404 (2008). https://doi.org/10.1103/PhysRevLett.100.016404
84. P. Hansmann, A. Toschi, X. Yang, O.K. Andersen, K. Held, Electronic structure of nickelates: from two-dimensional heterostructures to three-dimensional bulk materials. Phys. Rev. B **82**, 235,123 (2010). https://doi.org/10.1103/PhysRevB.82.235123
85. M.J. Han, X. Wang, C.A. Marianetti, A.J. Millis, Dynamical mean-field theory of nickelate superlattices. Phys. Rev. Lett. **107**, 206,804 (2011). https://doi.org/10.1103/PhysRevLett.107.206804
86. N. Parragh, G. Sangiovanni, P. Hansmann, S. Hummel, K. Held, A. Toschi, Effective crystal field and Fermi surface topology: a comparison of d- and dp-orbital models. Phys. Rev. B **88**, 195,116 (2013). https://doi.org/10.1103/PhysRevB.88.195116
87. O.E. Peil, M. Ferrero, A. Georges, Orbital polarization in strained LaNiO$_3$: structural distortions and correlation effects. Phys. Rev. B **90**, 045,128 (2014). https://doi.org/10.1103/PhysRevB.90.045128
88. E. Benckiser, M.W. Haverkort, S. Brück, E. Goering, S. Macke, A. Frañó, X. Yang, O.K. Andersen, G. Cristiani, H.U. Habermeier, A.V. Boris, I. Zegkinoglou, P. Wochner, H.J. Kim, V. Hinkov, B. Keimer, Orbital reflectometry of oxide heterostructures. Nat. Mater. **10**, 189 (2011). https://doi.org/10.1038/nmat2958
89. M. Wu, E. Benckiser, M.W. Haverkort, A. Frano, Y. Lu, U. Nwankwo, S. Brück, P. Audehm, E. Goering, S. Macke, V. Hinkov, P. Wochner, G. Christiani, S. Heinze, G. Logvenov, H.U. Habermeier, B. Keimer, Strain and composition dependence of orbital polarization in nickel oxide superlattices. Phys. Rev. B **88**, 125,124 (2013). https://doi.org/10.1103/PhysRevB.88.125124
90. A.S. Disa, D.P. Kumah, A. Malashevich, H. Chen, D.A. Arena, E.D. Specht, S. Ismail-Beigi, F.J. Walker, C.H. Ahn, Orbital engineering in symmetry-breaking polar heterostructures. Phys. Rev. Lett. **114**, 026,801 (2015). https://doi.org/10.1103/PhysRevLett.114.026801
91. H. Park, A.J. Millis, C.A. Marianetti, Influence of quantum confinement and strain on orbital polarization of four-layer LaNiO$_3$ superlattices: A DFT + DMFT study. Phys. Rev. B **93**, 235,109 (2016). https://doi.org/10.1103/PhysRevB.93.235109
92. M. Uchida, K. Ishizaka, P. Hansmann, Y. Kaneko, Y. Ishida, X. Yang, R. Kumai, A. Toschi, Y. Onose, R. Arita, K. Held, O.K. Andersen, S. Shin, Y. Tokura, Pseudogap of metallic layered nickelate $R_{2-x}Sr_xNiO_4$ (R=Nd,Eu) crystals measured using angle-resolved photoemission

spectroscopy. Phys. Rev. Lett. **106**, 027,001 (2011). https://doi.org/10.1103/PhysRevLett.106.027001
93. H. Chen, H. Park, A.J. Millis, C.A. Marianetti, Charge transfer across transition-metal oxide interfaces: emergent conductance and electronic structure. Phys. Rev. B **90**, 245,138 (2014). https://doi.org/10.1103/PhysRevB.90.245138
94. Z. Fang, N. Nagaosa, Quantum versus Jahn-Teller orbital physics in YVO_3 and $LaVO_3$. Phys. Rev. Lett. **93**, 176,404 (2004). https://doi.org/10.1103/PhysRevLett.93.176404
95. M. De Raychaudhury, E. Pavarini, O.K. Andersen, Orbital fluctuations in the different phases of YVO_3 and $LaVO_3$. Phys. Rev. Lett. **99**, 126,402 (2007). https://doi.org/10.1103/PhysRevLett.99.126402
96. Y. Hotta, T. Susaki, H.Y. Hwang, Polar discontinuity doping of the $LaVO_3/SrTiO_3$ interface. Phys. Rev. Lett. **99**, 236,805 (2007). https://doi.org/10.1103/PhysRevLett.99.236805
97. E. Assmann, P. Blaha, R. Laskowski, K. Held, S. Okamoto, G. Sangiovanni, Oxide heterostructures for efficient solar cells. Phys. Rev. Lett. **110**, 078–701 (2013). https://doi.org/10.1103/PhysRevLett.110.078701
98. E. Assmann, Spectral properties of strongly correlated materials. Ph.D. thesis, TU Wien (2015)
99. P. Werner, K. Held, M. Eckstein, Role of impact ionization in the thermalization of photoexcited mott insulators. Phys. Rev. B **90**, 235,102 (2014). https://doi.org/10.1103/PhysRevB.90.235102
100. L. Wang, Y. Li, A. Bera, C. Ma, F. Jin, K. Yuan, W. Yin, A. David, W. Chen, W. Wu, W. Prellier, S. Wei, T. Wu, Device performance of the Mott insulator $LaVO_3$ as a photovoltaic material. Phys. Rev. Applied **3**, 064,015 (2015). https://doi.org/10.1103/PhysRevApplied.3.064015
101. M. Nakamura, F. Kagawa, T. Tanigaki, H.S. Park, T. Matsuda, D. Shindo, Y. Tokura, M. Kawasaki, Spontaneous polarization and bulk photovoltaic effect driven by polar discontinuity in $LaFeO_3/SrTiO_3$ heterojunctions. Phys. Rev. Lett. **116**, 156,801 (2016). https://doi.org/10.1103/PhysRevLett.116.156801
102. G. Koster, L. Klein, W. Siemons, G. Rijnders, J.S. Dodge, C.B. Eom, D.H.A. Blank, M.R. Beasley, Structure, physical properties, and applications of $SrRuO_3$ thin films. Rev. Mod. Phys. **84**, 253–298 (2012). https://doi.org/10.1103/RevModPhys.84.253
103. W.E. Bell, M. Tagami, High-temperature chemistry of the ruthenium-osygen system. J. Phys. Chem. **67**, 2432–2436 (1963). https://doi.org/10.1021/j100805a042
104. D. Toyota, I. Ohkubo, H. Kumigashira, M. Oshima, T. Ohnishi, M. Lippmaa, M. Takizawa, A. Fujimori, K. Ono, M. Kawasaki, H. Koinuma, Thickness-dependent electronic structure of ultrathin $SrRuO_3$ films studied by in situ photoemission spectroscopy. Appl. Phys. Lett. **87**, 162,508 (2005). https://doi.org/10.1063/1.2108123
105. J. Xia, W. Siemons, G. Koster, M.R. Beasley, A. Kapitulnik, Critical thickness for itinerant ferromagnetism in ultrathin films of $SrRuO_3$. Phys. Rev. B **79**, 140,407 (2009). https://doi.org/10.1103/PhysRevB.79.140407
106. J.M. Rondinelli, N.M. Caffrey, S. Sanvito, N.A. Spaldin, Electronic properties of bulk and thin film $SrRuO_3$: Search for the metal-insulator transition. Phys. Rev. B **78**, 155,107 (2008). https://doi.org/10.1103/PhysRevB.78.155107
107. P. Mahadevan, F. Aryasetiawan, A. Janotti, T. Sasaki, Evolution of the electronic structure of a ferromagnetic metal: case of $SrRuO_3$. Phys. Rev. B **80**, 035,106 (2009). https://doi.org/10.1103/PhysRevB.80.035106
108. G. Rijnders, D.H.A. Blank, J. Choi, C.B. Eom, Enhanced surface diffusion through termination conversion during epitaxial $SrRuO_3$ growth. Appl. Phys. Lett. **84**, 505 (2004). https://doi.org/10.1063/1.1640472
109. E. Jakobi, S. Kanungo, S. Sarkar, S. Schmitt, T. Saha-Dasgupta, LDA + DMFT study of Ru-based perovskite $SrRuO_3$ and $CaRuO_3$. Phys. Rev. B **83**, 041,103 (2011). https://doi.org/10.1103/PhysRevB.83.041103
110. M. Kim, B.I. Min, Nature of itinerant ferromagnetism of $SrRuO_3$: A DFT+DMFT study. Phys. Rev. B **91**, 205,116 (2015). https://doi.org/10.1103/PhysRevB.91.205116

111. L. Si, Z. Zhong, J.M. Tomczak, K. Held, Route to room-temperature ferromagnetic ultrathin SrRuO$_3$ films. Phys. Rev. B **92**, 041,108 (2015). https://doi.org/10.1103/PhysRevB.92.041108
112. E. Assmann, P. Wissgott, J. Kuneš, A. Toschi, P. Blaha, K. Held, woptic: optical conductivity with Wannier functions and adaptive k-mesh refinements. Comput. Phys. Commun. **202**, 1 (2016). https://doi.org/10.1016/j.cpc.2015.12.010
113. S. Bhandary, E. Assmann, M. Aichhorn, K. Held, Charge self-consistency in density functional theory + dynamical mean field theory: k-space reoccupation and orbital order, (2016). Phys. Rev. B **94**, 155–131 (2016). https://doi.org/10.1103/PhysRevB.94.155131
114. D. Xiao, W. Zhu, Y. Ran, N. Nagaosa, S. Okamoto, Interface engineering of quantum Hall effects in digital transition metal oxide heterostructures. Nature Commun. **2**, 596 (2011). https://doi.org/10.1038/ncomms1602

Chapter 10
Spectroscopic Characterisation of Multiferroic Interfaces

M.-A. Husanu and C. A. F. Vaz

Abstract In this chapter we discuss the capabilities of X-ray photoemission and absorption spectroscopies for the investigation of the electronic, magnetic and electric properties of multiferroic materials and heterostructures. As complementary techniques providing element selective information on both occupied and empty states, their combination delivers a comprehensive picture of the chemical state of individual species, magnetic moments, bulk and surface band structure, and local atomic environment at the interface between dissimilar materials. By directly probing the electronic structure at the atomic level, unique insights can be learned about the mechanisms responsible for the magnetoelectric couplings in this fascinating class of materials.

10.1 Introduction

The term *multiferroic* designates material systems that are characterised by the presence of, and a coupling between, multiple order parameters, such as magnetism and ferroelectricity (in which case the coupling is called *magnetoelectric coupling*). What makes such systems attractive is the prospect of controlling one order parameter, such as magnetism, by an external excitation that drives the other order parameter, such as an electric field: in such a case, electric field control of magnetism would result, which could be of significant interest from an applied perspective. Such practical considerations have been a driving force for the intensive research interest which these systems have motivated, but equally important has been the prospect of gaining deeper insights into the role of electron correlations in determining mul-

---

M.-A. Husanu (✉) · C. A. F. Vaz (✉)
Paul Scherrer Institut, Villigen PSI, Villigen, Switzerland
e-mail: ahusanu@infim.ro

C. A. F. Vaz
e-mail: carlos.vaz@cantab.net

M.-A. Husanu
National Institute of Materials Physics, Magurele, Romania

tifunctional behaviour. In fact, multiferroic systems constitute an important class of materials where the physical mechanisms underlying spin, charge, orbital, and lattice order need to be directly brought together to explain their complex and multivariate behaviour, ranging from magnetic-driven ferroelectricity [1–3], competing magnetic ground states [4, 5], to charge-modulated magnetism [6–8] and coupled phonon-magnon collective excitation modes (electromagnons) [9–15].

While scientifically and intellectually stimulating, the complexity associated with these systems makes their study experimentally challenging: often such correlated behaviour manifests only at low temperatures; the synthesis of such materials tend to occur within narrow windows in the growth parameter space; and the interdependence between the order parameters often implies relatively small magnetoelectric couplings. One route to introducing some simplicity to this problem is that of constructing artificial structures that bring together materials which are not multiferroic per se, but which may develop a magnetoelectric coupling at the interface region, where the symmetry requirements for multiferroic behaviour may now be satisfied [16], as schematically illustrated in Fig. 10.1. Such approach for designing new functionalities is a particular case of a more general trend of exploring interfacial phenomena to induce new physical properties intrinsic to the interface [17–20].

The complexity of the physical phenomena underlying multiferroic behaviour in single phase compounds, which originates from the intricate details of the electron correlated behaviour, are now reduced to an interface region typically of a few unit cells in width. Such an approach has the added advantage that the choice of materials is significantly expanded and the process lends itself naturally to size scaling. In fact, a large body of work has been devoted to developing such *artificial multiferroic heterostructures*, largely made possible by the recent advances in epitaxial thin film growth and in the development of new tools capable of characterising systems with nano- and atomic scale resolution. The growth of high-quality epitaxial oxide thin films (by molecular beam epitaxy, pulsed laser deposition, and r.f. magnetron

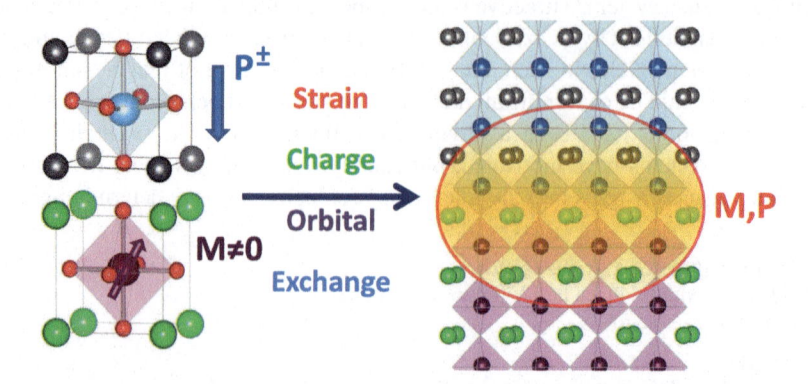

Fig. 10.1 Concept of artificial multiferroic heterostructures: by coupling the individual atomic behaviour at the interface, new functionalities emerge, including magnetoelectric coupling at ferroelectric-ferromagnetic interfaces

sputtering, to mention the most common methods for thin film growth) can presently be achieved routinely with an accurate control over the film thickness (in some instances, possible to monitor in real time during growth through the observation of oscillations in reflection high energy electron diffraction spots); over the epitaxial strain by deposition on suitably mismatched substrates [21]; over the growth of new crystalline phases not stable in bulk [22] and of artificial heterostructures with high degrees of crystalline perfection, including the growth of complex oxides on Si(001) without the formation of an amorphous oxide interface layer [23, 24]. The ability to fabricate sharp and abrupt interfaces is of particular importance for the case of multiferroic interfaces, since the origin of the magnetoelectric coupling is a strictly interfacial phenomenon. In tandem, new tools have been developed to probe matter down to the atomic scale, including the recent developments in high resolution transmission electron microscopy [25, 26] and in X-ray based techniques. Among such techniques, X-ray spectroscopy has emerged as a powerful tool for the electronic, magnetic, and ferroelectric characterisation of thin films and interfaces, including multiferroic heterostructures. In this chapter we discuss the unique capabilities of X-ray absorption and photoemission spectroscopy in advancing our understanding of the interfacial magnetoelectric coupling in various types of artificial multiferroic systems.

10.1.1 Intrinsic Multiferroics and Heterostructures

Systems that exhibit multiple order parameters are comparatively rare in nature. This is a consequence of the different requisites that are necessary for the simultaneous presence of the different order parameters. In the case of magnetoelectric multiferroics, these include break of inversion symmetry for electric order and break of time reversal symmetry for magnetism.

Hence, the class of materials that are simultaneously ferroelectric and magnetic are a much smaller subset of either class. Additionally, conflicting requirements for magnetism and ferroelectricity apply, such as a dielectric state for ferroelectricity (while a large number of ferromagnetic systems are electric conductors) or to a certain extent, the tendency for ferroelectricity to occur at cations with empty shells while magnetism requires partially filled electron shells [27, 28]. Nevertheless, a large number of single-phase, *intrinsic* multiferroic materials have been discovered, such as the ferroelectric and antiferromagnetic $BiFeO_3$ [29] and $BiMnO_3$ [30] perovskite oxides, the hexagonal or orthorhombic rare-earth manganites $YMnO_3$, $HoMnO_3$, $TbMnO_3$ [31], Fe_3O_4 magnetite [32, 33], or the barium fluorides, $BaMF_4$ with M = Mn, Fe, Co and Ni in octahedral coordination surrounded by F atoms [34]. In such materials, magnetic order arises solely through the spin moment and is stabilized by either direct or indirect (double or super) exchange interactions. Ferroelectricity, however, can have a number of physically distinct origins. In the canonical ferroelectrics, such as $BaTiO_3$ and $PbTiO_3$, ferroelectricity is related to structural distortions of the lattice and off-centering of cations with respect to a perfectly symmetric arrangement [35], while in a different class of materials the electric dipole moment is associated to particular configurations of s lone pair electrons, as in Bi in $BiFeO_3$; another source of ferroelectricity results from charge ordering, as in the case of magnetite [33]. In all these cases (called type I multiferroics [36]), the source of ferroelectricity and magnetism are largely unrelated, such that the magnetic and ferroelectric responses are largely decoupled, a fact which explains why in general, such intrinsic multiferroics have a weak magnetoelectric coupling coefficient. In the so-called type II multiferroics, one order parameter arises as a consequence of the onset of the other (mostly associated with the role of the Dzyaloshinskii-Moriya interaction, DMI [37, 38]) and the magnetoelectric coupling is therefore large. These are electronically complex materials characterised by strong electron correlation effects, but with ordering temperatures that tend to be well below room temperature. Such materials systems have been the subject of extensive attention and a number of excellent reviews have been made available recently [39–44].

Another approach to multiferroic behaviour is to combine dissimilar materials where the relevant degrees of freedom couple in the contact region in such a way as to generate an interfacial coupling between the different order parameters. In this approach, time and space symmetry are automatically broken at the interface and such cross-couplings are therefore allowed [45]. In fact, this concept has been applied already in a variety of physical systems, including granular composites [46–48], nanostructured combinations [49, 50] and layered multiferroic systems [51–55]. The various types of magnetoelectric coupling that can emerge at such interfaces will be described briefly below (see [56–58] for more extensive reviews).

10.1.2 Types of Interfacial Coupling

Since functionality in heterostructures comes from the cooperation of the intrinsic effects of the joining materials, it is instructive to inspect how the interface properties can be triggered by each of the degrees of freedom operating in the contact region. While ultimately all forms of interfacial coupling are a result of electron sharing, it is useful to consider the different mechanisms that give rise to magnetoelectric coupling, namely, strain, charge, and spin exchange processes.

10.1.2.1 Strain-Mediated Magnetoelectricity

Historically, the first effective realization of an artificial coupling between electric and magnetic ferroic phases aiming at designing large magnetoelectric effects relied on an elastic coupling between ferroelectrics with high piezoelectric response and spin-polarized systems with high magnetostrictive response in composite systems, where strain is transmitted through the interface to excite the magnetic or ferroelectric component with an applied electric or magnetic field, respectively. The coupling in such cases is indirect, with either the electric field piezoelectrically transforming first into strain followed by a magnetic response through the magnetoelastic effect by modifying the magnetic configuration and the total magnetisation, or conversely, the magnetic phase is altered first under an applied magnetic field which then triggers strain through magnetostriction, resulting in a modulated converse piezoelectric response. $BaTiO_3$, $PbTiO_3$, $PbZr_xTi_{1-x}O_3$ (PZT) are among the most common ferroelectrics used to tune the response of the magnetic component in multiferroic composites, while Ni [59, 60], ferrite materials ($NiFe_2O_4$, $CoFe_2O_4$) [49] or Fe nanoislands [61, 62] have been used successfully as magnetic counterparts. An additional requirement is that the overall structure must have a large resistivity in order to sustain the applied electric field.

10.1.2.2 Charge Mediated Multiferroicity

A more direct way of controlling the magnetic response in multiferroic heterostructures is to modulate the charge carrier density in the contact region by switching the ferroelectric between different polarisation states. A text-book case is the interplay between charge doping and magnetic order at the interface of ferroelectric $BaTiO_3$ or PZT with mixed valency manganites $La_{1-x}M_xMnO_3$ (M = Ca, Sr, Ba), the latter being associated with rich phase diagrams that reflect the role of strong electron correlations and the strong couplings between charge, spin, and lattice distortions [4, 5, 63–67]. In such heterostructures, screening of the ferroelectric polarisation by hole carriers in the manganite leads to states of hole accumulation or depletion near the interface, resulting in a change in the Mn^{3+}/Mn^{4+} ratio [8, 68] and to modifications in the electronic ground state of the system, including a change in the effective magnetic moment [6, 8, 68–71], a transition from metallic to insulating phase [72, 73], and from antiferromagnetic to ferromagnetic [6, 7] (see Fig. 10.2). Such mechanisms are equally interesting in the context of multiferroic tunnel junctions used for information storage and manipulation [74] as well as in spintronics, since such systems act as spin filters limiting or enhancing the interface conductivity depending on the spin orientation [75].

10.1.2.3 Orbital Reconstruction Mechanism

The interplay of spin, lattice, charge and orbital degrees of freedom in multiferroic oxide heterostructures manifests for example in the ability to control the magnitude of the spin polarisation in the metallic electrode when altering the occupancy of the orbitals involved in the chemical bond. In the case of the hole-doped $La_{1-x}Sr_xMnO_3$ (LSMO), the octahedral crystal field leads to a splitting of the Mn 3d states into 2-fold degenerate e_g states ($d_{3z^2-r^2}$, $d_{x^2-y^2}$) lying at the Fermi energy, and the occupied 3-fold degenerate t_{2g} states (d_{xy}, d_{yz}, d_{xz}) [76]. Further splitting, due to spin-orbit coupling and crystal-field interactions associated for example with the Jahn-Teller distortion, determine which orbitals are preferentially occupied, with a direct impact into charge transport [77], excited state dynamics [78] or magnetism. Strain [79],

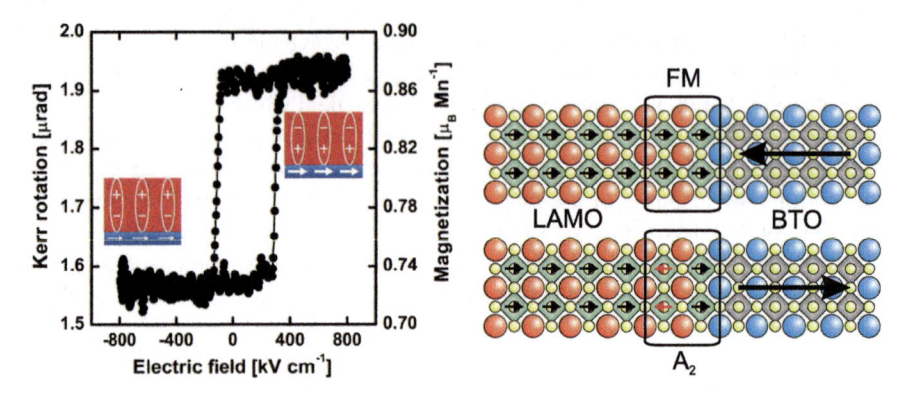

Fig. 10.2 Charge-driven spin modulation at a ferroelectric/ferromagnetic (PZT/LSMO) hole-doped interface (left, from [6]) and magnetic reconstruction in Sr, Ca-doped LaMnO$_3$/BaTiO$_3$ (LAMO/BTO) (right, from [7]), both driven by switching the direction of the ferroelectric polarisation

rotations of the octahedral cage [80] or modulations in the Mn-O bond length [19] have been shown to lead to preferential occupation within the e_g orbitals (*orbital polarisation*).

At an intuitive level one can understand the mechanism of interface orbital coupling as the alteration of the chemical bond lengths/angles by either strain, octahedral rotations or ferrodistortive displacements which change the orbital overlap and hybridisation, resulting in modified electronic properties. As an example, Cui et al. [81] showed experimentally how the ferroelectric instability [19] of BaTiO$_3$ in contact with an LSMO film propagates into the metallic electrode [82] as far as a few unit cells, modulating the Mn-O distance which in turn lifts the orbital degeneracy and e_g orbital occupancy: preferentially e_g ($d_{x^2-y^2}$) when the BaTiO$_3$ ferroelectric polarisation points away from the interface and preferentially e_g ($d_{3z^2-r^2}$) in the opposite polarisation state. This behaviour is responsible for a modulation in the transport [81] and magnetic properties [7, 68] as a function of the ferroelectric polarisation direction [19].

10.1.2.4 Interface Coupling Through Exchange Interaction

The exchange bias phenomenon refers to an horizontal shift of the *M-H* hysteresis loop when cooling a ferromagnet in contact with an antiferromagnet from above the Néel temperature under an applied magnetic field [83, 84]. Although the subject of investigation for many decades now, a full description of the processes giving rise to exchange bias is still not available due to the difficulty of fully accounting for the modified electronic and magnetic behaviour at the ferromagnetic-antiferromagnetic interface [85–89], including the structure of the uncompensated spins at the interface [85], the presence of interfacial spin structures [86, 90–92], or formation of domains with different boundary dynamics [93, 94].

Given that most single phase multiferroics are antiferromagnetic (including, BiFeO$_3$, one of the few systems that is multiferroic at room temperature), a natural extension is that of using exchange bias for achieving magnetoelectric coupling in multiferroic heterostructures [57, 95–98]. Typically, an antiferromagnetic ferroelectric intrinsic multiferroic is used as the electric actuator on a ferromagnetic adjacent layer. Bea et al. [99], studying a BiFeO$_3$/CoFeB bilayer, demonstrated the presence of exchange bias in the ferromagnet at room temperature due to interface coupling with the antiferromagnetic BiFeO$_3$, while Wu et al. [87] showed that, assisted by field effects, the magnitude of the exchange bias at a BiFeO$_3$/La$_{0.7}$Sr$_{0.3}$MnO$_3$ interface can be controlled by an electric field. Remarkably, it seems that in certain situations, such as in a Co/PZT/LSMO heterostructure, modulation of the exchange bias and magnetic anisotropy with the direction of the ferroelectric polarisation can appear even in the absence of a nominal antiferromagnetic layer [88], the effect appearing here presumably due to a reacted Co/PZT interface.

10.2 Introduction to PES and XAS: Differences and Complementarities

The magnetoelectric coupling mechanisms mentioned above have a common feature: their altered functionality arises at the interface and accessing the details of the processes responsible for such physical behaviour requires chemical sensitivity. X-ray absorption (XAS) and photoelectron spectroscopy (XPS or PES) give complementary information on the electronic structure and are particularly well suited for accessing buried interfaces. In the remainder of this chapter, we will consider how X-ray and photoemission spectroscopies can be applied to the characterisation and understanding of the interfacial mechanisms underlying magnetoelectric coupling phenomena. We start by providing a brief outline of the characteristics of both XPS and XAS.

10.2.1 X-ray Photoelectron Spectroscopy (XPS)

Photoelectron spectroscopy is a technique which gives information on material composition, the chemical state of atomic species within a given system, details of the local environment, and localization of emitting species. Conceptually, it is a photon-in electron-out method based on the photoelectric effect, where the electrons ejected from either core levels or from the valence band have a kinetic energy given by

$$E_k = h\nu - E_b, \qquad (10.1)$$

where $h\nu$ is the energy of incoming radiation and E_b the binding energy of the electron. The binding energy and, to a lesser extent, the peak shape, are the key physical parameters which can be further connected with:

- the chemical state of the atomic species by the so called "chemical shift" with respect to its neutral phase;
- the position of the emitting species through the background term associated to inelastic scattering of the photoelectrons on their way out the sample;
- the electronic correlations within the atomic shells which further induce broadening and shake-up/down satellites.

The chemical shift, in particular, provides invaluable information about the electronic state of the system, due to the high sensitivity of the energy levels (either core or valence states) to the slightest change in the screening of electron-electron or electron-nucleus interaction connected with the modification, formation and/or disappearance of chemical bonds and changes of the valence state.

If the radiation used as excitation source for the photoelectrons lies in the UV range, then one probes mostly states with small binding energy, i.e., located near the Fermi level or at the valence band maximum (VBM). In this case the technique is called ultraviolet photoemission spectroscopy (UPS) and its sensitivity to surface electronic states comes from the limited inelastic mean free path, λ, the distance over which electrons travel until their intensity decay to $1/e$ of the initial value,

$$I = I_0 e^{-d/\lambda(E)}, \tag{10.2}$$

with d the overlayer thickness. Hence, the probing depth in XPS is in the range of 3λ, with a dependence on the kinetic energy following a universal curve [100] with a minimum around $\sim$100 eV. For UPS, the probing depth associated with escaping photoelectrons is limited to the very first 1–3 surface layers. Given the large cross section for electron scattering and the extreme surface sensitivity, photoemission experiments are conducted in high vacuum in ultra-clean surfaces. UPS is characterised by an energy resolution in the meV range and high incoming photon flux when used with laboratory UV sources.

Soft X-ray (SXPES) at 100–1000 eV or hard X-rays (HAXPES), up to 10 keV energy range, allows one to extend the probing depth to buried interfaces or bulk-like regions, but with the price of a drop of 2–3 orders in the magnitude of the photoexcitation cross section. This is the reason why soft and hard X-ray photoemission are techniques particularly "photon hungry", especially when information on the valence band is required [101–103]. Moreover, the energy range of HAXPES is associated with a loss in momentum resolution due to electron-phonon coupling [103] or recoil effects [104] (with a few exceptions [103, 105]). In this context, SXPES is ideally suited for momentum resolved studies of buried interfaces as it combines a sharper momentum resolution compared with HAXPES and a higher penetration depth as compared to UPS.

10.2.2 X-ray Absorption Spectroscopy (XAS)

In X-ray absorption, an X-ray photon excites a core electron into unoccupied states in the system, a process which is seen as an abrupt increase in the absorption cross-section as the X-ray energy reaches the value corresponding to the binding energy of the core level electron (called absorption edge). The X-ray absorption process is driven by the electric field of the electromagnetic wave and the transition process is ruled predominantly by the electric dipole operator, hence it fulfils the electric dipole selection rules, including the requirement that the orbital quantum number changes by ± 1. The excited electron remains bound to the atom and the states which it can occupy lie above the Fermi energy, with the transition probability depending directly on the density of available final states, which in turn, depend intimately on the precise band structure of the solid as determined by chemical bonding, presence of magnetic order, crystalline environment, and so forth. The energy of the core level electron, on the other hand, may depend on the presence of core-level spin-orbit coupling, atomic number, and the ionic or valence state of the atom. Hence, the absorption spectrum of a material contains a wealth of information about the electronic structure and the use of such features to probe the latter is the purview of X-ray absorption spectroscopy (XAS) [106–108]. The absorption edges are often classified by assigning a letter K, L, M, N, O, P to transitions from $n = 1, 2, 3, 4, 5, 6$, levels, respectively, with a subscript to indicate the sublevel; alternatively, the electron transition can also be described by nl_s, where n is the atomic quantum number, $l = s, p, d, f, g$ corresponds to the orbital quantum number, and s is the spin quantum number. The X-ray absorption spectra can be divided into different energy regions where different electronic contributions dominate: the pre-edge region, mostly the result of orbital hybridisation between states of different orbital character (corresponding to transitions which otherwise would not be allowed by the dipole selection rules); the edge or near-edge region (also named X-ray near-edge absorption spectroscopy, or XANES, particularly at K edges); and the extended energy range above the absorption edge, where interference effects of the scattered wave with neighbouring atoms gives rise to a fine structure directly associated to local atomic bond lengths and bond angles (so-called extended X-ray absorption fine structure, EXAFS).

The fact that the wave function of core electrons tend to have well defined angular symmetries (as quantified by the orbital quantum numbers), means that transitions which originate from states with identical orbital states result in spectra that share some general common features, such as the K and L edges of the different elements. The transition energy, however, is specific to each element, giving XAS the element selectivity that is fully explored in combination with variable X-ray sources provided by X-ray synchrotrons. An important aspect of X-ray absorption is a consequence of the strongly localised nature of the radial wavefunction of the core electron, which overlaps only significantly with the radial wavefunctions of valence electrons in the immediate atomic neighbourhood, making XAS a very localised probe of the electronic structure [107]. The absorption process also depends strongly on the polarisation of the electromagnetic wave. In the case of circularly polarised light, the scattering process can result in a transfer of angular momentum from the photon to the electron, and becomes sensitive to the spin state of the core electron and to the presence of spin-asymmetries in the valence band (ferromagnetism). Linearly polarised light excites core electrons along directions parallel to the electric field of the X-ray light, providing sensitivity to asymmetric distributions in the electron charge density, including that originating from (anti)ferromagnetism and ferroelectricity.

XAS probes unoccupied states, which is different from XPS, where the excitation process ejects electrons from occupied states into the continuum, probing therefore occupied density of states. One advantage of XAS over XPS resides in the simplicity of measuring the X-ray absorption of materials. A direct measurement can be carried out by measuring the light transmitted across a specimen; such a process is often carried out, but for soft X-rays thin specimens need to be prepared due to the limited penetration depth of X-rays at the absorption edges (for example, ~70 nm for the Co L_3 edge at 779.7 eV) [109, 110]. Another method relies on measuring the intensity of photons emitted in the atomic transition to the ground state (fluorescence yield), which can be partial, when specific fluorescence lines are measured, or total, when the integrated fluorescence signal is measured. It probes a volume that is associated with the extinction length of the fluorescence X-rays in the material, which can be of several 100s nm. A third method relies on measuring the intensity of photoemitted electrons, called electron yield, which again can be either partial or total. Given that the inelastic mean free path of electrons in materials is very small, in the order of a few nm [100], the latter probing process is very surface sensitive. In the case of total electron yield, the bulk of the signal comes from secondary electrons that are generated in a cascade process from the de-excitation of high energy Auger electrons, providing an intrinsic signal amplification mechanism.

10.3 Photoemission Studies of Multiferroic Interfaces

10.3.1 Probing Interface Region with Core-Level PES

At its most basic level, the element selectivity and depth sensitivity of core-level spectroscopy can be explored for determining stoichiometry, intermixing, and the presence of defects and impurities: parameters which are critical in determining the physical properties of heterostructures. Moreover, in connection with the field of artificial multiferroic heterostructures, where the relevant mechanisms manifest at the interface, photoemission spectroscopy can be used as an effective tool in probing the electronic signature of the buried interface, including their evolution during film growth.

As an example, we consider the case of a thin Fe layer on top of ferroelectric $BaTiO_3$, which has been predicted to be multiferroic [111] with a coupling mechanism involving bond reconfiguration at the interface, driven by different hybridization of Ti 3d-Fe 3d states, which in turn depends on the direction of the ferroelectric

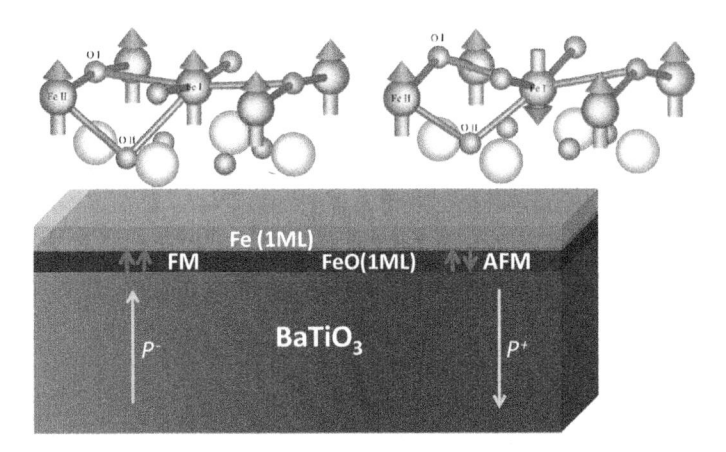

Fig. 10.3 Change of the magnetic order at the Fe/BaTiO$_3$ interface caused by switching the ferroelectric polarisation. Ferroelectric polarization towards the Fe/BaTiO$_3$ interface drives ferromagnetic ordering (**a**), while opposite polarization triggers antiferromagnetic ordering within FeO layer (**b**). Reproduced (in part) from [112], with permission

polarisation, P. The $P-$ state (defined here as pointing towards the ferromagnetic interface) is associated with a displacement of the Ti atoms closer to the Fe layer while the opposite behaviour is expected for the $P+$ state (see Fig. 10.3). These states are therefore associated with different lengths of the chemical bond between the interfacial Fe and Ti atoms, hence different overlapping of the electron clouds. Experimentally, such a system has been prepared in situ [112, 113] by depositing 2 atomic layers (ML) of Fe (1 ML = 1.4 Å) on top of ferroelectric BaTiO$_3$; an additional Co layer is deposited in order to preserve the magnetic state of Fe and the whole sample is capped with a protective Au layer. The region which is multiferroically active is the Fe/BaTiO$_3$ contact region, where the spin and ferroelectric degrees of freedom are coupled.

The mechanism for magnetoelectric coupling in this system has been determined experimentally to consist of a ON/OFF switching of the interface magnetic moment with the orientation of the ferroelectric polarisation, namely, of a transition from a ferromagnetic ground state in the interfacial layer to antiferromagnetic due to altered bond geometry in the contact region [112]. More specifically, the Fe$_1$-O-Fe$_2$ bond is different for the two ferroelectric configurations, where Fe$_1$ and Fe$_2$ are two inequivalent atoms in neighbouring sublattices connected through an oxygen atom. It is known that the indirect exchange parameters in spin-polarized systems are strongly affected by bond lengths and angles, and in this case it is found that it favours a ferromagnetic ground state in the FeO layer when the ferroelectric polarisation points towards the interface, while the opposite ferroelectric polarisation state favours an antiferromagnetic coupling (as schematically shown in Fig. 10.3). Surprisingly, the unintentional oxidation of the interfacial Fe is beneficial for the functionality of the heterostructure. In this case, the scenario for the magnetoelectric coupling is distinct from the canonical coupling of the magnetic and ferroelectric degrees of freedom in the ideal Fe/BaTiO$_3$ heterostructure with a sharp interface, which is characterised by a weaker magnetoelectric response with a modulation of the magnetic moment of only 0.06 μ_B per Fe atom [111].

The strong depth dependence of the photoemitted signal can be explored to study the structure and morphology of interfaces in thin film heterostructures. For the particular case mentioned above, of Fe deposited on BaTiO$_3$ substrate, the intensity of the Ti 2p line, attenuated by the Fe/FeO top layer is given by (10.2), while the intensity of the Fe 2p line can be expressed as:

$$I_{Fe} = N_{Fe}\lambda_{Fe}[1 - e^{-d_{Fe}/\lambda_{Fe}}] \tag{10.3}$$

where N_{Fe} is the atomic density and λ_{Fe} is the escape depth of Fe 2p electrons. The ratio

$$\frac{I_{Fe}}{I_{Ti}} = \frac{N_{Fe}}{N_{Ti}} \frac{\lambda_{Fe}}{\lambda_{Ti}} \frac{1 - e^{-d_{Fe}/\lambda_{Fe} \cos\theta}}{e^{-d_{Fe}/\lambda_{Ti} \cos\theta}}, \tag{10.4}$$

where θ is the angle between the electron analyser and the normal to the sample surface, describes the expected variation of the photoemitted intensity as a function of the Fe film thickness in the ideal case of flat and abrupt interfaces; interdiffusion, reactions at the interface or island-like growth instead of layer-by-layer will lead to deviations from the expected trend. In Fig. 10.4 (left panel) one can see how the intensity ratio I_{Fe}/I_{Ti} varies for Fe films deposited on BaTiO$_3$ (a) at room temperature and (b) at 373 K (also as a function of post-growth annealing). The fact that good agreement between the theoretical expression and the experimental data is observed only when the sample is prepared at higher temperature, indicates that deposition at 373 K favours a 2D growth with smooth and continuous Fe overlayer. Figure 10.4 (right panel) shows the XPS spectrum across the Fe L edge, where an additional component visible at higher binding energy in the Fe 2p spectrum that evolves with a post-annealing treatment indicates the presence of an oxide layer at the Fe/BaTiO$_3$ interface. In order to determine the thickness of the reacted layer, the previous model can be further elaborated to include an additional term which represents the contribution of the oxide layer beneath the Fe with thickness d_{FeO_x}. In this case the expression for the intensity ratios of Fe in the top layer and in the interface oxide layer becomes:

$$\frac{I_{FeO_x}}{I_{Fe}} = \frac{1 - e^{-d_{ox}/\lambda_{Fe}}}{e^{d_{Fe}/\lambda_{Fe}} - 1} \tag{10.5}$$

From (10.5) it is deduced that the thickness of the oxide layer is 0.14 nm (1ML). It is this layer which has been associated with the strong magnetoelectric response and ON/OFF switching of the Fe magnetisation due to switching of the magnetic ground state from ferromagnetic to antiferromagnetic, an effect which makes such system an efficient spin filter and of potential use for four-state memories.

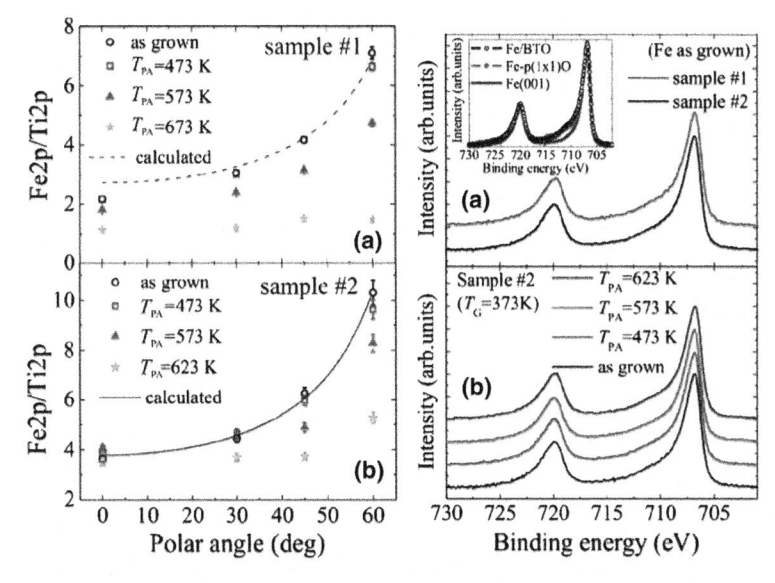

Fig. 10.4 Left: Variation of the XPS Fe/Ti intensity ratio as a function of polar angle for two Fe/BaTiO₃ samples grown at room temperature (**a**) and at 373 K (**b**), for various values of the post-growth annealing temperature. Lines are based on the theoretical curve for perfect 2D growth, which are found to provide a good agreement only for the sample prepared at higher deposition temperature. Right: Fe 2p core levels for the same two samples. Reproduced from [113], with permission

10.3.2 Schottky Barrier Height and Band Alignment at the Interface

One of the two building blocks of an artificial multiferroic is the ferroelectric component. Examples include PZT and BaTiO₃, which undergo a transition from a high-symmetry cubic structure to a low-symmetry state with a spontaneous electric polarisation below their critical temperature. They are also two of the most studied ferroelectrics, explored for applications in non-volatile memories [114], piezoelectric sensors, nanoactuators, lab-on-chips [115], pyroelectricity, solar energy and surface catalysis [116, 117]. In ferroelectric random access memories (FeRAMs), the capacitive readout is presently used. One disadvantage with such approach is that reading the data stored on a FeRAM requires applying an electric field to measure the amount of charge needed to flip the memory cell to the opposite state, destroying the information and requiring an additional re-writing operation to restore the bit. In addition, electrical fatigue is a well known problem that occurs with consecutive switching cycles, which leads to a drastic efficiency reduction due to reduced polarisation [118]. An alternative option is that of resistive readout in ferroelectric tunnel junctions [51, 119] based on tunneling electroresistance, a physical phenomenon associated with a modulation in the interface barrier height in response to polarisation reversal [120,

121]. Its functionality can be extended to four-state memory devices when multiferroic tunnel junctions and the additional tunneling magnetoresistive effect is used [74]. In both instances, a detailed knowledge of the mechanisms ruling the electronic behaviour of the interfaces is critical.

Of particular interest for the electrical properties of multiferroic heterostructures, including transport behaviour and conduction mechanisms operating in real devices, is the Schottky barrier, which may develop at metal/ferroelectric interfaces. When present, it leads to a rectifying behaviour with characteristics that can be parametrised in terms of the barrier width and height. The Schottky barrier height depends primarily on the difference between the work functions ($\phi_{M,S}$) of the two materials in contact; depending on the sign of $\phi_M - \phi_S$, the electrons tend to pass either from the conduction band of the semiconductor into the metal ($\phi_M > \phi_S$, corresponding to a bending of the semiconductor bands "upwards") and vice-versa. The resulting deformed band structure acts as a barrier that electrons have to overcome in order to transit the interface. In addition, other mechanisms may intervene and modify the height of the barrier, for example, structural defects at the interface which may act as recombination sites [120, 122, 123]. Since photoelectron spectroscopy probes the occupied density of states, it can be used to determine the relative band alignment of the different components of an heterostructure, from which the Schottky barrier height can be derived [124]. More specifically, with XPS one correlates the position of the Fermi level of the metal to that of the core-level shifts within the ferroelectric part; additionally, direct information on the local environment, composition and chemical state of the interface can be obtained as well.

The mechanism for band alignment and formation of a Schottky contact at multiferroic interfaces is far from trivial, since it involves a complex interplay between the electric field associated with polarisation charges, the depolarisation field which tends to restore charge neutrality, and imperfect screening of the interface charges near the metallic contact. Already in the case of free standing ferroelectric slabs, there is an inherent band bending [125–127] which, as illustrated in Fig. 10.5, depends on the direction of the ferroelectric polarisation due to the surface bound charge of the ferroelectric. This generates an interface built-in potential of the form $eP\delta/\varepsilon_r$, where δ is the distance the potential propagates into the ferroelectric, ε_r is the relative dielectric permittivity of the ferroelectric and P the ferroelectric polarization. This potential is further associated with a shift in the photoemission spectra towards lower ($P-$) or higher ($P+$) binding energies [127–129].

In addition to the band bending term due to differing work functions, the metal/ferroelectric contact leads to an additional modification of the band bending at the interface associated to the polarisation charges on a characteristic length scale that depends on the work function of the metal contact. In the particular case of the interface between ferroelectric PZT and Au, the metallic contact screens the polarisation charges at the interface, decreasing the band bending in the contact region to 0.45 eV and resulting in a Schottky barrier height of 1.12 eV for holes [127].

To determine experimentally the Schottky barrier height at the interface between two materials, say a ferroelectric A and a metal B, from the core-level position, it

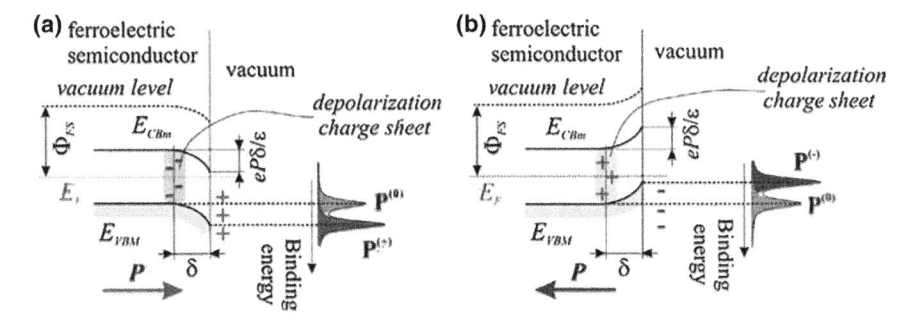

Fig. 10.5 Influence of the ferroelectric polarisation state on the photoemission spectra: $P+$ polarisation, pointing away from the surface creates a component at higher binding energy while $P-$ polarisation is seen at lower binding energies. Reproduced from [126], with permission

is useful to split the observed band offsets in terms of the ferroelectric and valence band offsets,

$$\Phi_p = \Delta E_v + \Delta V(z), \tag{10.6}$$

where $\Delta V(z)$ accounts for the skewing effect due to the ferroelectric potential, and ΔE_v, the valence band offset term, which has exclusively electronic origin and can be expressed as:

$$\Delta E_v = (E_B - E_A)_{A/B} - (E_B - E_F)_B - (E_A - \text{VBM})_A, \tag{10.7}$$

with E_F is the Fermi level of the metal and VMB is the valence band maximum of the ferroelectric.

Popescu et al. [130] calculated the Schottky barrier height for the BaTiO$_3$/LSMO interface for the hole-depletion state, where the top BaTiO$_3$ layer was made sufficiently thin in order to allow photoelectrons excited in the interface region to be detected (see Fig. 10.6). The terms of (10.7) are deduced based on the position of the Ti 2p and valence band maximum in BaTiO$_3$ and Sr 3d level in LSMO. The correction term of (10.6), $\Delta V(z)$, which stands for the skewing of the bands within the BaTiO$_3$ due to the ferroelectric polarisation, has been determined by comparing the position of the valence band maximum at normal incidence, which is more bulk-sensitive, and at grazing incidence for probing more of the surface. The modulation of the Schottky barrier height for the two directions of ferroelectric polarisation is then given by:

$$\Delta \Phi_B = 2\lambda_{\text{eff}} \frac{D_S}{\varepsilon_0} \tag{10.8}$$

with D_S the dielectric displacement in BaTiO$_3$, ε_0 the vacuum permittivity and λ_{eff}, the effective screening length which is a non-universal parameter that depends on the nature of the contacts and interface chemistry [19].

Fig. 10.6 Band alignment at the BaTiO3/LSMO multiferroic interface inferred from photoemission data. From [130], with permission

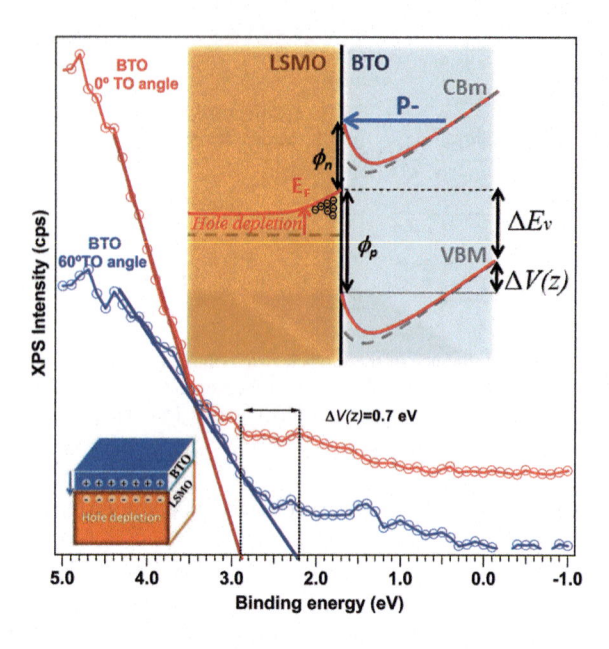

Table 10.1 Experimental estimates for the modulation of the Schottky barrier height (SBH) in ferroelectric-metallic structures with the direction of the ferroelectric polarisation ($P+$, $P-$ pointing away or towards the interface, respectively)

System	SBH ($P+$) (eV)	SBH ($P-$) (eV)	$\Delta\Phi_B$ (eV)	Reference
Pt/bulk-BaTiO3	0.3	0.95	0.6	[131]
RuO2/BaTiO3	0.35	1.45	1.1	[131]
Pt/thin-BaTiO3	0.4	0.85	0.45	[132]
LSMO/2.8 nm BaTiO3	1.35	1.05	0.3	[133]

A list of experimental estimates for the modulation of the Schottky barrier height in ferroelectric-metallic structures with the direction of the ferroelectric polarisation, reported in the literature from photoemission, is given in Table 10.1. One finds that the ferroelectric-induced modulation of the Schottky barrier height is large, reaching 1.1 eV in the case of RuO2/BaTiO3 [131]. In the case of Pt/thin-BaTiO3/Nb-SrTiO3, a smaller $\Delta\Phi_B = 0.45$ eV for the Pt/BaTiO3 interface is accompanied by a transition from Schottky to ohmic conduction at the bottom BaTiO3/Nb-SrTiO3 contact.

One aspect which has to be treated carefully concerns the dynamics of charge reorganization at multiferroic interfaces under X-ray illumination, which has implications to a proper understanding of the measurements performed with synchrotron light. Due to the high photon fluxes there is a high density of emitted secondary electrons which may screen the ferroelectric surface charge. Such an effect has been reported by Wu et al. [134] while probing the interface region of buried LSMO under

4 nm PZT. By following the Sr 3d core levels of the buried LSMO as a function of exposure time, they observed a variation of the position of the Sr lines with saturation behaviour in the range of minutes. In simple terms, the internal field associated with the ferroelectric polarisation acts as an effective bias which redistributes the photo-generated carriers, such that the depolarisation field of PZT is gradually suppressed to eliminate the polar discontinuity. This suggests that the best approach to investigating band alignment and cross coupling between ferroelectric and spin degrees of freedom might be in *operando* conditions (i.e., with bias applied on the metal/ferroelectric junction), i.e., in conditions that mimic the functioning of real devices.

10.3.3 Angle-Resolved Photoelectron Spectroscopy of Multiferroic Interfaces

The most direct way of accessing the electronic properties of a material is in angle resolved photoelectron spectroscopy (ARPES), since it gives information on the spectral function of a system, provides a mapping of the band structure, gives the carrier density at the Fermi level, and the topology of the reciprocal space. However, ARPES is generally limited to surfaces because of the small escape depth of the photoelectrons generated with conventional UV light sources. Probing deeper into the bulk with soft X-ray radiation comes at a price of losing 2–3 orders of magnitude in photoexcitation cross-section. The solution is to use synchrotron sources with high brilliance in order to compensate for the small signal in the valence band.

The first photoemission experiments with resolution in the k-space were performed for the LSMO ($x = 0.3$) interface buried under a thin ferroelectric layer at the Swiss Light Source [133]. LSMO is characterised by mixed Mn^{3+}/Mn^{4+} valence induced by the doping of the parent compound $LaMnO_3$ with Sr atoms. Knowledge of the bulk band structure, especially along the k_z direction, requires sharp momentum resolution achievable with higher probing depth in soft X-ray ARPES (SX-ARPES). In another study, the theoretical topology [135] of the reciprocal space for bare LSMO has been confirmed using ARPES [136]. The momentum space is characterised by electron spheres at the Γ point: Mn 3d e_g $d_{3z^2-r^2}$ and $d_{x^2-y^2}$ degenerate orbitals with equal occupancy and hole cuboids at the corners of the 3D Brillouin zone. The electron spheres are probed using 643 eV incoming photons and hole states using 708 eV. In each case, the signature of a rhombohedrally distorted cell is seen as "shadows" between the main features (Fig. 10.7). Additionally, element and chemical specificity of the states near the Fermi level has been achieved in resonant photoemission measurements at the Mn 2p edges (~643 eV) which, due to the intra-atomic 2p–3d resonance, disclose the exact contribution of Mn states at the Fermi level [137].

By taking advantage of recent developments in SX-ARPES, which provide unprecedented photon flux [101], the interfacial electronic properties of a 2.5 nm $BaTiO_3/LSMO/SrTiO_3(001)$ heterostructure could be visualized directly as well with

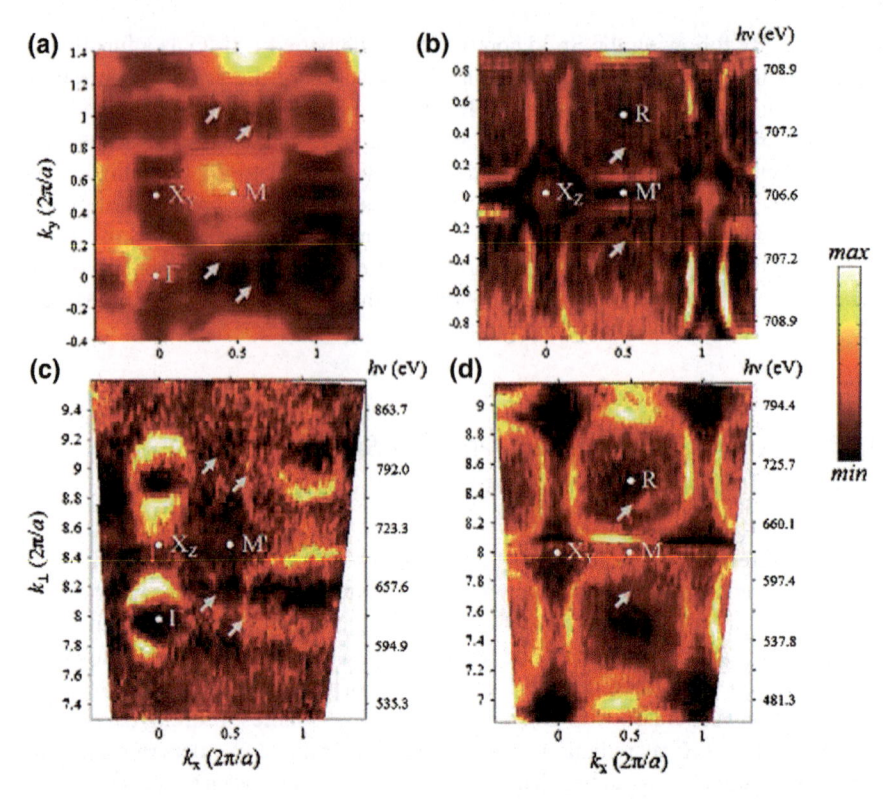

Fig. 10.7 **a** Experimental $k_\parallel$ cuts of the Fermi surface measured as a function of the polar angle θ, representing the ΓXM plane [**a**, measured at $h\nu = 643$ eV] and XZM'R plane [**b**, with $h\nu$ marked on the right varying around 708 eV as a function of k_y to keep $k_\perp$ constant]; experimental $k_\perp$ cuts measured under $h\nu$ variations (marked on the right for $k_x = 0$) representing the X_XM (**c**) and X_YMR (**d**) planes. For the latter, $h\nu$ tracked the sample rotation to keep constant $k_y = \pi/a$. The "shadow" Fermi surface contours are marked by arrows. From [136], with permission

momentum resolution using soft X-ray linear polarized light. The ferroelectric polarisation points towards the interface ($P-$), thus inducing a hole-depleted region near the interface [6–8, 68, 75]. The band structures recorded along the ΓX direction at 643 eV for a reference LSMO film and the LSMO/BaTiO$_3$ heterostructure are shown in Fig. 10.8(c) and (d), respectively, and show that the signature of hole depletion at the LSMO interface is the increase of the Fermi wavevector from $k_F = 0.25/\pi$ in the LSMO reference film to $k_F = 0.27/\pi$ for the buried interface. The Luttinger theorem [138] which links the number of particles in the ground state with the "volume" enclosed by its Fermi surface (*Luttinger volume*) confirms that this finding is consistent with the scenario of a hole depletion (electron doping) state. Another effect associated with the hole-depleted interface is a renormalisation of the effective mass of the electrons at the interface due to increased correlation effects by a factor of $\sim$2 compared with the bare surface. This value has been deduced from a parabolic

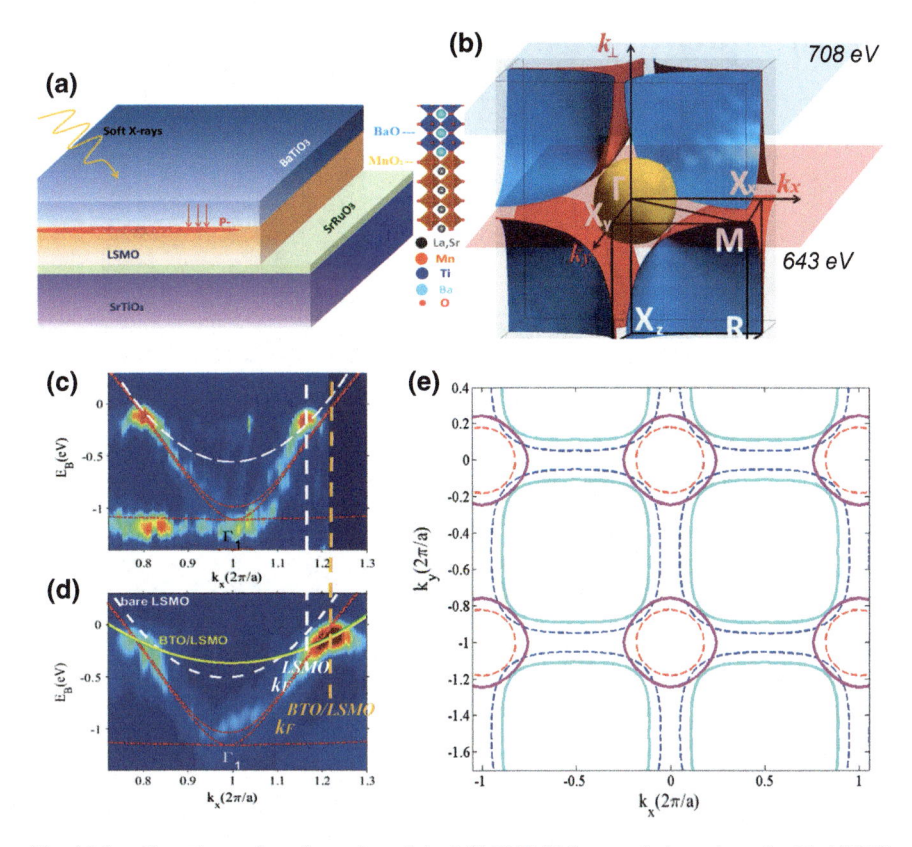

Fig. 10.8 a Experimental configuration of the LSMO/BaTiO$_3$ sample investigated with ARPES and **b** geometry of the reciprocal space. Experimental band structures recorded on **c** bare LSMO and **d** LSMO/BaTiO$_3$ interface along the ΓX direction showing the effect of electron doping which manifests in the increase of the k_F for the doped interface. **e** The four Fermi surfaces, calculated for states containing the electron spheres and hole cuboids (dashed lines for bare LSMO and continuous line for the hole-depleted interface) summarizes the effects of charge modulation due to ferroelectric effect. From [133], with permission

fit of the kink in the electronic bands ranging in the 0–0.3 eV interval below the Fermi level. The calculated Fermi surfaces in both planes containing Γ and R points for the LSMO film and hole-depleted interface are superimposed in the same image shown in Fig. 10.8e. They indicate that when the density of hole states (inferred from isoenergetic cuts in the ΓX$_Z$R plane) decreases, the electron density (extracted from Fermi surfaces in ΓX$_X$M plane) increases, so that the overall neutrality of the system is preserved.

In order to achieve a definitive picture of the charge modulation effect, the Fermi surfaces have been recorded at 643 eV to probe the ΓX$_X$M plane containing electron spheres and at 708 eV to probe the hole states at the LSMO/BaTiO$_3$ interface, shown in Fig. 10.9a, b, and for the bare LSMO (c, d). The resemblance with the bulk LSMO

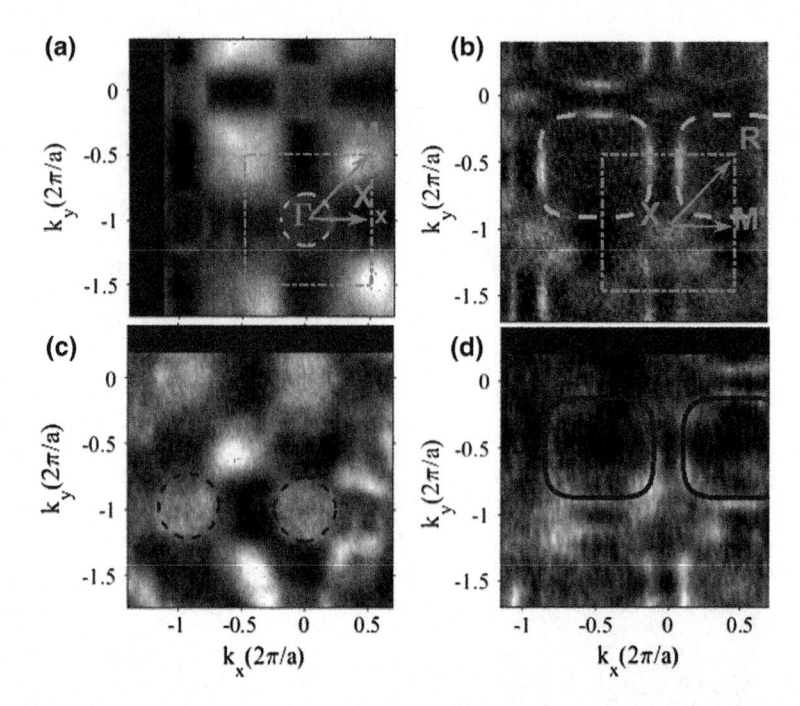

Fig. 10.9 a Fermi surfaces recorded at 643 eV revealing the electron spheres in Γ and **b** hole cuboids at 708 eV for the LSMO/BaTiO$_3$ interface and **c–d** and **a–b** bare LSMO taken as reference. From [133], with permission

is striking, with no major modification of the reciprocal space topology. The charge enrichment seen in the larger spheres centered at Γ accompanied by the shrinking of the hole cuboids in the corners of the Brillouin zone corresponds to a modulation of ~0.1 e/unit cell [7, 68]. Such direct measurements of the band structure modifications induced by the ferroelectric polarisation at the multiferroic interface measured using ARPES open the way for further studies involving metallic interfaces modified by the ferroelectric field effect, such as in PbTiO$_3$/SrRuO$_3$, PZT/LSMO or BiFeO$_3$/LSMO, which are of particular interest from both practical and basic science perspectives.

10.4 X-ray Absorption Spectroscopy Studies of Multiferroic Systems

X-ray absorption spectroscopy, with its ability to probe the electronic structure of matter and its capability for discriminating different aspects of the material properties, including magnetism and ferroelectricity, has been widely employed in the investigation of multiferroic systems. In this section we describe some of the unique contributions of this technique to our understanding of multiferroic phenomena, which largely explore the fact that by tuning the X-ray energy to the absorption edges of the different elements present in the system and by choosing the appropriate light polarisation, the different facets of multiferroic systems can be probed: either the electronic structure by probing the unoccupied density of states above the valence band, which reflect closely the local crystalline atomic environment; or by probing the magnetic moment using the X-ray magnetic circular dichroic effect (XMCD), or ferroelectricity by using linear dichroism, the latter two aspects being often associated with different atomic species in the system. In addition to standard spectroscopy, several XAS-based techniques have been developed that provide spatially resolved absorption maps of the sample, enabling local measurements of the magnetic and ferroelectric properties, including the determination of magnetic and ferroelectric configurations. In the following we aim at illustrating some of these capabilities by highlighting some of the work where the use of XAS has been employed to study and characterise interfacial phenomena at ferroelectric/ferromagnetic systems, including magnetoelectric coupling. For simplicity, we discuss first non-spatially resolved XAS and we consider separately L-edge spectroscopy, associated with p to d transitions in the 3d transition metals, and K-edge spectroscopy, associated with s to p transitions (X-ray absorption near edge spectroscopy, or XANES). We then consider techniques that can provide spatially-resolved XAS maps of the sample, with emphasis to X-ray photoemission electron microscopy (XPEEM), which has been widely used to the investigation of multiferroic heterostructures.

10.4.1 L-edge XAS Studies of Multiferroics and Heterostructures

X-ray absorption spectroscopy is a very powerful and mature technique that is routinely used for the electronic characterisation of materials, including ferromagnets, ferroelectrics and single phase multiferroics [139–142]. Its simplest application is that of identifying the oxidation or valence states, or the chemical bond and environment, by comparison with reference or calculated spectra, but more advanced use of X-ray absorption include probing ferromagnetic order through the XMCD effect (and in some instances, of the quantitative determination of spin and orbital contributions to the total moment through sum rules [143, 144]); the probing of antiferromagnetic and ferroelectric configuration through linear dichroic effects; to the determination of bond angles and bond lengths in EXAFS. In systems where ferromagnetic and ferroelectric order originate from different atomic species, the energy selectivity associated with X-ray absorption allows one to probe both aspects of the system largely independently in a single experiment. One example from the literature illustrating this capability is the report by Yi et al. [145] showing the control of magnetic order at the interface between $La_{0.5}Ca_{0.5}MnO_3$ and $BiFeO_3$ through the magnetoelectric coupling: by switching the $BiFeO_3$ ferroelectric polarisation, both a change in the magnetic moment of the Mn and of the Fe are observed, from zero to a finite value, as determined from XMCD measurements at the Fe and Mn L edges. The development of a net Fe magnetisation in this system was attributed to antiferromagnetic superexchange between Fe and Mn atoms at the interface, which is modulated by the changes in charge carrier density resulting from screening of the $BiFeO_3$ ferroelectric polarisation. Another instance where a net magnetisation in the Fe atoms of $BiFeO_3$ films interfaced with a doped manganite develops, has been reported for the $LSMO/BiFeO_3$ interface [146], also probed by XMCD; the effect in this instance was attributed to electronic orbital reconstruction at this interface [81] and was found additionally to be associated with the development of exchange bias in the system [146, 147]. The role of modified orbital occupancy at multiferroic interfaces has also been investigated for the LSMO ($x = 0.175$)/PZT interface using soft X-ray spectroscopy and the linear and circular dichroic effects [68]. In this work, it is found that switching the direction of the ferroelectric polarisation results in a energy shift in the Mn L-edge spectra associated to a change in Mn valency, in a change in the magnetic moment (higher for the depletion state), and in a change in the orbital occupancy, effects that are confined to the interface region. These results are in agreement with previous spectroscopy measurements on LSMO ($x = 0.2$)/PZT [8], where those changes were attributed to a change in the interfacial charge carrier density (see Sect. 10.4.2); in the work by Preziosi et al. [68], however, those changes are attributed to changes in the polarisation dependent structural distortions of the interfacial MnO_6 octahedra, as found in LSMO ($x = 0.2$)/$BaTiO_3$ [19].

XAS has been used extensively to determine the effect of strain on the electronic and magnetic structure [148]. Among the magnetic materials used in strain-mediated multiferroic heterostructures, we mention the metallic ferromagnets, including the doped manganites, where piezostrain induces changes in the magnetic anisotropy via magnetoelastic interaction, which gives rise to changes in the magnetisation easy axis and in the coercivity. A particularly strong manifestation of these effects is that of ferroelectric domain imprint on the magnetic configuration of the ferromagnetic film, as observed in Fe films deposited on a- and c-oriented $BaTiO_3$ crystals [149–151]. Ferrites, including magnetite, have also been widely investigated in this context, partly due to their high resistivity and partly due to their chemical stability vis-à-vis the ferroelectric perovskites. Ferrite-perovskite composites were the first artificial multiferroic composite materials synthesised with the aim of overcoming the small number of intrinsic multiferroic materials available in nature and their relatively small magnetoelectric coupling [152–154]. Such composite structures can be synthesised in a variety of ways, including euctetic growth in the bulk, sintering of nanopowders of the individual materials, self-organised growth induced by phase separation in co-deposition of thin films [49, 155–157], the growth of ferrite/perovskite layered structures [158, 159]; the various synthesis processes, materials combinations and materials characterisation have generated a vast literature aiming at maximising the elastic coupling between the composite phases and achieving larger magnetoelectric effects, and to reducing charge leakage (mostly in the ferromagnetic component) [40, 56, 160–162]. Spectroscopy studies of such systems have concentrated on the determination of the magnetic moment associated with the different elements of the magnetic component (the spin alignment in the Ni and Co ferrites is antiferromag-

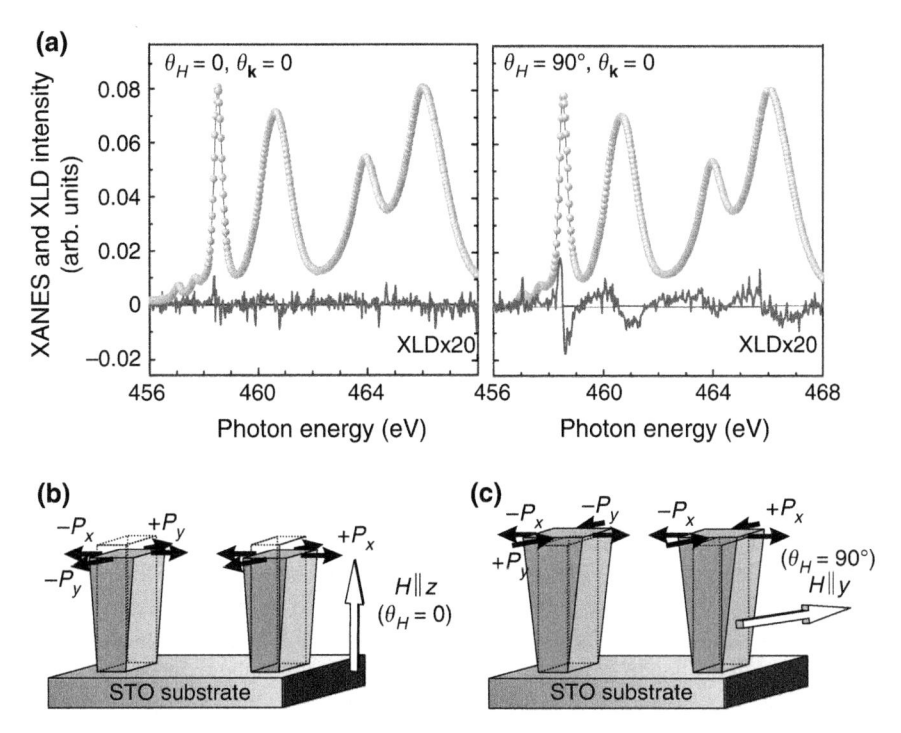

Fig. 10.10 a X-ray absorption spectra (circles) and linear dichroism (lines) at the Ti $L_{2,3}$ edge at normal incidence measured at remanence after applying a saturating magnetic field to the $CoFe_2O_4/BaTiO_3$ nanocomposite sample along the perpendicular direction (left) and in-plane (right). **b** and **c** show a schematic of the expected lattice distortion in the $BaTiO_3$ nanostructures induced by magnetostriction from the $CoFe_2O_4$. From [156], with permission

netic, with the non-zero total moment arising as a consequence of the different spin states of the cations composing the two antiferromagnetic sub-lattices) [163, 164] and with the identification of the relative site occupation of the different magnetic cations. The element specificity of XAS has been explored to obtain element-specific magnetic hysteresis loops, for example, for the Ni and Fe cations in a $NiFe_2O_4$-$BaTiO_3$ ceramic [165].

One example where the use of the natural dichroism that arises with changes in the crystal symmetry induced by strain is employed to understanding the nature of the magnetoelectric coupling in strain mediated multiferroic nanostructures is provided by Schmitz-Antoniak et al. [156, 165] in [001]-oriented $CoFe_2O_4/BaTiO_3$ multiferroic nanocomposite, where they consider the structural changes on the $BaTiO_3$ induced by the $CoFe_2O_4$ magnetostriction. This is seen as the emergence of an in-plane linear dichroism at the Ti $L_{2,3}$ edge going from the case where the magnetic field is applied out of plane (resulting in an in-plane cubic symmetry) to applying the magnetic field in-plane (resulting in a uniaxial distortion of the $BaTiO_3$ lattice), as shown in Fig. 10.10.

10.4.2 K-edge X-ray Absorption Spectroscopy of Multiferroics

The absorption K edges are characterised by excitation of core electron from s states, which are not subject to spin-orbit coupling (L = 0). In addition, the final p states in the 3d magnetic systems have only very small spin imbalances, such that absorption at K edges are not very sensitive to magnetic order.[1] The absorption K-edges associated with 1 s electrons are often characterised by an abrupt, step-like increase in the absorption spectrum and the sensitivity of the edge energy value to the atomic chemical state and environment, including crystal symmetry, makes K-edge spectroscopy very useful to assessing ionic and valence states in the system, as well as in identifying structural phases. The absorption K edges of the 3d transition metals lie in the 5–9 keV energy range, where the attenuation length of x-rays becomes relatively large (a few μm in the 3d elements and in the metre range in air). At these energies and for elements with atomic number $Z > 20$, fluorescence yield becomes a significant relaxation channel for the absorption process [139], making XAS in this energy range eminently bulk sensitive when measuring fluorescence yield. However, in heterostructures, it is possible to interrogate small regions of the sample structure when they are composed of, or contain a specific element, where the use of high brilliance synchrotron x-rays compensate for the reduced scattering volume. One key advantage of XAS at these energies is its ability to probe deeply buried layers, which is more challenging with other spectroscopic techniques, such as photoelectron spectroscopy or soft X-ray spectroscopy [109, 167].

The power of XANES to probe changes in the electronic state can be illustrated by the work carried out on multiferroic PZT/LSMO device structures. In this system, one exploits the surface bound charge of the ferroelectric layer to induce a charge modulation in the metallic LSMO channel layer (so-called ferroelectric field effect), which results in either depletion or accumulation of hole carriers at the LSMO/PZT interface for the two states of the ferroelectric polarisation direction. Using this approach, a large magnetoelectric effect was observed by direct magneto-optic Kerr effect measurements, seen both as a large change in the saturation magnetisation as a function of the direction of the ferroelectric polarisation and as a modulation in the magnetic critical temperature and coercivity [6, 168]. Such results indicate the presence of a charge-driven modulation of the magnetic moment and of the exchange interactions in the LSMO. To address the electronic nature of the magnetoelectric coupling in this system, the electronic state was directly probed with XANES by measuring the changes in the absorption K edge of Mn as a function of the ferroelectric polarisation direction, as shown in Fig. 10.11. The samples for these experiments consist of 10–12 u.c. LSMO ($x = 0.2$) film deposited on a $SrTiO_3$(001) substrate, where the thickness is right at the onset of electrical conductivity in the LSMO film [72]. A 200 nm thick

---

[1] However, due to electronic hybridisation of the final 3p states with 3d states in the magnetic transition metals, pre-edge features may contain magnetic information, and in fact, the first experimental observation of the XMCD effect was at the Fe K edge [166], where the largest spin asymmetry is observed just below the main absorption edge.

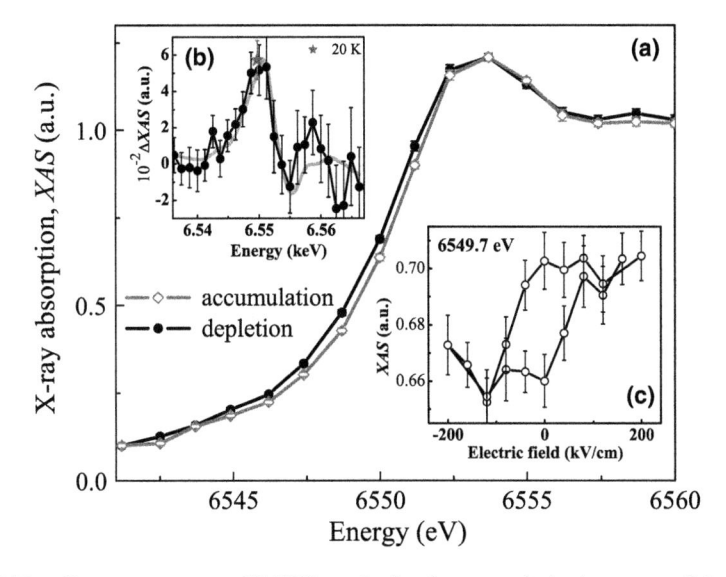

Fig. 10.11 a Room temperature XANES results for the two polarisation states of the PZT. b Difference in X-ray absorption for the two PZT polarisation states; the full line models this difference assuming a rigid shift in the Mn absorption edge. c Variation of the X-ray light absorption as a function of the applied gate voltage at a fixed energy of 6549.7 eV. From [8], with permission

PZT layer is then subsequently deposited, followed by a Au top contact, used with the LSMO film as electrodes for switching the ferroelectric polarisation. The sample is made in the form of a Hall bar device structure by lithographic methods, with the active area of the device sufficiently large ($160 \times 320\,\mu m^2$) to accommodate the X-ray beam. One finds that, as the direction of the ferroelectric polarisation switches the system from depletion to the accumulation state, the absorption edge of Mn shifts by 0.1 eV to higher energies.

It is well established that the position of the absorption edge is very sensitive to the cationic valency in a wide array of compounds [169, 170]. In the case of the manganites, the energy edge position has been found to vary with the doping level x between $x = 0$ (Mn^{3+}) and $x = 1$ (Mn^{4+}) by about 3.5 eV in $LaMn_{1-x}Co_xO_3$ [171], 3–4.2 eV in $La_{1-x}Ca_xMnO_3$ [172–174], 2.5–3 eV in $La_{1-x}Sr_xMnO_3$ [175–177], accompanied by other more subtle changes in the spectra, including peak amplitude and edge shape. Generally speaking, the edge energy position increases from the metallic state (Mn^0) to higher formal valence states. Several effects can contribute to this shift in the absorption edge, including changes in the Mn–O bonding distance (larger bonding distances favour ionic over covalent bonding and a reduction in the bandwidth of the 4p valence states), and a change in the nominal valency, which affects the position of both core and valence energy levels [177]. Hence, in the manganite series, the change in edge energy with doping can have both a ion-ligand bond length contribution as well as a contribution arising from the change in the valence state of the cations proper. In the present case, the observed changes in the absorption edge energy are expected to originate from the change in cation valency only, giving a change in valency of $\Delta x = 0.1$ between the accumulation and depletion states averaged over the whole of the LSMO film [175]. This corresponds to a total change in interfacial charge of 1.1 e per square unit cell, in good agreement with the $2P_s$ value found for the saturation ferroelectric polarisation, showing that the PZT polarisation is effectively screened by charge carriers from the LSMO and indicating also a low density of charge traps at this interface [178]. This change in valency is expected to take place mostly at the PZT/LSMO interface, to within the screening length of LSMO, of about 1 u.c. [72]. These results have been confirmed by more recent XAS measurements for LSMO ($x = 0.175$)/PZT heterostructures [68], where the probing of the Mn 3d states indicate a change in the orbital character at the interface (attributed to changes in the bond lengths at the interface atoms, instead of strictly to changes in occupancy arising from hole depletion or accumulation) [19].

Another direct application of 1s K edge XANES to the study of the magnetoelectric coupling is illustrated by the work of Park et al. [179] in $CoFe_2O_4$/PMN-PT multiferroic heterostructures. In this study, a 100 nm thick $CoFe_2O_4$ film is deposited on a ferroelectric PMN-PT(001) substrate, where a strain-induced magnetoelectric effect leads to an electric field modification of the coercive field and remanent magnetisation. The effect of piezo-strain on the electronic properties of the $CoFe_2O_4$ was studied using XANES at both the Fe and Co K edges with and without the applied electric field, where it is found that the applied electric field results in a shift of the X-ray absorption edge to higher energies, in the case of the Co K edge, and to a shift in the opposite direction at the Fe K edge. These changes in the cation valency are explained in terms of a reduction of the oxidation state of Fe^{3+} and increase in the oxidation state of Co^{2+} under the action of an in-plane compressive strain induced by the electric field, which is attributed to a partial charge redistribution induced by the lattice distortions, an effect that can be seen as the counterpart of the example shown in Fig. 10.10 [156].

10.4.3 X-ray Photoemission Electron Microscopy (XPEEM)

Spatially resolved maps of the X-ray absorption of materials can provide direct access to non-uniformities in the electronic, magnetic, and ferroelectric state and is implemented in some well established techniques, either based in full-field imaging (X-ray photoemission electron microscopy, PEEM, and X-ray holography) or in rastering modes (scanning transmission X-ray microscopy, STXM, and X-ray ptychography) [180, 181]. In STXM, the X-ray beam is focused to spot sizes down to about 10–15 nm using Fresnel zone plates, and rastered through the sample to provide a direct measurement of the X-ray absorption of the sample with nm spatial resolution. Due to the requirement that the sample be X-ray transparent, it has been less explored to studying multiferroic systems. A more recent development is that of X-ray ptychography [182–184], where the scattering patterns of overlapping areas of the sample are used to reconstruct both amplitude and phase of the scattered waves, and where spatial resolutions are limited by the largest scattered wavevector measured, reaching 8 nm in 2D and down to 16 nm in 3D tomography [185–187]. In X-ray holography, an X-ray beam transmitted through the sample interferes with a reference beam providing a scattering pattern that include information both about the amplitude and phase of the scattered beam, and in particular, variations in intensity caused by the XMCD effect, for example. It has a spacial resolution that is limited by the size of the reference beam, which can be made as small as 30 nm, and since the sample is also the optical element, it uses relatively simple experimental set-ups. The ultimate use of this technique combines the reference aperture with the sample in a single support, making the measurement largely vibration insensitive and providing extra spatial sensitivity well beyond the nominal spatial resolution, down to a few nm [188, 189]. One important constraint for this technique resides in the sample fabrication, which needs to be X-ray transparent at the sample and reference aperture regions only, often requiring an involved sample fabrication procedure [188]. In XPEEM, the sample is uniformly illuminated with X-rays and a magnified image of the locally emitted photoelectrons (whose intensity is proportional to the X-ray absorption) is obtained using electron optics. Since it relies on detecting photoemitted electrons, its probing depth is of the order of a few nm, making PEEM particularly suited to studying surface and interface phenomena [190, 191]. Its spatial resolution is of the order of 30 nm in XPEEM, limited mostly by aberrations introduced by the spread in the energy and emission angle of the photoemitted electrons, but also by the X-ray energy dispersion (aberration-corrected PEEM microscopes can theoretically improve the spatial resolution down to $\sim$4 nm [192, 193]). One strength of PEEM is its ability to provide direct maps of the magnetic and ferroelectric domain structure through X-ray dichroic effects; in particular, due to its high spatial resolution, the magnetic configuration of submicrometre-sized elements can be measured. An example of this capability is illustrated in Fig. 10.12, where the ferroelectric and antiferromagnetic domain structure of a $BiFeO_3$ sample is obtained over the same area by determining the linear dichroism at the O K edge (for the ferroelectric order) and at the Fe L edge (for the antiferromagnetic order) [194]. Also worth noting is that, since the XMCD effect does not rely on a direct magnetic interaction, the magnetic measurements with XPEEM have the advantage that the magnetic state of the system is not disturbed by the probe (unlike in magnetic force microscopy, for example, where the magnetic tip can perturb the magnetic state).

A different contrast mechanism in PEEM relies on differences in the surface workfunction; in particular, for ferroelectrics, it has been demonstrated that the presence of a ferroelectric polarisation modifies the surface potential and makes ferroelectric domains visible in PEEM, particularly in so-called threshold photoemission, which uses photon energies close to the surface work function [195]. Since most XPEEM units are also equipped with an electron gun and can therefore also double as a low energy electron microscope (LEEM), it is worth mentioning a different mechanism for probing ferroelectric domains based on operating the system in mirror mode; in this case, different intensity contrasts are obtained for low energy electrons feeling the electrostatic potential at the surface generated by the ferroelectric polarisation to produce high resolution images of the ferroelectric domain configuration of the

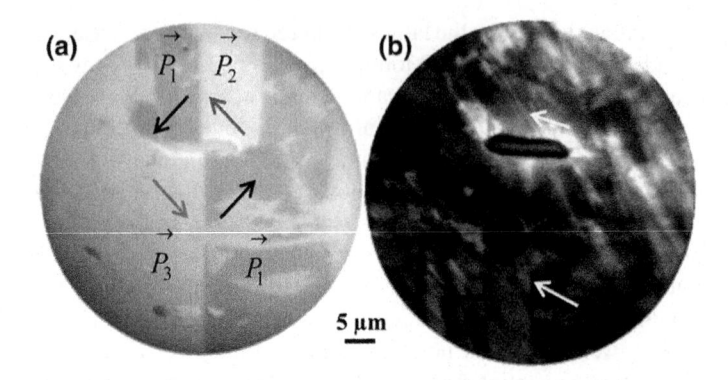

Fig. 10.12 XPEEM images taken on a BiFeO$_3$ single crystal showing **a** ferroelectric domains and **b** the antiferromagnetic domain structure obtained by using linear dichroism effects at the O 1s edge and Fe L$_2$ edge, respectively. Reproduced with permission from [194]

sample [196]. One important requirement for XPEEM is that the sample be slightly conducting and compatible with ultrahigh vacuum conditions; in fact, the sensitivity of the imaging quality to local charging of the sample tends to be the main difficulty in using XPEEM for studying multiferroics. Despite this, XPEEM has been successfully used to study such systems, and has provided unique information about the electronic and magnetic changes as a function of the applied electric field, and therefore of the magnetoelectric coupling mechanisms [95, 197].

A striking example of the use of XPEEM to study the magnetoelectric coupling in multiferroic systems is that provided by the work carried out by Zhao et al. [29] on BiFeO$_3$(001) thin films, where a direct correlation between the ferroelectric domain structure, measured by piezoelectric force microscopy (PFM), and the antiferromagnetic domain structure, measured by XPEEM, is found, by measuring the same area of the sample in both techniques. This study could demonstrate, in particular, that the switching of the ferroelectric polarisation also switches the direction of the antiferromagnetic domains, demonstrating therefore a direct coupling between the ferroelectric and magnetic order parameters in BiFeO$_3$ (a result also obtained independently from neutron scattering measurements [198]). The determination of the antiferromagnetic domain structure relies on the linear magnetic dichroic (XMLD) effect; the latter effect can be exploited in two different ways for imaging, one by measuring the difference in absorption for two different directions of vibration of the electric field of the light, and the other relying on changes in the linear dichroism as a function of energy (in particular, at two energies where the linear dichroism has opposite signs, to provide enhanced contrast). The first process is a true measure of the linear dichroism, but has the difficulty that, in XPEEM, to probe in-plane crystal orientations in a thin film, a physical rotation of the sample is required.[2] Hence, it is easier to probe the difference in absorption at two different energies: the acquisition

---

[2]The low angle of incidence typical of XPEEM implies that s-polarised light is fully in the plane of the sample, while p-polarised light has a dominant out of plane component. The latter geometry

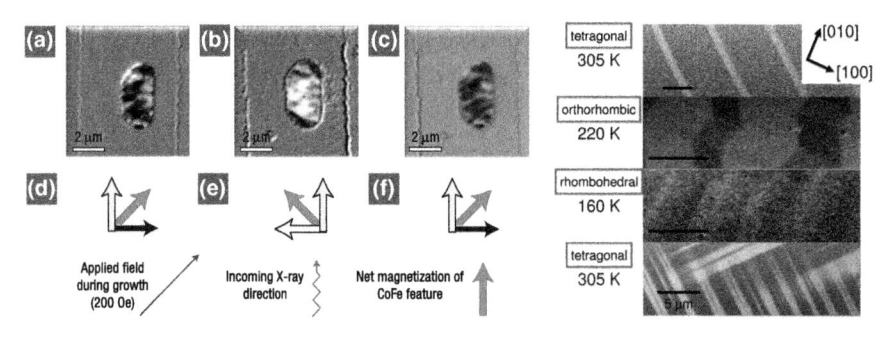

Fig. 10.13 Left: XMCD-XPEEM images taken at the Fe L edge showing switching of the magnetisation direction of the FeCo element with electric field applied in-plane. Right: XPEEM-XLD images of CoFe$_2$O$_4$/BaTiO$_3$ heterostructures at the Fe L edge showing the changes in the linear dichroism as a function of the temperature-driven phases transitions of the BaTiO$_3$ crystal that reflect the change in the magnetic domain structure. Reproduced with permission from [203] (left) and from [204] (right), respectively

time is faster since changing the energy is a faster process than physically rotating the sample or changing the light polarisation (unless the beamline is equipped with two twin undulators [199]); by choosing two energies at which the linear dichroism is inverted, a larger signal to noise ratio can be obtained [200–202]. However, the presence of linear dichroism can also be associated with the presence of a natural linear dichroism (XNLD) associated with the crystal structure. A separation of these two effects can be accomplished by increasing the temperature to above the Néel temperature of the system, where the XMLD signal vanishes [197].

In another instance, the electric field control of the magnetic state of a CoFe patterned element on a BiFeO$_3$ film was demonstrated by Chu et al. [203]. In this study, a local electric field applied close to a micrometre-sized CoFe element using lateral electrodes patterned into a bottom SrRuO$_3$ conducting layer was used to switch the ferroelectric polarisation. By directly measuring the magnetic domain structure using the XMCD effect at the Co L-edge, a change in the average magnetisation by about 90° could be observed, shown in Fig. 10.13 (left). In this case, the magnetoelectric effect is a consequence of a combination of two coupling effects, a first one related to an intrinsic coupling in the BiFeO$_3$ between the ferroelectric and the antiferromagnetic order parameters (as discussed above) and a second process corresponding to an exchange interaction between the antiferromagnetic order in BiFeO$_3$ and the ferromagnetic order in the CoFe (although effects related to strain may also be present).

XPEEM has also been used to investigate the elastic coupling between magnetic thin films and nanostructures deposited on ferroelectric substrates. Two ferroelectric substrate materials have been widely investigated, BaTiO$_3$ and [Pb(Mg$_x$Nb$_{1-x}$)O$_3$]$_y$-[PbTiO$_3$]$_{1-y}$ (PMN-PT), in the case of BaTiO$_3$ as a result of the presence of several

---

has associated with it a natural linear dichroic contribution due to the break of symmetry at the interface (in particular, taking into account the surface sensitivity of PEEM).

phase transitions in a very accessible temperature range (making it particularly useful for the investigation of magnetoelectric coupling in heterostructures [158, 205–207]), and for PMN-PT due to its very large piezoresponse. For example, $CoFe_2O_4$, $NiFe_2O_4$ [204] and LSMO ($x = 0.3$) [208] deposited on $BaTiO_3$ single crystals are shown to present local magnetic anisotropies that reflect the local ferroelectric domain structure of $BaTiO_3$, which when cooled down from above the critical temperature of 393 K to room temperature breaks into a multidomain structure composed of a- and c-domains that reflect its tetragonal crystal structure. In XPEEM these domains can be directly visualised by measuring the linear dichroism at the Ti L edge and, by measuring both the XMCD and linear dichroism at the Fe L edge for $CoFe_2O_4$ and $NiFe_2O_4$, both the magnetic domain structure and the crystal symmetry of the ferromagnetic layer are also revealed, showing that the magnetic film exhibits a cubic anisotropy (both magnetic and crystalline) at $BaTiO_3$ c-domains and uniaxial anisotropy at the $BaTiO_3$ a-domains, the latter as a consequence of the uniaxial strain induced by the $BaTiO_3$. Strikingly, the ferroelectric domain structure is made to reflect directly on the linear dichroism of the $CoFe_2O_4$ film, as seen as a function of the different crystal structures of $BaTiO_3$ with decreasing temperature (Fig. 10.13 (right)). Another advantage of XPEEM is that magnetic images can be obtained while applying electric fields to the sample [209]. Such capabilities have been explored, for example, to observe changes in the magnetic state of Ni structures deposited on PMN-PT(110) as a function of the applied electric field [210, 211]. In the case of submicrometre-sized Ni dots deposited on PMN-PT(110), a direct visualisation of a strain-induced rotation of the magnetisation by 90° with the applied electric field demonstrates the presence of a very large magnetoelectric coupling in such system as a result of the very large change in the total magnetic moment [210].

10.5 Conclusions and Outlook

In this chapter we gave a (necessarily brief) survey of some recent advances in the characterisation of single phase and artificial multiferroic heterostructures, to show how the underlying processes and the several key parameters which are involved in the magnetoelectric coupling can be inferred from X-ray absorption and photoelectron spectroscopy measurements. In particular, we aimed at illustrating how the interplay between spin, charge or orbital degrees of freedom couple to provide a new magnetoelectric effect that is absent in the parent compounds and how such coupling can be better understood by correlating the information on magnetic state, orbital occupancy, and local electronic environment provided by X-ray absorption techniques with the information acquired in photoemission on occupied states, by tapping directly into the chemical state, band structure and interface electrostatic potential of the system. We anticipate that novel technical developments, such as spin-resolved ARPES [101], high resolution spectromicroscopy enabled by X-ray ptychography, and the capability for probing the electronic structure at ultrafast timescales made possible by X-ray free-electron laser sources [58], will further enable the disclosure of new phenomena and provide new insights into the processes governing the magnetoelectric coupling in single phase and artificial multiferroics, a class of materials that has inspired and pointed new directions for designing novel materials through interface engineering.

Acknowledgements We are very thankful to the authors that allowed reproduction of their work here. M.-A.H. was supported by the Swiss Excellence Scholarship Grant ESKAS-No. 2015.0257 and in part by the Romanian UEFISCDI Agency under Contract No. PN-II-ID-JRP-2011-2.

References

1. T. Kimura, T. Goto, H. Shintani, K. Ishizaka, Y. Tokura, Nature **426**, 55 (2003)
2. G. Lawes, A.B. Harris, T. Kimura, N. Rogado, R.J. Cava, A. Aharony, O. Entin-Wohlman, T. Yildirim, M. Kenzelmann, C. Broholm, A.P. Ramirez, Phys. Rev. Lett. **95**, 087205 (2005)
3. T. Kimura, Annu. Rev. Mater. Res. **37**, 387 (2007)
4. Y. Tokura, Y. Tomioka, J. Magn. Magn. Mater. **200**, 1 (1999)
5. E. Dagotto, T. Hotta, A. Moreo, Phys. Rep. **344**, 1 (2001)
6. H.J.A. Molegraaf, J. Hoffman, C.A.F. Vaz, S. Gariglio, D. van der Marel, C.H. Ahn, J.-M. Triscone, Adv. Mater. **21**, 3470 (2009)
7. J.D. Burton, E.Y. Tsymbal, Phys. Rev. B **80**, 174406 (2009)
8. C.A.F. Vaz, J. Hoffman, Y. Segal, J.W. Reiner, R.D. Grober, Z. Zhang, C.H. Ahn, F.J. Walker, Phys. Rev. Lett. **104**, 127202 (2010)
9. G.A. Smolenskiĭ, I.E. Chupis, Sov. Phys. Usp. **25**, 475 (1982)
10. A. Pimenov, A.A. Mukhin, V.Yu. Ivanov, V.D. Travkin, A.M. Balbashov, A. Loidl, Nat. Phys. **2**, 97 (2006)
11. A.B. Sushkov, R.V. Aguilar, S. Park, S.-W. Cheong, H.D. Drew, Phys. Rev. Lett. **98**, 027202 (2007)
12. H. Katsura, A.V. Balatsky, N. Nagaosa, Phys. Rev. Lett. **98**, 027203 (2007)
13. A. Pimenov, A.M. Shuvaev, A.A. Mukhin, A. Loidl, J. Phys. Condens. Matter **20**, 434209 (2008)
14. A.M. Shuvaev, A.A. Mukhin, A. Pimenov, J. Phys. Condens. Matter **23**, 113201 (2011)
15. Y. Takahashi, R. Shimano, Y. Kaneko, H. Murakawa, Y. Tokura, Nat. Phys. **8**, 121 (2012)
16. W. Eerenstein, N.D. Mathur, J.F. Scott, Nature **442**, 759 (2006)
17. P. Zubko, S. Gariglio, M. Gabay, P. Ghosez, J.-M. Triscone, Ann. Rev. Condens. Matter Phys. **2**, 141 (2011)
18. Editorial. The interface is still the device. Nat. Mater. **11**, 91 (2012)
19. H. Chen, Q. Qiao, M.S.J. Marshall, A.B. Georgescu, A. Gulec, P.J. Phillips, R.F. Klie, F.J. Walker, C.H. Ahn, S. Ismail-Beigi, Nano Lett. **14**, 4965 (2014)
20. C.A.F. Vaz, F.J. Walker, C.H. Ahn, S.S. Ismail-Beigi, J. Phys. Condens. Matter **27**, 123001 (2015)
21. D.G. Schlom, L.-Q. Chen, C.-B. Eom, K.M. Rabe, S.K. Streiffer, J.-M. Triscone, Annu. Rev. Mater. Res. **37**, 589 (2007)
22. C.A.F. Vaz, J.A.C. Bland, G. Lauhoff, Rep. Prog. Phys. **71**, 056501 (2008)
23. R.A. McKee, F.J. Walker, M.F. Chisholm, Phys. Rev. Lett. **81**, 3014 (1998)
24. R.A. McKee, F.J. Walker, M.F. Chisholm, Science **293**, 468 (2001)
25. D.A. Muller, Nat. Mater. **8**, 263 (2009)
26. H. Tan, S. Turner, E. Yücelen, J. Verbeeck, G. Van Tendeloo, Phys. Rev. Lett. **107**, 107602 (2011)
27. N.A. Hill, J. Phys. Chem. B **104**, 6694 (2000)
28. R. Ramesh, N.A. Spaldin, Nat. Mater. **6**, 21 (2007)

29. T. Zhao, A. Scholl, F. Zavaliche, K. Lee, M. Barry, A. Doran, M.P. Cruz, Y.H. Chu, C. Ederer, N.A. Spaldin, R.R. Das, D.M. Kim, S.H. Baek, C.B. Eom, R. Ramesh, Nat. Mater. **5**, 823 (2006)
30. J.Y. Son, G.B. Kim, C.H. Kim et al., Appl. Phys. Lett. **846**, 4971 (2004)
31. E. Hanamura, Y. Tanabe, Phase Transit. **79**, 957 (2006)
32. Y. Miyamoto, K. Ishiyama, Solid State Commun. **87**, 581 (1993)
33. C. Medrano, M. Schlenker, J. Baruchel, J. Espeso, Y. Miyamoto, Phys. Rev. B **59**, 1185 (1999)
34. C. Ederer, N. Spaldin, Phys. Rev. B **74**, 024102 (2006)
35. B. Meyer, D. Vanderbilt, Phys. Rev. B **65**, 104111 (2002)
36. D.I. Khomskii, Physics **2**, 20 (2009)
37. I.E. Dzialoshinskii, Sov. Phys. JETP **5**, 1259 (1957)
38. T. Moriya, Phys. Rev. **120**, 91 (1960)
39. S.-W. Cheong, M. Mostovoy, Nat. Mater. **6**, 13 (2007)
40. K.F. Wang, J.M. Liu, Z.F. Ren, Adv. Phys. **58**, 321 (2009)
41. G. Catalan, J.F. Scott, Adv. Mater. **21**, 2463 (2009)
42. T. Kimura, Annu. Rev. Condens. Mater. Phys. **3**, 93 (2012)
43. J. Fontcuberta, C. R. Phys. **16**, 204 (2015)
44. S. Dong, J.-M. Liu, S.-W. Cheong, Z. Ren, Adv. Phys. **64**, 519 (2016)
45. J.M. Rondinelli, M. Stengel, N.A. Spaldin, Nat. Nanotechnol. **3**, 46 (2008)
46. G. Srinivasan, R.C.P. DeVreugd, C.S. Flattery, V.M. Laletsin, N. Paddubnaya, Appl. Phys. Lett. **85**, 2550 (2004)
47. J. Zhai, S. Dong, Z. Xing, J. Li, D. Viehland, Appl. Phys. Lett. **89**, 083507 (2006)
48. C. Miclea, C. Tanasoiu, L. Amarande, C.F. Miclea, C. Plavitu, M. Cioangher, L. Trupina, C.T. Miclea, T. Tanasoiu, M. Susu, J. Opt. Adv. Mater. **12**, 272 (2010)
49. H. Zheng, J. Wang, S.E. Lofland, Z. Ma, L. Mohaddes-Ardabili, T. Zhao, L. Salamanca-Riba, S.R. Shinde, S.B. Ogale, F. Bai, D. Viehland, Y. Jia, D.G. Schlom, M. Wuttig, A. Roytburd, R. Ramesh, Science **303**, 661 (2004)
50. C.-W. Nan, G. Liu, Y. Lin, H. Chen, Phys. Rev. Lett. **94**, 197203 (2005)
51. V. Garcia, M. Bibes, L. Bocher, S. Valencia, F. Kronast, A. Crassous, X. Moya, S. Enouz-Vedrenne, A. Gloter, D. Imhoff, C. Deranlot, N.D. Mathur, S. Fusil, K. Bouzehouane, A. Barthelemy, Science **327**, 1106 (2010)
52. S. Ryu, J.H. Park, H.M. Jang, Appl. Phys. Lett. **91**, 142910 (2007)
53. M.P. Singh, W. Prellier, L. Mechin, C. Simon, B. Raveau, J. Appl. Phys. **99**, 024105 (2006)
54. A.D. Caviglia, S. Gariglio, N. Reyren, D. Jaccard, T. Schneider, M. Gabay, S. Thiel, G. Hammerl, J. Mannhart, J.-M. Triscone, Nature **456**, 624 (2008)
55. W. Eerenstein, M. Wiora, J.L. Prieto, J.F. Scott, N.D. Mathur, Nat. Mater. **6**, 348 (2007)
56. C.A.F. Vaz, J. Hoffman, C.H. Ahn, R. Ramesh, Adv. Mater. **22**, 2900 (2010)
57. C.A.F. Vaz, J. Phys. Condens. Matter **24**, 333201 (2012)
58. C.A.F. Vaz, U. Staub, J. Mater. Chem. C **1**, 6731 (2013)
59. J.J. Wang, J.M. Hu, J. Ma, J.X. Zhang, L.Q. Chen, C.W. Nan, Sci. Rep. **4**, 7507 (2014)
60. M. Ghidini, R. Pellicelli, J.L. Prieto, X. Moya, J. Soussi, J. Briscoe, S. Dunn, N.D. Mathur, Nat. Commun. **4**, 1453 (2013)
61. L. Gerhard, T.K. Yamada, T. Balashov, A.F. Takacs, R.J.H. Wesselink, M. Dne, M. Fechner, S. Ostanin, A. Ernst, I. Mertig, W. Wulfhekel, Nat. Nanotechnol. **5**, 792 (2010)
62. L. Gerhard, R.J.H. Wesselink, S. Ostanin, A. Ernst, W. Wulfhekel, Phys. Rev. Lett. **111**, 167601 (2013)
63. M. Imada, A. Fujimori, Y. Tokura, Rev. Mod. Phys. **70**, 1039 (1998)
64. J.M.D. Coey, M. Viret, S. von Molnár, Adv. Phys. **48**, 167 (1999)
65. N. Tsuda, K. Nasu, A. Fujimori, K. Siratori, *Electronic Conduction in Oxides*, 2nd edn. (Springer, Berlin, 2000)
66. E.L. Nagaev, Phys. Rep. **346**, 387 (2001)
67. M. Ziese, Rep. Prog. Phys. **65**, 143 (2002)
68. D. Preziosi, M. Alexe, D. Hesse, M. Salluzzo, Phys. Rev. Lett **115**, 157401 (2015)

69. H. Lu, T.A. George, Y. Wang, I. Ketsman, J.D. Burton, C.-W. Bark, S. Ryu, D.J. Kim, J. Wang, C. Binek, P.A. Dowben, A. Sokolov, C.-B. Eom, E.Y. Tsymbal, A. Gruverman, Appl. Phys. Lett. **100**, 232904 (2012)
70. P.M. Leufke, R. Kruk, R.A. Brand, H. Hahn, Phys. Rev. B **87**, 094416 (2013)
71. T.L. Meyer, A. Herklotz, V. Lauter, J.W. Freeland, J. Nichols, S. Lee, E.-J. Guo, T.Z. Ward, N. Balke, S.V. Kalinin, M.R. Fitzsimmons, H.N. Lee, Phys. Rev. B **94**, 174432 (2016)
72. X. Hong, A. Posadas, C.H. Ahn, Appl. Phys. Lett. **86**, 142501 (2005)
73. J. Hoffman, X. Hong, C.H. Ahn, Nanotechnology **22**, 254014 (2011)
74. D. Pantel, S. Goetze, D. Hesse, M. Alexe, Nat. Mater. **11**, 289 (2012)
75. Y.W. Yin, J.D. Burton, Y.-M. Kim, A.Y. Borisevich, S.J. Pennycook, S.M. Yang, T.W. Noh, A. Gruverman, X.G. Li, E.Y. Tsymbal, Q. Li, Nat. Mater. **12**, 397 (2013)
76. Y. Tokura, N. Nagaosa, Science **288**, 462 (2000)
77. J.P. Velev, C.-G. Duan, J.D. Burton, A. Smogunov, M.K. Niranjan, E. Tosatti, S.S. Jaswal, E.Y. Tsymbal, Nano Lett. **9**, 427 (2009)
78. M. Frst, R.I. Tobey, S. Wall, H. Bromberger, V. Khanna, A.L. Cavalieri, Y.-D. Chuang, W.S. Lee, R. Moore, W.F. Schlotter, J.J. Turner, O. Krupin, M. Trigo, H. Zheng, J.-F. Mitchell, S.S. Dhesi, J.P. Hill, A. Cavalleri, Phys. Rev. B **84**, 241104(R) (2011)
79. D. Pesquera, G. Herranz, A. Barla, E. Pellegrin, F. Bondino, E. Magnano, F. Sanchez, J. Fontcuberta, Nat. Commun. **3**, 1189 (2012)
80. E.J. Moon, R. Colby, Q. Wang, E. Karapetrova, C.M. Schleputz, M.R. Fitzsimmons, S.J. May, Nature. Commun. **5**, 5710 (2014)
81. B. Cui, C. Song, H. Mao, H. Wu, F. Li, J. Peng, G. Wang, F. Zeng, F. Pan, Adv. Mater. **27**, 6651 (2015)
82. J.M. Pruneda, V. Ferrari, R. Rurali, P.B. Littlewood, N.A. Spaldin, E. Artacho, Phys. Rev. Lett. **99**, 226101 (2007)
83. W.H. Meiklejohn, C.P. Bean, Phys. Rev. **102**, 1413 (1956)
84. W.H. Meiklejohn, J. Appl. Phys. **33**, 1328 (1962)
85. J. Nogués, I.K. Schuller, J. Magn. Magn. Mater. **192**, 203 (1999)
86. S. Picozzi, A. Continenza, A.J. Freeman, J. Appl. Phys. **94**, 4723 (2003)
87. S.M. Wu, S.A. Cybart, D. Yi, J.M. Parker, R. Ramesh, R.C. Dynes, Phys. Rev. Lett. **110**, 067202 (2013)
88. A. Quindeau, I. Fina, X. Marti, G. Apachitei, P. Ferrer, C. Nicklin, E. Pippel, D. Hesse, M. Alexe, Sci. Rep. **5**, 9749 (2015)
89. C.A.F. Vaz, E.I. Altman, V.E. Henrich, Phys. Rev. B **81**, 104428 (2010)
90. S. Picozzi, A. Continenza, A., J. Freeman. Phys. Rev. B **69**, 09423 (2004)
91. J. Curiale, M. Granada, H.E. Troiani, R.D. Sanchez, A.G. Leyva, P. Levy, K. Samwer, Appl. Phys. Lett. **95**, 043106 (2009)
92. K. Nakamura, Y. Kato, T. Akiyama, T. Ito, Phys. Rev. Lett. **96**, 047206 (2006)
93. P. Borisov, A. Hochstrat, X. Chen, W. Kleemann, C. Binek, Phys. Rev. Lett. **94**, 117203 (2005)
94. X. He, N. Wu, A.N. Caruso, E. Vescovo, K.D. Belashchenko, P.A. Dowben, C. Binek, Nat. Mater. **9**, 579 (2010)
95. M.B. Holcomb, S. Polisetty, A.F. Rodríguez, V. Gopalan R. Ramesh, Int. J. Mod. Phys. B **26**, 1230004 (2012)
96. R. Ramesh, Phil. Trans. R. Soc. A **372**, 20120437 (2014)
97. M. Guennou, M. Viret, J. Kreisel, C. R. Phys. **16**, 182 (2015)
98. V. Garcia, M. Bibes, A. Barthélémy, C. R. Phys. **16**, 168 (2015)
99. H. Bea, M. Bibes, S. Cherifi, F. Nolting, B. Warot-Fonrose, S. Fusil, G. Herranz, C. Deranlot, E. Jacquet, K. Bouzehouane, A. Barthélémy, Appl. Phys. Lett. **89**, 242114 (2006)
100. S. Tanuma, C.J. Powel, D.R. Penn, Surf. Interface Anal. **17**, 911 (1991)
101. V.N. Strocov, M. Kobayashi, X. Wang, L.L. Lev, J. Krempasky, V.V. Rogalev, T. Schmitt, C. Cancellieri, M.L. Reinle-Schmitt, Synchr. Rad. News **27**, 31 (2014)
102. S. Mukherjee, P. K. Santra, and D. D. Sarma. Depth profiling and internal structure determination of low dimensional materials using X-ray photoelectron spectroscopy. In *Hard X-ray Photoelectron Spectroscopy (HAXPES)*, number 59 in Springer Series in Surface Sciences (Springer, Heidelberg, 2015), p. 309

103. A.X. Gray, C. Papp, S. Ueda, B. Balke, Y. Yamashita, L. Plucinski, J. Minr, J. Braun, E.R. Ylvisaker, C.M. Schneider, W.E. Pickett, H. Ebert, K. Kobayashi, C.S. Fadley, Nat. Mater. **10**, 759 (2011)
104. Y. Kayanuma, Recoil effects in X-ray photoelectron spectroscopy, in *Hard X-ray Photoelectron Spectroscopy (HAXPES)*, number 59 in Springer Series in Surface Sciences (Springer, Heidelberg, 2015), p. 175
105. A.X. Gray, J. Minar, S. Ueda, P.R. Stone, Y. Yamashita, J. Fujii, J. Braun, L. Plucinski, C.M. Schneider, G. Panaccione, H. Ebert, O.D. Dubon, K. Kobayashi, C.S. Fadley, Nat. Mater. **11**, 957 (2012)
106. F. de Groot, J. Vogel, Fundamentals of X-ray absorption and dichroism: the multiple approach, in *Neutron and X-ray Spectroscopy*, ed. by F. Hippert, E. Geissler, J.L. Hodeau, E. Lelièvre-Berna, J.-R. Regnard (Springer, Dordrecht, 2006), p. 3
107. J. Stöhr, H.C. Siegmann, *Magnetism* (Springer, Berlin, 2006)
108. M. Salluzzo, G. Ghiringhelli, Soft X-ray absorption spectroscopy and resonant X-ray inelastic scattering on oxides and their heterostructures, in *Spectroscopy of TMO interfaces*, ed. by C. Cancellieri, V.N. Strocov (Springer, Berlin, 2017), p. XXX
109. C.A.F. Vaz, C. Moutafis, C. Quitmann, J. Raabe. Appl. Phys. Lett. **101**, 083114 (2012)
110. B.L. Henke, E.M. Gullikson, J.C. Davis, At. Data Nucl. Data Tables **54**, 181 (1993)
111. C.G. Duan, S.S. Jaswal, E.Y. Tsymbal, Phys. Rev. Lett. **97**, 047201 (2006)
112. G. Radaelli, D. Petti, E. Plekhanov, I. Fina, P. Torelli, B.R. Salles, M. Cantoni, C. Rinaldi, D. Gutierrez, G. Panaccione, M. Varela, S. Picozzi, J. Fontcuberta, R. Bertacco, Nat. Commun. **5**, 3404 (2013)
113. G. Radaelli, M. Cantoni, L. Lijun, M. Espabodi, R. Bertacco, J. Appl. Phys. **115**, 063501 (2014)
114. F.J. Scott. Ferroelectric Memories (Springer, Heidelberg, 2000)
115. A.J. Moulson, J.M. Herbert, *Electroceramics: Materials and Properties Applications*, 2nd edn. (Wiley, New York, 2003)
116. L. Li, P.A. Salvador, G.S. Rother, Nanoscale **6**, 24 (2014)
117. D. Cao, C. Wang, F. Zheng, L. Fang, W. Dong, A.K. Tagantsev, J. Appl. Phys. **90**, 1287 (2001)
118. A. Chanthbouala, A. Crassous, V. Garcia, K. Bouzehouane, S. Fusil, X. Moya, J. Allibe, B. Dlubak, J. Grollier, S. Xavier, C. Deranlot, A. Moshar, R. Proksch, N.D. Mathur, M. Bibes, A. Barthlmy, Nat. Nanotechnol. **7**, 101 (2012)
119. A. Gruverman, D. Wu, H. Lu, Y. Wang, H.W. Jang, C.M. Folkman, M.Y. Zhuravlev, D. Felker, M. Rzchowski, C.-B. Eomm, E.Y. Tsymbal, Nano Lett. **9**, 3539 (2009)
120. S. Takatani, M. Hiroshi, K.-A. Keiko, K. Torii, J. Appl. Phys. **85**, 7784 (1999)
121. J.P. Velev, C.-G. Duan, K.D. Belashchenko, S.S. Jaswal, E.Y. Tsymbal, Phys. Rev. Lett. **98**, 137201 (2007)
122. R. Schafranek, S. Payan, M. Maglione, A. Klein, Phys. Rev. B **77**, 195310 (2008)
123. A.G. Boni, I. Pintilie, L. Pintilie, D. Preziosi, H. Deniz, M. Alexe, J. Appl. Phys. **113**, 224103 (2013)
124. S.A. Chambers, Y. Liang, Z. Yu, R. Droopad, J. Ramdani, K. Eisenbeiser, Appl. Phys. Lett. **77**, 1662 (2000)
125. N.G. Apostol, C.M. Teodorescu, Band bending at metal-semiconductor interfaces. ferroelectric surfaces and metal-ferroelectric interfaces investigated by photoelectron spectroscopy, in Surface Science Tools for Nanomaterials Characterization (Springer, Heidelberg, 2015), p. 405
126. L.E. Stoflea, N.G. Apostol, L. Trupina, C.M. Teodorescu, J. Mater. Chem. A **2**, 14386 (2014)
127. N.G. Apostol, L.E. Stoflea, G.A. Lungu, C. Chirila, L. Trupina, R.F. Negrea, C. Ghica, L. Pintilie, C.M. Teodorescu, Appl. Surf. Sci. **273**, 415 (2013)
128. E. Kroger, A. Petraru, A. Quer, R. Soni, M. Kallane, N.A. Pertsev, H. Kohlstedt, K. Rossnagel, Phys. Rev. B **93**, 235415 (2016)
129. D.G. Popescu, M.A. Husanu, L. Trupina, L. Hrib, L. Pintilie, A. Barinov, S. Lizzit, P. Lacovig, C.M. Teodorescu, Phys. Chem. Chem. Phys. **17**, 509 (2015)
130. D.G. Popescu, N. Barrett, C. Chirila, I. Pasuk, M.A. Husanu, Phys. Rev. B **92**, 235442 (2015)

131. F. Chen, A. Klein, Phys. Rev. B **86**, 094105 (2012)
132. J.E. Rault, G. Agnus, T. Maroutian, V. Pillard, P. Lecoeur, G. Niu, B. Vilquin, M.G. Silly, A. Bendounan, F. Sirotti, N. Barrett, Phys. Rev. B **87**, 155146 (2013)
133. M.A. Husanu, D.G. Popescu, L. Hrib, C. Chirila, C.M. Teodorescu, L. Pintilie, F. Bisti, V.N. Strocov. unpublished
134. S.M. Wu, S.A. Cybart, P. Yu, M.D. Rossell, J.X. Zhang, R. Ramesh, R.C. Dynes, Nat. Mater. **9**, 756 (2010)
135. J. Krempaský, V.N. Strocov, L. Patthey, P.R. Willmott, R. Herger, M. Falub, P. Blaha, M. Hoesch, V. Petrov, M.C. Richter, O. Heckmann, K. Hricovini, Phys. Rev. B **77**, 165120 (2008)
136. L.L. Lev, J. Krempaský, U. Staub, V.A. Rogalev, T. Schmitt, M. Shi, P. Blaha, A.S. Mishchenko, A.A. Veligzhanin, Y.V. Zubavichus, M.B. Tsetlin, H. Volfov, J. Braun, J. Minr, V.N. Strocov, Phys. Rev. Lett. **114**, 237601 (2015)
137. M. Kobayashi, I. Muneta, Y. Takeda, Y. Harada, A. Fujimori, J. Krempasky, T. Schmitt, S. Ohya, M. Tanaka, M. Oshima, V.N. Strocov, Phys. Rev. B **89**, 205204 (2014)
138. J.M. Luttinger, Phys. Rev. **119**, 1153 (1960)
139. B.K. Agarwal, *X-ray Spectroscopy* (Springer, Berlin, 1991)
140. P.M. Bertsch, D.B. Hunter, Chem. Rev. **101**, 1809 (2001)
141. F.M.F. de Groot, Chem. Rev. **101**, 1779 (2001)
142. H. Wende, Rep. Prog. Phys. **67**, 2105 (2004)
143. B.T. Thole, P. Carra, F. Sette, G. van der Laan, Phys. Rev. Lett. **68**, 1943 (1992)
144. P. Carra, B.T. Thole, M. Altarelli, X. Wang, Phys. Rev. Lett. **70**, 694 (1993)
145. D. Yi, J. Liu, S. Okamoto, S. Jagannatha, Y.-C. Chen, P. Yu, Y.-H. Chu, E. Arenholz, R. Ramesh, Phys. Rev. Lett. **111**, 127601 (2013)
146. P. Yu, J.-S. Lee, S. Okamoto, M.D. Rossell, M. Huijben, C.-H. Yang, Q. He, J.X. Zhang, S.Y. Yang, M.J. Lee, Q.M. Ramasse, R. Erni, Y.-H. Chu, D.A. Arena, C.-C. Kao, L.W. Martin, R. Ramesh, Phys. Rev. Lett. **105**, 027201 (2010)
147. S.S. Rao, J.T. Prater, F. Wu, C.T. Shelton, J.-P. Maria, J. Narayan, Nano Lett. **13**, 5814 (2013)
148. C.A.F. Vaz, J.A. Moyer, D. Arena, C.H. Ahn, V.E. Henrich, Phys. Rev. B **90**, 024414 (2014)
149. T.H.E. Lahtinen, J.O. Tuomi, S. van Dijken, Adv. Mater. **23**, 3187 (2011)
150. T.H.E. Lahtinen, K.J.A. Franke, S. van Dijken, Sci. Rep. **2**, 258 (2012)
151. K.J.A. Franke, B. Van de Wiele, Y. Shirahata, S.J. H"am"al"ainen, T. Taniyama, S. van Dijken, Phys. Rev. X **5**, 011010 (2015)
152. J. van Suchtelen, Philips Res. Repts **27**, 28 (1972)
153. J. van den Boomgaard, D.R. Terrel, R.A.J. Born, H.F.J.I. Giller, J. Mater. Sci. **9**, 1705 (1974)
154. A.M.J.G. van Run, D.R. Terrell, J.H. Scholing, J. Mater. Sci. **9**, 1710 (1974)
155. F. Zavaliche, H. Zheng, L. Mohaddes-Ardabili, S.Y. Yang, Q. Zhan, P. Shafer, E. Reilly, R. Chopdekar, Y. Jia, P. Wright, D.G. Schlom, Y. Suzuki, R. Ramesh, Nano Lett. **5**, 1793 (2005)
156. C. Schmitz-Antoniak, D. Schmitz, P. Borisov, F.M.F. de Groot, S. Stienen, A. Warland, B. Krumme, R. Feyerherm, E. Dudzik, W. Kleemann, H. Wende, Nat. Commun. **4**, 2051 (2013)
157. Y.-J. Chen, Y.-H. Hsieh, S.-C. Liao, Z. Hu, M.-J. Huang, W.-C. Kuo, Y.-Y. Chin, T.-M. Uen, J.-Y. Juang, C.-H. Lai, H.-J. Lin, C.-T. Chena, Y.-H. Chu, Nanoscale **5**, 4449 (2013)
158. C.A.F. Vaz, J. Hoffman, A.-B. Posadas, C.H. Ahn, Appl. Phys. Lett. **94**, 022504 (2009)
159. V.H. Babu, R.K. Govind, K.-M. Schindler, M. Welke, R. Denecke, J. Appl. Phys. **114**, 113901 (2013)
160. C.-W. Nan, M.I. Bichurin, S. Dong, D. Viehland, G. Srinivasan, J. Appl. Phys. **103**, 031101 (2008)
161. G. Srinivasan, Annu. Rev. Mater. Res. **40**, 153 (2010)
162. R.C. Kambale, D.-Y. Jeong, J. Ryu, Adv. Condens. Matter Phys. **2012**, 824643 (2012)
163. B.Y. Wang, H.T. Wang, S.B. Singh, Y.C. Shao, Y.F. Wang, C.H. Chuang, P.H. Yeh, J.W. Chiou, C.W. Pao, H.M. Tsai, H.J. Lin, J.F. Lee, C.Y. Tsai, W.F. Hsieh, M.-H. Tsaif, W.F. Pong, RSC Adv. **3**, 7884 (2013)
164. V.K. Verma, V.R. Singh, K. Ishigami, G. Shibata, T. Harano, T. Kadono, A. Fujimori, F.-H. Chang, H.-J. Lin, D.-J. Huang, C.T. Chen, Y. Zhang, J. Liu, Y. Lin, C.-W. Nan, A. Tanaka, Phys. Rev. B **89**, 115128 (2014)

165. M. Etier, C. Schmitz-Antoniak, S. Salamon, H. Trivedi, Y. Gao, A. Nazrabi, J. Landers, D. Gautam, M. Winterer, D. Schmitz, H. Wende, V.V. Shvartsmana, D.C. Lupascu, Acta Mater. **90**, 1 (2015)
166. G. Schütz, W. Wagner, W. Wilhelm, P. Kienle, R. Zeller, R. Frahm, G. Materlik, Phys. Rev. Lett. **58**, 737 (1987)
167. C.A.F. Vaz, C. Moutafis, M. Buzzi, J. Raabe, J. Electron Spectrosc. Rel. Phenom. **189**, 1 (2013)
168. C.A.F. Vaz, Y. Segal, J. Hoffman, R.D. Grober, F.J. Walker, C.H. Ahn, Appl. Phys. Lett. **97**, 042506 (2010)
169. P.P. Kirichok, G.S. Podval'nykh, L.M. Letyuk, Russ. Phys. J. **14**, 983 (1971)
170. P.P. Kirichok, A.V. Kopaev, V.P. Pashchenko, Rus. Phys. J. **28**, 849 (1985)
171. M. Sikora, C. Kapusta, K. Knížek, Z. Jirák, C. Autret, M. Borowiec, C.J. Oates, V. Procházka, D. Rybicki, D. Zajac, Phys. Rev. B **73**, 094426 (2006)
172. G. Subías, J. García, M.G. Proietti, J. Blasco. Phys. Rev. B **56**, 8183 (1997)
173. M. Croft, D. Sills, M. Greenblatt, C. Lee, S.-W. Cheong, K.V. Ramanujachary, D. Tran, Phys. Rev. B **55**, 8726 (1997)
174. F. Bridges, C.H. Booth, M. Anderson, G.H. Kwei, J.J. Neumeier, J. Snyder, J. Mitchell, J.S. Gardner, E. Brosha, Phys. Rev. B **63**, 214405 (2001)
175. T. Shibata, B.A. Bunker, J.F. Mitchell, Phys. Rev. B **68**, 024103 (2003)
176. R. Bindu, S.K. Pandey, A. Kumar, S. Khalid, A.V. Pimpale, J. Phys. Condens. Matter **17**, 6393 (2005)
177. S.K. Pandey, R. Bindu, A. Kumar, S. Khalid, A.V. Pimpale, Pramana—J. Phys. **70**, 359 (2008)
178. C.A.F. Vaz, J. Hoffman, Y. Segal, M.S.J. Marshall, J.W. Reiner, Z. Zhang, R.D. Grober, F.J. Walker, C.H. Ahn, J. Appl. Phys. **109**, 07D905 (2011)
179. J.H. Park, J.-H. Lee, M.G. Kim, Y.K. Jeong, M.-A. Oak, H.M. Jang, H.J. Choi, J.F. Scott, Phys. Rev. B **81**, 134401 (2010)
180. J. Stöhr, H.A. Padmore, S. Anders, T. Stammler, M.R. Scheinfein, Surf. Rev. Lett. **5**, 1297 (1998)
181. F. Nolting, Magnetic imaging with X-rays, in *Magnetism and Synchrotron Radiation*, volume 133, of Proceedings in Physics, ed. by E. Beaurepaire, et al. (Springer, Berlin, 2010), p. 345
182. J. Miao, P. Charalambous, J. Kirz, D. Sayre, Nature **400**, 342 (1999)
183. M. Dierolf, A. Menzel, P. Thibault, P. Schneider, C.M. Kewish, R. Wepf, O. Bunk, F. Pfeiffer, Nature **467**, 436 (2010)
184. A.M. Maiden, G.R. Morrison, B. Kaulich, A. Gianoncelli, J.M. Rodenburg, Nat. Commun. **4**, 1669 (2013)
185. M. Holler, A. Diaz, M. Guizar-Sicairos, P. Karvinen, E. Färm, E. Härkönen, M. Ritala, A. Menzel, J. Raabe, O. Bunk, Sci. Rep. **4**, 3857 (2014)
186. D.A. Shapiro, Y.-S. Yu, T. Tyliszczak, J. Cabana, R. Celestre, W. Chao, K. Kaznatcheev, A.L.D. Kilcoyne, F. Maia, S. Marchesini, Y.S. Meng, T. Warwick, L.L. Yang, H.A. Padmore, Nat. Photonics **8**, 765 (2014)
187. X. Shi, P. Fischer, V. Neu, D. Elefant, J.C.T. Lee, D.A. Shapiro, M. Farmand, T. Tyliszczak, H.-W. Shiu, S. Marchesini, S. Roy, S.D. Kevan, Appl. Phys. Lett. **108**, 094103 (2016)
188. F. Buettner, M. Schneider, C.M. Guenther, C.A.F. Vaz, B. Laegel, D. Berger, S. Selve, M. Klaeui, S. Eisebitt, Opt. Express **21**, 30563 (2013)
189. F. Büttner, C. Moutafis, M. Schneider, B. Krüger, C.M. Günther, J. Geilhufe, C.V. Korff, C.K. Schmising, J. Mohanty, B. Pfau, S. Schaffert, A. Bisig, M. Foerster, T. Schulz, C.A.F. Vaz, J.H. Franken, H.J.M. Swagten, M. Kläui, S. Eisebitt, Nat. Phys. **11**, 225 (2015)
190. A. Locatelli, E. Bauer, J. Phys. Condes. Matter **20**, 093002 (2008)
191. X.M. Cheng, D.J. Keavney, Rep. Prog. Phys. **75**, 026501 (2012)
192. R.M. Tromp, J.B. Hannonand, A.W. Ellis, W. Wan, A. Berghaus, O. Schaff, Ultramicroscopy **110**, 852 (2010)
193. R.M. Tromp, J.B. Hannon, W. Wan, A. Berghaus, O. Schaff, Ultramicroscopy **127**, 25 (2013)
194. R. Moubah, M. Elzo, S. El Moussaoui, D. Colson, N. Jaouen, R. Belkhou, M. Viret, Appl. Phys. Lett. **100**, 042406 (2012)

195. A. Sander, M. Christl, C.-T. Chiang, M. Alexe, W. Widdra, J. Appl. Phys. **118**, 224102 (2015)
196. S. Cherifi, R. Hertel, S. Fusil, H. Béa, K. Bouzehouane, J. Allibe, M. Bibes, A. Barthélémy, Phys. Status Solidi RRL **4**, 22 (2010)
197. Q. He, E. Arenholz, A. Scholl, Y.-H. Chu, R. Ramesh, Curr. Opin. Solid State Mater. Sci. **16**, 216 (2012)
198. D. Lebeugle, D. Colson, A. Forget, M. Viret, A.M. Bataille, A. Gusakov, Phys. Rev. Lett. **100**, 227602 (2008)
199. U. Flechsig, F. Nolting, A. Fraile Rodríguez, J. Krempaský, C. Quitmann, T. Schmidt, S. Spielmann, D. Zimoch, AIP Conf. Proc. **1234**, 705 (2010)
200. A. Scholl, J. Stohr, J. Luning, J.W. Seo, J. Fompeyrine, H. Siegwart, J.P. Locquet, F. Nolting, S. Anders, E.E. Fullerton, M.R. Scheinfein, H.A. Padmore, Science **287**, 1014 (2000)
201. F. Nolting, A. Scholl, J. Stöhr, J.W. Seo, J. Fompeyrine, H. Siegwart, J.-P. Locquet, S. Anders, J. Lüning, E.E. Fullerton, M.F. Toney, M.R. Scheinfein, H.A. Padmore, Nature **405**, 767 (2000)
202. J. Lüning, F. Nolting, A. Scholl, H. Ohldag, J.W. Seo, J. Fompeyrine, J.-P. Locquet, J. Stöhr. Phys. Rev. B **84**, 1174 (2004)
203. Y.-H. Chu, L.W. Martin, M.B. Holcomb, M. Gajek, S.-J. Han, Q. He, N. Balke, C.-H. Yang, D. Lee, W. Hu, Q. Zhan, P.-L. Yang, A. Fraile-Rodríguez, A. Scholl, S.X. Wang, R. Ramesh, Nat. Mater. **7**, 478 (2008)
204. R.V. Chopdekar, V.K. Malik, A. Fraile Rodríguez, L. Le Guyader, Y. Takamura, A. Scholl, D. Stender, C.W. Schneider, C. Bernhard, F. Nolting, L.J. Heyderman, Phys. Rev. B **86**, 014408 (2012)
205. M.K. Lee, T.K. Nath, C.B. Eom, M.C. Smoak, F. Tsui, Appl. Phys. Lett. **77**, 3547 (2000)
206. R.V. Chopdekar, Y. Suzuki, Appl. Phys. Lett. **89**, 182506 (2006)
207. H.F. Tian, T.L. Qu, L.B. Luo, J.J. Yang, S.M. Guo, H.Y. Zhang, Y.G. Zhao, J.Q. Li, Appl. Phys. Lett. **92**, 063507 (2008)
208. R.V. Chopdekar, J. Heidler, C. Piamonteze, Y. Takamura, A. Scholl, S. Rusponi, H. Brune, L.J. Heyderman, F. Nolting, Eur. Phys. J. B **86**, 241 (2013)
209. M. Buzzi, C.A.F. Vaz, J. Raabe, F. Nolting, Electric field stimulation set-up for photoemission electron microscopes. Rev. Sci. Instrum. **86**, 083702 (2015)
210. M. Buzzi, R.V. Chopdekar, J.L. Hockel, A. Bur, T. Wu, N. Pilet, P. Warnicke, G.P. Carman, L.J. Heyderman, F. Nolting, Phys. Rev. Lett. **111**, 027204 (2013)
211. S. Finizio, M. Foerster, M. Buzzi, B. Krüger, M. Jourdan, C.A.F. Vaz, J. Hockel, T. Miyawaki, A. Tkach, S. Valencia, F. Kronast, G.P. Carman, F. Nolting, M. Kläui, Phys. Rev. Appl. **1**, 021001 (2014)

Chapter 11
Oxides and Their Heterostructures Studied with X-Ray Absorption Spectroscopy and Resonant Inelastic X-Ray Scattering in the "Soft" Energy Range

M. Salluzzo and G. Ghiringhelli

Abstract Soft X-ray absorption spectroscopy (XAS) and resonant inelastic X-ray scattering (RIXS) have become essential experimental tools for the investigations the complex physics of transition metal oxide (TMO) heterostructures. XAS has been long used to determine the valence, the orbital and magnetic properties of transition metals. More recently, linear and circular dichroism in XAS have been widely applied to determine the crystal field splitting, the atomic orbital and spin moments, and the magnetic order of 3d-states, in bulk sample, in thin films and at atomically-sharp interfaces. Although less common, RIXS is also gaining popularity for its capability of accessing local and collective excitations at a time; the recent technical advances have been established RIXS as an important method for the determination of the electronic and magnetic properties of TMOs. This chapter is a brief review of the salient XAS and RIXS results on TMO and TMO heterostructures published in the last 15 years.

11.1 Introduction

In the last decades, the study of transition metal oxides (TMO) attracted the interest of the condensed matter community due to their intriguing physical phenomena, including colossal magnetoresistance in manganites [1] and high critical temperature (T_c) superconductivity in copper oxides compounds (cuprates) [2]. A complete understanding of the extraordinary physics of these materials remains elusive in many cases. At the same time, the physics of TMO continues to provide new surprises. Researches in artificial epitaxial TMO heterostructures demonstrated the creation

---

M. Salluzzo (✉)
CNR-SPIN, Complesso MonteSantangelo via Cinthia, 80126 Naples, Italy
e-mail: marco.salluzzo@spin.cnr.it

G. Ghiringhelli
CNR-SPIN and Dipartimento di Fisica, Politecnico di Milano, piazza Leonardo da Vinci 32, 20133 Milan, Italy

© Springer International Publishing AG, part of Springer Nature 2018
C. Cancellieri and V. Strocov (eds.), *Spectroscopy of Complex Oxide Interfaces*, Springer Series in Materials Science 266,
https://doi.org/10.1007/978-3-319-74989-1_11

of new systems at atomically sharp interfaces, characterized by physical properties substantially different from the bulk properties of each layer. These results are facing the condensed matter community with new challenges [3], which require the development and the use of experimental techniques able to selectively address the electronic and magnetic properties of few atomic layers at the interfaces between different TMOs.

The latest technological improvements in X-ray absorption spectroscopy (XAS) and especially resonant inelastic X-ray scattering (RIXS) are establishing these techniques as essential methods to study both the bulk and interfacial physics of TMOs and their heterostructures. XAS, together with its polarization dependent related spectroscopies [X-ray non-magnetic and magnetic linear dichroisms (XLD, XMLD) and X-ray magnetic circular dichroism (XMCD)], is able to provide selective information about the crystal field splitting and the orbital symmetry of the ground state, the magnitude of the orbital and spin moments, and about the magnetic order of transition metal 3d-states with extraordinary sensitivity, down to fraction of a monolayer. RIXS, on the other hand, is able to provide detailed d-d (crystal field), charge transfer and intra-band electron-hole pair excitations spectral distributions; and, when performed with sufficient resolution, RIXS can be used to map the dispersion of low energy collective excitation, like magnetic related excitations (magnons, bimagnons, spinons), phonon related excitations, orbital and charge order collective phenomena (orbitons, charge density waves). The advent of new, brilliant synchrotron radiation sources in the last few years has been extending the study of these phenomena to samples of few atomic layers and to the interfaces of TMO heterostructures, opening new directions in the exploration of the physics of TMOs also in nanostructures.

This chapter represents a review of the state of art of research in the study of TMO heterostructures using XAS and RIXS techniques. The main aim is to demonstrate the unique capabilities of these methods by giving few selected examples of outstanding achievements obtained in the field in last few years of research.

The chapter is organized as follows: Sect. 11.2 is devoted to an overview of X-ray absorption spectroscopy, with a short theoretical introduction allowing the basic assessment of the measurable quantities with particular emphasis on XLD and XMCD and relative sum-rules. Section 11.3 is a brief introduction to resonant inelastic X-ray scattering. Section 11.4 is dedicated to XAS and RIXS studies of hole doped copper-oxide high Tc materials (cuprates) and related heterostructures, including cuprate/manganite heterostructures, cuprate/$SrTiO_3$ as well as cuprate/$LaAlO_3$ superlattices, and to XAS, RIXS, XLD and XMCD studies on the LAO/STO quasi 2D-electron gas (q2DEG), as other important example of novel system where the interface physics dominate the electronic and magnetic properties. Finally, Sect. 11.5 present some future directions of XAS and RIXS for the study of the physics of transition metal oxides.

11.2 Introduction to X-Ray Absorption Spectroscopy and Resonant Inelastic X-Ray Scattering

Soft X-ray XAS and RIXS spectroscopy techniques are based on the inelastic scattering process of soft X-rays with matter, characterized by energies in the 300–1200 eV range. In the resonant X-ray absorption process, the incoming photon (photon-in) promotes the excitation of a core-electron of the absorbing ion into empty excited states. This process is the basis of X-ray absorption spectroscopy, where the absorption cross section is measured as function of the incident photon energy. As described in Fig. 11.1, a XAS spectra can be measured by looking at different sources of information as function of the incoming photon energy, including the intensity of emitted photoelectrons, of secondary and Auger fluorescent X-rays (called Total Fluorescence Yield, TFY), and the current of electrons which neutralize the sample after the absorption process (Total Electron Yield, TEY). The Resonant Inelastic X-ray Scattering (RIXS) technique, on the other hand, studies the photons which are scattered-out of the sample after the decay of the resonant excited core-electron into the initial state. In the following section we will give a brief outline of XAS and RIXS that will serve as guideline for the understanding of the main results on oxide heterostructures presented later. For more detailed description we refer to textbooks [4] and excellent review papers [5].

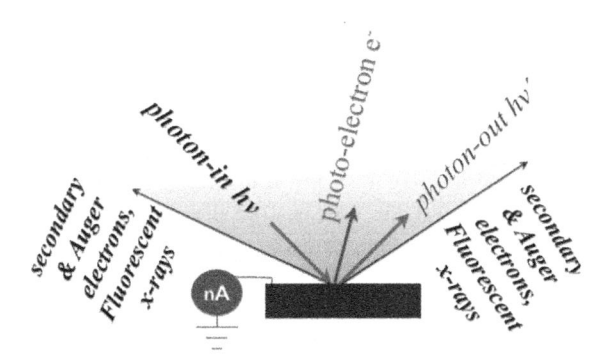

Fig. 11.1 A schematic of the X-ray scattering process. The XAS spectra can be measured by using several schemes, including the total electron yield method, which consist in the measurement of the electron current which is generate between the ground and sample due to the exit of secondary, Auger and photoelectrons, due to the absorption process, as function of the incoming photon energy. In the total fluorescence yield, a photodiode measures the intensity of fluorescent emitted X-rays. In RIXS, a special spectrometer is used to measure the number, the energy and, if possible, the polarization of outgoing photons generated by the scattering process

11.2.1 X-Ray Absorption Spectroscopy

In the X-ray absorption process, a core electron is excited to an empty state and, as such, XAS spectroscopy is a probe of the unoccupied part of the electronic structure of the system. The electronic and magnetic properties of transition metal oxides are dominated by the occupation of the 3d-orbitals, which can be probed at the $L_{2,3}$ absorption edge of the absorbing ion. From XAS it is possible to obtain information related to the electronic configuration and magnetic properties of the selected ion, making the technique elemental and orbital selective. In the $L_{2,3}$ edge XAS process, a 2p core electron is resonantly excited to the empty 3d states. The final state is characterized by an extra 3d electron and a core-hole in the 2p orbital. The core-hole and the 3d electron have a strong coulomb interaction, which strongly perturb the ground state of the system.

It is possible to show that, with good approximation, a XAS spectra, I_{XAS}, is given by:

$$I_{XAS}(E) \propto |\langle f|\hat{e} \cdot r|i\rangle|^2 \rho(E), \tag{11.1}$$

where $\langle f|\hat{e} \cdot r|i\rangle$ is the expectation probability of the dipole moment operator. The initial state is the ground state of the system, while the final state can be described as the initial state with a core-electron excited to the unoccupied states of the system. $\rho(E)$ is the density of unoccupied electronic states in the continuum. The final states that can effectively be reached are determined by the well-known dipole operator selection rules. At the L-edge, the large spin-orbit splitting of the 2p core level gives rise to two absorption resonances, i.e. the L_3 and the L_2 absorption edges, as shown in Fig. 11.2. The separation between L_2 and L_3 is relatively small for light 3d transition metals (e.g. ~6 eV for Ti) and larger for late transition metals like Cu (20 eV).

In spite of the simplicity of the process, the theoretical calculation of L-edge XAS spectra of TMOs cannot be performed using a single-particle approximation approach because the interaction between the core-hole and excited electron is very strong. This give rise to so-called multiplet effects which has to be properly treated. It turns out that one of the most successful method to treat multiplet effects is based on a ligand-field multiplet model, an atomistic, localized, approach which is able to treat exactly the presence of a core-hole. In this model, the dipole matrix element is calculated assuming a scattering process from an isolated ion in a given crystal field environment with opportune symmetry. For example, in perfectly cubic perovskites the crystal field symmetry is octahedral, which splits the 3d levels in t_{2g} ($3d_{xy}$, $3d_{xz,yz}$) and e.g. ($3d_{x2-y2}$ and $3d_{z2}$) multiplets, separated in energy by a quantity named $10Dq$ (Fig. 11.2a). Several codes have been developed to calculate the atomic multiplet spectrum of transition metals and rare earths with very good success, like Missing [6] and CTM4XAS [7].

The shape of the XAS spectra depends crucially on the formal oxidation state, i.e. the valence, of the transition metal, as shown for example in Fig. 11.3 for the case of Mn with (mainly) Mn^{2+} and Mn^{3+} configurations. Thus, XAS is an extremely

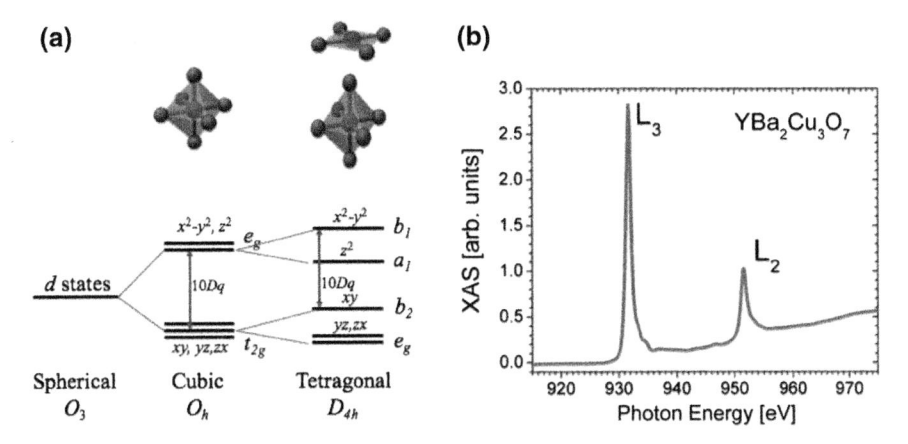

Fig. 11.2 a Effective energy diagram of a TM 3d orbitals in the case isolated ion, of a cubic and tetragonal crystal field splitting. **b** Typical XAS spectrum acquired in the TEY mode of a YBaCuO$_7$ high Tc superconductor, showing the two main absorption L$_3$ and L$_2$ edges

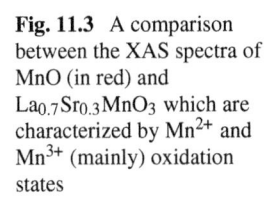

Fig. 11.3 A comparison between the XAS spectra of MnO (in red) and La$_{0.7}$Sr$_{0.3}$MnO$_3$ which are characterized by Mn^{2+} and Mn^{3+} (mainly) oxidation states

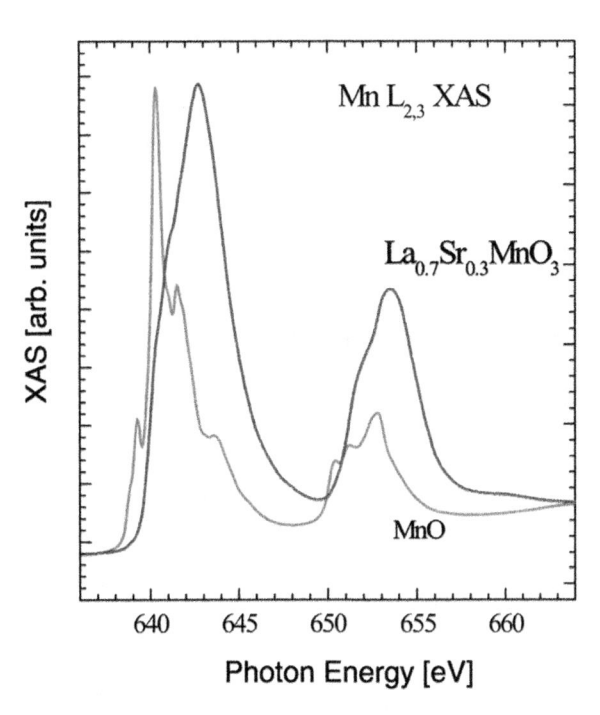

sensitive probe of the valence state of the transition metal in a crystal, which is often intimately related to the electronic and magnetic properties of the material.

In many TMO compounds the hybridization between the oxygen-2p and the transition-metal 3d states is so strong that additional features appear in the XAS

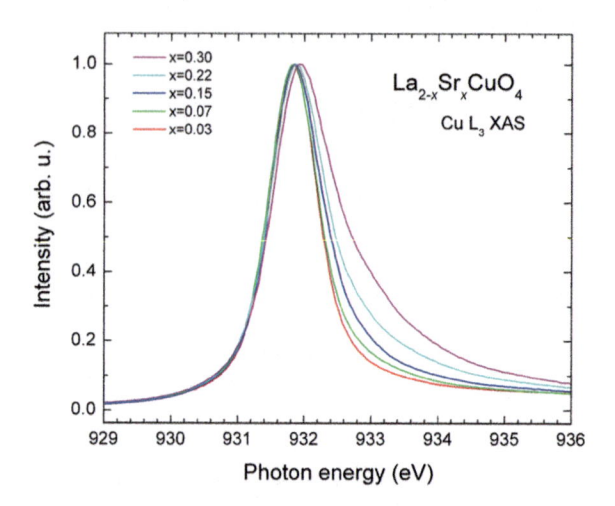

Fig. 11.4 XAS spectra as function of the Sr-doping of La$_{2-x}$Sr$_x$CuO$_4$ single crystals

spectra not reproduced by the crystal field ligand theory. In Fig. 11.4, we show the important case of High-Tc superconductors. The L$_3$-edge XAS spectra of undoped, antiferromagnetic insulating, compounds, are characterized by a sharp peak, assigned to a 2p^{6}3d^9 to 2p^{5}3d^{10} transition of Cu in formal Cu^{2+} configuration at 932 eV (Fig. 11.4). By doping, holes are introduced into the CuO$_2$ planes, and consequently the formal Cu oxidation state increases. Experimentally, the main peak becomes broader and asymmetric. Both broadening and the asymmetry are related to the appearance of a satellite 1.4 eV above the main peak, whose spectral weight is proportional to the amount of holes introduced into the CuO$_2$ planes. This feature is related to the 2p-3d hybridization between a fraction of copper and oxygen ions, forming Zhang-Rice singlets [8].

These effects can be introduced in the multiplet atomic model by explicitly considering the charge-transfer (CT) among the TM and the oxygen ions in small cluster calculations. CT considers the possibility that a valence 3d electron is shared with the neighboring oxygen ions, forming a so called ligand 3d$^n\underline{L}$ state. In the case of cuprates the additional satellite in XAS absorption is assigned to a 3d$^9\underline{L}$ configuration in the system, where $\underline{L}$ denotes a ligand hole in the O2p state(s). Within the multiplet model, charge transfer effects are parameterized by a charge transfer energy Δ_{CT}, i.e. the energy between the 3d^n and 3d$^{n+1}\underline{L}$ configuration, an on site U_{dd} (Hubbard) coulomb potential among d-electrons and an intersite core-hole coulomb repulsion U_{pd}.

11.2.2 X-Ray Linear Dichroism

The X-ray Linear Dichroism (XLD) is the dependence of the absorption cross section on the angle between the linear polarization vector and the sample lattice. It can be

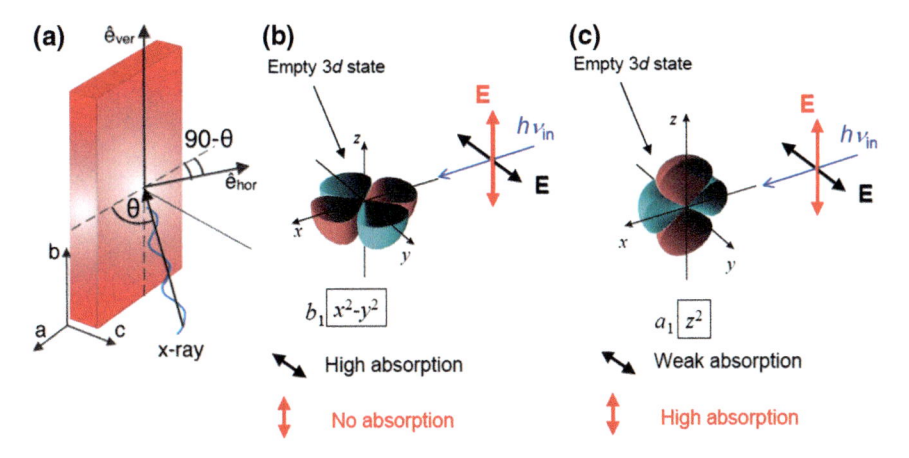

Fig. 11.5 a Schematic of an XLD measurement: the incident angle of X-rays, with respect to the sample surface (ab-plane), placed in the vertical laboratory plane, is θ. The linear polarization can be either in the horizontal plane or in the vertical plane. **b** and **c** show the anisotropic absorption in cases of final 3d states having x^2-y^2 or z^2 symmetry

used to obtain information about the crystal field splitting within the t_{2g} or e.g. multiplets in case of symmetry lower than cubic; and/or about the spin orientation in ferro- or antiferro-magnetic materials. In particular it has been widely used in the case of (001) TMO heterostructures to study the splitting between orbitals characterized by pure in-plane ($3d_{xy}$, $3d_{x2-y2}$) and out-of-plane symmetries ($3d_{xz,yz}$, $3d_{z2}$) and their consequent uneven occupation. Note that a description of the ground state in term of atomic 3d orbitals is valid only in the case of octahedral (Oh) and tetragonal (D4h) crystal-field symmetry; however lower symmetries (e.g. trigonal D3h) can be treated as well by dealing with other orbitals that result from linear combinations of the usual atomic 3d orbitals.

In Fig. 11.5, we show the general principle of the XLD measurements in the case of (001) TMO oxides which retain a cubic or a tetragonal symmetry. The X-rays are sent on the sample at grazing incidence with respect to the sample surface (θ). Assuming the sample mounted vertically in the laboratory frame (the usual geometrical configuration), one can measure XAS spectra using different linear polarizations of the photons, i.e. along the vertical direction, I_v, and along the horizontal direction perpendicular to the beam, I_h. Because the sample surface of 001 oriented crystals (ab-plane) contains the vertical laboratory direction, the XAS associated to radiation with electric field in the plane, I_{ab}, is equal to I_v, while I_h contains information on both in plane and out of plane (I_c) absorption. In particular, I_c can be determined from I_v and I_h through the simple formula:

$$I_c = (I_H - I_V \sin^2(\theta))/\cos^2(\theta) \tag{11.2}$$

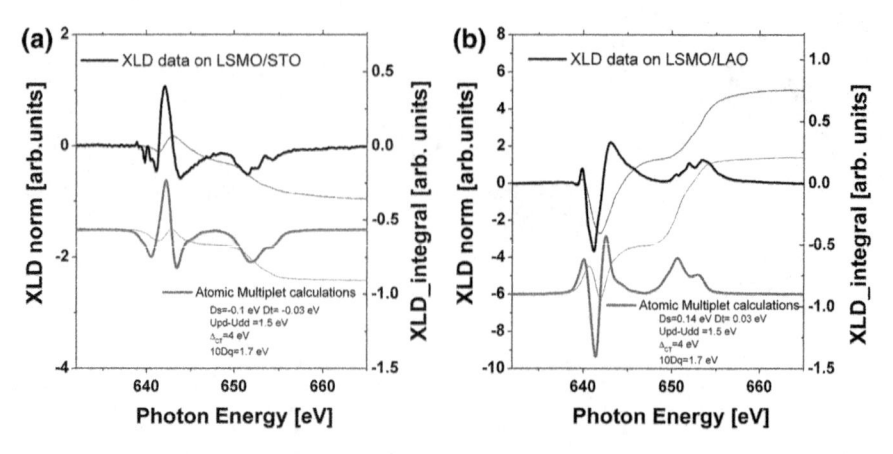

Fig. 11.6 XLD data (black lines, right axes) on $La_{0.7}Sr_{0.3}MnO_3$ thin films deposited on **a** $SrTiO_3$ and $LaAlO_3$ taken from [9]. Red lines are atomic multiplet splitting calculations (red lines, right axes) including charge transfer in the case of **a** 3d $_{x2-y2}$ and **b** $3d_{z2}$ predominant occupation. Thin black and red lines are the integrated intensities related to the effective anisotropic occupations (see (11.4))

The XLD spectrum can be defined as the difference between out-of-plane and in-plane XAS, $XLD = I_c - I_{ab}$, which is obviously related to the out-of-plane and in-plane splitting and anisotropic occupancy.

In Fig. 11.6 we show two examples of dichroism spectra acquired at room temperature for the cases of an in-plane tensile strained $La_{0.7}Sr_{0.3}MnO_3$ (LSMO) film (deposited on STO) and a compressively strained LSMO (deposited on LAO) which show respectively predominant electron occupation of the 3d $_{x2-y2}$ and of $3d_{z2}$ e.g. states [9]. The XLD data are reproduced by atomic multiplet scattering calculations in D4h symmetry, which corresponds to a tetragonal crystal field. The crystal field parameters are Ds and Dt. The energy position of each individual 3d orbital are directly related to the parameters $10Dq$, Ds and Dt. Another way to get semi-quantitative information about the occupation of different 3d orbitals, is by using the sum rules, which relate the integral of the XLD spectrum normalized to integral of the total absorption, D_L, to the expectation value of the quadrupole moment operator, Q_{zz}, associated to the charge distribution of the incomplete 3d-shell. It can be shown that such integral is given by [10]:

$$D_L = \frac{\int_{L_3+L_2} (I_{ab} - I_c)dE}{\int_{L_3+L_2} (2I_{ab} + I_c)dE} = \frac{\langle Q_{zz} \rangle}{h(2l-1)l} = \frac{1}{2}\frac{\langle \sum_i [3l_z^2 - l(l+1)]_i \rangle}{h(2l-1)l}, \quad (11.3)$$

where l is the angular momentum ($l = 2$ for 3d orbitals), l_z is the projected angular momentum along the z-axis.

For example, in the case of manganites ($3d^4$), assuming that the t_{2g} orbitals are equally occupied, we get [11]:

$$D_L = \frac{\int_{L_3+L_2} (I_{ab} - I_c)dE}{\int_{L_3+L_2} (2I_{ab} + I_c)dE} = \frac{1}{2} \frac{[6n_{xy} - 3n_{xz} - 3n_{yz} + 6n_{x2-y2} - 6n_{z2}]}{36} = \frac{n_{x2-y2} - n_{z2}}{12}$$

(11.4)

From (11.4) one can directly derive the anisotropic occupation of e.g. orbitals and the orbital polarizations of the 3d states.

11.2.3 X-Ray Magnetic Circular Dichroism

X-ray magnetic circular dichroism is a very powerful method to study magnetism in metallic, insulating and molecular systems, giving access to atomic moments with element and site selectivity. The exceptional sensitivity of XMCD allows its use not only on bulk samples, but also on ultra-thin films, interfaces and ultra-dilute systems. In Fig. 11.7 we show the typical experimental setup of an XMCD experiment. The dichroic signal is due to the different absorption cross sections of circularly polarized photons whether the sample magnetization is parallel or antiparallel to the direction of propagation of the beam. Therefore, to orient the magnetization of the sample a magnetic field is applied, usually directed along the X-ray beam but variable in amplitude and sign. In addition, the sample can be rotated around the vertical axis (**y** in Fig. 11.7), allowing the alignment of different crystallographic axes along the magnetizing field for the determination of magneto-crystalline anisotropy. The XMCD spectrum is the difference between XAS spectra measured, at fixed magnetization, with left (LCP, I^+) and right (RCP, I^-) circularly polarized photons. In order to eliminate or reduce systematic errors, usually the measurements are made with both opposite magnetizations: the resulting XMCD spectra are then reversed in signs and their average cancels most of the instrumental asymmetries due to differences in the beam intensity or spot size with the two polarizations.

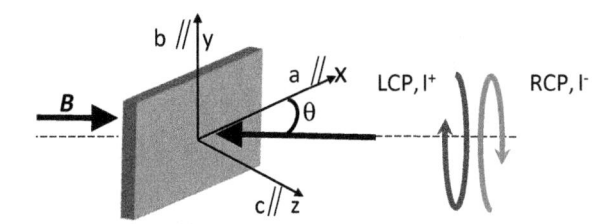

Fig. 11.7 Typical experimental setup of an XMCD experiment. The beam impinges on the sample at an angle θ respect the **ab** sample surface mounted in the **xy** plane (**y** is the vertical direction in the laboratory frame). LCP and RCP are respectively the left and right circular polarizations. **B** is the magnetic field, parallel or antiparallel to the beam

Fig. 11.8 Upper panel: Sum of average I^+ and I^- XAS spectra, normalized to the maximum at L_3 (black line), acquired in normal incidence at 5 T on a superconducting, optimally doped, $YBa_2Cu_3O_7$ thin film deposited on $SrTiO_3$. Bottom panel: XMCD spectra obtained from the difference of I^+ and I^- spectra. Data are taken from [12]

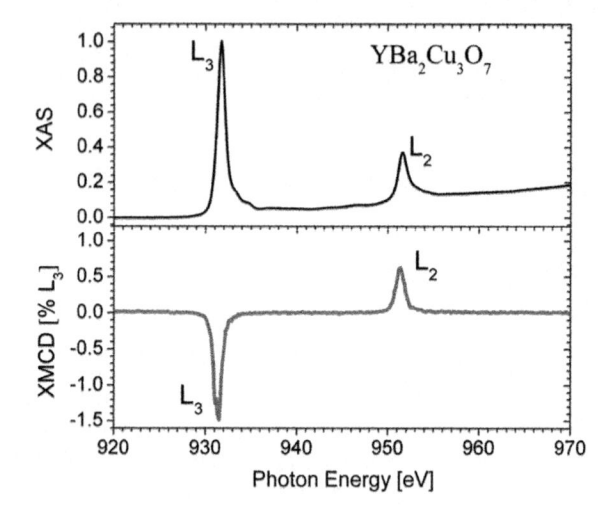

To illustrate the sensitivity of the technique in Fig. 11.8 we show the XMCD due to weak ferromagnetism of a layered cuprate superconductor. In these materials each Cu^{2+} site is characterized by a magnetic moment associated to the spin $S = 1/2$ of one electron. These spin are anti-ferromagnetically ordered at low temperature, with the atomic moments lying mainly in the ab plane. This configuration would give a zero XMCD signal. However, a strong field (5 T) along the c-axis at low temperature (5 K) can orient the small out of plane component of the spins leading to a measurable XMCD effect. Following the convention, the XMCD is negative at L_3 and positive at L_2. This case would bring 50% XMCD at L_3 for a fully magnetized sample; the 1.2% measured here can be attributed to an average out-of-plane canting of the Cu moments of ~1.4° [12].

The most powerful application of XMCD is the determination of the spin and of the orbital moment of TMs directly from the experimental spectra, using the sum rules [13]:

$$m_{orb} = -\frac{4\int_{L_3+L_2}(I^+ - I^-)dE}{3\int_{L_3+L_2}(I^+ + I^-)dE}(10 - n_{3d}) = \mu_B \langle L_z \rangle$$

$$m_{Effspin} = -\frac{6\int_{L_3}(I^+ - I^-)dE - 4\int_{L_3+L_2}(I^+ - I^-)dE}{\int_{L_3+L_2}(I^+ + I^-)dE}(10 - n_{3d}) = g\mu_B \langle S_{z_{eff}} \rangle$$

(11.5)

where n_{3d} is the total number of electrons in the 3d-orbitals (known from the nominal valence of the absorbing atom) and I^+ and I^- are the XAS spectra acquired with left- and right-handed polarizations, respectively. It must be noted that only the orbital moment can be exactly determined using the sum rules [14]. In fact, the effective spin moment, $\langle S_{z,eff} \rangle$, differs from the spin-moment $\langle S_z \rangle$ by the contribution of the magnetic dipole operator, T_z, related to the orbital anisotropy and to the spin-orbit coupling in the final 3d states:

$$\langle S_{zEff} \rangle = \langle Sz \rangle + \frac{7}{2} \langle Tz \rangle \qquad (11.6)$$

Although in most cases <T_z> cannot be determined experimentally, it can be calculated using atomic multiplet calculations as shown in [14].

11.3 Introduction to Resonant Inelastic X-Ray Scattering

Resonant Inelastic X-ray Scattering is a "photon in-photon out" synchrotron-based spectroscopic technique suitable for the study of elementary excitations in solids. The coherent absorption and re-emission of a photon can be seen as an inelastic process where the photon transfers energy, momentum and polarization to the scatterer. Therefore, an excitation takes place, characterized by given energy, momentum and symmetry. RIXS is based upon a two-steps process, as shown in Fig. 11.9a: the first step is nothing else that the XAS resonant absorption process, where a core electron is excited into an intermediate empty state above the Fermi level by the absorption of an X-ray photon with energy $h\nu_{in}$ and wave vector $\mathbf{k}_{in}$. This state is unstable and within its lifetime it rapidly decays. Although the main de-excitation mechanism is self-ionization (Auger emission), in a small but measurable fraction of cases the decay is radiative, i.e. another X-ray photon of energy $h\nu_{out}$ and momentum $\mathbf{k}_{out}$ is emitted. We are interested here in the transitions where a valence electron fills back the core-hole: if the electron is the same one originally promoted from the core level we have an elastic or quasi-elastic process, otherwise an electronic (electron-hole pair, charge transfer) or orbital (crystal field) excitation is left at the end of the process. The final state is thus characterized by energy $E = h\nu_{in} - h\nu_{out}$ and a momentum $\mathbf{q} = \mathbf{k}_{out} - \mathbf{k}_{in}$.

In the experiments the ingoing photons are known, as a monochromatic beam impinges on the sample at a given orientation with respect to the crystalline axes (Fig. 11.9b), while the parameters of the outgoing photons have to be measured: the spectral distribution of the scattered photons $I(h\nu_{out})$ is measured at fixed ($\mathbf{k}_{in}$, $h\nu_{in}$) and fixed emission direction ($\mathbf{k}_{out}$), so that it can be directly interpreted as $I(\mathbf{q}, E)$. In addition, one can consider that photons carry a spin ($S_{ph} = 1$) that can be transferred in the scattering process: knowing the angular momentum transferred would allow a full characterization of the excited state, including its "symmetry". This latter property is the hardest to determine experimentally: if the polarization of incident photons is usually known, as synchrotron radiation is usually fully polarized, the measurement of the scattered photons polarization is very difficult.

In RIXS the energy of the incoming photon is tuned on purpose at an absorption edge of a transition metal, which ensures not only a selection of the chemical species to be probed, but also a more favorable scattering cross-section with respect to the

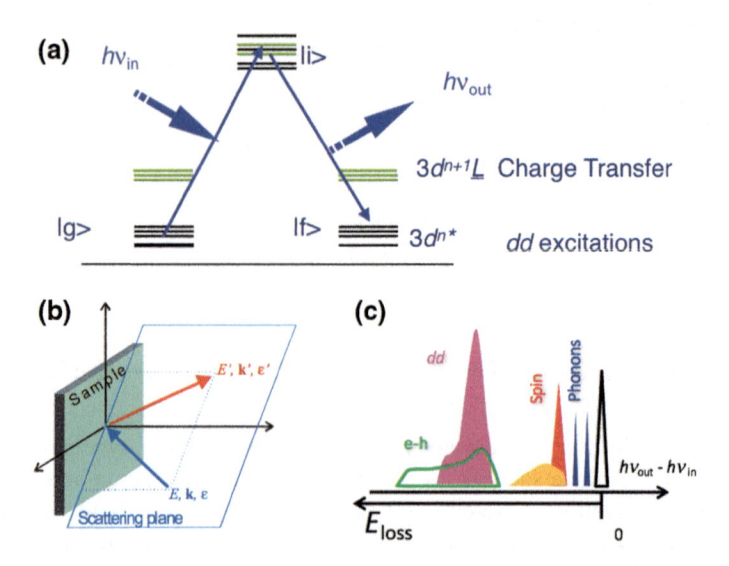

Fig. 11.9 a Schematic of the photon-in photon-out scattering process in a typical RIXS experiment. The photon-in with energy $E = h\nu_{in}$ excite a core electron (e.g. a 2p electron) in the ground state |g> to an intermediate |i> state (e.g. 3d), before decaying into the core hole, the excited electron can undergo' several type of intermediate excitation processes, leaving the system in another |f> final state, like those involving a charge transfer to a ligand $3d^{n+1}L$ state, or those concerning an internal d-d transition (d-d excitations), or involving spin and phonon excitations. As consequence of this process a photon-out with energy $E' = h\nu_{out}$ is emitted. b Geometrical configuration in a typical RIXS experiment. c Schematic of a RIXS spectrum characterized by various kind of excitations

non-resonant case and to other techniques. For a number of reasons RIXS is a unique method to study elementary excitations in condensed matter physics. First of all, no charge is added to or removed from the studied sample as a consequence of the scattering process. Therefore, the overall neutrality of the system is preserved. Moreover, photons carry a momentum that is inversely proportional to their wavelength, λ, which becomes not-negligible for X-rays, where λ is of the order of 0.1 nm, i.e. comparable to the interatomic distances in solids. This means that a sizable fraction of the Brillouin Zone of solids can be typically probed, contrarily to experiments using optical photons, which are restricted to a region very close to the Γ-point.

The excitations probed by RIXS are related to the charge, spin, orbital and lattice degrees of freedom of the studied system (Fig. 11.9c). Figure 11.10 shows a typical RIXS spectrum on a La_2CuO_4 (LCO) undoped cuprate. The high energy loss feature in TMO are charge transfer excitations, involving the transfer of a 3d electron to the ligand oxygen state. The intermediate region, between 0.5 and 2.5 eV are related to d-d excitations, corresponding to a reshuffling of the occupied and unoccupied 3d orbitals without changes in total 3d occupation number. The latter are forbidden by selection rules in the one photon absorption processes such as optical absorption or photoconductivity, but they are allowed in RIXS that is composed by two dipole allowed p-d, d-p transitions. A comparison between d-d excitations of various

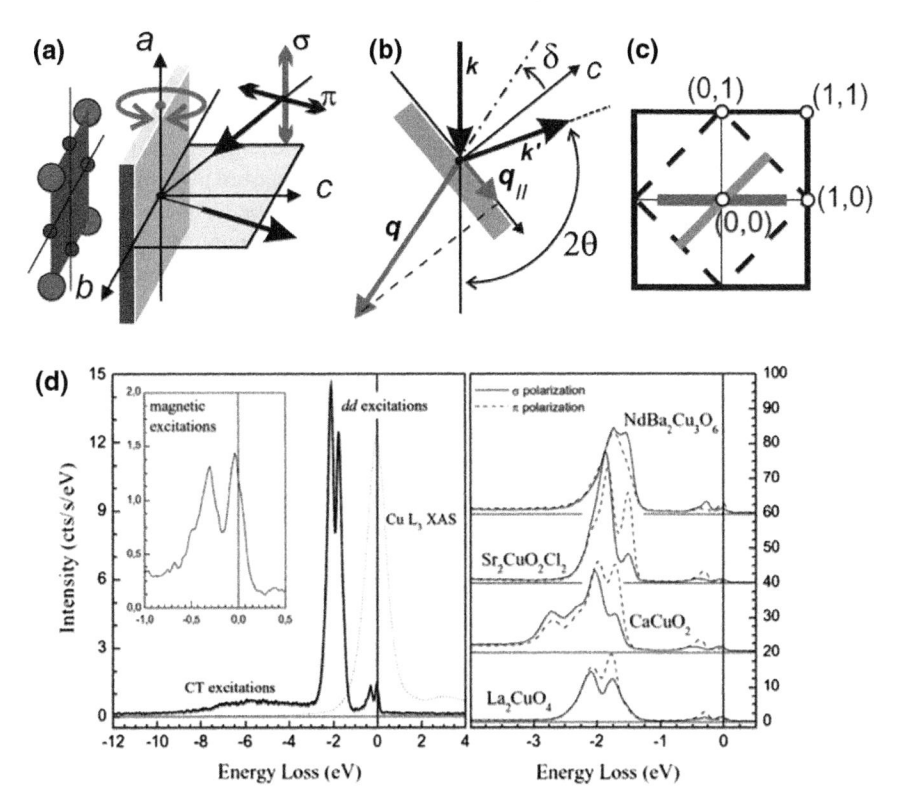

Fig. 11.10 a, b A layout of the experimental geometry of a RIXS experiment: the beam hits the sample surface (parallel to the ab plane) at incident angles θ_i and ϕ_i. The outgoing beam is collected at the angles θ_0 and ϕ_0. The scattering angle 2θ is fixed, whereas the incident angle and the azimuthal angle, which define the angle between the c-axis of the sample and the transferred momentum q (red arrow), δ can be varied. The projection of **q** onto the sample ab-plane, $q\|$, is changed by rotating the sample around a vertical axis, in order to access regions of the 2D reciprocal space far from the Γ-point, as indicated by the thick lines in panel (**c**). The [1, 0] and [1, 1] directions correspond to $\phi_i = 0$ and $45°$, respectively. **d** (Left panel) Example of Cu L_3 absorption (dashed) and RIXS (solid) spectra of LCO acquired with σ-polarization (Vertical linear polarization), and scattering and emission angles of $130°$ and $20°$ ($2\theta = 130°$, $\delta = 45°$). Charge transfer, d-d and magnetic excitations are highlight in the figure. A closer look at the mid-infrared energy region is given in the inset. (Right panel) RIXS spectra for LCO, SCOC, CCO and NBCO in the same experimental geometry. The data are taken from [15]

undoped cuprates is shown in Fig. 11.10b. These data allow the determination of the Cu-3d crystal field orbital configuration, a crucial parameter in the physics high Tc cuprates, as shown in [15].

Finally, the low energy part of the RIXS spectrum is related to presence of collective excitations, like magnetic excitations as in cuprates (magnons, bimagnons) and if resolution permits, lattice modes (phonons), whose intensity in the RIXS spectra is strictly related to the electron-phonon interaction.

A significant example of the potentialities of RIXS is measuring collective excitations in cuprates [16]. In Fig. 11.11a we show a typical RIXS spectrum measured at the ADRESS beamline of the Swiss Light Source (Paul Scherrer Institut) using the high-resolution SAXES spectrometer. The combined energy resolution for this experiment, performed in 2009, was 140 meV. The raw spectrum is composed by an elastic peak A, a single magnon peak B, a high-energy feature C, and a low-energy (90 meV) peak D. Peak D is related to a well-known optical phonon [17], and feature C is due to higher-order magnetic excitations, namely, bimagnons. In Fig. 11.11b we show a comparison between the dispersion of the single magnon feature compared to inelastic neutron scattering data [18], demonstrating a perfect agreement between the two techniques.

The dispersion of magnetic excitations has been measured in a number of other undoped cuprates. One of the most spectacular data-set was obtained on antiferromagnetic $Sr_2CuO_2Cl_2$ (SCOC), and is shown in Fig. 11.12a [19]. These results confirm earlier inelastic neutron scattering on La_2CuO_4, showing that in undoped cuprates the 2D antiferromagnetic structure and related spin-wave excitations cannot be explained without taking into account a significant electronic hopping beyond nearest-neighbor Cu ions, indicative of extended magnetic interactions.

RIXS has also been used to detect dispersing orbital excitations, as shown in the case of Sr_2CuO_3, a quasi-1D Mott insulator [20]. As shown in Fig. 11.12 (right panel), the orbital part of the RIXS spectrum shows a substantial dispersion along the CuO 1D-chains. This dispersion is interpreted as the evidence of a special quasiparticle, the orbiton, i.e. a non-local orbital excitation. The orbiton can be excited alone or together with a spin excitation, and the complex dispersion pattern is due to the combination of their individual propagation properties.

11.4 XAS and RIXS of Oxide Heterostructures

XAS and RIXS have been extensively used to investigate the magnetic, electronic properties of artificial oxide heterostructures obtained by the combination of different functional materials. The list of interesting examples is too long to be presented here, and includes nickelate based heterostructures [21, 22], multiferroic heterostructures [11, 23], ruthenates [24], titanates [25], manganites [26], and cuprates, in form of superlattices and bilayers.

In this chapter we will present the cases of cuprate and titanate heterostructures, which have triggered a broad interest in the oxide community. In the first part, we will give examples of the application of XAS, XLD and XMCD to the study of cuprate and in particular cuprate/manganite heterostructures. Then we will show the recent progresses of RIXS in the study of d-d excitation and magnon dispersion in artificial cuprate superlattices. Finally, we will present a short overview of the studies on the quasi two dimensional electron gas (q2DEG) formed at the interface between band insulating $LaAlO_3$ and $SrTiO_3$ oxides.

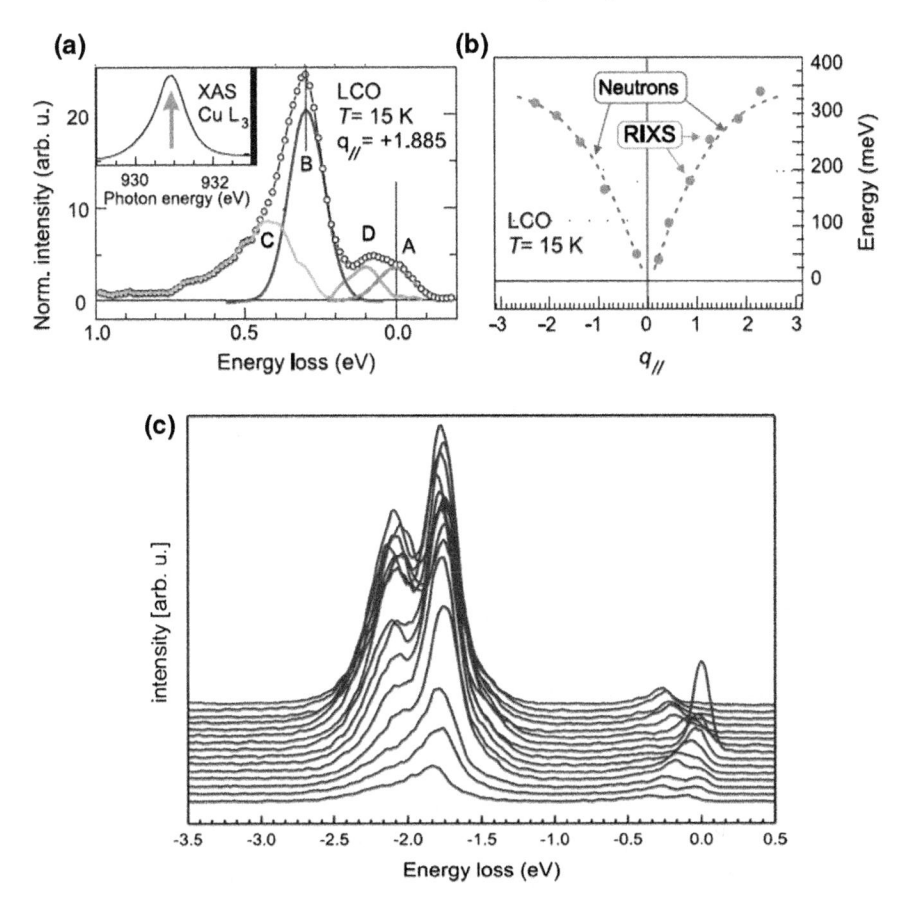

Fig. 11.11 a RIXS spectra on undoped La_2CuO_4: decomposition of the LCO spectrum at $ql = 1.85/a = 0.29$ rlu along the (100) direction: elastic (A) and single magnon peaks (B); multiple magnon (C) and optical phonons (D) spectral features. In the inset the L_3 absorption spectrum, the red arrow indicates excitation energy. Panel **b** single magnon dispersion determined by RIXS (blue dots) and by inelastic neutron scattering ([18], dashed purple line). Reprinted from [16]. Copyright 2010 American Physical Society. **c** RIXS spectrum as function of q∥ showing the magnon dispersion in the raw data

11.4.1 XAS on Cuprate Heterostructures

We have shown in Sects. 11.2 and 11.3 that XAS and RIXS can be used to obtain important information about the crystal field symmetry and strength, about the occupation of 3d orbitals, about the magnetic moment of a transition metal, about the collective excitations, including magnons and multi-magnons. Historically Cu-$L_{2,3}$ edge has been used to correlate the electronic signature of the so-called Zhang-Rice singlet to the hole density p in the CuO_2 planes of high Tc superconductors (HTS), as shown in Sect. 11.2.1 and Fig. 11.4. Another important result of XAS spectroscopy about

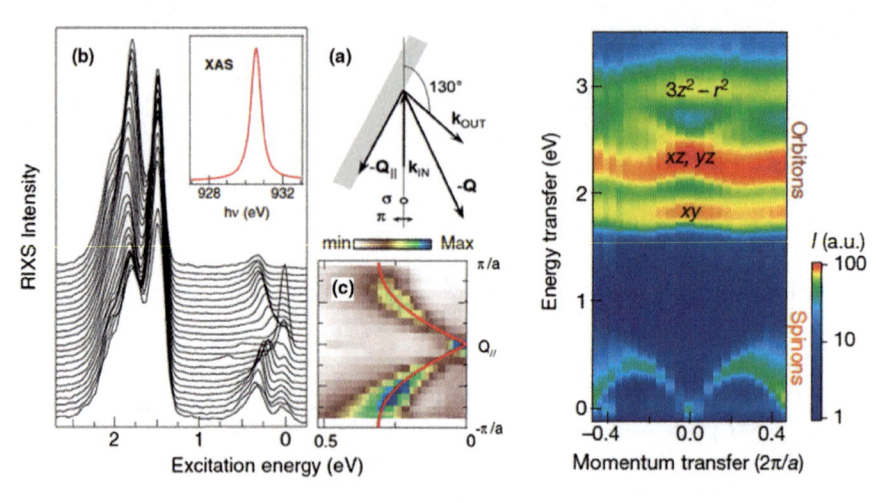

Fig. 11.12 Left Panels **a–c** RIXS on $Sr_2CuO_2Cl_2$ compound. **a** Schematics of the scattering geometry. **b** Cu L_3 RIXS spectra (T = 15 K) along the (100) direction. The incident energy is set at the maximum of the absorption (XAS) (inset). **c** Intensity map extracted from (**b**). The red line is the spin-wave dispersion for the NN Heisenberg model and J = 130 meV. Reprinted with permission from [19]. Copyright 2010 American Physical Society. Right Panel: RIXS intensity map of the dispersing spin and orbital excitations in Sr_2CuO_3 as functions of photon momentum transfer along the chains and photon energy transfer. Reproduced with permission from [20]. Copyright 2012 Nature Publishing Group

cuprates is related to the strong anisotropy of the absorption. Starting from the electronic configuration of the undoped parent compounds, e.g. La_2CuO_4, $YBa_2Cu_3O_6$, where copper has a Cu^{2+} formal valence with a $2p^63d^9$ configuration, due to the tetragonal crystal-field, the 3d-degeneracy is fully removed and the only empty 3d orbital has a x^2-y^2 symmetry, which determines the strong 2D-character of the electronic band structure of these materials. The symmetry of the empty 3d orbital was determined by measuring the angular dependence of the XAS spectra, in particular in the case of $La_{2-x}Sr_xCuO_4$ (LSCO) [27]. The XAS absorption is completely different in the case of linear polarization parallel (I_{ab}) or perpendicular (I_c) to the CuO_2 planes, as shown in Fig. 11.13a for a LSCO thin film with a doping $p = 0.1$ holes/CuO_2 plane. The intensity of the I_c XAS L_3 peak is only 6.5% of the I_{ab} absorption. The residual absorption observed in the I_c spectrum is due to the partial hybridization between apical $O2p_z$ and $Cu3d_{z2}$ orbitals. Thus, the degree of anisotropy of the Cu-3d orbitals is roughly related to the differences between the distances of planar and out-of-plane (apical) oxygen ions from the Cu-ion, or more precisely on the degree of hybridization between Cu and O in the two directions. By doping the charge reservoir, i.e. by ionic substitution or by changing the oxygen content, a fraction of copper sites changes their formal valence by adding holes to the CuO_2 planes. For each hole added, the local copper electronic configuration changes from $3d^9$ to a $3d^9\underline{L}$ ligand state, where an electron, instead of being localized at the Cu site, is shared among the Cu 3d $_{x2-y2}$ orbitals and the neighbor oxygen ions, forming the so called Zhang Rice singlet.

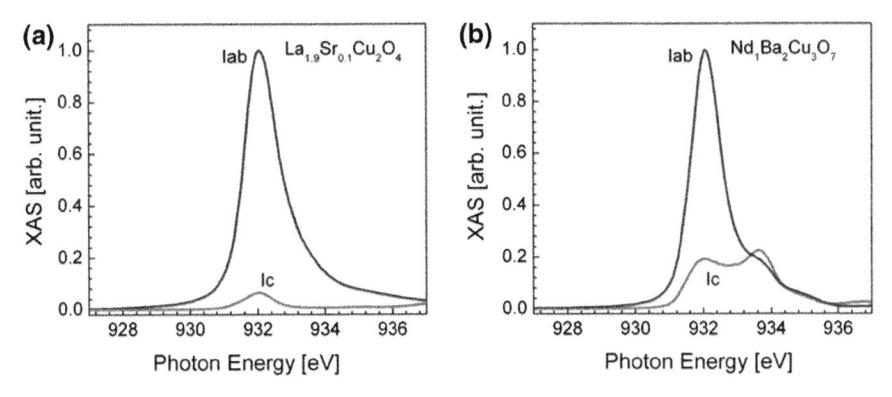

Fig. 11.13 In plane (Iab) and out of plane (Ic) XAS spectra normalized to the maximum of Iab on **a** underdoped $La_{1.9}Sr_{0.1}CuO_4$ and **b** optimally doped $Nd_1Ba_2Cu_3O_7$ (belonging to the $Y_1Ba_2Cu_3O_7$ 123 family) thin films

However, a similar hybridization can take place also among oxygen $2p_z$ and Cu $3d_{z2}$ orbitals. As matter of fact both I_{ab} and I_c Cu-L$_3$ spectra of doped cuprates are typically characterized by an absorption peak, associated to a $2p^63d^9 \rightarrow 2p^53d^{10}$ excitation, and a satellite which corresponds to the additional ligand configuration of the Cu ions. We have seen that the spectral weight of this satellite in the I_{ab} spectra is directly proportional to the density of holes in the CuO_2 planes, and from this feature one can effectively determine the carrier density in the CuO_2 planes. On the other hand, the $3d^9$ and $3d^9\underline{L}$ features in the I_c spectra should have a different meaning. In LSCO the I_c XAS is much smaller than in YBCO compounds, for example, and the satellite is barely visible. In YBCO, on the other hand, the $3d^9$ and $3d^9\underline{L}$ features have a similar order of magnitude intensity and quite stronger than the LSCO case (Fig. 11.13b). This is due to the particular structure of 123 cuprates (YBCO family) where Cu is present in the CuO_2 planes and in the chains. The hybridization between Cu$3d_{z2}$ and apical O$2p_z$ orbitals, Cu belonging to the CuO_2 plane and to the CuO chains, then determine the effective lower XAS anisotropy of these compounds compared to LSCO. In particular, the $3d^9\underline{L}$ peak in the I_c spectra of YBCO is a direct consequence of the charge transfer process of holes from the Cu(1)O chains to the CuO_2 planes.

Beyond these fundamental results, XAS has been used to study several other properties of cuprates. For example, the charge-transfer between Cu and both in-plane and out-of plane oxygen 2p orbitals, which can be on the other hand correlated to superconductivity. Namely the degree of localization of holes in the charge reservoir is crucial to understand the superconducting-insulating transition (SIT) in the YBCO family as directly measured by XAS. One of these studies has been performed on ultra-thin NdBCO films, which undergo a SIT as function of the thickness, without changing the composition. XAS spectroscopy demonstrated a SIT induced by the localization of a fraction of holes in the charge reservoir layers, as shown in Fig. 11.14 [28]. The main difference between the XAS spectra of superconducting and insulating samples is the anomalous increase of the $3d^9$ peak in the I_c XAS spectra, which probes

Fig. 11.14 In plane (Iab, upper panel) and out of plane (Ic, bottom panel) XAS spectra normalized to the maximum of Iab on 9 unit cells superconducting and 6 unit cells insulating $Nd_{1.2}Ba_{1.8}Cu_3O_7$ thin films

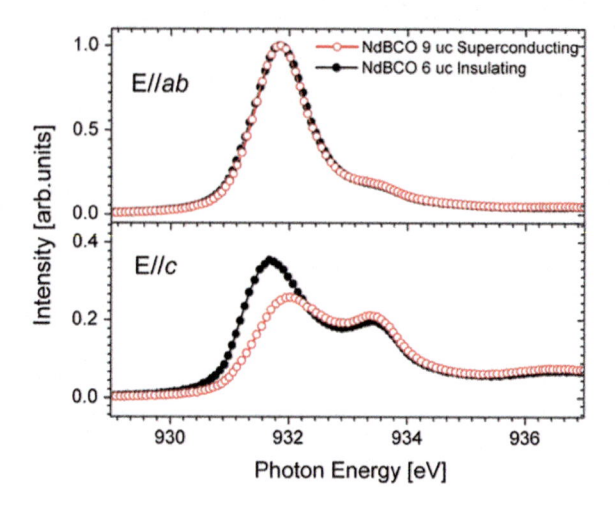

the out-of-plane Cu3d unoccupied states. This result is interpreted as the effect of localization of holes in the $3d_{z^2}$ orbitals of Cu ions belonging to the CuO_2 planes and to the CuO chains.

Another example concerns the orbital reconstruction phenomena in cuprate/manganite (superconducting/ferromagnetic) heterostructures, which show an anomalous degradation of the critical temperature attributed to the magnetic proximity effect induced by the manganite layer. A proper description of the phenomena in the case of these systems is still lacking, due to the variety and characteristic intriguing phenomena taking place at these interfaces. In [29], Chakhalian et al. reported an extreme reduction of the Cu-3d anisotropy at the interface between YBCO and LCMO thin films (Fig. 11.15a), which was ascribed to the formation of an anomalous interfacial Mn3d-O2p-Cu3d molecular orbital. The idea was that only the Cu(2)-sites belonging to the superconducting CuO_2 planes, where involved in this interfacial bonding, whereas Cu(1) sites, lying in the charge-reservoir and able of accommodating a variable amount of oxygen in the Cu(1)O chains, were supposed to play no direct role.

However, the characteristic molecular bonding along the c-axis can be equally interpreted as localization of holes in the interfacial charge reservoir layer, involving not only the Cu(2) ions but also the Cu(1)-O chains, as suggested by studies on isolated ultrathin films [28]. In order to clarify this issue, recently, a similar study has been performed on a different type of cuprate/manganite interface, namely in the case of $La_{1.85}Sr_{0.15}CuO_4/La_{0.75}Sr_{0.25}MnO_3$ superlattices, where a direct Mn-O-Cu bonding at the interface is not formed. Additionally, LSCO is characterized by only one copper site in the unit cell belonging exclusively to the CuO_2 planes [30]. One of the main result is that the interfacial Cu-3d states retain a strong anisotropy, as shown by the comparison between XAS spectra acquired with different linear polarization of the light (Fig. 11.15b). However, the bulk sensitive FY and the interface sensitive TEY signals show remarkable differences in the E//c component, as shown by the

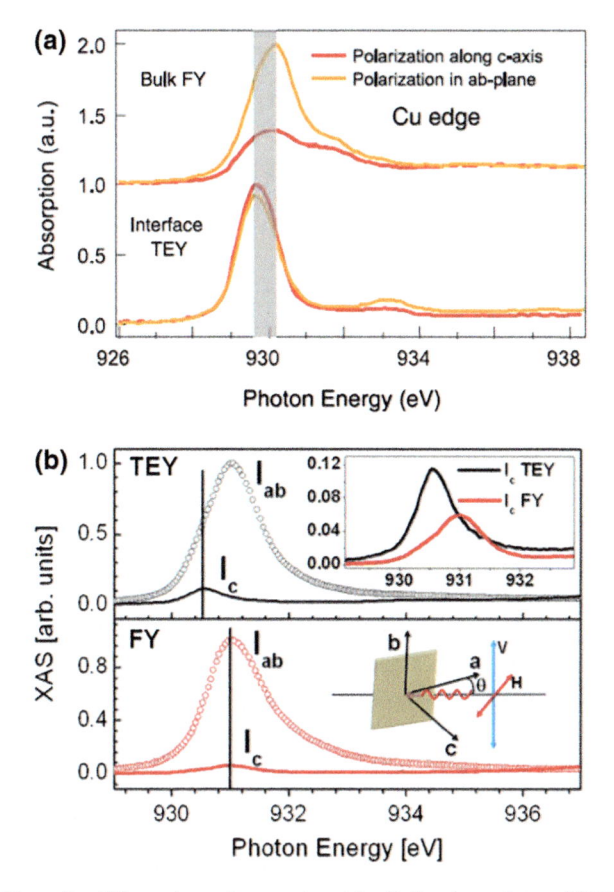

Fig. 11.15 a Normalized X-ray absorption spectra at the Cu L_3 absorption on YBCO/LCMO SLs, taken in bulk sensitive (FY, top panel) and interface sensitive (TEY, bottom panel) detection modes with varying photon polarization as indicated in the legend. Reproduced with permission from [29]. Copyright 2007 Science Publishing Group. **b** Cu L_3 absorption XAS spectra on LSCO/LSMO SLs, acquired by interface sensitive TEY (top panel) and bulk-sensitive FY (bottom panel); Iab (circles) is the normalized intensity for vertical (V) linear polarization, parallel to the ab plane; Ic (solid lines) is the corresponding intensity for horizontal (H) polarization, that is, almost parallel to the c-axis The inset in the top panel shows a direct comparison between TEY (black) and FY (red) with H polarization. In the bottom panel the experimental setup for XLD is sketched ($\theta = 70°$). Reproduced with permission from [30]. Copyright 2014 Nature Publishing Group

inset of Fig. 11.15b: after normalization of the respective in-plane spectra, the TEY XAS is two times stronger than the one measured in the bulk-sensitive FY mode. It is also shifted to lower energy, indicating the formation of a new state mainly composed by Cu $3d_{z^2}$ orbitals. The additional state is related to localization of holes in Cu$3d_{z^2}$ orbitals at the interface, and not the formation of a new molecular ordering as it is the case of YBCO/LCMO heterostructure. A consequence of the absence of a direct Cu-O-Mn bonding is that the orbital anisotropy of the Cu(2)-3d states is strong

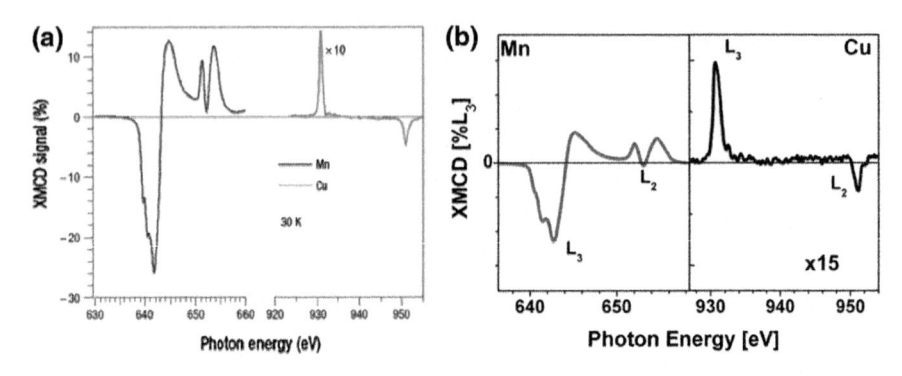

Fig. 11.16 XMCD signals obtained from the core-level absorption spectra for Cu (red multiplied by 10) and Mn (blue) in YBCO/LCMO SLs. Reproduced with permission from [33]. Copyright 2006 Nature Publishing Group. XMCD spectra for Cu (Black multiplied by 15) and Mn in LSCO/LSMO SLs [30]

also at the interface. Thus the formation of a particular interfacial molecular bonding cannot explain per se the suppression of superconductivity in both YBCO/LCMO and LSCO/LSMO SLs.

Another very intriguing, and related, phenomenon emerging at the cuprate/manganite heterostructures is the weak-ferromagnetism of the small out-of-plane Cu-$3d^9$ spin component, due to the proximity with the ferromagnetic LXMO (X = Ca, Sr) layer. While the emergence of a weak out-of-plane ferromagnetic ordering of the canted Cu-$3d^9$ spins is very well known in the case of undoped cuprates [31], its discovery in doped and superconducting compounds [32] was somewhat surprising in view of the well-known antagonism between superconductivity and magnetism. Actually many cuprates, both films and single crystals, exhibit an out-of-plane spin-moment due to the progressive alignment of canted Cu$3d^9$ spin, due to the Dzyaloshinskii-Moriya (DM) interaction, by a magnetic field perpendicular to the CuO$_2$ planes [32]. However, very surprisingly, Chakhalian et al. found that even at very low magnetic field the interfacial out-of-plane Cu-$3d^9$ magnetic moment are FM ordered and antiparallel to the adjacent manganite layer (Fig. 11.16a and [33]). This result was interpreted by Salafranca and Okamoto [34] as the consequence of an antiferromagnetic super-exchange interaction between Cu and Mn ions at the interface, mediated by the oxygen ions. The subsequent discovery of a Cu3d-O2p-Mn3d molecular ordering discussed above was interpreted in same framework, thus suggesting a crucial role of the specific Cu-O-Mn bonding in the magnetic and electronic reconstruction phenomena occurring at cuprate/manganite interfaces.

Following the same rationale, a different magnetic behavior would be expected in cuprate/manganite interfaces which do not exhibit a direct Cu-O-Mn bonding. Actually LSCO/LSMO interfaces show, on the other hand, the same antiferromagnetic coupling between Cu and Mn out-of-plane spins without the presence of Cu-O-Mn bonding and with negligible change of the interfacial Cu3d anisotropy

(Fig. 11.16b and [30]). Extending the theoretical model of Salafranca and Okamoto by introducing a two band description of the cuprate layer, which include both $Cu3d_{x2-y2}$ and $Cu3d_{z2}$ derived bands and the additional DM-interaction term in the Hamiltonian, it was shown that the Cu and Mn spins at the interface are always anti-ferromagnetically coupled. Moreover, spin-polarized electrons, transferred at the equilibrium from the manganite to the cuprate layer, perturb the magnetization of the Cu-3d moments along the interface over several layers inside the cuprate. This result suggests that the main driving force for the suppression of the superconductivity in cuprates is the perturbation of magnetic correlations among the CuO_2 planes occurring on several layer in cuprate/manganite heterostructures in general. The interfacial magnetic and orbital structure is more or less perturbed by the presence or absence of a direct molecular Cu-O-Mn bonding. However, the suppression of Tc in LSCO/LSMO heterostructures, which is as strong as the Tc suppression in YBCO/LCMO, is due to the perturbation over several layers of the magnetization profile of the Cu-3d moments, and not to the exclusive role of the interfaces.

11.4.2 RIXS of Cuprate Heterostructures

The interest of RIXS over other X-ray spectroscopies, like ARPES and scanning probe spectroscopy, is the capability to probe the bulk electronic structure of the material. For this reason, the main breakthroughs of the application of soft-X-ray RIXS has been mainly in the study of the electronic and magnetic collective excitation of bulk cuprates where, due to the characteristic momentum exchange of RIXS, this technique has been complementary to inelastic neutron scattering. In particular, RIXS is able to probe a different and larger portion of the Brillouin zone which is inaccessible by neutrons for the weakness of the signal, in turn due to the low flux and to the smallness of scattering cross sections. Moreover, RIXS can be used also for the study of very small single crystals and thin films, due to very high efficiency of the resonant scattering process. This property makes RIXS one of the few techniques able to measure collective excitations of ultrathin films and buried heterostructures, possibly down to the single unit cell, and single interfaces.

Notable examples are RIXS studies on $La_2CuO_4/LaAlO_3$ (LCO/LAO) [35] and $CaCuO_2/SrTiO_3$ (CCO/STO) superlattices [36] composed by the repetition of few unit cells cuprate layers, down to the smallest number of two CuO_2 planes separated by insulating $LaAlO_3$ and $SrTiO_3$ band insulators, respectively. In spite of the reduced probed volume, RIXS has been capable to measure the crystal-field (through the study of the d-d excitation) and the magnon dispersion of few, atomically sharp, interfaces. In the case of non-superconducting LCO/LAO SLs, the spin-excitations were observed down to 1 uc thick LCO layers, each containing two CuO_2 planes (Fig. 11.17a). In [36], M. Minola et al. have found that the dynamic of magnon dispersion in superconducting CCO/STO SLs is progressively reduced decreasing the thickness of the CCO layer, but even in the case of SLs composed by only two CCO unit cells, which contain two CuO_2 atomic planes, the spin-excitations

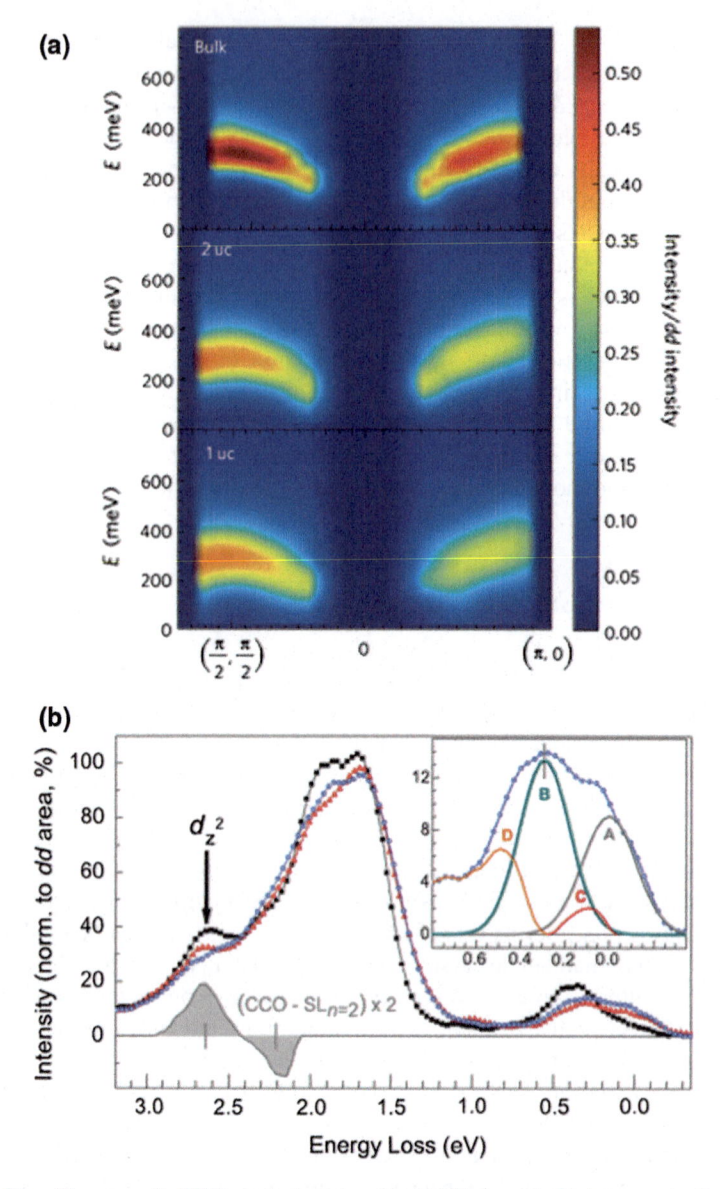

Fig. 11.17 a The magnetic RIXS scattering intensity at 15 K along the high-symmetry lines in the Brillouin zone for a bulk film (upper panel), 2 uc (15x[2LaAlO$_3$ + 2La$_2$CuO$_4$]) (middle panel) and 1 uc (25x[LaAlO$_3$ + La$_2$CuO$_4$]) (bottom panel). Reproduced with permission from [35]. Copyright 2012 Nature Publishing Group. **b** RIXS raw spectra at Cu L$_3$ edge of bulk CaCuO$_2$ (black filled squares) and (CaCuO$_2$)n/(SrTiO$_3$)n SLs, with n = 3 (red filled triangles) and n = 2 (blue open circles), at q|| = 0.375 r.l.u. and T = 20 K using π incident polarization. Intensities are normalized to the dd area put to 100. The shaded area under the gray solid line represents the difference between the bulk CaCuO2 and the SL with n = 2 multiplied by 2. The inset shows the decomposition of SL n = 2 spectrum in the low energy region: elastic (A) and magnon (B) peaks; optical phonon (C) and multiple magnons (D) spectral features. Reproduced with permission from [36]. Copyright © 2012 American Physical Society

survive (Fig. 11.17b), as in the case of LCO/LAO SLs. More recently, a similar result was found in the case of SCO/STO superlattices [37], again demonstrating that the engineering of novel heterostructures is a feasible method to tune the electronic and magnetic properties of cuprate thin films.

These experiments, show that RIXS can be successfully applied to the study of single interfaces, and single layer oxides.

11.4.3 XAS and RIXS of LaAlO$_3$/SrTiO$_3$ Heterostructures

The interface between band insulating LaAlO$_3$ (LAO) thin films and bulk SrTiO$_3$ (STO) [38], host a quasi two-dimensional electron system (q2DES) stands as a model of oxide heterostructures characterized by electronic properties which differ substantially from the bulk constituent layers. It shows quite unique properties, including remarkable mobility up to 5×10^4 cm^2 V^{-1} s^{-1} (at 4.2 K) [39] and low temperature superconductivity [40]. Moreover, its electronic properties can be largely tuned by electric field effect, which allows, for example, a control of metal to insulating transition [41, 42] (even at room temperature) and a large range modulation of Rashba spin-orbit coupling [43]. Among the models explaining the formation of the q2DES, an electronic reconstruction scenario has been put forward by several groups. This model has some precise predictions, based on the idea that the polar instability of the LAO film requires a transfer of 0.5 electrons from the AlO$_2$ surface to Ti-3d states. This mechanism would be favoured by the variable valence of transition metals. In particular, to make the system stable, 50% of interfacial titanium ions are expected to change valence at the interface from Ti^{4+} to Ti^{3+}. This change of valence can be detected by XAS. However, as shown in Fig. 11.18, the XAS spectra of LAO/STO samples is very similar to that one conducting (Nb-doped) and even insulating (stoichiometric) SrTiO$_3$ single crystals. As matter of fact, the XAS spec-

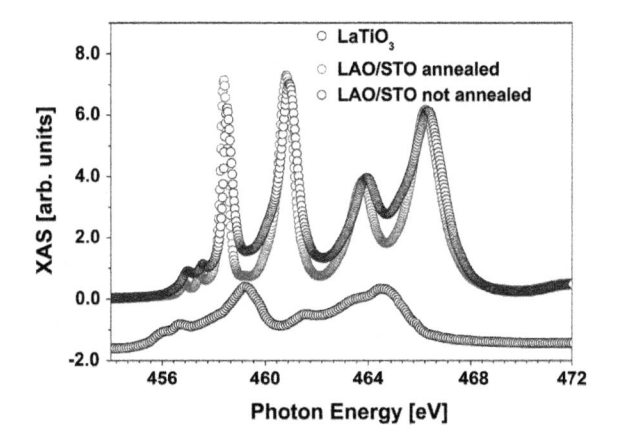

Fig. 11.18 XAS data on LAO/STO samples annealed (red circles) and non-annealed (blue circles) in oxygen after the deposition (from [44]). Data for LTO are also shown as reference

tra can be reproduced by atomic multiplet splitting calculations assuming titanium ions in a Ti^{4+} oxidation state, i.e. a $3d^0$ ground state configuration, and crystal field parameters typical of insulating STO. The Ti L-edge XAS spectra, are characterized by four main peaks (**a1, b1, a2, b2**), related to empty t_{2g} and e.g. 3d-orbitals. This is at odds from the case of $LaTiO_3$, which is an antiferromagnetic Mott insulator and represents the closest realization of a $3d^1$ Ti^{3+} configuration, with one electron per site localized in 3d-Ti states. The $LaTiO_3$ XAS spectrum is characterized by much broader peaks located at energies intermediate between the **a1, b1** and **a2, b2** in STO.

This result apparently contradicts the metallic character of the LAO/STO system, which implies the presence of delocalized electrons in the Ti-3d bands. However, titanium $L_{2,3}$-edge XAS is not very sensitive to the delocalized electrons. Indeed, in the XAS process the electrons excited into the 3d states are not efficiently screened.

Consequently, the conducting electrons in Ti-bands do not show any signature, unless they are so many that the effective oxidation state of more than 10% of titanium ions changes from 4+ to 3+. On the contrary, XAS can measure the presence of electrons localized in Ti^{3+} states, as in the case of $LaTiO_3$. In LAO/STO signatures of localized electrons at Ti-sites (Ti^{3+}) are observed in non-stoichiometric samples which contain oxygen vacancies, for example, in standard LAO/STO interfaces deposited by pulsed laser deposition in $P_{O2} = 10^{-4}$ mbar of oxygen pressure and at deposition temperatures of 780 °C, without high oxygen pressure post-annealing [44]. In these cases, we can notice that (Fig. 11.18) the XAS spectra show a transfer of spectral weight from **a1, b1** (and from **a2, b2**) features, typical of a Ti^{4+} valence, to the middle regions where contribution from Ti^{3+} is expected.

The presence of both delocalized and localized electrons in LAO/STO superlattices, was demonstrated by RIXS in [45] (see Fig. 11.19) on samples with different degrees of oxygen vacancies content. The data show a weak signature of d-d excitations around -1.2 eV energy loss, which is attributed to delocalized electrons due to their dispersive character, and at about -3 eV, attributed to localized Ti^{3+} electrons; both features are strongly suppressed in samples heavily annealed in oxygen to reduce the oxygen vacancies content.

Beside the presence of electrons in the conduction bands, one of the most important difference between bulk STO and LAO/STO heterostructures is the fact that the there is an inversion of hierarchy between states characterized by in-plane and out-of-plane orbitals, predicted by several theoretical calculations. This inversion has been demonstrated by using X-ray linear dichroism. The XLD data (Fig. 11.20, [25]) of TiO_2 terminated STO single crystals can be perfectly reproduced by atomic multiplet calculations of Ti^{4+} $3d^0$ configuration in D4h symmetry with crystal field splitting $\Delta t_{2g} = 3d_{xz,yz} - 3d_{xy} = -25$ meV and $\Delta e_{2g} = 3d_{z2} - 3d_{x2-y2} = -40$ meV. Both the sign and the shape of the dichroism change in LAO/STO bilayer.

In particular, the XLD spectra of LAO/STO can be perfectly reproduced by atomic multiplet calculations with $\Delta t_{2g} = 3d_{xz,yz} - 3d_{xy} = +50$ meV and $\Delta e_{2g} = 3d_{z2} - 3d_{x2-y2} = +90$ meV [25]. However, also in the LAO/STO case, the XLD spectra is calculated assuming a Ti^{4+} ionic configuration, and not the mixed Ti^{4+} and Ti^{3+} scenario expected in the electronic reconstruction scenario. These results can be explained by the fact that the fraction of electrons in the interfacial conduction band is much lower than 0.5 electrons/2D unit cell.

Fig. 11.19 a Total electron yield (TEY) Ti 2p-XAS spectra of a LAO/STO SL(LAO3). Arrows label the incident energies used for RIXS. **b** Series of RIXS spectra of LAO3 SL with out-of-plane polarization. RIXS spectra from STO substrate (solid black line) and STO film (dotted gray line) references are also displayed. Reproduced with permission from [45]. Copyright © 2012 American Physical Society

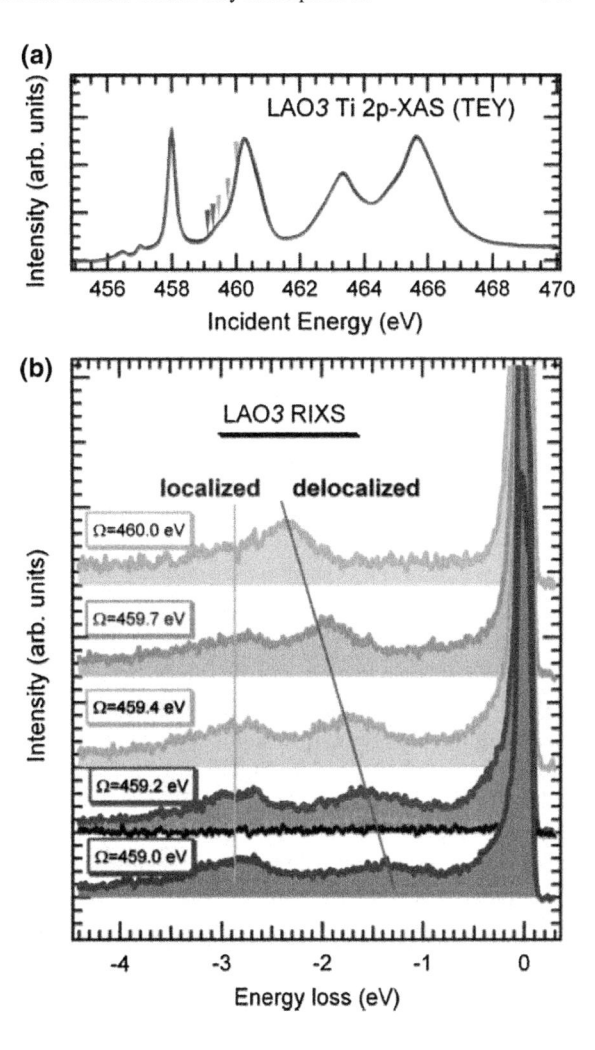

The real ground state of the LaAlO$_3$/SrTiO$_3$ system has also been subject of controversy due to observations of superconductivity of the q2DES below 0.3 K in some samples and of magnetism in others. Theoretically, titanate heterostructures can become ferromagnetic and metallic by an interface orbital/magnetic reconstruction. In principle, 3d electrons, carrying not only charge but also spin, can order ferromagnetically in the presence of electron correlation. Coexistence of magnetism and superconductivity at low temperatures were reported by scanning SQUID (Superconducting Quantum Interference device) [46], SQUID [47], and torque magnetometry [48].

Unfortunately, these techniques are not elemental and orbital selective, do not probe the Ti-magnetism alone, and are not sensitive only to the interface states. Thus, the data reported could be also attributed to a magnetic signal coming from

Fig. 11.20 Ti L_{23} edge XLD (Ic-Iab) spectra of STO (red symbols) and conducting LAO/STO (green symbols) (from [25]). Black lines are atomic multiplet calculations assuming a Ti^{4+} configuration

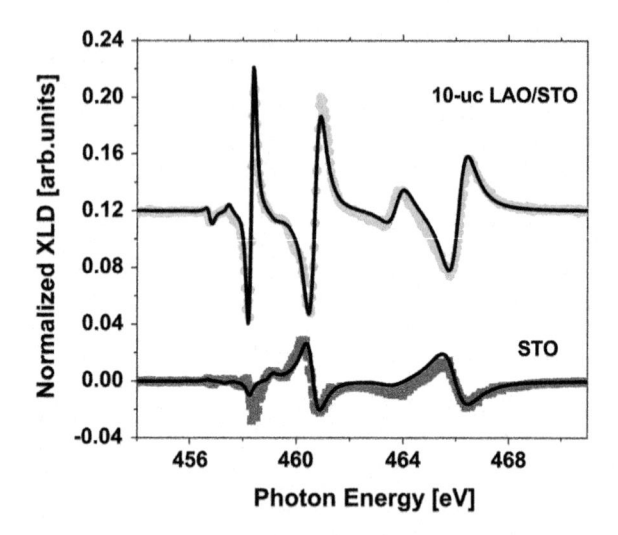

inner STO layers and from the topmost (LAO) films. Besides these studies, other experimental reports did not find any signature of interfacial magnetism, including polarized neutron reflectometry [49] and β-detected nuclear magnetic resonance [50]. These studies show that any Ti^{3+}-magnetic moment in this system is very small, incompatible with the large values (up to 0.3 μ_B/Ti at the interface) estimated by torque magnetometry and by SQUID.

One of the few experimental techniques able to address the Ti-magnetism with both elemental, orbital and interface selectivity is X-ray magnetic circular dichroism at the Ti $L_{2,3}$ edge. According to atomic multiplet splitting calculations (Fig. 11.21), the simulated XMD spectra of Ti^{3+} and Ti^{4+} ionic configurations are very different both quantitatively and qualitatively. The XMCD spectra are characterized by main features which correlate with the maximum of intensity in the corresponding XAS spectra. The splitting of spin-up and spin-down final states due to the addition of magnetic exchange field (here 10 meV), gives rise to a finite XMCD signal, but a corresponding null effective magnetic moment. On the other hand, the Ti^{3+} XMCD signal is much larger and corresponds to the presence of the effective magnetic moment of the Ti^{3+} electrons.

Lee et al. performed XMCD experiments on 3 uc LAO/STO sample and found, in magnetic field of 0.2 T almost parallel to the interface, a magnetic signal attributed to Ti^{3+} $3d_{xy}$ electrons in the interfacial layer [51]. While the XAS spectrum is similar to that one of STO, the XMCD exhibits features at energies corresponding to **a1**, **b1** and **a2**, **b2**, typical of a titanium in a Ti^{4+} oxidation state. The data can be reproduced assuming that the sample is characterized by a fraction of Ti^{3+} magnetic moments. However, even including charge transfer effects [Fig. 11.21, red line bottom panel], in order to get a good matching between the data and model, one has to assume rigid energy shift of the calculated Ti^{3+} spectrum.

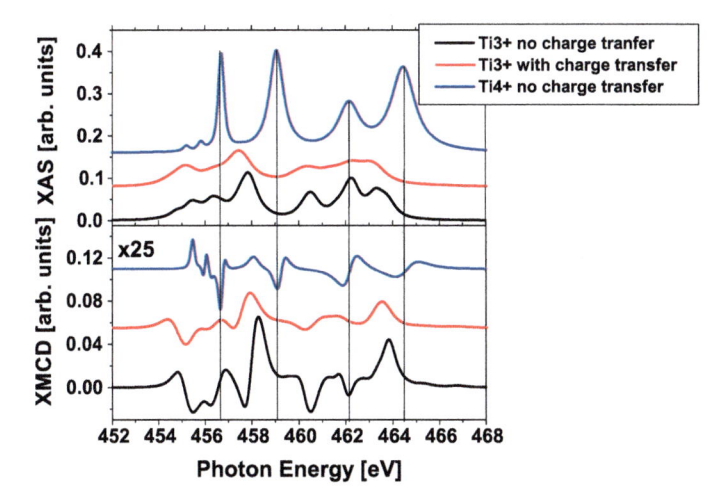

Fig. 11.21 Simulated XAS (upper panel) and XMCD (bottom panel) spectra of L_{23} Ti-edge in Ti^{3+} and Ti^{4+} configurations using the atomic multiplet scattering code CTM4XAS [7]. Blue lines are obtained assuming a Ti^{4+} oxidation state of titanium, red and black data are obtained assuming a Ti^{3+} (with and without charge transfer respectively) oxidation state. The calculations have been performed in C4 symmetry (equivalent to octahedral configuration) and an effective exchange field of +10 meV. The charge transfer parameters, 10Dq and slater integral are taken from [44, 51] in the case of Ti^{4+} and Ti^{3+} configurations respectively

By performing a systematic experimental work on samples grown by different groups in both strongly and poorly oxidizing conditions, we have found different titanium XMCD results [44]. We have studied standard LAO/STO samples, grown in $P_{O2} = 8 \times 10^{-5}$ mbar and annealed in 200 mbar of O_2 after the deposition, and non-standard LAO/STO, without post-annealing containing large amounts of oxygen vacancies at the interface and localized Ti^{3+} electrons. The experiments were performed on the same samples at two X-ray synchrotron facilities, i.e. the beam-line ID08 of the European Synchrotron Radiation Facility and the X-TREME beam-line of the Swiss Light Source.

As shown in Fig. 11.22a, we found that the magnetic (both orbital and spin) moment is negligible in optimally oxygenated LAO/STO interfaces. On the other hand, LAO/STO samples containing oxygen vacancies show Ti^{3+} localized spins, as seen by XMCD features resonating at the t_{2g} Ti^{3+} XAS main peaks. This study shows that localized and magnetic Ti^{3+} 3d-electrons are introduced by oxygen vacancies. Together with a Ti^{3+} signal, the XMCD of these non-annealed LAO/STO samples exhibit features resonating also at the main Ti^{4+} peaks. This result suggests that the presence of localized and magnetic Ti^{3+} 3d-electrons induces a small spin-splitting of the $3d^0$ states. Very strikingly, according to calculations, the XMCD sign related to the these Ti^{4+} related signal is opposite to expected sign in case of a positive exchange interaction (e.g. for trivial Zeeman splitting, Fig. 11.21 blue line), and consequently cannot be attributed to the effect of the external magnetic field [44].

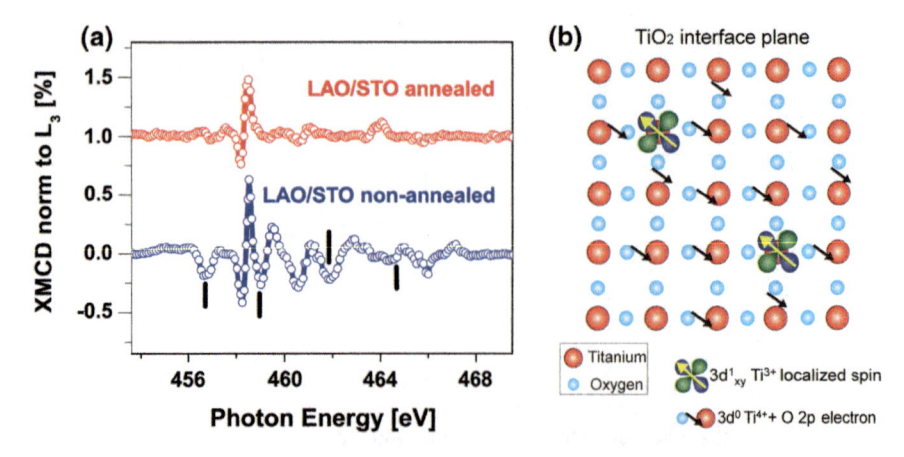

Fig. 11.22 **a** XMCD data measured at 3 K and 6 T on annealed (red symbols) and non-annealed (blue symbols) LAO/STO interfaces. Black sticks indicate features associated to localized Ti^{3+} magnetic moments. **b** A cartoon illustration of the proposed mechanism of magnetism at the titanate heterostructures. In LAO/STO oxygen vacancies lead to localized Ti^{3+} sites, holding sizable spin and orbital magnetic moments thanks to their 3dxy electron. The q2DES gets polarized by these magnetic impurities and can mediate a long-range magnetic interaction among them at low temperatures. Reproduced with permission from [44]. Copyright © 2012 American Physical Society

This result suggests that a negative exchange interaction between delocalized and localized (magnetic) 3d-electrons takes places, which induce a spin-polarization in the 2DEG, as shown schematically in Fig. 11.22b. Since a purely Ti^{3+} magnetic moment is detected only in the case of samples containing oxygen defects it is possible to rule out ferromagnetism as an intrinsic property of this system. In the case of annealed LAO/STO interfaces, the XMCD data show a tiny (Ti^{4+}) XMCD signal. Here, the Ti-XMCD and XAS spectra cannot be explained by the presence of Ti^{3+} magnetic moments. Yet, at L$_3$ and for t$_{2g}$ states, the XMCD is different from zero even at 0.1 T. This is an intriguing result, which requires further investigations, also in view of theoretical proposal of spiral magnetism in the q2DES [52].

A robust Ti-ferromagnetism has been recently found in the case of delta-doped LaAlO$_3$/EuTiO$_3$/SrTiO$_3$ interfaces, where 2 unit cells of EuTiO$_3$ are intercalated between the STO single crystal and the LAO film. This is shown in Fig. 11.23, were we compare the normalized XMCD spectra of several LAO/ETO/STO and LAO/STO heterostructures measured in the same conditions (T = 3 K, H = 6 T). One can see that, while the shape of the XMCD show some similarities among the samples, a sizable signal is observed only in the case of the delta-doped heterostructures. Moreover, as shown in [53], only delta-doped heterostructures are characterized by a remanent XMCD signal at zero field, a magnetic field dependence of the XMCD typical of a FM, together with a Ti-XMCD temperature dependence which show a FM transition at around Tc = 7 K, the same temperature at which the Eu4f ions order ferromagnetically. Eu and Ti FM in this heterostructure are directly coupled [53].

11 Oxides and Their Heterostructures Studied with X-Ray Absorption ...

Fig. 11.23 XMCD data, normalized to the maximum of the XAS spectrum at L_3, measured at 3 K and 6 T on LAO/ETO/STO (green line), non-annealed LAO/STO (dark-cyan line), annealed (blue line) and LAO/STO(2uc)/STO (orange line) interfaces

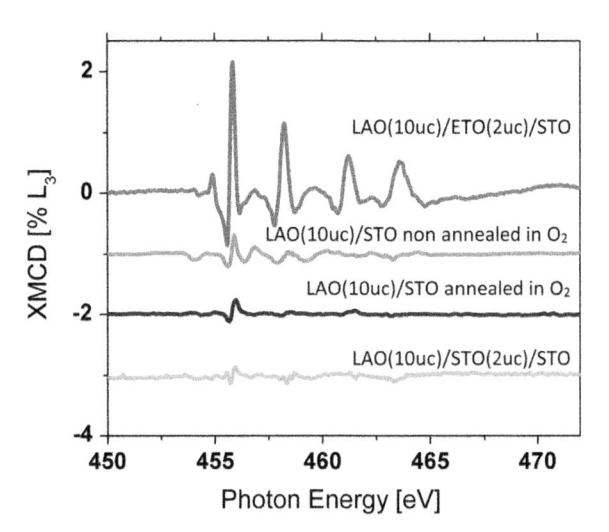

11.5 Perspectives and Conclusions

Core level soft X-ray spectroscopies have emerged as ideal tools for the determination of the electronic and magnetic properties of ultrathin TMO films and TMO heterostructures. Their strength is in the chemical and site selectivity, which permits to probe the relevant sample volume cancelling the signal from the rest of the sample or from the substrate. The resonant enhancement boosts the sensitivity of these bulk sensitive techniques, allowing them to probe not only buried interfaces but also individual unit cells at the surface. Finally, the strong transition selection rules and the large spin-orbit interaction of the 2p core levels provide unambiguous interpretation of the main phenomenology in polarization dependence XAS (XMCD, XLD) and in RIXS (d-d excitations, spin excitations).

The clarity of the results on cuprates and STO/LAO systems presented in this chapter are pushing a growing number of researcher in the field of TMO heterostructures to use XAS and RIXS. A limitation has been up to now the difficult access to beam lines with sufficient quality: stability, reproducibility, absence of spurious asymmetries are key requirements for high level XMCD and XLD, high flux and energy resolution are necessary for measuring significant RIXS spectra. But the situation is improving, and new ambitious facilities are being built at several synchrotrons around the world. A wider access to polarization dependent XAS and high resolution RIXS will allow more systematic studies of the subtle and intriguing electronic properties of TMO heterostructures, so that we will be able to nail down intrinsic properties and discard defect-related phenomena. This is the necessary step towards a true engineering of TMO interface properties targeting their eventual utilization in the future devices of oxide electronics.

Acknowledgements The Authors are grateful to C. Aruta, L. Braicovich, N. B. Brookes, G. Dellea, G. M. De Luca, D. Di Castro, M. Minola, M. Moretti Sala, Y. Y. Peng, for all the discussions and heavy-work during the experiments at the ID08-ESRF and ADDRESS (SLS) beamlines. GG acknowledges the support of the grant "P-ReXS" of the Fondazione CARIPLO and Regione Lombardia. MS acknowledges the support of Regione Campania through Legge Regionale 5 funding.

References

1. M.B. Salamon, M. Jaime, Rev. Mod. Phys. **73**, 583 (2001)
2. E. Fradkin, S.A. Kivelson, Nat. Phys. **8**, 864 (2012)
3. J. Mannhart, D.G. Schlom, Science **327**, 1607 (2010)
4. F. de Groot, A. Kotani, *Core Level Spectroscopy of Solids* (CRC Press, 2010)
5. L.J.P. Ament, M. van Veenendaal, T.P. Devereaux, J.P. Hill, J. van den Brink, Rev. Mod. Phys. **83**, 705 (2011)
6. MISSING package, based on Cowan's code, http://www.esrf.eu
7. E. Stavitski, F.M.F. de Groot, Micron **41**, 687 (2010)
8. F.C. Zhang, T.M. Rice, Phys. Rev. B **37**, 3759 (1988)
9. C. Aruta, G. Ghiringhelli, V. Bisogni, L. Braicovich, N. Brookes, A. Tebano, G. Balestrino, Phys. Rev. B **80**, 014431 (2009)
10. M.N. Regueiro, M. Altarelli, C.T. Chen, Phys. Rev. B **51**, 629 (1995)
11. D. Preziosi, M. Alexe, D. Hesse, M. Salluzzo, Phys. Rev. Lett. **115**, 157401 (2015)
12. C.T. Chen, Y.U. Idzerda, H.J. Lin, N.V. Smith, G. Meigs, E. Chaban, G.H. Ho, E. Pellegrin, F. Sette, Phys. Rev. Lett. **75**, 152 (1995)
13. G.M. De Luca, G. Ghiringhelli, M. Moretti Sala, S. Di Matteo, M.W. Haverkort, H. Berger, V. Bisogni, J.C. Cezar, N.B. Brookes, M. Salluzzo, Phys. Rev. B **82**, 214504 (2010)
14. C. Piamonteze, P. Miedema, F. de Groot, Phys. Rev. B **80**, 184410 (2009)
15. M. Moretti Sala, V. Bisogni, C. Aruta, G. Balestrino, H. Berger, N.B. Brookes, G.M. de Luca, D. Di Castro, M. Grioni, M. Guarise, P.G. Medaglia, F. Miletto Granozio, M. Minola, P. Perna, M. Radović, M. Salluzzo, T. Schmitt, K.J. Zhou, L. Braicovich, G. Ghiringhelli, New J. Phys. **13**, 043026 (2011)
16. L. Braicovich, J. van den Brink, V. Bisogni, M.M. Sala, L.J.P. Ament, N.B. Brookes, G.M. De Luca, M. Salluzzo, T. Schmitt, V.N. Strocov, G. Ghiringhelli, Phys. Rev. Lett. **104**, 077002 (2010)
17. W.J. Padilla, M. Dumm, S. Komiya, Y. Ando, D.N. Basov, Phys. Rev. B **72**, 205101 (2005)
18. R. Coldea, S.M. Hayden, G. Aeppli et al., Phys. Rev. Lett. **86**, 5377 (2001)
19. M. Guarise, B.D. Piazza, M.M. Sala, G. Ghiringhelli, L. Braicovich, H. Berger, J.N. Hancock, D. van der Marel, T. Schmitt, V.N. Strocov, L.J.P. Ament, J. van den Brink, P.H. Lin, P. Xu, H.M. Rønnow, M. Grioni, Phys. Rev. Lett. **105**, 157006 (2010)
20. J. Schlappa, K. Wohlfeld, K.J. Zhou, M. Mourigal, M.W. Haverkort, V.N. Strocov, L. Hozoi, C. Monney, S. Nishimoto, S. Singh, A. Revcolevschi, J.S. Caux, L. Patthey, H.M. Rønnow, J. van den Brink, T. Schmitt, Nature **485**, 82 (2012)
21. E. Benckiser, M.W. Haverkort, S. Brück, E. Goering, S. Macke, A. Frañó, X. Yang, O.K. Andersen, G. Cristiani, H.-U. Habermeier, A.V. Boris, I. Zegkinoglou, P. Wochner, H.-J. Kim, V. Hinkov, B. Keimer, Nat. Mater. **10**, 189 (2011)
22. A.S. Disa, D.P. Kumah, A. Malashevich, H. Chen, D.A. Arena, E.D. Specht, S. Ismail-Beigi, F.J. Walker, C.H. Ahn, Phys. Rev. Lett. **114**, 026801 (2015)
23. D. Yi, J. Liu, S. Okamoto, S. Jagannatha, Y.-C. Chen, P. Yu, Y.-H. Chu, E. Arenholz, R. Ramesh, Phys. Rev. Lett. **111**, 127601 (2013)
24. J.W. Seo, W. Prellier, P. Padhan, P. Boullay, J.Y. Kim, H. Lee, C.D. Batista, I. Martin, E.E.M. Chia, T. Wu, B.G. Cho, C. Panagopoulos, Phys. Rev. Lett. **105**, 167206 (2010)

25. M. Salluzzo, J.C. Cezar, N.B. Brookes, V. Bisogni, G.M. De Luca, C. Richter, S. Thiel, J. Mannhart, M. Huijben, A. Brinkman, Phys. Rev. Lett. **102**, 166804 (2009)
26. C. Aruta, C. Adamo, A. Galdi, P. Orgiani, V. Bisogni, N.B. Brookes, J.C. Cezar, P. Thakur, C.A. Perroni, G. De Filippis, V. Cataudella, D.G. Schlom, L. Maritato, G. Ghiringhelli, Phys. Rev. B **80**, 140405 (2009)
27. C.T. Chen, L.H. Tjeng, J. Kwo, H.L. Kao, P. Rudolf, F. Sette, R.M. Fleming, Phys. Rev. Lett. **68**, 2543 (1992)
28. M. Salluzzo, G. Ghiringhelli, N. Brookes, G. De Luca, F. Fracassi, R. Vaglio, Phys. Rev. B **75**, 054519 (2007)
29. J. Chakhalian, J.W. Freeland, H.U. Habermeier, G. Cristiani, G. Khaliullin, M. van Veenendaal, B. Keimer, Science **318**, 1114 (2007)
30. G.M. De Luca, G. Ghiringhelli, C.A. Perroni, V. Cataudella, F. Chiarella, C. Cantoni, A.R. Lupini, N.B. Brookes, M. Huijben, G. Koster, G. Rijnders, M. Salluzzo, Nature Communications **5**, 5626 (2014)
31. T. Thio, T.R. Thurston, N.W. Preyer, P.J. Picone, M.A. Kastner, H.P. Jenssen, D.R. Gabbe, C.Y. Chen, R.J. Birgeneau, A. Aharony, Phys. Rev. B **38**, 905 (1988)
32. G.M. De Luca, G. Ghiringhelli, M. Moretti Sala, S. Di Matteo, M.W. Haverkort, H. Berger, V. Bisogni, J.C. Cezar, N.B. Brookes, M. Salluzzo, Phys. Rev. B **82**, 214504 (2010)
33. J. Chakhalian, J.W. Freeland, G. Srajer, J. Strempfer, G. Khaliullin, J.C. Cezar, T. Charlton, R. Dalgliesh, C. Bernhard, G. Cristiani, H.U. Habermeier, B. Keimer, Nat. Phys. **2**, 244 (2006)
34. J. Salafranca et al., Competition between covalent bonding and charge transfer at complex-oxide interfaces. Phys. Rev. Lett. **112**, 196802 (2014)
35. M.P.M. Dean, R.S. Springell, C. Monney, K.J. Zhou, J. Pereiro, I. Bozovic, B.D. Piazza, H.M. Rønnow, E. Morenzoni, J. van den Brink, T. Schmitt, J.P. Hill, Nat. Mater. **11**, 850 (2012)
36. M. Minola, D. Di Castro, L. Braicovich, N.B. Brookes, D. Innocenti, M. Moretti Sala, A. Tebano, G. Balestrino, G. Ghiringhelli, Phys. Rev. B **85**, 235138 (2012)
37. M. Dantz, J. Pelliciari, D. Samal, V. Bisogni, Y. Huang, P. Olalde-Velasco, V.N. Strocov, G. Koster, T. Schmitt, Sci. Rep. **6**, 32896 (2016)
38. A. Ohtomo, H.Y. Hwang, Nature **427**, 423 (2004)
39. M. Huijben, G. Koster, M.K. Kruize, S. Wenderich, J. Verbeeck, S. Bals, E. Slooten, B. Shi, H.J.A. Molegraaf, J.E. Kleibeuker, S. van Aert, J.B. Goedkoop, A. Brinkman, D.H.A. Blank, M.S. Golden, G. Van Tendeloo, H. Hilgenkamp, G. Rijnders, Adv. Funct. Mater. **23**, 5240 (2013)
40. N. Reyren, S. Thiel, A.D. Caviglia, L.F. Kourkoutis, G. Hammerl, C. Richter, C.W. Schneider, T. Kopp, A.S. Ruetschi, D. Jaccard, M. Gabay, D.A. Muller, J.M. Triscone, J. Mannhart, Science **317**, 1196 (2007)
41. A.D. Caviglia, S. Gariglio, N. Reyren, D. Jaccard, T. Schneider, M. Gabay, S. Thiel, G. Hammerl, J. Mannhart, J.M. Triscone, Nature **456**, 624 (2008)
42. S. Thiel, G. Hammerl, A. Schmehl, C.W. Schneider, J. Mannhart, Science **313**, 1942 (2006)
43. A.D. Caviglia, M. Gabay, S. Gariglio, N. Reyren, C. Cancellieri, J.M. Triscone, Phys. Rev. Lett. **104**, 126803 (2010)
44. M. Salluzzo, S. Gariglio, D. Stornaiuolo, V. Sessi, S. Rusponi, C. Piamonteze, G.M. De Luca, M. Minola, D. Marrè, A. Gadaleta, H. Brune, F. Nolting, N.B. Brookes, G. Ghiringhelli, Phys. Rev. Lett. **111**, 087204 (2013)
45. K.-J. Zhou, M. Radovic, J. Schlappa, V. Strocov, R. Frison, J. Mesot, L. Patthey, T. Schmitt, Phys. Rev. B **83**, 201402 (2011)
46. J.A. Bert, B. Kalisky, C. Bell, M. Kim, Y. Hikita, H.Y. Hwang, K.A. Moler, Nat. Phys. **7**, 767 (2011)
47. Ariando, X. Wang, G. Baskaran, Z.Q. Liu, J. Huijben, J.B. Yi, A. Annadi, A.R. Barman, A. Rusydi, S. Dhar, Y.P. Feng, J. Ding, H. Hilgenkamp, T. Venkatesan, Nat. Commun. **2**, 188 (2011)
48. L. Li, C. Richter, J. Mannhart, R.C. Ashoori, Nat. Phys. **7**, 762 (2011)
49. M.R. Fitzsimmons, N.W. Hengartner, S. Singh, M. Zhernenkov, F.Y. Bruno, J. Santamaria, A. Brinkman, M. Huijben, H.J.A. Molegraaf, J. de la Venta, I.K. Schuller, Phys. Rev. Lett. **107**, 217201 (2011)

50. Z. Salman, O. Ofer, M. Radović, H. Hao, M. Ben Shalom, K.H. Chow, Y. Dagan, M.D. Hossain, C.D.P. Levy, W.A. MacFarlane, G.M. Morris, L. Patthey, M.R. Pearson, H. Saadaoui, T. Schmitt, D. Wang, R.F. Kiefl, Phys. Rev. Lett. **109**, 257207 (2012)
51. J.S. Lee, Y.W. Xie, H.K. Sato, C. Bell, Y. Hikita, H.Y. Hwang, C.C. Kao, Nat. Mater. **12**, 703 (2013)
52. S. Banerjee, O. Erten, M. Randeria, Nat. Phys. **9**, 626 (2013)
53. D. Stornaiuolo, C. Cantoni, G.M. De Luca, R. Di Capua, E. Di Gennaro, G. Ghiringhelli, B. Jouault, D. Marrè, D. Massarotti, F. Miletto Granozio, I. Pallecchi, C. Piamonteze, S. Rusponi, F. Tafuri, M. Salluzzo, Nat. Mater. **15**, 278 (2016)

Index

A
Ab-initio, 181
Ab-initio calculations, 182
Absorption edge, 269
Adsorbate(s), 22, 44, 47
Amorphous layer, 45
Analytical continuation, 224
Anatase titanium dioxide (TiO_2), 73
Anderson impurity problem, 223
Angle-resolved photoelectron spectroscopy (ARPES), 261
Antiferromagnetic, 186
ARPES, 1, 55, 77
ARPES analyzer, 12
ARPES experiments, 190
Artificial multiferroic, 246, 257, 266, 274
Atomic force microscope, 46
Atomic multiplet spectrum
Atomic-plane Bragg reflection, 170
Atomic-plane reflection, 160

B
Band alignment, 257, 258
Band bending, 88, 91, 93, 258
Band filling, 201
Band offset, 90–92, 259
Band structure, 120, 219, 261
Band structure calculations, 190
Band velocities, 197
Band-bending, 58
Berezinskii-Kosterlitz-Thouless transition, 43
$BiFeO_3/(Ca_{1-x}Ce_x)MnO_3$, 169
Bloch-Boltzmann Theory, 196
BoltZTraP code, 196

Bond reconfiguration, 254
Boundary conditions, 66
Bulk-doping, 57
Buried interface(s), 8, 254

C
$CaCuO_2/SrTiO_3$ (CCO/STO, 303
Capacitance, 38, 47
Capacitive readout, 257
Carrier density(ies), 64, 76
Cation mixing, 186, 188
Charge mobility, 28
Charge modulation, 263, 268
Charge transfer, 20, 21
Charge-transfer mechanism, 184
Chemical shift, 88, 89, 252
Chemical species, 167, 293
Chemical termination, 19
Circular dichroism, 120
Clogston-Chandrasekhar limit, 44
Composite charge carrier, 123
Confinement potential, 40
Confining potential, 58
Constant initial state spectra, 98, 99
Copper electronic configuration, 298
Core level spectra, 90
 Al $1s$, 91
 Sr $3d$, 94
 Ti $2p$, 88, 94, 95, 99
Coulomb repulsion
 on-site, 101
Coupling constant, 76, 125
Critical thickness, 38, 90, 91, 93, 101–103, 184
Cross-section problem, 109

Crystal field splitting, 228, 234
Crystal momentum conservation, 204
CuO2 planes, 288, 297–300, 302, 303

D
2D growth, 256
2DEL, 55, 56, 77
2DES mobility, 130
3D electronic structure, 113
D-orbital, 57
d -p model, 228
Debye model, 158
Debye wavelength, 204
Defects, 22
Deformation potential, 197
Delta-doping, 190
Density functional theory (DFT), 216, 218, 220
Density of states, 184
DFT+DMFT, 217, 218, 220, 235
DFT+U, 216, 219
Diagonal hopping, 195
Dielectric constant, 66
Diffusive thermopower, 203
Diluted magnetic semiconductor, 114
Diode, 47
Direct transitions (DTs), 157
Dispersion kink, 127, 130
Doping, 234
Doping effect, 121
Doping equivalence, 203
Double-exchange model, 116
Drude model, 199
Dynamical mean field theory (DMFT), 216, 217, 222
Dzyaloshinskii-Moriya (DM) interaction, 248, 302

E
Effective mass, 65, 191
  electron, 262
Effective mass tensor, 197
Electrical fatigue, 257
Electrolyte, 167
Electromigration, 141
Electron correlation(s), 133, 193, 245
Electron inelastic mean free path, 87, 89
Electron mean free path, 8
Electron mobility, 199
Electron-phonon scattering, 109
Electron screening, 136
Electron yield, 254
Electron-phonon coupling, 203
Electron-phonon interaction, 75, 123
Electron-phonon scattering, 204

Electronic, 47
Electronic correlations, 215
Electronic phase separation, 138
Electronic reconstruction, 88, 90, 95, 102, 226
  built-in potential, 90, 91, 102
  polar discontinuity, 95, 102
Electronic spectral function, 129
Electrons
  localized, 101
  mobile, 101
Electrostatic field, 183
Electrostatic field effect, 38, 44
Electrostatic modeling, 181
Electrostatic potential, 183
Element-resolved Bloch spectral functions, 176
Elemental sensitivity, 173
Epitaxial oxide, 17
Exchange bias, 250
Exchange interaction, 247
Experimental geometry, 110
Extended X-ray absorption fine structure (EXAFS), 253, 265

F
Fermi energy, 191
Fermi surface, 61, 69, 73, 74, 120
Ferroelastic, 39
Ferroelectric field effect, 264, 268
Ferromagnetic, 187
Ferromagnetism, 232
Field effect, 141
Fluorescence yield, 254, 268
Formation energy, 187
Fröhlich interaction, 75, 124
Fractured surfaces, 64
Free-electron final state (FEFS), 173

G
γ-Al$_2$O$_3$, 93, 94
GaAs, 10
Gas thickness, 203
Gate dielectric, 44
Gate field, 192
GdTiO$_3$/SrTiO$_3$, 163
Generalized-gradient approximation, 182
Green's function, 223
Ground state, 41, 45
Growth temperature, 192

H
Hall coefficient, 197
Hall effect, 40, 42
Hall mobility, 199
Hall resistivity, 184, 198

HAXPES, 252
Heavy subband, 62
Heterojunction, 20
High temperature superconductors, 130
High-resolution, 56
Holes, 6, 90, 135, 186, 188, 199, 230, 231, 235, 288, 298, 299, 301
Holstein interaction, 124
Holstein-type polaron, 132

I
Impact ionization, 231
Impurity state, 115
Impurity-assisted trapping, 126
In-gap state(s), 59, 98–101, 135
Independent Boson model, 133
Indirect exchange, 255
Inelastic mean free path, 252
In-plane and out-of-plane orbitals, 39, 306
Insulating-metal transition, 188
Insulator to metal transition, 26
Interface
  conducting, 88
  contrast, 88
  electronic and chemical structure, 87
  intermixing, 95
Interface conductivity, 249
Interface coupling, 246
Interface magnetism, 251
Interface quantum well, 117
Interfacial charge carriers, 133
Interfacial ferromagnetism, 137
Interfacial magnetism, 141
Interference effect, 154
Intrinsic multiferroic, 248
Inversion symmetry, 247

J
Jahn-Teller distortion, 161

K
Potassium tantalate, $KTaO_3$ (KTO), 72

L
La_2CuO_4 (LCO), 294
$La_{0.67}Sr_{0.33}MnO_3/SrTiO_3$, 160
$LaAlO_3$, 37
$LaAlO_3/SrTiO_3$, 38, 88, 95, 97, 99, 101
$LaCrO_3/SrTiO_3$, 91
LAO stoichiometry, 29
LAO/STO, 38
LAO/STO interface, 12
Large polarons, 75, 130
Lifshitz transition, 41, 141

Light-induced, 59
Liquid/solid interface, 167
Lithography, 45
Local density approximation, 182
Localization of the 2DES, 119
Localized states, 182
Localized electrons at Ti-sites, 306
Luttinger count, 121, 139
Luttinger volume, 262

M
Magnetic anisotropy, 251
Magnetic moment
  local, 101
Magnetic tunnel junction, 42, 159
Magnetism, 41, 42, 217, 226
Magnetization, 42, 47, 160, 176, 291, 303
Magnetoelastic, 249
Magnetoelectric coupling, 245, 255, 265, 266, 272, 274
Magnetoelectric effect, 270, 273
Magnetoelectric response, 256
Magnetoresistance, 40, 42
Magnetoresistivity, 184
Magnetostrictive, 248
Magnetotransport, 40
Magnons, 284, 295–297, 304
Many-body interactions, 5, 74
Mass enhancement, 71, 76
Mass renormalization, 125
Matrix element, 5, 66
Maximally localized Wannier orbitals, 222, 229
Metal contact, 188
Metal-insulator transition, 249
Molecular bonding, 300, 302
Momentum, 4
Momentum resolution, 113
Momentum-resolved electronic structure, 172
Mössbauer spectra, 159
Mott insulator, 216, 225
Multi-band behavior, 202
Multi-orbital Hubbard model, 222
Multiferroic, 245, 264
  heterostructure, 258, 265
  interface, 254, 260, 261, 264
Multiferroic manganites, 186
Multifunctional, 246
Multilayer heterostructure, 154

N
Nanoelectronic applications, 207
Near-total-reflection (NTR), 157
Nernst effect, 40

Non-linear Hall resistance, 202
Non-local correlations, 217

O
(111), (011) surface 2DELs (SrTiO$_3$), 69
Orbital character, 73, 120
Orbital coupling, 250
Orbital ordering, 62
Orbital polarization, 193, 229, 250
Orbital reconstruction, 70, 266
Ox-deficient samples, 139
Oxidation state, 88, 89
Oxide heterostructures, 215, 296
  LaTiO$_3$/SrTiO$_3$ (LTO/STO), 226
  SrRuO$_3$/SrTiO$_3$ (SRO/STO), 232–234, 236
  SrTiO$_3$/LaVO$_3$ (STO/LVO), 230–232, 236
  SrVO$_3$/LaVO$_3$ (SVO/LVO), 230
  SrVO$_3$/SrTiO$_3$ (SVO/STO), 218–226, 230
Oxide interfaces, 38
Oxitronics, 47
Oxygen depletion, 99
Oxygen doping, 186
Oxygen dosing, 99, 101
Oxygen pressure, 26
Oxygen vacancies, 27, 41, 42, 58, 95, 98, 99, 134, 226, 234
  cluster, 101
  electron doping, 95, 99, 101, 102
Oxygen vacancy, 59, 188

P
Patterning, 45
Peak-dip-hump (PDH) structure, 127
Periodic boundary conditions, 182
Perovskites, 17
Perturbation expansion, 125
Phonon modes, 124, 130
Phonon sidebands, 127
Phonon-drag, 198
Phonon-drag peak, 204
Phonon-phonon scattering, 204
Phonons, 76
Photoabsorption cross sections, 88, 96, 101
Photoelectron, 2
Photoelectron analyzer, 110
Photoelectron spectroscopy, 258
  angle-dependence, 89
  angle-integrated, 88
  core levels, 88, 96
  damping of photoelectrons, 90, 93
  depth profiling, 88
  hard X-rays, 87, 96, 100
  probing depth, 87–89
  resonant, 96–100
  soft X-rays, 87, 96, 100
  valence band, 90, 96, 97
Photoemission, 245
Photoemission spectroscopy, 216
Photoemitted intensity, 256
Piezoelectric, 248
Poisson distribution, 129
Polar catastrophe, 181
Polar discontinuity, 20, 183, 261
Polar field, 232
Polar oxide interface, 21
Polar structural phase transition, 39
Polaronic states, 182
Polaronic weight, 136
Polarons, 123
Probing depth, 9, 111
Pseudo-self-interaction corrected energy functional, 182
Pulsed laser deposition, 24

Q
Quantum confinement, 55, 71, 203
Quantum critical point, 45
Quantum Monte Carlo (QMC), 223, 235
Quantum well, 63
Quasiparticle, 6, 98–101
Quasiparticle renormalization, 216

R
Rare-earth nickelates RNiO$_3$, 227
Rashba, 41, 42, 45, 46, 67, 69
Rashba effect, 195
Reacted interface, 251
Reacted layer, 256
Reconstructions, 60, 71, 74
Relaxation-time approximation, 196
Replica band, 77
Resistive readout, 257
Resistivity, 232
Resistivity oscillations, 196
Resonant excitation, 109, 154
Resonant inelastic X-ray scattering (RIXS), 284
Resonant photoemission, 114, 118
RHEED, 25
Rigid band approximation, 190
RKKY interaction, 101
Rumpling, 184

S
SARPES, 69
Scanning transmission X-ray microscopy, 270
  STXM, 270
Schottky barrier, 258

Index

height, 257
Seebeck coefficient, 202
Seebeck effect, 39
self-energy Σ, 219, 223
Self-trapping, 126
Sheet conductance, 26
Sheet resistivity, 201
Shubnikov-de Haas, 195
Shubnikov-de Haas oscillations, 40
SiO$_x$/EuO/Si heterostructure, 112
Soft X-ray ARPES, 56, 77
Soft X-ray reflectivity, 229
Solid solution, 189
Sound velocity, 197
Spatial extension, 136
Spatial extent, 66
Spectral function, 261
Spectral function of polarons, 127
Splitting, 38, 39, 67, 69, 73, 101, 161, 163, 191, 193, 234
Spin currents, 46
Spin filter, 256
spin-moment, 292
Spin polarization, 142
Spin-orbit, 194
Spin-orbit coupling, 38
Spin-orbit interaction, 67, 72
Spin-resolved, 69
Spin-resolved ARPES, 137
Spin-splitting, 67
Spin to charge current conversion, 47
Spintronics, 46
Sr$_2$CuO$_2$Cl$_2$(SCOC), 296
SrRuO$_3$ (SRO), 232
SrTiO$_3$, 37
Standing-wave photoemission, 153
Step-and-terrace morphology, 25
STO, 56
Strontium titanate, SrTiO3 (STO), 61
Stoichiometric interface, 207
Strain, 234
Strontium titanate, 57
Subband(s), 61, 73
Subband structure, 66
Superconducting cuprates, 186
Superconductivity, 41, 42, 45, 76
Superexchange, 266
Super-exchange interaction, 302
Superstructures, 142
Surface polarity, 188
Surface reconstruction, 207
Surfaces, 55
SX-ARPES, 9

Symmetry breaking, 192
Synchrotron, 10
Synchrotron radiation beamlines, 110

T
t_{2g} splitting, 191
TEM/EELS, 160
Tetragonal crystal-field, 298
Thermal conductivity, 202
Thermoelectric efficiency, 202
Thermopower, 39
Tight-binding, 66
Ti-magnetism, 308
Time reversal, 247
TMO heterostructures, 135, 283, 284, 311
Transition metal dichalcogenide, 113
Transition metal oxides, 87, 215
Transverse electric field, 198
Tunneling electroresistance, 257
Tunneling magnetoresistance, 258
Two-dimensional, 62
Two-dimensional electron system, 89, 95, 97, 117
 charge carrier density, 89
 coexistence with ferromagnetism, 101
 thickness, 89
Two-dimensional superconductivity, 43
Two-dimensional system, 37

U
Ultraviolet photoemission spectroscopy, 252
Unintentional oxidation, 255
Universal curve, 87, 89, 108
Universal curve of photoemission, 8
UV valence band, 3

V
Vacancy concentration, 188
Valence band spectra, 90

W
W2dynamics, 235
Wavefunctions, 63, 66, 71
Weak localization, 38
Wien2K, 221, 235
Wien2wannier, 222

X
X-ray absorption, 271
X-ray absorption near edge spectroscopy, 253, 265
 XANES, 253, 265, 268, 270
X-ray absorption spectra, 98, 99

X-ray absorption spectroscopy, 245, 253, 264, 265
X-ray holography, 270, 271
X-ray irradiation, 139
X-ray Linear Dichroism (XLD), 65, 288, 306
X-ray magnetic circular dichroism (XMCD), 265, 268, 271, 273, 274, 284
X-ray magnetic linear dichroism (XMLD), 272, 273
X-ray natural linear dichroism (XNLD), 273
X-ray photoelectron spectroscopy, 251
X-ray photoemission electron microscopy (XPEEM), 265, 270, 271, 273, 274
X-ray ptychography, 270
XLD spectra of LAO/STO, 306
XPS depth profiling, 118

Z
Zener breakdown, 21, 183

Printed by Printforce, the Netherlands